EDITION

10

FUNDAMENTALS OF
Mathematics

JAMES VAN DYKE
JAMES ROGERS
HOLLIS ADAMS

Portland Community College

BROOKS/COLE
CENGAGE Learning™

Australia • Brazil • Japan • Korea • Mexico • Singapore • Spain • United Kingdom • United States

Fundamentals of Mathematics, **Tenth Edition**
James Van Dyke, James Rogers, Hollis Adams

Acquisitions Editor: Marc Bove

Development Editor: Stefanie Beeck

Assistant Editor: Shaun Williams

Editorial Assistant: Zachary Crockett

Media Editors: Guanglei Zhang

Marketing Manager: Gordon Lee

Marketing Assistant: Angela Kim

Marketing Communications Manager: Darlene Macanan

Content Project Manager: Jennifer Risden

Design Director: Rob Hugel

Art Director: Vernon Boes

Print Buyer: Judy Inouye

Rights Acquisitions Specialist: Don Schlotman

Production Service: MPS Limited, a Macmillan Company

Text Designer: Kim Rokusek

Photo Researcher: Bill Smith Group

Copy Editor: Martha Williams

Illustrator: MPS Limited, a Macmillan Company

Cover Designer: Roger Knox

Cover Image: Jean-Pierre Lescourret/Corbis

Compositor: MPS Limited, a Macmillan Company

Library of Congress Control Number: 2010922324

ISBN-13: 978-0-538-49797-8

ISBN-10: 0-538-49797-1

Brooks/Cole
20 Davis Drive
Belmont, CA 94002-3098
USA

Cengage Learning is a leading provider of customized learning solutions with office locations around the globe, including Singapore, the United Kingdom, Australia, Mexico, Brazil, and Japan. Locate your local office at **www.cengage.com/global**.

Cengage Learning products are represented in Canada by Nelson Education, Ltd.

To learn more about Brooks/Cole, visit **www.cengage.com/brookscole**

Purchase any of our products at your local college store or at our preferred online store **www.cengagebrain.com**

Printed in the United States of America
1 2 3 4 5 6 7 14 13 12 11 10

To

Carol Van Dyke

Elinore Rogers

Scott Huff

TABLE OF CONTENTS

GOOD ADVICE FOR *Studying*

New Habits from Old 166

CHAPTER 3 Fractions and Mixed Numbers 167

GOOD ADVICE FOR *Studying*

Preparing for Tests 276

CHAPTER 4 Decimals 277

GOOD ADVICE FOR
Studying

CHAPTER 5

GOOD ADVICE FOR
Studying

CHAPTER 6

TO THE STUDENT

"It looks so easy when you do it, but when I get home . . . " is a popular lament of many students studying mathematics.

The process of learning mathematics evolves in stages. For most students, the first stage is listening to and watching others. In the middle stage, students experiment, discover, and practice. In the final stage, students analyze and summarize what they have learned. Many students try to do only the middle stage because they do not realize how important the entire process is.

Here are some steps that will help you to work through all the learning stages:

1. Go to class every day. Be prepared, take notes, and most of all, think actively about what is happening. Ask questions and keep yourself focused. This is prime study time.

2. Begin your homework as soon after class as possible. Start by reviewing your class notes and then read the text. Each section is organized in the same manner to help you find information easily. The objectives tell you what concepts will be covered, and the vocabulary lists all the new technical words. There is a **How & Why** section for each objective that explains the basic concept, followed by worked sample problems. As you read each example, make sure you understand every step. Then work the corresponding **Warm-Up** problem to reinforce what you have learned. You can check your answer at the bottom of the page. Continue through the whole section in this manner.

3. Now work the exercises at the end of the section. The A group of exercises can usually be done in your head. The B group is harder and will probably require pencil and paper. The C group problems are more difficult, and the objectives are mixed to give you practice at distinguishing the different solving strategies. As a general rule, do not spend more than 15 minutes on any one problem. If you cannot do a problem, mark it and ask someone (your teacher, a tutor, or a study buddy) to help you with it later. Do not skip the **Maintain Your Skills** problems. They are for review and will help you practice earlier procedures so you do not become "rusty." The answers to the odd exercises are in the back of the text so you can check your progress.

4. In this text, you will find **State Your Understanding** exercises in every section. Taken as a whole, these exercises cover *all* the basic concepts in the text. You may do these orally or in writing. Their purpose is to encourage you to analyze or summarize a skill and put it into words. We suggest that you do these in writing and keep them all together in a journal. Then they are readily available as a review for chapter tests and exams.

5. When preparing for a test, work the material at the end of the chapter. The **True/False Concept Review** and the **Chapter Test** give you a chance to review the concepts you have learned. You may want to use the chapter test as a practice test.

If you have never had to write in a math class, the idea can be intimidating. Write as if you are explaining to a classmate who was absent the day the concept was discussed. Use your own words—*do not copy out of the text*. The goal is that you understand the concept, not that you can quote what the authors have said. Always use complete sentences, correct spelling, and proper punctuation. Like everything else, writing about math is a learned skill. Be patient with yourself and you will catch on.

Since we have many students who do not have a happy history with math, we have included **Good Advice for Studying**—a series of eight checklists that address various problems that are common for students. They include advice on time management, organization, test taking, and reducing math anxiety. We talk about these things with our own students, and hope that you will find some useful tips.

We really want you to succeed in this course. If you go through each stage of learning and follow all the steps, you will have an excellent chance for success. But remember, you are in control of your learning. The effort that you put into this course is the single biggest factor in determining the outcome. Good luck!

James Van Dyke
James Rogers
Hollis Adams

TO THE INSTRUCTOR

Fundamentals of Mathematics, **Tenth Edition,** continues this text's tradition of organizing exposition around a short list of objectives in each section. Clear, accessible writing explains concepts in the context of "how" and "why," and then carefully matches those concepts with a variety of well-paced exercises. It's a formula that has worked for hundreds of thousands of students, including those who are anxious about the course. And it's a formula that's appropriate for individual study or for lab, self-paced, lecture, group, or combined formats.

New to the Tenth Edition

- Classroom Activities have been added at the end of each chapter. These activities (2 per chapter) are designed to be done in a group using $\frac{1}{2}$–1 hour of class time. Many are activities used by the authors in their own classes. These replace the Group Work problems in each section.
- New examples appear in each section's *Examples* and *Warm-Ups* (which place examples and related exercises side by side).
- Approximately 30% of the text's section exercises are new.
- New and updated application problems reflect the emphasis on real-world data.
- The topic of compound interest has been rewritten and simplified and there is a new emphasis on credit cards in Chapter 6.
- *Good Advice for Studying* has been completely rewritten and reorganized, with new topics added. A directory is included at each site.

A Textbook of Many Course Formats

Fundamentals of Mathematics is suitable for individual study or for a variety of course formats: lab, both supervised and self-paced; lecture; group; or combined formats. For a lecture-based course, for example, each section is designed to be covered in a standard 50-minute class. The lecture can be interrupted periodically so that students individually can work the **Warm-Up** exercises or work in small groups on the group work. In a self-paced lab course, **Warm-Up** exercises give students a chance to practice while they learn, and get immediate feedback since warm-up answers are printed on the same page. Using the text's ancillaries, instructors and students have even more options available to them. Computer users, for example, can take advantage of complete electronic tutorial and testing systems that are fully coordinated with the text.

Pedagogy

The pedagogical system of *Fundamentals of Mathematics* meets two important criteria: coordinated purpose and consistency of presentation.

Each section begins with numbered **Objectives**, followed by definitions of new **Vocabulary** to be encountered in the section. Following the vocabulary, **How & Why** segments, numbered to correspond to the objectives, explain and demonstrate concepts and skills. Throughout the **How & Why** segments, **skill boxes** clearly summarize and outline the skills in step-by-step form. Also throughout the segments, **concept boxes** highlight appropriate properties, formulas, and theoretical facts underlying the skills. Following each **How & Why** segment are **Examples** and **Warm-Ups**. Each example of an objective is paired with a warm-up, with workspace provided. Solutions to the warm-ups are given at the bottom of the page, affording immediate feedback. The examples also include, where suitable, a relevant application of the objective. Examples similar to each other are linked by common **Directions** and a common **Strategy** for solution. Directions and strategies are closely related to the skill boxes. Connecting examples by a common solution method helps students recognize the similarity of problems and their solutions, despite their specific differences. In this way, students may improve their problem-solving skills. In both **How & Why** segments and in the **Examples**, **Caution** remarks help to forestall common mistakes.

Teaching Methodology

As you examine the Tenth Edition of *Fundamentals of Mathematics,* you will see distinctive format and pedagogy that reflect these aspects of teaching methodology:

Teaching by Objective Each section focuses on a short list of objectives, stated at the beginning of the section. The objectives correspond to the sequence of exposition and tie together other pedagogy, including the highlighted content, the examples, and the exercises.

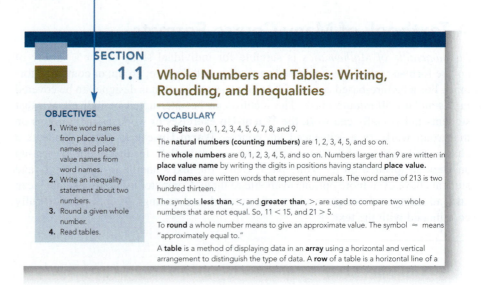

SECTION 1.1

Whole Numbers and Tables: Writing, Rounding, and Inequalities

OBJECTIVES

1. Write word names from place value names and place value names from word names.
2. Write an inequality statement about two numbers.
3. Round a given whole number.
4. Read tables.

VOCABULARY

The **digits** are 0, 1, 2, 3, 4, 5, 6, 7, 8, and 9.

The **natural numbers (counting numbers)** are 1, 2, 3, 4, 5, and so on.

The **whole numbers** are 0, 1, 2, 3, 4, 5, and so on. Numbers larger than 9 are written in **place value name** by writing the digits in positions having standard **place value.**

Word names are written words that represent numerals. The word name of 213 is two hundred thirteen.

The symbols **less than**, <, and **greater than**, >, are used to compare two whole numbers that are not equal. So, 11 < 15, and 21 > 5.

To **round** a whole number means to give an approximate value. The symbol ≈ means "approximately equal to."

A **table** is a method of displaying data in an **array** using a horizontal and vertical arrangement to distinguish the type of data. A **row** of a table is a horizontal line of a

Teaching by Application Each chapter leads off with an application that uses the content of the chapter. Exercise sets have applications that use this material or that are closely related to it. Applications are included in the examples for most objectives. Other applications appear in exercise sets. These cover a diverse range of fields, demonstrating the utility of the content in business, environment, personal health, sports, and daily life.

Whole Numbers

APPLICATION

The ten top-grossing movies in the United States are given in Table 1.1.

TABLE 1.1 The Ten Top-Grossing Movies in the United States

Name	Year Produced	Earnings
Avatar	2009	$696,000,000
Titanic	1997	$600,788,188
The Dark Knight	2008	$533,184,219
Star Wars: Episode IV—A New Hope	1977	$460,998,007
Shrek 2	2004	$437,212,000
E.T. The Extra-Terrestrial	1982	$434,974,579
Star Wars: Episode I—The Phantom Menace	1999	$431,088,301
Pirates of the Caribbean: Dead Man's Chest	2006	$423,416,000
Spider-Man	2002	$407,681,000
Star Wars: Episode III—Revenge of the Sith	2005	$380,270,577

SOURCE: MovieWeb.com

GROUP DISCUSSION

1. Which movie in the table is the oldest?
2. How is the table organized?
3. What is the difference in earnings between the top-grossing movie and the tenth one?

Emphasis on Language New words of each section are explained in the vocabulary segment that precedes the exposition. Exercise sets include questions requiring responses written in the students' own words.

Adding and Subtracting Whole Numbers

VOCABULARY

Addends are the numbers that are added. In $9 + 20 + 3 = 32$, the addends are 9, 20, and 3.

The result of adding is called the **sum**. In $9 + 20 + 3 = 32$, the sum is 32.

The result of subtracting is called the **difference**. So in $62 - 34 = 28$, 28 is the difference.

A **polygon** is any closed figure whose sides are line segments.

The **perimeter** of a polygon is the distance around the outside of the polygon.

OBJECTIVES

1. Find the sum of two or more whole numbers.
2. Find the difference of two whole numbers.
3. Estimate the sum or difference of whole numbers.
4. Find the perimeter of a polygon.

Emphasis on Skill, Concept, and Problem Solving Each section covers concepts and skills that are fully explained and demonstrated in the exposition for each objective.

HOW & WHY

■ **OBJECTIVE 1** Find the sum of two or more whole numbers.

When Jose graduated from high school he received cash gifts of $50, $20, and $25. The total number of dollars received is found by adding the individual gifts. The total number of dollars he received is 95. In this section we review the procedure for adding and subtracting whole numbers.

The addition facts and place value are used to add whole numbers written with more

---

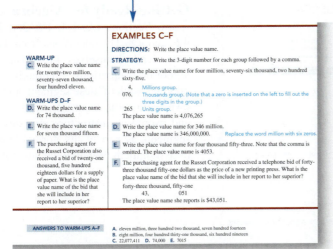

EXAMPLES C–F

DIRECTIONS: Write the place value name.

STRATEGY: Write the 3-digit number for each group followed by a comma.

WARM-UP
C. Write the place value name for twenty-two million, seventy-seven thousand, four hundred eleven.

C. Write the place value name for four million, seventy-six thousand, two hundred sixty-five.
4,   Millions group.
076, Thousands group. (Note that a zero is inserted on the left to fill out the three digits in the group.)
265   Units group.
The place value name is 4,076,265.

WARM-UPS D–F
D. Write the place value name for 74 thousand.

D. Write the place value name for 346 million.
The place value name is 346,000,000.   Replace the word *million* with six zeros.

E. Write the place value name for seven thousand fifteen.

E. Write the place value name for four thousand fifty-three. Note that the comma is omitted. The place value name is 4053.

F. The purchasing agent for the Russet Corporation also received a bid of twenty-one thousand, five hundred eighteen dollars for a supply of paper. What is the place value name of the bid that she will include in her report to her superior?

F. The purchasing agent for the Russet Corporation received a telephone bid of forty-three thousand fifty-one dollars as the price of a new printing press. What is the place value name of the bid that she will include in her report to her superior?
forty-three thousand, fifty-one
43,   051
The place value name she reports is $43,051.

ANSWERS TO WARM-UPS A–F
A. eleven million, three hundred two thousand, seven hundred fourteen
B. eight million, four hundred thirty-one thousand, six hundred nineteen
C. 22,077,411 D. 74,000 E. 7015

Carefully constructed examples for each objective are connected by a common strategy that reinforces both the skill and the underlying concepts. Skills are not treated as isolated feats of memorization but as the practical result of conceptual understanding. Skills are strategies for solving related problems. Students see the connections between problems that require similar strategies.

Emphasis on Success and Preparation

Integrated throughout the text, the following features focus on study skills, math anxiety, calculators, and simple algebraic equations.

Good Advice for Studying is continued from the previous editions, but reorganized as *checklists*. Taken as a whole, they address the unique study problems that students of *Fundamentals of Mathematics* experience. Students learn college survival skills, general study skills, and study skills specific to mathematics and to the pedagogy and ancillaries of *Fundamentals of Mathematics*. Special techniques are described to overcome the pervasive problems of math anxiety. Though a checklist begins each chapter, students may profit by reading all the checklists at once and then returning to them as the need arises. A directory of all the checklists is included at each site.

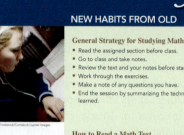

GOOD ADVICE FOR *Studying*

NEW HABITS FROM OLD

© Thinkstock/Comstock/Jupiter Images

GOOD ADVICE FOR STUDYING

General Strategy for Studying Math
- Read the assigned section before class.
- Go to class and take notes.
- Review the text and your notes before starting the exercises.
- Work through the exercises.
- Make a note of any questions you have.
- End the session by summarizing the techniques you have learned.

How to Read a Math Text
- Begin by reading the objectives and new vocabulary.
- Read with a pencil/highlighter and mark up the text with notes to yourself.
- After reading an example, try the corresponding warm-up exercise.
- If you don't remember an earlier concept, go back and review before continuing.

Keep a Positive Attitude
- Begin each study session by expecting that you will be able to learn the skills.
- Do not spend more than 15 minutes on any one problem. If you can't solve it, mark it to ask about later and go on.
- Follow up on all your questions with your instructor or in the tutoring center or with a study buddy.
- Always end a study session with a problem that you have successfully solved—even if it means going backward. Your brain will remember that you were successful.
- Do not allow negative self-talk. Remind yourself that you have a new chance to be successful and that you are capable of learning this material.

CALCULATOR EXAMPLE:

E. Find the value of 11^6.

Calculators usually have an exponent key marked $\boxed{y^x}$ or $\boxed{\wedge}$. If your calculator doesn't have such a key, you can multiply the repeated factors.

The value is 1,771,561.

F. Mabel needs 125 square yards of carpet to carpet her house wall to wall. Write the amount of carpet needed using an exponent to show the measure.

125 square yards = 125 yd^2 Square yards is written yd^2.

Mabel needs 125 yd^2 of carpet.

Calculator examples, marked by the symbol 🖩, demonstrate how a calculator may be used, though the use of a calculator is left to the discretion of the instructor. Nowhere is the use of a calculator required. Appendix A reviews the basics of operating a scientific calculator.

GETTING READY FOR ALGEBRA

OBJECTIVES

Solve an equation of the form $x + a = b$ or $x - a = b$, where a, b, and x are whole numbers.

VOCABULARY

An **equation** is a statement about numbers that says that two expressions are equal. Letters, called **variables** or **unknowns**, are often used to represent numbers.

HOW & WHY

■ **OBJECTIVE 1** Solve an equation of the form $x + a = b$ or $x - a = b$, where a, b, and x are whole numbers.

Examples of equations are

$9 = 9$ $13 = 13$ $123 = 123$ $30 + 4 = 34$ $52 - 7 = 45$

When variables are used an equation can look like this:

$x = 7$ $x = 10$ $y = 18$ $x + 5 = 13$ $y - 8 = 23$

An equation containing a variable can be true only when the variable is replaced by a specific number. For example,

$x = 7$ is true only when x is replaced by 7.
$x = 10$ is true only when x is replaced by 10.
$y = 18$ is true only when y is replaced by 18.
$x + 5 = 13$ is true only when x is replaced by 8, so that $8 + 5 = 13$.
$y - 8 = 23$ is true only when y is replaced by 31, so that $31 - 8 = 23$.

Getting Ready for Algebra segments follow Sections 1.2, 1.4, 1.6, 3.4, 3.9, 4.3, 4.6, and 4.8. The operations from these sections lend themselves to solving simple algebraic equations. Though entirely optional, each of these segments includes its own exposition, examples with warm-ups, and exercises. Instructors may cover these segments as part of the normal curriculum or assign them to individual students.

Exercises, Reviews, Tests

Thorough, varied, properly paced, and well-chosen exercises are a hallmark of *Fundamentals of Mathematics*. Exercise sets are provided at the end of each section and a review set at the end of each chapter.

Section exercises are paired so that virtually each odd-numbered exercise, in Sections A and B, is paired with an even-numbered exercise that is equivalent in type and difficulty. Since answers for odd-numbered exercises are in the back of the book, students can be assigned odd-numbered exercises for practice and even-numbered exercises for homework. Section exercises are categorized to satisfy teaching and learning aims. Exercises for estimation, mental computation, pencil and paper computation, application, and calculator skills are provided, as well as opportunities for students to challenge their abilities, master communications skills, and participate in group problem solving.

- **Category A** exercises, organized by section objective, are those that most students should be able to solve mentally, without pencil, paper, or calculator. Mentally working problems improves students' estimating abilities. These can often be used in class as oral exercises.
- **Category B** exercises, also organized by objective, are similar except for level of difficulty. All students should be able to master Category B.
- **Category C** exercises contain applications and more difficult exercises. Since these are not categorized by objective, the student must decide on the strategy needed to set up and solve the problem. These applications are drawn from business, health and nutrition, the environment, consumerism, sports, and various science fields. Both professional and daily-life uses of mathematics are incorporated.

■ **OBJECTIVE 1** Divide whole numbers. (See page 47.)

A *Divide.*

1. $8\overline{)72}$ 2. $8\overline{)88}$ 3. $6\overline{)78}$ 4. $4\overline{)84}$

5. $5\overline{)435}$ 6. $3\overline{)327}$ 7. $5\overline{)455}$ 8. $9\overline{)549}$

9. $136 \div 8$ 10. $180 \div 5$ 11. $880 \div 22$ 12. $850 \div 17$

13. $492 \div 6$ 14. $1668 \div 4$ 15. $36 \div 7$ 16. $79 \div 9$

17. $81 \div 17$ 18. $93 \div 29$

19. The quotient in division has no remainder when the last difference is _____.

20. For $360 \div 12$, in the partial division $36 \div 12 = 3$, 3 has place value _____.

B *Divide.*

21. $18,306 \div 6$ 22. $21,154 \div 7$ 23. $\dfrac{768}{24}$ 24. $\dfrac{558}{62}$

25. $46\overline{)2484}$ 26. $38\overline{)2546}$ 27. $46\overline{)4002}$ 28. $56\overline{)5208}$

29. $542\overline{)41,192}$ 30. $516\overline{)31,992}$ 31. $355\overline{)138,805}$ 32. $617\overline{)124,017}$

33. $43\overline{)7822}$ 34. $56\overline{)7288}$ 35. $57\overline{)907}$ 36. $39\overline{)797}$

37. $(78)(?) = 1872$ 38. $(?)(65) = 4225$ 39. $27\overline{)345,672}$ 40. $62\overline{)567,892}$

41. $55,892 \div 64$. Round quotient to the nearest ten. 42. $67,000 \div 43$. Round quotient to the nearest hundred.

43. $225,954 \div 415$. Round quotient to the nearest hundred. 44. $535,843 \div 478$. Round quotient to the nearest hundred.

C *Exercises 45–48. The revenue department of a state had the following collection data for the first 3 weeks of April.*

Taxes Collected

Number of Returns	Total Taxes Paid
Week 1—4563	$24,986,988
Week 2—3981	$19,315,812
Week 3—11,765	$48,660,040

45. Find the taxes paid per return during week 1.

46. Find the taxes paid per return during week 2.

STATE YOUR UNDERSTANDING

74. Explain how to find the product of $\dfrac{35}{24}$ and $\dfrac{40}{14}$.

75. Explain how to find the quotient of $\dfrac{35}{24}$ and $\dfrac{40}{14}$.

196 3.3 Multiplying and Dividing Fractions

State Your Understanding exercises require a written response, usually no more than two or three sentences. These responses may be kept in a journal by the student. Maintaining a journal allows students to review concepts as they have written them. These writing opportunities facilitate student writing in accordance with standards endorsed by AMATYC and NCTM.

Challenge exercises stretch the content and are more demanding computationally and conceptually.

Maintain Your Skills exercises continually reinforce mastery of skills and concepts from previous sections. The problems are specially chosen to review topics that will be needed in the next section.

Classroom Activities and **Group Projects** provide opportunities for small groups of students to work together to solve problems and create reports. While the use of these is optional, the authors suggest the assignment of two or three of these per semester or term to furnish students with an environment for exchanging ideas. Classroom Activities encourage cooperative learning as recommended by AMATYC and NCTM guidelines.

KEY CONCEPTS

SECTION 1.1 Whole Numbers and Tables: Writing, Rounding, and Inequalities

Definitions and Concepts	Examples	
The whole numbers are 0, 1, 2, 3, and so on.	238	two hundred thirty-eight
	6,198,349	six million, one hundred ninety-eight thousand, three hundred forty-nine
One whole number is smaller than another if it is to the left on the number line.	$3 < 6$	
One whole number is larger than another if it is to the right on the number line.	$14 > 2$	
To round a whole number, • Round to the larger number if the digit to the right is 5 or more	$6,745 \approx 7,000$	(nearest thousand)
• Round to the smaller number if the digit to the right is 4 or less	$6,745 \approx 6,700$	(nearest hundred)

Key Concepts recap the important concepts and skills covered in the chapter. The Key Concepts can serve as a quick review of the chapter material.

REVIEW EXERCISES

SECTION 1.1

Write the word name for each of these numbers.

1. 607,321

2. 9,070,800

Write the place value name for each of these numbers.

3. Sixty-two thousand, three hundred thirty-seven

4. Five million, four hundred forty-four thousand, nineteen

Chapter Review Exercises provide a student with a set of exercises, usually 8–10 per section, to verify mastery of the material in the chapter prior to taking an exam.

TEST

	Answers
1. Divide: $72\overline{)15{,}264}$	**1.** _____
2. Subtract: $9615 - 6349$	**2.** _____
3. Simplify: $55 \div 5 + 6 \cdot 4 - 7$	**3.** _____
4. Multiply: $37(428)$	**4.** _____
5. Insert $<$ or $>$ to make the statement true: 368 371	**5.** _____
6. Multiply: 55×10^6	**6.** _____
7. Multiply: $608(392)$	**7.** _____

Chapter Test exercises end the chapter. Written to imitate a 50-minute exam, each chapter test covers all of the chapter content. Students can use the chapter test as a self-test before the classroom test.

Chapter True/False Concept Review exercises require students to judge whether a statement is true or false and, if false, to rewrite the sentence to make it true. Students evaluate their understanding of concepts and also gain experience using the vocabulary of mathematics.

TRUE/FALSE CONCEPT REVIEW

Check your understanding of the language of basic mathematics. Tell whether each of the following statements is true (always true) or false (not always true). For each statement you judge to be false, revise it to make a statement that is true.

	Answers
1. All whole numbers can be written using nine digits.	**1.** _____
2. In the number 8425, the digit 4 represents 400.	**2.** _____
3. The word *and* is not used when writing the word names of whole numbers.	**3.** _____

Other Teaching and Learning Resources

FOR INSTRUCTORS

Complete Solutions Manual
1-111-42949-9
The Complete Solutions Manual provides worked-out solutions to all of the problems in the text.

Enhanced Webassign with eBook (One-Term Access Card)
Enhanced WebAssign, used by over one million students at more than 1,100 institutions, allows you to assign, collect, grade, and record homework assignments via the web. This proven and reliable homework system includes thousands of algorithmically generated homework problems, links to relevant textbook sections, video examples, problem-specific tutorials, and more.

Diagnostic quizzing for each chapter identifies concepts that students still need to master, and directs them to the appropriate review material. Students will appreciate the interactive Premium eBook, which offers search, highlighting, and note-taking functionality, as well as links to multimedia resources, all available to students when you choose Enhanced WebAssign.

Note that the WebAssign problems for this text are highlighted by a ▶.

PowerLecture with ExamView®
1-111-42950-2
This CD-ROM provides the instructor with dynamic media tools for teaching. Create, deliver, and customize tests (both print and online) in minutes with ExamView® Computerized Testing Featuring Algorithmic Equations. Easily build solution sets for homework or exams using Solution Builder's online solutions manual. Microsoft® PowerPoint® lecture slides and figures from the book are also included on this CD-ROM.

Text-Specific DVDs
1-111-43002-0
These DVDs cover selected sections and are ideal for promoting individual study and review.

Solution Builder
This online solutions manual allows instructors to create customizable solutions that they can print out to distribute or post as needed. This is a convenient and expedient way to deliver solutions to specific homework sets.
   Visit www.cengage.com/solutionbuilder.

www.cengage.com/math/vandyke
This site features a range of teaching and learning resources, including a variety of chapter-specific materials for students.

FOR STUDENTS

Student Solutions Manual
1-111-42948-0
The **Student Solutions Manual** provides worked-out solutions to the odd-numbered problems in the text.

Text-Specific DVDs
1-111-43002-0
These DVDs cover selected sections and are ideal for promoting individual study and review.

Enhanced WebAssign with eBook
Enhanced WebAssign, used by over one million students at more than 1,100 institutions, allows you to assign, collect, grade, and record homework assignments via the web. This proven and reliable homework system includes thousands of algorithmically generated homework problems, links to relevant textbook sections, video examples, problem-specific tutorials, and more.

For more information

Ask your Cengage Learning sales representative how to package any of the above student resources with the text.

Due to contractual reasons, certain ancillaries are available only in higher education or U.S. domestic markets. Minimum purchases may apply to receive the ancillaries at no charge. For more information, please contact your local Cengage Learning sales representative.

Acknowledgments

The authors appreciate the unfailing and continuous support of their families who made the completion of this work possible. We are also grateful to Marc Bove of Brooks/Cole for his suggestions during the preparation and production of the text. We also want to thank the following professors and reviewers for their many excellent contributions to the development of the text: Kinley Alston, Trident Technical College; Carol Barner, Glendale Community College; Karen Driskell, Calhoun Community College; Beverlee Drucker, Northern Virginia Community College; Dale Grussing, Miami-Dade Community College, North Campus; Dianne Hendrickson, Becker College; Eric A. Kaljumagi, Mt. San Antonio College; Joanne Kendall, College of the Mainland; Christopher McNally, Tallahassee Community College; Michael Montano, Riverside Community College; Kim Pham, West Valley College; Ellen Sawyer, College of Dupage; Leonard Smiglewski, Penn Valley Community College; Brian Sucevic, Valencia Community College; Stephen Zona, Quinsigamond Community College.

Special thanks to Gloria Langer for her careful reading of the text and for the accuracy review of all the problems and exercises in the text.

James Van Dyke
James Rogers
Hollis Adams

FUNDAMENTALS OF
Mathematics

GOOD ADVICE FOR *Studying*

STRATEGIES FOR SUCCESS

© Ana Blazic/Shutterstock.com

What Is Math Anxiety?

- A learned fear response to math that causes disruptive, debilitating reactions to tests.
- A dread of doing **anything** that involves numbers.
- Worrisome thoughts and the inability to concentrate and recall what you've learned.
- Imagined catastrophic consequences for your failure to be successful in math.
- Physical signs include muscle tightness, stomach upset, sweating, headache, shortness of breath, shaking, or rapid heartbeat.

Math Anxiety Can Be Unlearned

- You can choose to learn behaviors that are more useful to achieve success in math.
- You can learn and choose the ways that work best for you.

Strategies to Overcome Math Anxiety

- Study math in ways **proven** to be effective in learning mathematics and taking tests.
- Learn to physically and mentally **relax**, to manage your anxious feelings, and to think rationally and positively.
- Make a time commitment to practice relaxation techniques, study math, and record your thought patterns.

How the Text Will Help

- Read To the Student at the beginning of this book to understand the authors' organization or "game plan" for your math experience in this course.
- Find more Good Advice for Studying sections at the beginning of each chapter. You may want to read ahead so you can improve even more quickly.

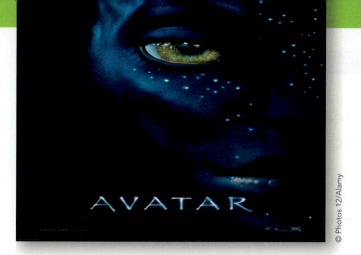

© Photos 12/Alamy

Whole Numbers

APPLICATION

The ten top-grossing movies in the United States are given in Table 1.1.

TABLE 1.1 The Ten Top-Grossing Movies in the United States

Name	Year Produced	Earnings
Avatar	2009	$696,000,000
Titanic	1997	$600,788,188
The Dark Knight	2008	$533,184,219
Star Wars: Episode IV—A New Hope	1977	$460,998,007
Shrek 2	2004	$437,212,000
E.T. The Extra-Terrestrial	1982	$434,974,579
Star Wars: Episode I—The Phantom Menace	1999	$431,088,301
Pirates of the Caribbean: Dead Man's Chest	2006	$423,416,000
Spider-Man	2002	$407,681,000
Star Wars: Episode III—Revenge of the Sith	2005	$380,270,577

SOURCE: MovieWeb.com

GROUP DISCUSSION

1. Which movie in the table is the oldest?
2. How is the table organized?
3. What is the difference in earnings between the top-grossing movie and the tenth one?

SECTION 1.1

Whole Numbers and Tables: Writing, Rounding, and Inequalities

OBJECTIVES

1. Write word names from place value names and place value names from word names.
2. Write an inequality statement about two numbers.
3. Round a given whole number.
4. Read tables.

VOCABULARY

The **digits** are 0, 1, 2, 3, 4, 5, 6, 7, 8, and 9.

The **natural numbers (counting numbers)** are 1, 2, 3, 4, 5, and so on.

The **whole numbers** are 0, 1, 2, 3, 4, 5, and so on. Numbers larger than 9 are written in **place value name** by writing the digits in positions having standard **place value.**

Word names are written words that represent numerals. The word name of 213 is two hundred thirteen.

The symbols **less than**, $<$, and **greater than**, $>$, are used to compare two whole numbers that are not equal. So, $11 < 15$, and $21 > 5$.

To **round** a whole number means to give an approximate value. The symbol $\approx$ means "approximately equal to."

A **table** is a method of displaying data in an **array** using a horizontal and vertical arrangement to distinguish the type of data. A **row** of a table is a horizontal line of a table and reads left to right across the page. A **column** of a table is a vertical line of a table and reads up or down the page. For example, in Table 1.2 the number "57" is in row 3 and column 2.

	Column 2		
134	56	89	102
14	116	7	98
65	57	12	67
23	56	7	213

TABLE 1.2

HOW & WHY

■ **OBJECTIVE 1** Write word names from place value names and place value names from word names.

In our written whole number system (called the Hindu-Arabic system), digits and commas are the only symbols used. This system is a positional base 10 (decimal) system. The location of the digit determines its value, from right to left. The first three place value names are one, ten, and hundred. See Figure 1.1.

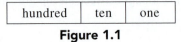

hundred	ten	one

Figure 1.1

For the number 764,

4 is in the ones place, so it contributes 4 ones, or 4, to the value of the number.

6 is in the tens place, so it contributes 6 tens, or 60, to the value of the number.

7 is in the hundreds place, so it contributes 7 hundreds, or 700, to the value of the number.

So 764 is 7 hundreds + 6 tens + 4 ones, or 700 + 60 + 4. These are called **expanded forms** of the number. The word name is seven hundred sixty-four.

For numbers larger than 999, we use commas to separate groups of three digits. The first four groups are unit, thousand, million, and billion (Figure 1.2). The group on the far left may have one, two, or three digits. All other groups must have three digits. Within each group the names are the same (hundred, ten, and one).

hundred ten one	hundred ten one	hundred ten one	hundred ten one
billion	million	thousand	(unit)

Figure 1.2

For 63,506,345,222 the group names are

63	506	345	222
billion	million	thousand	unit

The number is read "63 billion, 506 million, 345 thousand, 222." The word *units* for the units group is not read. The complete word name is sixty-three billion, five hundred six million, three hundred forty-five thousand, two hundred twenty-two.

To write the word name from a place value name

1. From left to right, write the word name for each set of three digits followed by the group name (except units).
2. Insert a comma after each group name.

CAUTION

The word *and* is not used to write names of whole numbers. So write three hundred ten, **NOT** three hundred. and ten, also one thousand, two hundred twenty-three, **NOT** one thousand and two hundred twenty-three.

To write the place value name from the word name of a number, we reverse the previous process. First identify the group names and then write each group name in the place value name. Remember to write a 0 for each missing place value. Consider

three billion, two hundred thirty-five million, nine thousand, four hundred thirteen

three *billion,* two hundred thirty-five *million,* nine *thousand,* four hundred thirteen	Identify the group names. (*Hint:* Look for the commas.)
3 billion, 235 million, 9 thousand, 413	Write the place value name for each group.
3,235,009,413	Drop the group names. Keep all commas. Zeros must be inserted to show that there are no hundreds or tens in the thousands group.

To write a place value name from a word name

1. Identify the group names.
2. Write the three-digit number before each group name, followed by a comma. (The first group, on the left, may have fewer than three digits.) It is common to omit the comma in 4-digit numerals.

Numbers like 81,000,000,000, with all zeros following a single group of digits, are often written in a combination of place value notation and word name. The first set of digits on the left is written in place value notation followed by the group name. So 81,000,000,000 is written 81 billion.

EXAMPLES A–B

DIRECTIONS: Write the word name.

STRATEGY: Write the word name of each set of three digits, from left to right, followed by the group name.

A. Write the word name for 31,781,564.

31	Thirty-one million,
781	seven hundred eighty-one thousand,
564	five hundred sixty-four

Write the word name for each group followed by the group name.

> ### CAUTION
> Do not write the word *and* when reading or writing a whole number.

The word name is thirty-one million, seven hundred eighty-one thousand, five hundred sixty-four.

B. Write the word name for 9,382,059.
Nine million, three hundred eight-two thousand, fifty-nine.

WARM-UP
A. Write the word name for 11,302,714.

WARM-UP
B. Write the word name for 8,431,619.

EXAMPLES C–F

DIRECTIONS: Write the place value name.

STRATEGY: Write the 3-digit number for each group followed by a comma.

C. Write the place value name for four million, seventy-six thousand, two hundred sixty-five.

4,	Millions group.
076,	Thousands group. (Note that a zero is inserted on the left to fill out the three digits in the group.)
265	Units group.

The place value name is 4,076,265

D. Write the place value name for 346 million.
The place value name is 346,000,000. Replace the word *million* with six zeros.

E. Write the place value name for four thousand fifty-three. Note that the comma is omitted. The place value name is 4053.

F. The purchasing agent for the Russet Corporation received a telephone bid of forty-three thousand fifty-one dollars as the price of a new printing press. What is the place value name of the bid that she will include in her report to her superior?

forty-three thousand, fifty-one

| 43, | 051 |

The place value name she reports is $43,051.

WARM-UP
C. Write the place value name for twenty-two million, seventy-seven thousand, four hundred eleven.

WARM-UPS D–F
D. Write the place value name for 74 thousand.

E. Write the place value name for seven thousand fifteen.

F. The purchasing agent for the Russet Corporation also received a bid of twenty-one thousand, five hundred eighteen dollars for a supply of paper. What is the place value name of the bid that she will include in her report to her superior?

ANSWERS TO WARM-UPS A–F

A. eleven million, three hundred two thousand, seven hundred fourteen
B. eight million, four hundred thirty-one thousand, six hundred nineteen
C. 22,077,411 D. 74,000 E. 7015
F. The place value name she reports is $21,518

HOW & WHY

■ **OBJECTIVE 2** Write an inequality statement about two numbers.

If two whole numbers are not equal, then the first is either *less than* or *greater than* the second. Look at the number line (or ruler) in Figure 1.3.

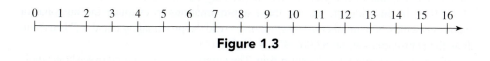

Figure 1.3

Given two numbers on a number line or ruler, the number on the right is the larger. For example,

$9 > 7$	9 is to the right of 7, so 9 is greater than 7.
$11 > 1$	11 is to the right of 1, so 11 is greater than 1.
$14 > 8$	14 is to the right of 8, so 14 is greater than 8.
$13 > 0$	13 is to the right of 0, so 13 is greater than 0.

Given two numbers on a number line or ruler, the number on the left is the smaller. For example,

$3 < 9$	3 is to the left of 9, so 3 is less than 9.
$5 < 12$	5 is to the left of 12, so 5 is less than 12.
$1 < 9$	1 is to the left of 9, so 1 is less than 9.
$10 < 14$	10 is to the left of 14, so 10 is less than 14.

For larger numbers, imagine a longer number line. Notice how the points in the symbols $<$ and $>$ point to the smaller of the two numbers. For example,

$$181 < 715$$
$$87 > 56$$
$$5028 > 5026$$

To write an inequality statement about two numbers

1. Insert $<$ between the numbers if the number on the left is smaller.
2. Insert $>$ between the numbers if the number on the left is larger.

EXAMPLES G–H

DIRECTIONS: Insert $<$ or $>$ to make a true statement.

STRATEGY: Imagine a number line. The smaller number is on the left. Insert the symbol that points to the smaller number.

G. Insert the appropriate inequality symbol: 62   83

$62 < 83$

H. Insert the appropriate inequality symbol: 7909   7099

$7909 > 7099$

WARM-UPS G–H

G. Insert the appropriate inequality symbol: 164   191

H. Insert the appropriate inequality symbol: 6318   6269

ANSWERS TO WARM-UPS G–H

G. $<$ **H.** $>$

HOW & WHY

■ **OBJECTIVE 3** Round a given whole number.

Many numbers that we see in daily life are approximations. These are used to indicate the approximate value when it is believed that the exact value is not important to the discussion. So attendance at a political rally may be stated at 15,000 when it was actually 14,783. The amount of a deficit in the budget may be stated as $2,000,000 instead of $2,067,973. In this chapter, we use these approximations to estimate the outcome of operations with whole numbers. The symbol ≈, read "approximately equal to," is used to show the approximation. So $2,067,973 ≈ $2,000,000.

We approximate numbers by **rounding**. The number line can be used to see how whole numbers are rounded. Suppose we wish to round 57 to the nearest ten. See Figure 1.4

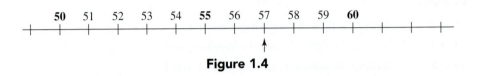

Figure 1.4

The arrow under the 57 is closer to 60 than to 50. We say "to the nearest ten, 57 rounds to 60."

We use the same idea to round any number, although we usually make only a mental image of the number line. The key question is, Is this number closer to the smaller rounded number or to the larger one? Practically, we need to determine only if the number is more or less than half the distance between the rounded numbers.

To round 47,472 to the nearest thousand without a number line, draw an arrow under the digit in the thousands place.

47,472
↑

Because 47,472 is between 47,000 and 48,000, we must decide which number it is closer to. Because 47,500 is halfway between 47,000 and 48,000 and because 47,472 < 47,500, we conclude that 47,472 is less than halfway to 48,000.

Whenever the number is less than halfway to the larger number, we choose the smaller number.

47,472 ≈ 47,000   47,472 is closer to 47,000 than to 48,000.

> ### To round a number to a given place value
>
> 1. Draw an arrow under the given place value.
> 2. If the digit to the right of the arrow is 5, 6, 7, 8, or 9, add one to the digit above the arrow. (Round to the larger number.)
> 3. If the digit to the right of the arrow is 0, 1, 2, 3, or 4, do not change the digit above the arrow. (Round to the smaller number.)
> 4. Replace all the digits to the right of the arrow with zeros.

EXAMPLES I–J

DIRECTIONS: Round to the indicated place value.

STRATEGY: Choose the larger number if the digit to the right of the round-off place is 5 or greater, otherwise, choose the smaller number.

I. Round 344,599 to the nearest ten thousand.

344,599   Draw an arrow under the ten-thousands place.
↑

340,000   The digit to the right of the arrow is 4.
           Because 4 < 5, choose the smaller number.

So 344,599 ≈ 340,000.

J. Round the numbers to the indicated place value.

Number	Ten	Hundred	Thousand
862,548	862,550	862,500	863,000
35,632	35,630	35,600	36,000

HOW & WHY

■ **OBJECTIVE 4**   Read tables.

Data are often displayed in the form of a *table*. We see tables in the print media, in advertisements, and in business presentations. Reading a table involves finding the correct *column* and *row* that describes the needed information, and then reading the data at the intersection of that column and that row.

TABLE 1.3   **Student Course Enrollment**

Class	Mathematics	English	Science	Humanities
Freshman	950	1500	500	1200
Sophomore	600	700	650	1000
Junior	450	200	950	1550
Senior	400	250	700	950

For example, in Table 1.3, to find the number of sophomores who take English, find the column headed English and the row headed Sophomore and read the number at the intersection.

The number of sophomores taking English is 700.

We can use the table to compare enrollments by class. For instance, are more seniors or sophomores taking science? From the table we see that 650 sophomores are taking science and 700 seniors are taking science. Since 700 > 650, more seniors than sophomores are taking science.

EXAMPLE K

DIRECTIONS: Answer the questions associated with the table.

STRATEGY: Examine the rows and columns of the table to determine the values that are related.

WARM-UP

K. Use the table in Example K to answer the questions.

1. Which location has the lowest-priced home sold?
2. Round the average price of a home sold in S.E. Portland to the nearest thousand.
3. Which location has the higher price for a home sold, N.E. Portland or S.E. Portland?

K. This table shows the value of homes sold in the Portland metropolitan area for a given month.

Values of Houses Sold

Location	Lowest	Highest	Average
N. Portland	$86,000	$258,500	$184,833
N.E. Portland	$78,000	$220,000	$165,091
S.E. Portland	$82,000	$264,000	$173,490
Lake Oswego	$140,000	$1,339,000	$521,080
W. Portland	$129,500	$799,000	$354,994
Beaverton	$98,940	$665,000	$293,737

1. In which location was the highest-priced home sold?
2. Which area has the highest average sale price?
3. Round the highest price of a house in Beaverton to the nearest hundred thousand.

1. We look at the Highest column for the largest entry. It is $1,339,000, which is in the fourth row. So Lake Oswego is the location of the highest-priced home sold.
2. Looking at the Average column for the largest entry, we find $521,080 in the fourth row. So Lake Oswego has the largest average sale price.
3. Looking at the Highest column and the sixth row, we find $665,000. So the highest price of a house in Beaverton is $700,000, rounded to the nearest hundred thousand.

ANSWER TO WARM-UP K

K. 1. N.E. Portland has the lowest-priced home sold. 2. The rounded price is $173,000. 3. S.E. Portland has the higher price.

EXERCISES 1.1

■ **OBJECTIVE 1** Write word names from place value names and place value names from word names. (See page 4.)

A *Write the word names of each of these numbers.*

1. 843

2. 196

3. 460

4. 710

5. 7020

6. 66,086

Write the place value name.

7. Eighty-seven

8. Thirty-nine

9. Nine thousand, five hundred

10. Nine thousand, five

11. One hundred one million

12. 493 thousand

B *Write the word name of each of these numbers.*

13. 27,680

14. 27,068

15. 207,690

16. 270,069

17. 54,000,000

18. 780,000

Write the place value name.

19. Two hundred forty-three thousand, seven hundred

20. Three hundred fifty-nine thousand, eight

21. Twenty-two thousand, five hundred seventy

22. Twenty-three thousand, four hundred seventy-seven

23. Nineteen billion

24. Nine hundred thousand, five

◼ **OBJECTIVE 2** Write an inequality statement about two numbers. (See page 7.)

A *Insert < or > between the numbers to make a true statement.*

25. 12   22

26. 53   49

27. 61   54

28. 62   71

B

29. 246   251

30. 212   208

31. 7470   7850

32. 2751   2693

◼ **OBJECTIVE 3** Round a given whole number. (See page 8.)

A *Round to the indicated place value.*

33. 742 (ten)

34. 794 (ten)

35. 2655 (hundred)

36. 8250 (hundred)

B

	Number	Ten	Hundred	Thousand	Ten Thousand
37.	607,546				
38.	467,299				
39.	6,545,742				
40.	4,309,498				

■ **OBJECTIVE 4** Read tables. (See page 9.)

A *Exercises 41– 45. The percent of people who do and do not exercise regularly, broken down by income levels, is shown in the following table. (Source: Centers for Disease Control and Prevention)*

Regular Exercises by Income Level

Income	Does Exercise	Does Not Exercise
$0–$14,999	35%	65%
$15–$24,999	40%	60%
$25–$50,000	45%	55%
Over $50,000	52%	48%

41. What percent of people in the income level of $15–$24,999 exercise regularly?

42. Which income level has the lowest percent of regular exercisers?

43. Which income level has the lowest percent of nonexercisers?

44. Which income level(s) have less than 50% nonexercisers?

45. Use words to describe the trend indicated in the table.

B *Exercises 46–50. The following table gives the number of homeless in selected cities according to data collected by the National Alliance to End Homelessness.*

Number of Homeless in Selected Cities

City	2005	2007
Atlanta	6832	6840
Boston	5819	5104
Chicago	6680	5979
Minneapolis	3415	2984
New Orleans	2051	1619
San Diego	4268	3485
Seattle	7315	7902

46. Which city in the table had essentially no change in the number of homeless?

47. Which cities had an increase in the number of homeless? Which cities had a decrease?

48. Write an inequality comparing the homeless populations of Boston and San Diego in 2007.

49. Write an inequality comparing the homeless populations of Chicago and Atlanta in 2007?

50. Which city had the largest population of homeless in 2005? Which had the smallest?

C *Write the place value name.*

51. Six hundred fifty-six million, seven hundred thirty-two thousand, four hundred ten.

52. Nine hundred five million, seven hundred seventy-seven

Exercises 53–54. The following chart gives the average income of the top 5% of families and the bottom 20% of families in Iowa. (Source: CNNMoney.com)

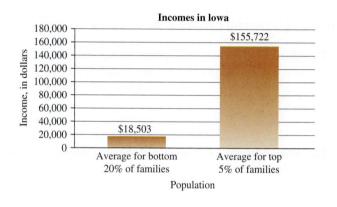

53. Write the word name for the average salary for the poor in Iowa.

54. Write the word name for the average salary of the rich in Iowa.

Insert < or > between the numbers to make a true statement.

55. 4553   4525

56. 21,186   21,299

57. What is the largest 4-digit number?

58. What is the smallest 5-digit number?

Round to the indicated place value.

59. 74,571,111 (hundred thousand)

60. 64,009,880 (ten thousand)

61. Round 63,749 to the nearest hundred. Round 63,749 to the nearest ten and then round your result to the nearest hundred. Why did you get a different result the second time? Which method is correct?

62. Hazel bought a plasma flat screen television set for $2495. She wrote a check to pay for it. What word name did she write on the check?

63. Kimo bought a used Toyota Camry for $11,475 and wrote a check to pay for it. What word name did he write on the check?

Exercises 64–65. A biologist with the Columbia River Inter-Tribal Fish Commission estimates the sockeye salmon run on the Columbia River at 25,400 in 2007 and 230,000 in 2008.

64. Write the word name for the number of salmon in 2007.

65. Write an inequality comparing the salmon runs in 2007 and 2008.

66. Some sources estimate that the world population will exceed 7 billion, 302 million, 186 thousand by 2015. Write the place value name for this world population.

67. The purchasing agent for Print-It-Right received a telephone bid of thirty-six thousand, four hundred seven dollars as the price for a new printing press. What is the place value name for the bid?

68. The Oak Ridge Missionary Baptist Church in Kansas City took out a building permit for $2,659,500. Round the building permit price of the church to the nearest hundred thousand dollars.

69. Ten thousand shares of the Income Fund of America sold for $185,200. What is the value of the sale, to the nearest thousand dollars?

Exercises 70–72. The following table gives the total carbon emissions for the state of California, measured in metric tons.

Metric Tons of Carbon Emissions in California

1970	1980	1990	2000
80,661,690	90,729,685	94,073,529	100,655,263

70. Write the place value name for the number of metric tons of carbon emissions in California in 1980.

71. Round the number of metric tons of carbon emissions in California in 2000 to the nearest million metric tons.

72. What is the general trend in carbon emissions in California over the past 30 years?

Exercises 73–76. The following table gives the per capita income in the New England states in a recent year, according to the U.S. Bureau of Economic Analysis.

Per Capita Income in New England

State	Per Capita Income
Connecticut	$50,762
Maine	$32,095
Massachusetts	$46,299
New Hampshire	$39,753
Rhode Island	$37,523
Vermont	$34,871

73. Write the word name of the per capita income in Maine.

74. Round the per capita income in Massachusetts to the nearest thousand.

75. Which state has the smallest per capita income?

76. Which state has the largest per capita income?

77. The distance from Earth to the sun was measured and determined to be 92,875,328 miles. To the nearest million miles, what is the distance?

78. According to the National Cable Television Association, the top six pay-cable networks for 2008 are shown in the table to the right. To what place value do these numbers appear to have been rounded?

Network	Subscribers
Discovery	98,000,000
TNT	98,000,000
ESPN	97,800,000
CNN	97,500,000
USA	97,500,000
Lifetime	97,300,000

Exercises 79–80. The number of marriages each month for a recent year, according to data from the U.S. Census Bureau is given in the following table.

Number of Marriages per Month

Month	Number of Marriages, in Thousands	Month	Number of Marriages, in Thousands
January	110	July	235
February	155	August	239
March	118	September	225
April	172	October	231
May	241	November	171
June	242	December	184

79. Rewrite the information ordering the months from most number of marriages to least number of marriages. Use place value notation when writing the number of marriages.

80. Do you think the number of marriages has been rounded? If so to what place value?

Exercises 81–82. The six longest rivers in the United States are as follows:

Arkansas	1459 miles
Colorado	1450 miles
Mississippi	2340 miles
Missouri	2315 miles
Rio Grande	1900 miles
Yukon	1079 miles

81. List the rivers in order of increasing length.

82. Do you think any of the river lengths have been rounded? If so, which ones?

83. The state motor vehicle department estimated the number of licensed automobiles in the state to be 2,376,000, to the nearest thousand. A check of the records indicated that there were actually 2,376,499. Was their estimate correct?

84. The total land area of Earth is approximately 52,425,000 square miles. What is the land area to the nearest million square miles?

Exercises 85–86. The following figure lists some nutritional facts about two brands of peanut butter.

Skippy® Super Chunk

Nutrition Facts
Serving Size 2 tbsp (32g)
Servings Per Container about 15

Amount Per Serving

Calories 190   **Calories from Fat**   140

	% Daily Values
Total Fat 17g	26%
Saturated Fat 3.5g	17%
Cholesterol 0mg	0%
Sodium 140mg	6%
Total Carbohydrate 7g	2%
Dietary Fiber 2g	8%
Sugars 3g	
Protein 7g	

Jif® Creamy
Simply Jif contains 2g sugar per serving.
Regular Jif contains 3g sugar per serving

Nutrition Facts
Serving Size 2 tbsp (31g)
Servings Per Container about 16

Amount Per Serving

Calories 190   **Calories from Fat**   130

	% Daily Values
Total Fat 16g	25%
Saturated Fat 3g	16%
Cholesterol 0mg	0%
Sodium 65mg	3%
Total Carbohydrate 6g	2%
Dietary Fiber 2g	9%
Sugars 2g	
Protein 8g	

85. List the categories of nutrients for which Jif has fewer of the nutrients than Skippy.

86. Round the sodium content in each brand to the nearest hundred. Do the rounded numbers give a fair comparison of the amount of sodium in the brands?

Exercises 87–90 relate to the chapter application. See Table 1.1, page 3.

87. Write the name for the dollar amount taken in by *The Dark Knight*.

88. Round the amount taken in by *Spider-Man* to the nearest hundred thousand.

89. Round the amount taken in by *Shrek 2* to the nearest million.

90. Do the numbers in Table 1.1 appear to be rounded?

STATE YOUR UNDERSTANDING

91. Explain why "base ten" is a good name for our number system.

92. Explain what the digit 9 means in 295,862.

93. What is rounding? Explain how to round 87,452 to the nearest thousand and to the nearest hundred.

CHALLENGE

94. What is the place value for the digit 5 in 3,456,709,230,000?

95. Write the word name for 5,326,901,570,000.

96. Arrange the following numbers from smallest to largest: 1234, 1342, 1432, 1145, 1243, 1324, and 1229.

97. What is the largest value of X that makes 2X56 > 2849 false?

98. Round 967,345 to the nearest hundred thousand.

99. Round 49,774 to the nearest hundred thousand.

100. Two other methods of rounding are called the "odd/even method" and "truncating." Research these methods. (*Hint:* Try the library or talk to science and business instructors.)

SECTION
1.2

Adding and Subtracting Whole Numbers

VOCABULARY

Addends are the numbers that are added. In 9 + 20 + 3 = 32, the addends are 9, 20, and 3.

The result of adding is called the **sum**. In 9 + 20 + 3 = 32, the sum is 32.

The result of subtracting is called the **difference**. So in 62 − 34 = 28, 28 is the difference.

A **polygon** is any closed figure whose sides are line segments.

The **perimeter** of a polygon is the distance around the outside of the polygon.

OBJECTIVES

1. Find the sum of two or more whole numbers.
2. Find the difference of two whole numbers.
3. Estimate the sum or difference of whole numbers.
4. Find the perimeter of a polygon.

HOW & WHY

■ **OBJECTIVE 1**   Find the sum of two or more whole numbers.

When Jose graduated from high school he received cash gifts of $50, $20, and $25. The total number of dollars received is found by adding the individual gifts. The total number of dollars he received is 95. In this section we review the procedure for adding and subtracting whole numbers.

The addition facts and place value are used to add whole numbers written with more than one digit. Let's use this to find the sum of the cash gifts that Jose received. We need to find the sum of

50 + 20 + 25

By writing the numbers in expanded form and putting the same place values in columns it is easy to add.

$$
\begin{array}{rl}
50 = & 5 \text{ tens} + 0 \text{ ones} \\
20 = & 2 \text{ tens} + 0 \text{ ones} \\
+25 = & \underline{2 \text{ tens} + 5 \text{ ones}} \\
& 9 \text{ tens} + 5 \text{ ones} = 95
\end{array}
$$

So, 50 + 20 + 25 = 95. Jose received $95 in cash gifts.

Because each place can contain only a single digit, it is often necessary to rewrite the sum of a column.

$$
\begin{array}{rl}
77 = & 7 \text{ tens} + 7 \text{ ones} \\
+16 = & \underline{1 \text{ tens} + 6 \text{ ones}} \\
& 8 \text{ tens} + 13 \text{ ones}
\end{array}
$$

Because 13 ones is a 2-digit number it must be renamed:

$$8 \text{ tens} + 13 \text{ ones} = 8 \text{ tens} + 1 \text{ ten} + 3 \text{ ones}$$
$$= 9 \text{ tens} + 3 \text{ ones}$$
$$= 93$$

So the sum of 77 and 16 is 93.

The common shortcut is shown in the following sum. To add $497 + 307 + 135$, write the numbers in a column.

$$\begin{array}{r} 497 \\ 307 \\ +135 \\ \hline \end{array}$$ Written this way, the digits in the ones, tens, and hundreds places are aligned.

$$\begin{array}{r} \overset{1}{497} \\ 307 \\ +135 \\ \hline 9 \end{array}$$ Add the digits in the ones column: $7 + 7 + 5 = 19$. Write 9 and carry the 1 (1 ten) to the tens column.

$$\begin{array}{r} \overset{11}{497} \\ 307 \\ +135 \\ \hline 39 \end{array}$$ Add the digits in the tens column: $1 + 9 + 0 + 3 = 13$. Write 3 and carry the 1 (10 tens = 1 hundred) to the hundreds column.

$$\begin{array}{r} \overset{11}{497} \\ 307 \\ +135 \\ \hline 939 \end{array}$$ Add the digits in the hundreds column: $1 + 4 + 3 + 1 = 9$

To add whole numbers

1. Write the numbers in a column so that the place values are aligned.
2. Add each column, starting with the ones (or units) column.
3. If the sum of any column is greater than nine, write the ones digit and "carry" the tens digit to the next column.

EXAMPLES A–C

DIRECTIONS: Add.

STRATEGY: Write the numbers in a column. Add the digits in the columns starting on the right. If the sum is greater than 9, "carry" the tens digit to the next column.

A. Add: $684 + 537$

$$\begin{array}{r} \overset{11}{684} \\ +537 \\ \hline 1221 \end{array}$$ Add the numbers in the ones column. $4 + 7 = 11$. Because the sum is greater than 9, write 1 in the ones column and carry the 1 to the tens column. Add the numbers in the tens column. $1 + 8 + 3 = 12$. Write 2 in the tens column and carry the 1 to the hundreds column. Add the numbers in the hundreds column. $1 + 6 + 5 = 12$. Because all columns have been added there is no need to carry.

WARM-UP
A. Add: $851 + 379$

ANSWER TO WARM-UP A

A. 1230

B. Add 68, 714, 7, and 1309. Round the sum to the nearest ten.

$$
\begin{array}{r}
{\scriptstyle 1\ 2} \\
68 \\
714 \\
7 \\
+1309 \\
\hline
2098
\end{array}
$$

When writing in a column, make sure the place values are aligned properly.

$2098 \approx 2100$ Round to the nearest ten.

CALCULATOR EXAMPLE:

C. Add: $7659 + 518 + 7332 + 4023 + 1589$

Calculators are programed to add numbers just as we have been doing by hand. Simply enter the exercise as it is written horizontally and the calculator will do the rest.

The sum is 21,121.

HOW & WHY

■ **OBJECTIVE 2**   Find the difference of two whole numbers.

Marcia went shopping with $78. She made purchases totaling $53. How much money does she have left? Finding the difference in two quantities is called subtraction. When we subtract $53 from $78 we get $25.

Subtraction can also be thought of as finding the missing addend in an addition exercise. For instance, $9 - 5 = ?$ asks $5 + ? = 9$. Because $5 + 4 = 9$, we know that $9 - 5 = 4$. Similarly, $47 - 15 = ?$ asks $15 + ? = 47$. Because $15 + 32 = 47$, we know that $47 - 15 = 32$.

For larger numbers, such as $875 - 643$, we take advantage of the column form and expanded notation to find the missing addend in each column.

$$
\begin{array}{rl}
875 = & 8 \text{ hundreds} + 7 \text{ tens} + 5 \text{ ones} \\
-643 = & 6 \text{ hundreds} + 4 \text{ tens} + 3 \text{ ones} \\
\hline
& 2 \text{ hundreds} + 3 \text{ tens} + 2 \text{ ones} = 232
\end{array}
$$

Check by adding:
$$
\begin{array}{r}
643 \\
+232 \\
\hline
875
\end{array}
$$

So, $875 - 643 = 232$.

Now consider the difference $672 - 438$. Write the numbers in column form.

$$
\begin{array}{rl}
672 = & 6 \text{ hundreds} + 7 \text{ tens} + 2 \text{ ones} \\
-438 = & 4 \text{ hundreds} + 3 \text{ tens} + 8 \text{ ones}
\end{array}
$$

Here we cannot subtract 8 ones from 2 ones, so we rename by "borrowing" one of the tens from the 7 tens (1 ten = 10 ones) and adding the 10 ones to the 2 ones.

$$
\begin{array}{rl}
& \quad\quad\quad {\scriptstyle 6\ tens} \quad\ {\scriptstyle 12\ ones} \\
672 = & 6 \text{ hundreds} + 7 \text{ tens} + 2 \text{ ones} \\
-438 = & 4 \text{ hundreds} + 3 \text{ tens} + 8 \text{ ones} \\
\hline
& 2 \text{ hundreds} + 3 \text{ tens} + 4 \text{ ones} = 234
\end{array}
$$

Check by adding:
$$
\begin{array}{r}
{\scriptstyle 1} \\
438 \\
+234 \\
\hline
672
\end{array}
$$

We generally don't bother to write the expanded form when we subtract. We show the shortcut for borrowing in the examples.

> **To subtract whole numbers**
>
> 1. Write the numbers in a column so that the place values are aligned.
> 2. Subtract in each column, starting with the ones (or units) column.
> 3. When the numbers in a column cannot be subtracted, borrow 1 from the next column and rename by adding 10 to the upper digit in the current column and then subtract.

EXAMPLES D–H

DIRECTIONS: Subtract and check.

STRATEGY: Write the numbers in columns. Subtract in each column. Rename by borrowing when the numbers in a column cannot be subtracted.

D. Subtract: $78 - 27$

$$\begin{array}{r} 78 \\ -27 \\ \hline 51 \end{array}$$

Subtract the ones column: $8 - 7 = 1$.
Subtract the tens column: $7 - 2 = 5$.

CHECK:

$$\begin{array}{r} 51 \\ +27 \\ \hline 78 \end{array}$$

So $78 - 27 = 51$.

E. Find the difference: $836 - 379$

$$\begin{array}{r} {}^{2\ 16} \\ 8\,\cancel{3}\,6 \\ -3\,7\,9 \\ \hline 7 \end{array}$$

In order to subtract in the ones column we borrow 1 ten (10 ones) from the tens column and rename the ones ($10 + 6 = 16$).

$$\begin{array}{r} {}^{7\ 12} \\ {}^{2\ 16} \\ \cancel{8}\,\cancel{3}\,6 \\ -3\,7\,9 \\ \hline 4\,5\,7 \end{array}$$

Now in order to subtract in the tens column, we must borrow 1 hundred (10 tens) from the hundreds column and rename the tens ($10 + 2 = 12$).

CHECK:

$$\begin{array}{r} 379 \\ +457 \\ \hline 836 \end{array}$$

So $836 - 379 = 457$.

F. Subtract 759 from 7300.

$$\begin{array}{r} 7300 \\ -\ 759 \end{array}$$

We cannot subtract in the ones column, and since there are 0 tens, we cannot borrow from the tens column.

$$\begin{array}{r} {}^{2\ 10} \\ 7\,\cancel{3}\,0\,0 \\ -\ 7\,5\,9 \end{array}$$

We borrow 1 hundred (1 hundred = 10 tens) from the hundreds place.

WARM-UP
D. Subtract: $78 - 35$

WARM-UP
E. Find the difference:
$823 - 476$

WARM-UP
F. Subtract 495 from 7100.

ANSWERS TO WARM-UPS D–F

D. 43   **E.** 347   **F.** 6605

$$\begin{array}{r} {\scriptstyle 9\,10} \\[-2pt] {\scriptstyle 2\,\cancel{1}0} \\[-2pt] 7\,3\,\cancel{0}\,\cancel{0} \\ -\ \ 7\,5\,9 \end{array}$$ Now borrow 1 ten (1 ten = 10 ones). We can now subtract in the ones and tens columns but not in the hundreds column.

$$\begin{array}{r} {\scriptstyle 6\ 12\ 9\,10} \\[-2pt] {\scriptstyle 2\,\cancel{1}0} \\[-2pt] 7\,\cancel{3}\,\cancel{0}\,\cancel{0} \\ -\ \ 7\,5\,9 \\ \hline 6\,5\,4\,1 \end{array}$$ Now borrow 1 thousand (1 thousand = 10 hundreds). We can now subtract in every column.

CHECK:
$$\begin{array}{r} 6541 \\ +\ \ 759 \\ \hline 7300 \end{array}$$

Let's try Example F again using a technique called "reverse adding." Just ask your-self, "What do I add to 759 to get 7300?"

$$\begin{array}{r} 7300 \\ -\ \ 759 \end{array}$$ Begin with the ones column. 9 is larger than 0, so ask "What do I add to 9 to make 10?"

$$\begin{array}{r} 7300 \\ -\ \ 759 \\ \hline 1 \end{array}$$ Because $1 + 9 = 10$, we write the 1 in the ones column and carry the 1 over to the 5 to make 6. Now ask "What do I add to 6 to make 10?"

$$\begin{array}{r} 7300 \\ -\ \ 759 \\ \hline 41 \end{array}$$ Write 4 in the tens column and carry the 1 over to the 7 in the hundreds column. Now ask "What do I add to 8 to make 13?"

$$\begin{array}{r} 7300 \\ -\ \ 759 \\ \hline 6541 \end{array}$$ Write the 5 in the hundreds column. Finally, ask "What do I add to the carried 1 to make 7?"

The advantage of this method is that 1 is the largest amount carried, so most people can do this process mentally.

So $7300 - 759 = 6541$.

CALCULATOR EXAMPLE:

G. Subtract 58,448 from 75,867.

> **CAUTION**
>
> When a subtraction exercise is worded "Subtract A from B," it is necessary to reverse the order of the numbers. The difference is B − A.

Enter $75{,}867 - 58{,}448$.
The difference is 17,419.

H. Maxwell Auto is advertising a $986 rebate on all new cars priced above $15,000. What is the cost after rebate of a car originally priced at $16,798?

STRATEGY: Because the price of the car is over $15,000, we subtract the amount of the rebate to find the cost.

$$\begin{array}{r} 16{,}798 \\ -\ \ \ 986 \\ \hline 15{,}812 \end{array} \qquad \textbf{CHECK:}\quad \begin{array}{r} 15{,}812 \\ +\ \ \ 986 \\ \hline 16{,}798 \end{array}$$

The car costs $15,812.

WARM-UP

G. Subtract: 59,677 from 68,143.

WARM-UP

H. Maxwell Auto is also advertising a $2138 rebate on all new cars priced above $32,000. What is the cost after rebate of a car originally priced at $38,971?

ANSWERS TO WARM-UPS G–H

G. 8466 **H.** The cost of the car is $36,833.

HOW & WHY

■ **OBJECTIVE 3** Estimate the sum or difference of whole numbers.

The sum or difference of whole numbers can be estimated by rounding each number to a specified place value and then adding or subtracting the rounded values. Estimating is useful to check to see if a calculated sum or difference is reasonable or when the exact sum is not needed. For instance, estimate the sum by rounding to the nearest thousand.

6359	6000	Round each number to the nearest thousand.
3790	4000	
9023	9000	
4825	5000	
+ 899	+ 1000	
	25,000	

The estimate of the sum is 25,000.

Another estimate can be found by rounding each number to the nearest hundred.

6359	6400	Round each number to the nearest hundred.
3790	3800	
9023	9000	
4825	4800	
+ 899	+ 900	
	24,900	

We can use the estimate to see if we added correctly. If a calculated sum is not close to the estimated sum, you should check the addition by re-adding. In this case the calculated sum, 24,896, is close to the estimated sums of 25,000 and 24,900.

Estimate the difference of two numbers by rounding each number. Subtract the rounded numbers.

8967	9000	Round each number to the nearest hundred.
−5141	−5100	Subtract.
	3900	

The estimate of the difference is 3900. We use the estimate to see if the calculated difference is correct. If the calculated difference is not close to 3900, you should check the subtraction. In this case, the calculated difference is 3826, which is close to the estimate.

WARM-UP

I. Estimate the sum by rounding each number to the nearest hundred:

643 + 72 + 422 + 875 + 32 + 91

ANSWER TO WARM-UP I

I. The estimated sum is 2100.

EXAMPLES I–M

DIRECTIONS: Estimate the sum or difference.

STRATEGY: Round each number to the specified place value. Then add or subtract.

I. Estimate the sum by rounding each number to the nearest hundred:

475 + 8795 + 976 + 6745 + 5288 + 12

475	500	Round each number to the nearest hundred.
8795	8800	With practice, this can be done mentally.
976	1000	
6745	6700	
5288	5300	
+ 12	+ 0	
	22,300	

The estimated sum is 22,300.

J. Estimate the difference of 56,880 and 28,299 by rounding to the nearest thousand.

$$\begin{array}{r} 57{,}000 \\ -28{,}000 \\ \hline 29{,}000 \end{array}$$ Round each number to the nearest thousand.

The estimated difference is 29,000.

K. Petulia subtracts 756 from 8245 and gets a difference of 685. Estimate the difference by rounding to the nearest hundred to see if Petulia is correct.

$$\begin{array}{r} 8200 \\ -\ 800 \\ \hline 7400 \end{array}$$ Round each to the nearest hundred.

The estimated answer is 7400 so Petulia is not correct. Apparently she did not align the place values correctly. Subtracting we find the correct answer.

$$\begin{array}{r} 8\,2\,4\,5 \\ -\ \ 7\,5\,6 \\ \hline 7\,4\,8\,9 \end{array}$$

Petulia is not correct; the correct answer is 7489.

L. Joan and Eric have a budget of $1200 to buy new furniture for their living room. They like a sofa that costs $499, a love seat at $449, and a chair at $399. Round the prices to the nearest hundred dollars to estimate the cost of the items. Will they have enough money to make the purchases?

Sofa: $499	500
Love seat: $449	400
Chair: $399	+ 400
	1300

Round each price to the nearest hundred.

The estimated cost, $1300, is beyond their budget, so they will have to rethink the purchase.

M. The population of Alabama is about 4,661,900 and the population of Mississippi is about 2,938,600. Estimate the difference in the populations by rounding each to the nearest hundred thousand.

Alabama: 4,661,900	4,700,000
Mississippi: 2,938,600	−2,900,000
	1,800,000

Round each population to the nearest hundred thousand.

So the estimated difference in populations is 1,800,000.

HOW & WHY

OBJECTIVE 4 Find the perimeter of a polygen.

A polygon is a closed figure whose sides are line segments, such as rectangles, squares, and triangles (Figure 1.5). An expanded discussion of polygons can be found in Section 7.3.

Common polygons

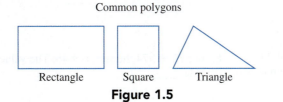

Rectangle Square Triangle

Figure 1.5

The perimeter of a polygon is the distance around the outside. To find the perimeter we add the lengths of the sides.

EXAMPLES N–O

DIRECTIONS: Find the perimeter of the polygon.

STRATEGY: Add up the lengths of all the sides.

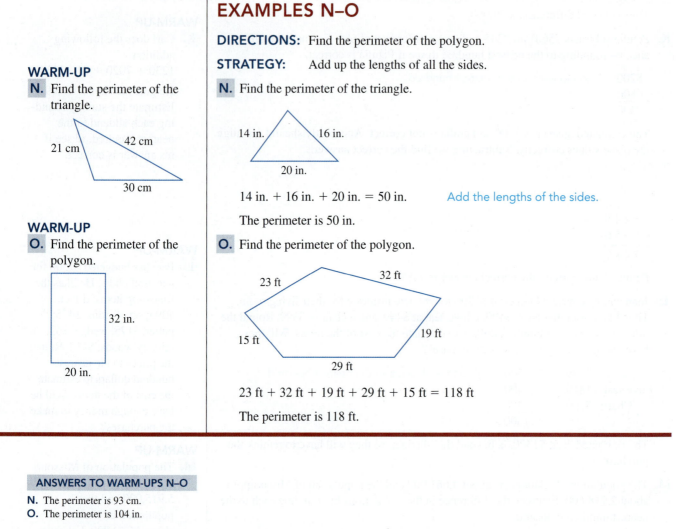

N. Find the perimeter of the triangle.

14 in. + 16 in. + 20 in. = 50 in.   Add the lengths of the sides.

The perimeter is 50 in.

O. Find the perimeter of the polygon.

23 ft + 32 ft + 19 ft + 29 ft + 15 ft = 118 ft

The perimeter is 118 ft.

WARM-UP
N. Find the perimeter of the triangle.

21 cm   42 cm   30 cm

WARM-UP
O. Find the perimeter of the polygon.

32 in.   20 in.

ANSWERS TO WARM-UPS N–O

N. The perimeter is 93 cm.
O. The perimeter is 104 in.

EXERCISES 1.2

■ **OBJECTIVE 1** Find the sum of two or more whole numbers. (See page 17.)

A *Add.*

1. 75 + 38

2. 23 + 85

3. 724 + 218

4. 765 + 127

5. 212
      +495

6. 467
      +324

7. When you add 26 and 39, the sum of the ones column is 15. You must carry the _____ to the tens column.

8. In 572 + 374 the sum is X46. The value of X is _____.

B *Add.*

9. 515 + 2908 + 387

10. 874 + 7052 + 418

11. 7 + 85 + 607 + 5090

12. 3 + 80 + 608 + 7050

13. 2795 + 3643 + 7055 + 4004 (Round sum to the nearest hundred.)

14. 6732 + 9027 + 5572 + 3428 (Round sum to the nearest hundred.)

■ **OBJECTIVE 2** Find the difference of two whole numbers. (See page 19.)

A *Subtract.*

15.   8 hundreds + 7 tens + 4 ones
     −5 hundreds + 7 tens + 2 ones

16.   5 hundreds + 4 tens + 8 ones
     −2 hundreds + 2 tens + 5 ones

17. 406 − 72

18. 764 − 80

19. 876 − 345

20. 848 − 622

21. When subtracting 73 − 18, you can borrow 1 from the 7. The value of the borrowed 1 is _____ ones.

22. When subtracting 526 − 271, you can borrow from the _____ column to subtract in the _____ column.

B *Subtract.*

23. 944 − 458

24. 861 − 468

25. 300 − 164

26. 600 − 388

27. 8769 − 4073 (Round difference to the nearest hundred.)

28. 9006 − 6971 (Round difference to the nearest hundred.)

■ **OBJECTIVE 3** Estimate the sum or difference of whole numbers. (See page 22.)

A *Estimate the sum by rounding each number to the nearest hundred.*

29. 546 + 577

30. 495 + 912

31.   2044
      4550
     +3449

32.   5467
      3811
     +2199

Estimate the difference by rounding each number to the nearest hundred.

33. 675 − 349

34. 768 − 571

35.   9765
     −4766

36.   5479
     −2599

B *Estimate the sum by rounding each number to the nearest thousand.*

37.   3209
      7095
      4444
      2004
     +3166

38.   5038
      4193
      2121
      5339
     +6560

39.   45,902
      33,333
      57,700
     +23,653

40.   12,841
      29,671
      21,951
     +73,846

Estimate the difference by rounding each number to the nearest thousand.

41. 7822
−3098

42. 9772
−4192

43. 65,808
−32,175

44. 92,150
−67,498

■ **OBJECTIVE 4** Find the perimeter of a polygon. (See page 23.)

A *Find the perimeter of the following polygons.*

45.

11 cm

11 cm

46.

8 ft   13 ft

17 ft

47.

56 in.

40 in.   36 in.

89 in.

48.

2 km

24 km

B

49.

7 ft

50.

4 cm   4 cm

10 cm

51.

4 in.

10 in.

40 in.   18 in.

52.

5 m   5 m

6 m   8 m

11 m

6 m   14 m

25 m

C *Exercises 53–58 refer to the sales chart, which gives the distribution of car sales among dealers in Wisconsin.*

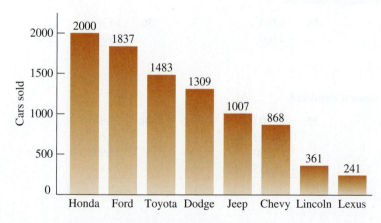

53. What is the total number of Fords, Toyotas, and Lexuses sold?

54. What is the total number of Chevys, Lincolns, Dodges, and Hondas sold?

55. How many more Hondas are sold than Fords?

56. How many more Toyotas are sold than Jeeps?

57. What is the total number sold of the three best-selling cars?

58. What is the difference in cars sold between the best-selling car and the least-selling car?

59. The biologist at the Bonneville fish ladder counted the following number of coho salmon during a one-week period: Monday, 1046; Tuesday, 873; Wednesday, 454; Thursday, 1156; Friday, 607; Saturday, 541; and Sunday, 810. How many salmon went through the ladder that week? How many more salmon went through the ladder on Tuesday than on Saturday?

60. Michelle works the following addition problem, 345 + 672 + 810 + 921 + 150, and gets a sum of 1898. Estimate the answer by rounding each addend to the nearest hundred to see if Michelle's answer is reasonable. If not, find the correct sum.

61. Ralph works the following subtraction problem, 10,034 − 7959, and gets a difference of 2075. Estimate the answer by rounding each number to the nearest thousand to see if Ralph's answer is reasonable. If not, find the correct difference.

62. The state of Alaska has an area of 570,374 square miles, or 365,039,104 acres. The state of Texas has an area of 267,277 square miles, or 171,057,280 acres. Estimate the difference in the areas using square miles rounded to the nearest ten thousand. Estimate the difference in the areas using acres rounded to the nearest million.

63. Philipe buys a refrigerator for $376, an electric range for $482, a dishwasher for $289, and a microwave oven for $148. Estimate the cost of the items by rounding each cost to the nearest hundred dollars.

Exercises 64–66. The table gives the number of offences reported to law enforcement in Miami, Florida, in 2007, according to the FBI's Uniform Crime Reports.

Violent Crimes	
Murder	77
Forcible rape	57
Robbery	253
Aggravated assault	3446

Property Crimes	
Burglary	482
Larceny-theft	12,480
Motor vehicle theft	3876
Arson	176

64. Find the total number of reported violent crimes.

65. Find the total number of reported property crimes.

66. How many more reported burglaries were there than robberies?

Exercises 67–69. A home furnace uses natural gas, oil, or electricity for the energy needed to heat the house. We humans get our energy for body heat and physical activity from calories in our food. Even when resting we use energy for muscle actions such as breathing, heartbeat, digestion, and other functions. If we consume more calories than we use up, we gain weight. If we consume less calories than we use, we lose weight. Some nutritionists recommend about 2270 calories per day for women aged 18–30 who are reasonably active.

Sasha, who is 22 years old, sets 2250 calories per day as her goal. She plans to have pasta with marinara sauce for dinner. The product labels shown here give the number of calories in each food.

Pasta

Nutrition Facts

Serving Size 2 oz (56g) dry
(1/8 of the package)
Servings Per Container 8

Amount Per Serving

Calories 200   Calories from Fat 10

	% Daily Value*
Total Fat 1 g	2%
Saturated Fat 0g	0%
Cholesterol 0mg	0%
Sodium 0mg	0%
Total Carbohydrate 41g	14%
Dietary Fiber 2g	8%
Sugars 3g	
Protein 7g	

Vitamin A 0%	•	Vitamin C 0%
Calcium 0%	•	Iron 10%
Thiamin 35%	•	Riboflavin 15%
Niacin 15%	•	

*Percent Daily Values are based on a 2,000 calorie diet. Your daily values may be higher or lower depending on your calorie needs.

Marinara Sauce

Nutrition Facts

Serving Size 1/2 cup (125g)
Servings per Container approx 6

Amount Per Serving

Calories 60   Calories from Fat 20

	% Daily Value*
Total Fat 2g	3%
Saturated Fat 3g	0%
Cholesterol 0mg	0%
Sodium 370mg	15%
Total Carbohydrate 7g	2%
Dietary Fiber 2g	8%
Sugars 4g	
Protein 3g	

Vitamin A 15%	•	Vitamin C 40%
Calcium 0%	•	Iron 4%

67. If she eats two servings each of pasta and sauce, how many calories does she consume?

68. If Sasha has 550 more calories in bread, butter, salad, drink, and desert for dinner, how many total calories does she consume at dinner?

69. If Sasha keeps to her goal, how many calories could she have eaten at breakfast and lunch?

70. Super Bowl XIV was the highest attended Super Bowl, with a crowd of 103,985. Super Bowl XVII was the second highest attended, with a crowd of 103,667. The third highest attendance occurred at Super Bowl XI, with 103,438. What was the total attendance at all three Super Bowls? How many more people attended the highest attended game than the third highest attended game?

71. A forester counted 31,478 trees that are ready for harvest on a certain acreage. If Forestry Service rules require that 8543 mature trees must be left on the acreage, how many trees can be harvested?

72. The new sewer line being installed in downtown Chehalis will handle 475,850 gallons of refuse per minute. The old line handled 238,970 gallons per minute. How many more gallons per minute will the new line handle?

73. Fong's Grocery owes a supplier $36,450. During the month, Fong's makes payments of $1670, $3670, $670, and $15,670. How much does Fong's still owe, to the nearest hundred dollars?

74. In the spring of 1989, an oil tanker hit a reef and spilled 10,100,000 gallons of oil off the coast of Alaska. The tanker carried a total of 45,700,000 gallons of oil. The oil that did not spill was pumped into another tanker. How many gallons of oil were pumped into the second tanker? Round to the nearest million gallons.

© Christopher Poliquin/Shutterstock.com

75. The median family income of a region is a way of estimating the middle income. Half the families in the region make more than the median income and the other half of the families make less. In a recent year, the Department of Housing and Urban Development (HUD) estimated that the median family income for San Francisco was $95,000, and for Seattle it was $72,000. What place value were these figures rounded to and how much higher was San Francisco's median income than Seattle's?

76. The Grand Canyon, Zion, and Bryce Canyon parks are found in the southwestern United States. Geologic changes over a billion years have created these formations and canyons. The chart shows the highest and lowest elevations in each of these parks. Find the change in elevation in each park. In which park is the change greatest and by how much?

Elevations at National Parks

	Highest Elevation	Lowest Elevation
Bryce Canyon	8500 ft	6600 ft
Grand Canyon	8300 ft	2500 ft
Zion	7500 ft	4000 ft

Exercises 77–80. The average number of murder victims per year in the United States who are related to the murderer, according to statistics from the FBI, is given in the table.

Murder Victims Related to the Murderer

Wives	Husbands	Sons	Daughters	Fathers	Brothers	Mothers	Sisters
913	383	325	235	169	167	121	42

77. In an average year, how many more husbands killed their wives than wives killed their husbands?

78. In an average year, how many people killed their child?

79. In an average year, how many people killed a sibling?

80. In an average year, did more people kill their child or their parent?

Exercises 81–83. The table lists Ford brand vehicles for 2008, according to Blue Oval News.

Model	Units Sold
Ford cars	35,940
Crossover utility vehicles	24,310
Trucks and vans	58,640
Sport utility vehicles	12,180

81. How many more cars were sold than sport utility vehicles?

82. What were the combined sales of the trucks, vans, and utility vehicles?

83. How many more utility vehicles were sold than cars?

84. In the National Football League, the salary cap is the maximum amount that a club can spend on player salaries. For the 2009 season the salary cap was $127,000,000, and for the 2008 season it was $116,700,000. By how much did the salary cap increase from 2008 to 2009?

Exercises 85–86. In sub-Saharan Africa, 5 out of every 100 adults are living with HIV/AIDS. The table gives statistics for people in the region in 2007. (SOURCE: AVERT.org)

Total	22 million
Women	12 million
Children	1,800,000

85. How many adults in sub-Saharan Africa have HIV/AIDS?

86. How many men in sub-Saharan Africa have HIV/AIDS?

87. Find the perimeter of a rectangular house that is 62 ft long and 38 ft wide.

88. A farmer wants to put a fence around a triangular plot of land that measures 5 km by 9 km by 8 km. How much fence does he need?

89. Blanche wants to sew lace around the edge of a rectangular tablecloth that measures 64 in. by 48 in. How much lace does she need, ignoring the corners and the seam allowances?

90. Annisa wants to trim a picture frame in ribbon. The outside of the rectangular frame is 25 cm by 30 cm. How much ribbon does she need, ignoring the corners?

STATE YOUR UNDERSTANDING

91. Explain to a 6-year-old child why $15 - 9 = 6$.

92. Explain to a 6-year-old child why $8 + 7 = 15$.

93. Define and give an example of a sum.

94. Define and give an example of a difference.

95. Add the following numbers, round the sum to the nearest hundred, and write the word name for the rounded sum: one hundred sixty; eighty thousand, three hundred twelve; four hundred seventy-two thousand, nine hundred fifty-two; and one hundred forty-seven thousand, five hundred twenty-three.

96. How much greater is six million, three hundred fifty-two thousand, nine hundred seventy-five than four million, seven hundred six thousand, twenty-three? Write the word name for the difference.

97. Peter sells three Honda Civics for $15,488 each, four Accords for $18,985 each, and two Acuras for $30,798 each. What is the total dollar sales for the nine cars? How many more dollars were paid for the four Accords than the three Civics?

Complete the sum or difference by writing in the correct digit wherever you see a letter.

98.

```
  5A68
   241
+ 10A9
  B64C
```

99.

```
  4A6B
− C251
  15D1
```

100. Add and round to the nearest hundred.

```
   14,657
    3,766
  123,900
      569
   54,861
+ 346,780
```

Now round each addend to the nearest hundred and then add. Tell why the answers are different. Explain why this happens.

VOCABULARY

An **equation** is a statement about numbers that says that two expressions are equal. Letters, called **variables** or **unknowns**, are often used to represent numbers.

HOW & WHY

Examples of equations are

$$9 = 9 \qquad 13 = 13 \qquad 123 = 123 \qquad 30 + 4 = 34 \qquad 52 - 7 = 45$$

When variables are used an equation can look like this:

$$x = 7 \qquad x = 10 \qquad y = 18 \qquad x + 5 = 13 \qquad y - 8 = 23$$

An equation containing a variable can be true only when the variable is replaced by a specific number. For example,

$x = 7$ is true only when x is replaced by 7.

$x = 10$ is true only when x is replaced by 10.

$y = 18$ is true only when y is replaced by 18.

$x + 5 = 13$ is true only when x is replaced by 8, so that $8 + 5 = 13$.

$y - 8 = 23$ is true only when y is replaced by 31, so that $31 - 8 = 23$.

The numbers that make equations true are called *solutions*. Solutions of equations, such as $x - 7 = 12$, can be found by trial and error, but let's develop a more practical way.

Addition and subtraction are inverse, or opposite, operations. For example, if 14 is added to a number and then 14 is subtracted from that sum, the difference is the original number. So

$23 + 14 = 37$	Add 14 to 23.
$37 - 14 = 23$	Subtract 14 from the sum, 37. The difference is the original number, 23.

We use this idea to solve the following equation:

$x + 21 = 35$	21 is added to the number represented by x.
$x + 21 - 21 = 35 - 21$	To remove the addition and have only x on the left side of the
$x = 14$	equal sign, we subtract 21. To keep a true equation, we must subtract 21 from both sides.

To check, replace x in the original equation with 14 and see if the result is a true statement:

$x + 21 = 35$	
$14 + 21 = 35$	
$35 = 35$	The statement is true, so the solution is 14.

We can also use the idea of inverses to solve an equation in which a number is subtracted from a variable (letter):

$b - 17 = 12$	Since 17 is subtracted from the variable, we eliminate the
$b - 17 + 17 = 12 + 17$	subtraction by adding 17 to both sides of the equation. Recall that addition is the inverse of subtraction.
$b = 29$	The equation will be true when b is replaced by 29.

CHECK:	$b - 17 = 12$	
	$29 - 17 = 12$	Substitute 29 for b.
	$12 = 12$	True.

So the solution is $b = 29$.

EXAMPLES A–E

DIRECTIONS: Solve and check.

STRATEGY: Isolate the variable by adding or subtracting the same number to or from each side.

A.　　$x + 7 = 23$

$x + 7 = 23$

$x + 7 - 7 = 23 - 7$　　Because 7 is added to the variable, eliminate the addition by subtracting 7 from both sides of the equation.

$x = 16$　　Simplify.

CHECK:　$x + 7 = 23$

$16 + 7 = 23$　　Check by substituting 16 for x in the original equation.

$23 = 23$　　The statement is true.

The solution is $x = 16$.

B.　　$a - 24 = 50$

$a - 24 = 50$

$a - 24 + 24 = 50 + 24$　　Because 24 is subtracted from the variable, eliminate the subtraction by adding 24 to both sides of the equation.

$a = 74$　　Simplify.

CHECK:　$a - 24 = 50$

$74 - 24 = 50$　　Check by substituting 74 for a in the original equation.

$50 = 50$　　The statement is true.

The solution is $a = 74$.

C. $45 = b + 22$

In this example we do the subtraction vertically.

$$\begin{array}{r} 45 = b + 22 \\ -\,22 \quad\ -\,22 \\ \hline 23 = b \end{array}$$

Subtract 22 from both sides to eliminate the addition of 22.
Simplify.

CHECK:　$45 = b + 22$

$45 = 23 + 22$　　Substitute 23 for b.

$45 = 45$　　The statement is true.

The solution is $b = 23$.

D.　　$z - 33 = 41$

$z - 33 = 41$

$z - 33 + 33 = 41 + 33$　　Add 33 to both sides.

$z = 74$　　Simplify.

CHECK:　$z - 33 = 41$

$74 - 33 = 41$　　Substitute 74 for z.

$41 = 41$　　The statement is true.

The solution is $z = 74$.

E. The selling price for a pair of Nike "Air Deluxe" shoes is $139. If the markup on the shoes is $43, what is the cost to the store?

Cost + markup = selling price.

$C + M = S$	Since cost + markup = selling price.
$C + 43 = 139$	Substitute 43 for the markup and 139 for the selling price.
$C + 43 - 43 = 139 - 43$	Subtract 43 from both sides.
$C = 96$	

CHECK: Does the cost + the markup equal $139?

$96	Cost
+ 43	Markup
$139	Selling price

So the cost of the shoes to the store is $96.

ANSWER TO WARM-UP E

E. The golf clubs cost the store $438.

EXERCISES

■ **OBJECTIVE**   Solve an equation of the form $x + a = b$ or $x - a = b$, where a, b, and x are whole numbers. (See page 32.)

Solve and check.

1. $x + 12 = 24$ **2.** $x - 11 = 14$ **3.** $x - 6 = 17$

4. $x + 10 = 34$ **5.** $z + 13 = 27$ **6.** $b - 21 = 8$

7. $c + 24 = 63$ **8.** $y - 33 = 47$ **9.** $a - 40 = 111$

10. $x + 75 = 93$ **11.** $x + 91 = 105$ **12.** $x - 76 = 43$

13. $y + 67 = 125$ **14.** $z - 81 = 164$ **15.** $k - 56 = 112$

16. $c + 34 = 34$ **17.** $73 = x + 62$ **18.** $534 = a + 495$

19. $87 = w - 29$ **20.** $373 = d - 112$

21. The selling price for a computer is $1265. If the cost to the store is $917, what is the markup?

23. The length of a rectangular garage is 2 meters more than the width. If the width is 7 meters, what is the length?

22. The selling price of a trombone is $675. If the markup is $235, what is the cost to the store?

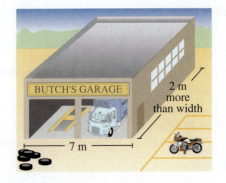

24. The width of a rectangular fish pond is 6 feet shorter than the length. If the length is 27 feet, what is the width?

25. A Saturn with manual transmission has an EPA highway rating of 5 miles per gallon more than the EPA highway rating of a Subaru Impreza. Write an equation that describes this relationship. Be sure to define all variables in your equation. If the Saturn has an EPA highway rating of 35 mpg, find the highway rating of the Impreza.

26. In a recent year in the United States, the number of deaths by drowning was 1700 less than the number of deaths by fire. Write an equation that describes this relationship. Be sure to define all variables in your equation. If there were approximately 4800 deaths by drowning that year, how many deaths by fire were there?

Exercises 27–28. A city treasurer made the following report to the city council regarding monies allotted and dispersed from a city parks bond.

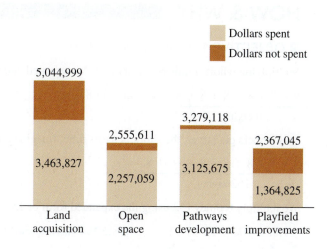

27. Write an equation that relates the total money budgeted per category to the amount of money spent and the amount of money not yet spent. Define all the variables.

28. Use your equation from Exercise 27 to calculate the amount of money not yet spent in each of the four categories.

1.3 Multiplying Whole Numbers

OBJECTIVES

1. Multiply whole numbers.
2. Estimate the product of whole numbers.
3. Find the area of a rectangle.

VOCABULARY

There are several ways to indicate multiplication. Here are examples of most of them, using 28 and 41.

$$28 \times 41 \qquad 28 \cdot 41 \qquad \begin{array}{r} 28 \\ \times 41 \end{array}$$

$$(28)(41) \qquad 28(41) \qquad (28)41$$

The **factors** of a multiplication exercise are the numbers being multiplied. In $7(9) = 63$, 7 and 9 are the factors.

The **product** is the answer to a multiplication exercise. In $7(9) = 63$, the product is 63.

The **area** of a rectangle is the measure of the surface inside the rectangle.

HOW & WHY

■ **OBJECTIVE 1** Multiply whole numbers.

Multiplying whole numbers is a shortcut for repeated addition:

$$\underbrace{8 + 8 + 8 + 8 + 8 + 8}_{6 \text{ eights}} = 48 \quad \text{or} \quad 6 \cdot 8 = 48$$

As numbers get larger, the shortcut saves time. Imagine adding 152 eights.

$$\underbrace{8 + 8 + 8 + 8 + 8 + \cdots + 8}_{152 \text{ eights}} = ?$$

We multiply 8 times 152 using the expanded form of 152.

$$\begin{aligned} 152 &= 100 + 50 + 2 \qquad &\text{Write 152 in expanded form.} \\ \times \ 8 &= \ 8 \\ &= 800 + 400 + 16 \qquad &\text{Multiply 8 times each addend.} \\ &= 1216 \qquad &\text{Add.} \end{aligned}$$

The exercise can also be performed in column form without expanding the factors.

$$\begin{array}{r} 152 \\ \times \ 8 \\ \hline 16 \\ 400 \\ 800 \\ \hline 1216 \end{array} \qquad \begin{aligned} 8(2) &= 16 \\ 8(50) &= 400 \\ 8(100) &= 800 \end{aligned} \qquad \begin{array}{r} {\color{red}4\,1} \\ 152 \\ \times \ 8 \\ \hline 1216 \end{array}$$

The form on the right shows the usual shortcut. The carried digit is added to the product of each column. Study this example.

$$\begin{array}{r} 635 \\ \times \ 47 \end{array}$$

First multiply 635 by 7.

```
  2 3
  6 3 5
×   4 7
  4 4 4 5
```

7(5) = 35. Carry the 3 to the tens column.
7(3 tens) = 21 tens. Add the 3 tens that were carried:
(21 + 3) tens = 24 tens. Carry the 2 to the hundreds column.
7(6 hundreds) = 42 hundreds. Add the 2 hundreds that were
carried: (42 + 2) hundreds = 44 hundreds.

Now multiply 635 by 40.

```
  1 2
  2 3
  6 3 5
×   4 7
  4 4 4 5
2 5 4 0 0
```

40(5) = 200 or 20 tens. Carry the 2 to the hundreds column.
40(30) = 1200 or 12 hundreds. Add the 2 hundreds that were
carried. (12 + 2) hundreds = 14 hundreds. Carry the 1 to the
thousands column. 40(600) = 24,000 or 24 thousands. Add
the 1 thousand that was carried: (24 + 1) thousands = 25 thousands.
Write the 5 in the thousands column and the 2 in the ten-thousands column.

```
  1 2
  2 3
  6 3 5
×   4 7
  4 4 4 5
2 5 4 0 0
2 9 8 4 5
```

Add the products.

Two important properties of arithmetic and higher mathematics are the *multiplication
property of zero* and the *multiplication property of one*.

As a result of the multiplication property of zero, we know that

$$0 \cdot 23 = 23 \cdot 0 = 0 \qquad \text{and} \qquad 0(215) = 215(0) = 0$$

As a result of the multiplication property of one, we know that

$$1 \cdot 47 = 47 \cdot 1 = 47 \qquad \text{and} \qquad 1(698) = 698(1) = 698$$

Multiplication Property of Zero

Multiplication property of zero:

$$a \cdot 0 = 0 \cdot a = 0$$

Any number times zero is zero.

Multiplication Property of One

Multiplication property of one:

$$a \cdot 1 = 1 \cdot a = a$$

Any number times 1 is that number.

EXAMPLES A–F

DIRECTIONS: Multiply.

STRATEGY: Write the factors in columns. Start multiplying with the ones digit. If the product is 10 or more, carry the tens digit to the next column and add it to the product in that column. Repeat the process for every digit in the second factor. When the multiplication is complete, add to find the product.

A. Multiply: 1(932)

1(932) = 932   Multiplication property of one.

B. Find the product: 7(4582)

```
  451
  4582
×    7
 32,074
```

Multiply 7 times each digit, carry when necessary, and add the number carried to the next product.

WARM-UP
C. Multiply: 76 · 63

C. Multiply: 54 · 49

$$
\begin{array}{r}
4 \\
3 \\
49 \\
\times\ 54 \\
\hline
196 \\
2450 \\
\hline
2646
\end{array}
$$

When multiplying by the 5 in the tens place, write a 0 in the ones column to keep the places lined up.

WARM-UP
D. Find the product of 826 and 307.

D. Find the product of 528 and 109.

STRATEGY: When multiplying by zero in the tens place, rather than showing a row of zeros, just put a zero in the tens column. Then multiply by the 1 in the hundreds place.

$$
\begin{array}{r}
528 \\
\times\ 109 \\
\hline
4752 \\
52800 \\
\hline
57552
\end{array}
$$

CALCULATOR EXAMPLE:

WARM-UP
E. 763(897)

E. 3465(97)

Most graphing calculators recognize implied multiplication but most scientific calculators do not. Be sure to insert a multiplication symbol between two numbers written with implied multiplication.

The product is 336,105.

WARM-UP
F. General Electric ships 88 cartons of lightbulbs to Lowe's. If each carton contains 36 lightbulbs, how many lightbulbs are shipped to Lowe's?

F. Hewlett-Packard ships 136 cartons of printer ink cartridges to an Office Depot warehouse. Each carton contains 56 cartridges. What is the total number of cartridges shipped to Office Depot?

STRATEGY: To find the total number of cartridges, multiply the number of cartons by the number of cartridges in each carton.

$$
\begin{array}{r}
136 \\
\times\ 56 \\
\hline
816 \\
6800 \\
\hline
7616
\end{array}
$$

Hewlett-Packard shipped 7616 cartridges to Office Depot.

HOW & WHY

■ **OBJECTIVE 2** Estimate the product of whole numbers.

The product of two whole numbers can be estimated by using **front rounding.** With front rounding we round to the highest place value so that all the digits become 0 except the first one. For example, if we front round 7654, we get 8000.

So to estimate the following product of 78 and 432, front round each factor and multiply.

$$
\begin{array}{r}
432 \\
\times\ 78
\end{array}
\qquad
\begin{array}{r}
400 \\
\times\ 80 \\
\hline
32,000
\end{array}
$$

Front round each factor and multiply.

The estimated product is 32,000, that is, $(432)(78) \approx 32,000$.

One use of the estimate is to see if the product is correct. If the calculated product is not close to 32,000, you should check the multiplication. In this case the actual product is 33,696, which is close to the estimate.

EXAMPLES G–J

DIRECTIONS: Estimate the product.

STRATEGY: Front round both factors and multiply.

G.
$$
\begin{array}{r} 298 \\ \times\ 46 \end{array} \qquad \begin{array}{r} 300 \\ \times\ 50 \\ \hline 15{,}000 \end{array} \qquad \text{\color{blue}{Front round and multiply.}}
$$

So, $(298)(46) \approx 15{,}000$.

H.
$$
\begin{array}{r} 3{,}792 \\ \times\ 412 \end{array} \qquad \begin{array}{r} 4{,}000 \\ \times\ \ \ 400 \\ \hline 1{,}600{,}000 \end{array}
$$

So $(3792)(412) \approx 1{,}600{,}000$.

I. Paul finds the product of 230 and 47 to be 1081. Estimate the product by front rounding, to see if Paul is correct. If not, find the actual product.

$$
\begin{array}{r} 230 \\ \times\ 47 \end{array} \qquad \begin{array}{r} 200 \\ \times\ 50 \\ \hline 10{,}000 \end{array}
$$

The estimate is 10,000, so Paul is not correct.

$$
\begin{array}{r} 230 \\ \times\ \ 47 \\ \hline 1610 \\ 9200 \\ \hline 10{,}810 \end{array}
$$

Paul was not correct; the correct product is 10,810.

J. John wants to buy seven shirts that cost $42 each. He has $300 in cash. Estimate the cost of the shirts to see if John has enough money to buy them.

$$
\begin{array}{r} \$42 \\ \times\ 7 \end{array} \qquad \begin{array}{r} \$40 \\ \times\ 7 \\ \hline \$280 \end{array} \qquad \text{\color{blue}{Front round the price of one shirt and multiply by the number of shirts.}}
$$

The estimated cost of the seven shirts is $280, so it looks as if John has enough money.

HOW & WHY

■ **OBJECTIVE 3** Find the area of a rectangle.

The area of a polygon is the measure of the space inside the polygon. We use area when describing the size of a plot of land, the living space in a house, or an amount of carpet. Area is measured in square units such as square feet or square meters. A square foot is literally a square with sides of 1 foot. The measure of the surface inside the square is 1 square foot.

When measuring the space inside a polygon, we divide the space into squares and count them. For example, consider a rug that is 2 ft by 3 ft (Figure 1.6).

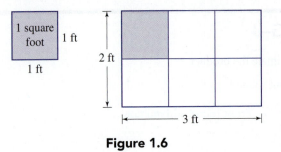

Figure 1.6

There are six squares in the subdivided rug, so the area of the rug is 6 square feet.

Finding the area of a rectangle, such as the area rug in the example, is relatively easy because a rectangle has straight sides and it is easy to fit squares inside it. The length of the rectangle determines how many squares will be in each row, and the width of the rectangle determines the number of rows. In the rug shown in Figure 1.6, there are two rows of three squares each because the width is 2 ft and the length is 3 ft. The product of the length and width gives the number of squares inside the rectangle.

Area of a rectangle = length · width

Finding the area of other shapes is a little more complicated, and is discussed in Section 7.4.

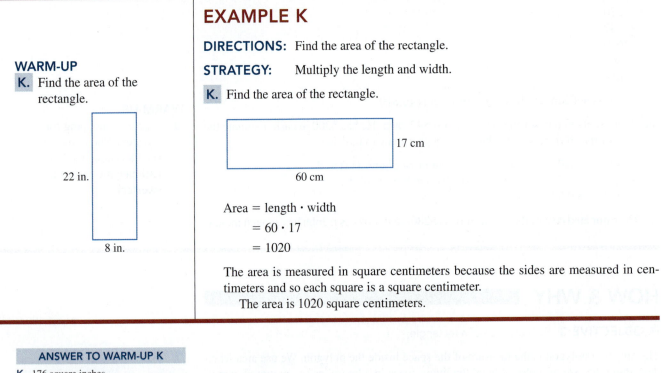

EXAMPLE K

DIRECTIONS: Find the area of the rectangle.

STRATEGY: Multiply the length and width.

K. Find the area of the rectangle.

17 cm

60 cm

Area = length · width

= 60 · 17

= 1020

The area is measured in square centimeters because the sides are measured in centimeters and so each square is a square centimeter.

The area is 1020 square centimeters.

WARM-UP

K. Find the area of the rectangle.

22 in.

8 in.

EXERCISES 1.3

■ **OBJECTIVE 1** Multiply whole numbers. (See page 36.)

A *Multiply.*

1.	83	2.	55	3.	97	4.	35
	× 7		× 4		× 3		× 7

5. 76
    × 4

6. 46
    × 6

7. 93
    × 7

8. 39
    × 8

9. 8 · 37

10. 6 · 55

11. (239)(0)

12. (1)(345)

13. 76
    × 40

14. 17
    × 50

15. In 326 × 52 the place value of the product of 5 and 3 is _____.

16. In 326 × 52 the product of 5 and 6 is 30 and you must carry the 3 to the _____ column.

B *Multiply.*

17. 646
    × 7

18. 562
    × 6

19. 804
    × 7

20. 408
    × 8

21. (53)(67)

22. (49)(55)

23. (94)(37)

24. (83)(63)

25. 416
    × 300

26. 582
    × 700

27. 904
    × 74

28. 608
    × 57

29. 747
    × 48

30. 534
    × 75

31. (87)(252) Round product to the nearest hundred.

32. (48)(653) Round product to the nearest thousand.

33. 312
    × 50

34. 675
    × 40

35. 527
    × 73

36. 265
    × 57

37. 738
    × 47

38. 684
    × 76

39. (4321)(76)

40. (6230)(94)

■ **OBJECTIVE 2** Estimate the product of whole numbers. (See page 38.)

A *Estimate the product using front rounding.*

41. 43 × 84

42. 68 × 22

43. 528 × 48

44. 693 × 38

45. 4510 × 53

46. 6328 × 27

B

47. 83 × 3046

48. 34 × 6290

49. 17,121 × 39

50. 52,812 × 81

51. 610 × 34,560

52. 555 × 44,991

■ **OBJECTIVE 3** Find the area of a rectangle. (See page 39.)

A *Find the area of the following rectangles.*

53.

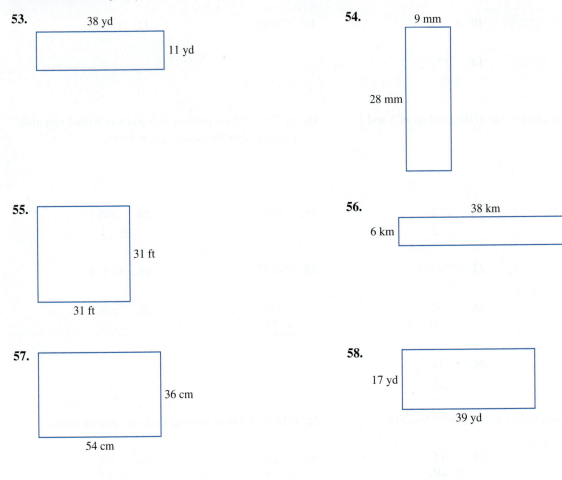

38 yd

11 yd

54.

9 mm

28 mm

55.

31 ft

31 ft

56.

38 km

6 km

57.

36 cm

54 cm

58.

17 yd

39 yd

59. What is the area of a rectangle that has a length of 17 ft and a width of 6 ft?

60. What is the area of a rectangle that measures 30 cm by 40 cm?

B *Find the area of the following.*

61.

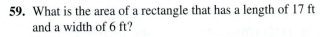

512 cm

102 cm

62.

176 in.

235 in.

63.

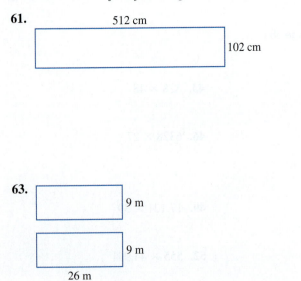

9 m

9 m

26 m

64.

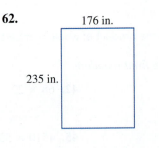

13 mi

8 mi   8 mi

65.

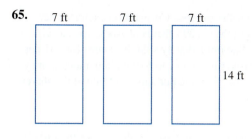

7 ft   7 ft   7 ft

14 ft

66.

25 mm

3 mm

3 mm

3 mm

3 mm

67. What is the area of three goat pens that each measure 6 ft by 11 ft?

68. What is the area of five bath towels that measure 68 cm by 140 cm?

C

69. Find the product of 505 and 773.

70. Find the product of 505 and 886.

71. Multiply and round to the nearest thousand: (744)(3193)

72. Multiply and round to the nearest ten thousand: (9007)(703)

73. Maria multiplies 59 times 482 and gets a product of 28,438. Estimate the product by front rounding to see if Maria's answer is reasonable.

74. John multiplies 791 by 29 and gets a product of 8701. Estimate the product by front rounding to see if John's answer is reasonable.

75. During the first week of the Rotary Club rose sale, 341 dozen roses are sold. The club estimates that a total of 15 times that number will be sold during the sale. What is the estimated number of dozens of roses that will be sold?

Exercises 76–79. Use the information on the monthly sales at Jeff's Used Greene Cars.

Monthly Sales at Jeff's Used Greene Cars

Car Model	Number of Cars Sold	Average Price per Sale
Civic	23	$15,844
Prius	31	$17,929
Smart Car	18	$15,237

76. Find the gross receipts from the sale of the Civics.

77. What are the gross receipts from the sale of the Prius?

78. Find the gross receipts from the sale of Smart Cars.

79. Find the gross receipts for the month (the sum of the gross receipts for each model) rounded to the nearest thousand dollars.

80. An average of 452 salmon per day are counted at the Bonneville fish ladder during a 17-day period. How many total salmon are counted during the 17-day period?

81. One year, the population of Washington County grew at a rate of 1874 people per month. What was the total growth in population for the year?

82. The CEO of Apex Corporation exercised his option to purchase 2355 shares of Apex stock at $13 per share. He immediately sold the shares for $47 per share. If broker fees came to $3000, how much money did he realize from the purchase and sale of the shares?

83. The comptroller of Apex Corporation exercised her option to purchase 1295 shares of Apex stock at $16 per share. She immediately sold the shares for $51 per share. If broker fees came to $1050, how much money did she realize from the purchase and sale of the shares?

84. Nyuen wants to buy mp3 players for his seven grand-children for Christmas. He has budgeted $500 for these presents. The mp3 player he likes costs $79. Estimate the total cost, by front rounding, to see if Nyuen has enough money in his budget for these presents.

85. Carmella needs to purchase 12 blouses for the girls in the choir at her church. The budget for the purchases is $480. The blouse she likes costs $38.35. Estimate the total cost, by front rounding, to see if Carmella has enough money in her budget for these blouses.

86. A certain bacteria culture triples its size every hour. If the culture has a count of 265 at 10 A.M., what will the count be at 2 P.M. the same day?

© YellowPixel/Shutterstock.com

Exercises 87–88. The depth of water is often measured in fathoms. There are 3 feet in a yard and 2 yards in a fathom.

87. How many feet are in a fathom?

88. How many feet are in 25 fathoms?

Exercises 89–91. A league is an old measure of about 3 nautical miles. A nautical mile is about 6076 feet.

89. How many feet are in a league?

90. There is famous book by Jules Verne entitled *20,000 Leagues Under the Sea*. How many feet are in 20,000 leagues?

91. The Mariana Trench in the Pacific Ocean is the deepest point of all the world's oceans. It is 35,840 ft deep. Is it physically possible to be 20,000 leagues under the sea?

Exercises 92–94. Because distances between bodies in the universe are so large, scientists use large units. One such unit is the light-year, which is the distance traveled by light in one year, or 5880 billion miles.

92. Write the place value notation for the number of miles in a light-year.

93. The star Sirius is recognized as the brightest star in the sky (other than the sun). It is 8 light-years from Earth. How many miles is Sirius from Earth?

94. The star Rigel in the Orion constellation is 545 light-years from Earth. How many miles away is Rigel from Earth?

Exercises 95–96. One model of an inkjet printer can produce 20 pages per minute in draft mode, 8 pages per minute in normal mode, and 2 pages per minute in best-quality mode.

95. Skye is producing a large report for her group. She selects normal mode and is called away from the printer for 17 minutes. How many pages of the report were printed in that time?

96. How many more pages can be produced in 25 minutes in draft mode than in 25 minutes in normal mode?

Exercises 97–98. In computers, a byte is the amount of space needed to store one character. Knowing something about the metric system, one might think a kilobyte is 1000 bytes, but actually it is 1024 bytes.

97. A computer has 256 KB (kilobytes) of RAM. How many bytes is this?

98. A megabyte is 1024 KB. A writable CD holds up to 700 MB (megabytes). How many bytes can the CD hold?

Exercises 99–100. A gram of fat contains about 9 calories, as does a gram of protein. A gram of carbohydrate contains about 4 calories.

99. A tablespoon of olive oil has 14 g of fat. How many calories is this?

100. One ounce of cream cheese contains 2 g of protein and 10 grams of fat. How many calories from fat and protein are in the cream cheese?

101. The water consumption in Hebo averages 534,650 gallons per day. How many gallons of water are consumed in a 31-day month, rounded to the nearest thousand gallons?

102. Ms. Munos orders 225 4G iPods shuffles for sale in her discount store. If she pays $55 per iPod and sells them for $76 each, how much do the iPods cost her and what is the net income from their sale? How much are her profits from the sale of the iPods?

103. Ms. Perta orders four hundred sixty-four studded snow tires for her tire store. She pays $48 per tire and plans to sell them for $106 each. What do the tires cost Ms. Perta and what is her gross income from their sale? What net income does she receive from the sale of the tires?

104. In 2008, Bill Gates of Microsoft was the richest person in the United States, with an estimated net worth of $57 billion. Write the place value name for this number. A financial analyst made the observation that the average person has a hard time understanding such large amounts. She gave the example that in order to spend $1 billion, one would have to spend $40,000 per day for 69 years, ignoring leap years. How much money would you spend if you did this?

Exercises 105–106 relate to the chapter application. See Table 1.1, page 3.

105. If the revenue from *Star Wars: Episode III* had doubled, would it have been the top-grossing film?

106. Which would result in more earnings—if *Titanic*'s earnings doubled or if *Spider-Man*'s earnings tripled?

 STATE YOUR UNDERSTANDING

107. Explain to an 8-year-old child that 3(8) = 24.

108. When 65 is multiplied by 8, we carry 4 to the tens column. Explain why this is necessary.

109. Define and give an example of a product.

 CHALLENGE

110. Find the product of twenty-four thousand, fifty-five and two hundred thirteen thousand, two hundred seventy-six. Write the word name for the product.

111. Tesfay harvests 82 bushels of wheat per acre from his 1750 acres of grain. If Tesfay can sell the grain for $31 a bushel, what is the crop worth, to the nearest thousand dollars?

Complete the problems by writing in the correct digit wherever you see a letter,

112.
$$\begin{array}{r} 51A \\ \times\ \underline{B2} \\ 10B2 \\ \underline{154C} \\ 1A5E2 \end{array}$$

113.
$$\begin{array}{r} 1A57 \\ \times\ \underline{42} \\ B71C \\ \underline{D428} \\ 569E4 \end{array}$$

Dividing Whole Numbers

VOCABULARY

There are a variety of ways to indicate division. These are the most commonly used:

$$72 \div 6 \qquad 6\overline{)72} \qquad \frac{72}{6}$$

The **dividend** is the number being divided, so in $54 \div 6 = 9$, the dividend is 54.

The **divisor** is the number that we are dividing by, so in $54 \div 6 = 9$, the divisor is 6.

The **quotient** is the answer to a division exercise, so in $54 \div 6 = 9$, the quotient is 9.

When a division exercise does not come out even, as in $61 \div 7$, the quotient is not a whole number.

$$\begin{array}{r} 8 \\ 7\overline{)61} \\ \underline{56} \\ 5 \end{array}$$

We call 8 the **partial quotient** and 5 the **remainder**. The quotient is written 8 R 5.

HOW & WHY

◼ **OBJECTIVE**   Divide whole numbers.

The division exercise $144 \div 24 = ?$ (read "144 divided by 24") can be interpreted in one of two ways.

How many times can 24 be subtracted from 144?   This is called the "repeated subtraction" version.

What number times 24 is equal to 144?   This is called the "missing factor" version.

All division problems can be done using repeated subtraction.

In 144 ÷ 24 = ?, we can find the missing factor by repeatedly subtracting 24 from 144:

```
  144
 − 24
  120
 − 24
   96        Six subtractions, so 144 ÷ 24 = 6.
 − 24
   72
 − 24
   48
 − 24
   24
 − 24
    0
```

The process can be shortened using the traditional method of guessing the number of 24s and subtracting from 144:

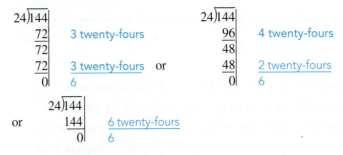

In each case, 144 ÷ 24 = 6.

We see that the missing factor in (24)(?) = 144 is 6. Because 24(6) = 144, consequently 144 ÷ 24 = 6.

This leads to a method for checking division. If we multiply the divisor times the quotient, we will get the dividend. To check 144 ÷ 24 = 6, we multiply 24 and 6.

(24)(6) = 144

So 6 is correct.

This process works regardless of the size of the numbers. If the divisor is considerably smaller than the dividend, you will want to guess a rather large number.

```
63)19,593
    6 300    100
   13 293
    6 300    100
    6 993
    6 300    100
      693
      630     10
       63
       63      1
        0     311
```

So, 19,593 ÷ 63 = 311.

All divisions can be done by this method. However, the process can be shortened by finding the number of groups, starting with the largest place value on the left, in the dividend, and then working toward the right. Study the following example. Note that the answer is written above the problem for convenience.

$$31\overline{)17{,}391}$$

Working from left to right, we note that 31 does not divide 1, and it does not divide 17. However, 31 does divide 173 five times. Write the 5 above the 3 in the dividend.

$$
\begin{array}{r}
561 \\
31\overline{)17{,}391} \\
\underline{155} \\
189 \\
\underline{186} \\
31 \\
\underline{31} \\
0
\end{array}
$$

5(31) = 155. Subtract 155 from 173. Because the difference is less than the divisor, no adjustment is necessary. Bring down the next digit, which is 9. Next, 31 divides 189 six times. The 6 is placed above the 9 in the dividend. 6(31) = 186. Subtract 186 from 189. Again, no adjustment is necessary, since 3 < 31. Bring down the next digit, which is 1. Finally, 31 divides 31 one time. Place the 1 above the one in the dividend. 1(31) = 31. Subtract 31 from 31, the remainder is zero. The division is complete.

CHECK:
$$
\begin{array}{r}
561 \\
\times\ 31 \\
\hline
561 \\
16{,}830 \\
\hline
17{,}391
\end{array}
$$

Check by multiplying the quotient by the divisor.

So $17{,}391 \div 31 = 561$.

Not all division problems come out even (have a zero remainder). In

$$
\begin{array}{r}
4 \\
21\overline{)94} \\
\underline{84} \\
10
\end{array}
$$

we see that 94 contains 4 twenty-ones and 10 toward the next group of twenty-one. The answer is written as 4 remainder 10. The word *remainder* is abbreviated "R" and the result is 4 R 10.

Check by multiplying (21)(4) and adding the remainder.

$$(21)(4) = 84$$
$$84 + 10 = 94$$

So $94 \div 21 = 4$ R 10.

The division $61 \div 0 = ?$ can be restated: What number times 0 is 61? $0 \times ? = 61$. According to the multiplication property of zero we know that $0 \times$ (any number) $= 0$, so it cannot equal 61.

> **CAUTION**
>
> Division by zero is not defined. It is an operation that cannot be performed.

When dividing by a single-digit number the division can be done mentally using "short division."

$$
\begin{array}{r}
423 \\
3\overline{)1269}
\end{array}
$$

Divide 3 into 12. Write the answer, 4, above the 2 in the dividend. Now divide the 6 by 3 and write the answer, 2, above the 6. Finally divide the 9 by 3 and write the answer, 3, above the 9.

The quotient is 423.

If the "mental" division does not come out even, each remainder is used in the next division.

$$\begin{array}{r} 4\,5\,2\,\text{R}\,2 \\ 3\overline{)13^{1}58} \end{array}$$

$13 \div 3 = 4$ R 1. Write the 4 above the 3 in the dividend. Now form a new number, 15, using the remainder 1 and the next digit 5. Divide 3 into 15. Write the answer, 5, above 5 in the dividend. Because there is no remainder, divide the next digit, 8, by 3. The result is 2 R 2. Write this above the 8.

The quotient is 452 R 2.

EXAMPLES A–E

DIRECTIONS: Divide and check.

STRATEGY: Divide from left to right. Use short division for single-digit divisors.

WARM-UP
A. $7\overline{)4249}$

A. $6\overline{)4854}$

STRATEGY: Because there is a single-digit divisor, we use short division.

$$\begin{array}{r} 809 \\ 6\overline{)4854} \end{array}$$

6 divides 48 eight times. 6 divides 5 zero times with a remainder of 5. Now form a new number, 54, using the remainder and the next number 4. 6 divides 54 nine times.

The quotient is 809.

> **CAUTION**
>
> A zero must be placed in the quotient so that the 8 and the 9 have the correct place values.

WARM-UP
B. Divide: $13\overline{)2028}$

B. Divide: $23\overline{)5635}$

STRATEGY: Write the partial quotients above the dividend with the place values aligned.

$$\begin{array}{r} 245 \\ 23\overline{)5635} \\ \underline{46} \\ 103 \\ \underline{92} \\ 115 \\ \underline{115} \\ 0 \end{array}$$

$23(2) = 46$

$23(4) = 92$

$23(5) = 115$

CHECK:
$$\begin{array}{r} 245 \\ \times\,23 \\ \hline 735 \\ 4900 \\ \hline 5635 \end{array}$$

The quotient is 245.

ANSWERS TO WARM-UPS A–B

A. 607   **B.** 156

C. Find the quotient: $\dfrac{127{,}257}{482}$

WARM-UP
C. Find the quotient: $\dfrac{233{,}781}{482}$

STRATEGY: When a division is written as a fraction, the dividend is above the fraction bar and the divisor is below.

```
              264
    482)127,257        482 does not divide 1.
        96 4           482 does not divide 12.
        30 85          482 does not divide 127.
        28 92          482 divides 1272 two times.
         1 937         482 divides 3085 six times.
         1 928         482 divides 1937 four times.
             9         The remainder is 9.
```

CHECK: Multiply the divisor by the partial quotient and add the remainder.

$$264(482) + 9 = 127{,}248 + 9$$
$$= 127{,}257$$

The answer is 264 with a remainder of 9, or 264 R 9.

You may recall other ways to write a remainder using fractions or decimals. These are covered in a later chapter.

CALCULATOR EXAMPLE:

D. Divide 73,965 by 324.
Enter the division: $73{,}965 \div 324$

$$73{,}965 \div 324 \approx 228.28703$$

The quotient is not a whole number. This means that 228 is the partial quotient and there is a remainder. To find the remainder, multiply 228 times 324. Subtract the product from 73,965. The result is the remainder.

$$73{,}965 - 228(324) = 93$$

So $73{,}965 \div 324 = 228$ R 93.

WARM-UP
D. Divide 47,753 by 415.

E. When planting Christmas trees, the Greenfir Tree Farm allows 64 square feet per tree. How many trees will they plant in 43,520 square feet?

STRATEGY: Because each tree is allowed 64 square feet, we divide the number of square feet by 64 to find out how many trees will be planted.

```
           680
    64)43,520
       38 4
        5 12
        5 12
          00
           0
           0
```

There will be a total of 680 trees planted in 43,520 square feet.

WARM-UP
E. The Greenfir Tree Farm allows 256 square feet per large spruce tree. If there are 43,520 square feet to be planted, how many trees will they plant?

ANSWERS TO WARM-UPS C–E
C. 485 R 11   **D.** 115 R 28   **E.** They will plant 170 trees.

OBJECTIVE 1 Divide whole numbers. (See page 47.)

A *Divide.*

1. $8\overline{)72}$ **2.** $8\overline{)88}$ **3.** $6\overline{)78}$ **4.** $4\overline{)84}$

5. $5\overline{)435}$ **6.** $3\overline{)327}$ **7.** $5\overline{)455}$ **8.** $9\overline{)549}$

9. $136 \div 8$ **10.** $180 \div 5$ **11.** $880 \div 22$ **12.** $850 \div 17$

13. $492 \div 6$ **14.** $1668 \div 4$ **15.** $36 \div 7$ **16.** $79 \div 9$

17. $81 \div 17$ **18.** $93 \div 29$

19. The quotient in division has no remainder when the last difference is _____.

20. For $360 \div 12$, in the partial division $36 \div 12 = 3$, 3 has place value _____.

B *Divide.*

21. $18,306 \div 6$ **22.** $21,154 \div 7$ **23.** $\dfrac{768}{24}$ **24.** $\dfrac{558}{62}$

25. $46\overline{)2484}$ **26.** $38\overline{)2546}$ **27.** $46\overline{)4002}$ **28.** $56\overline{)5208}$

29. $542\overline{)41,192}$ **30.** $516\overline{)31,992}$ **31.** $355\overline{)138,805}$ **32.** $617\overline{)124,017}$

33. $43\overline{)7822}$ **34.** $56\overline{)7288}$ **35.** $57\overline{)907}$ **36.** $39\overline{)797}$

37. $(78)(?) = 1872$ **38.** $(?)(65) = 4225$ **39.** $27\overline{)345,672}$ **40.** $62\overline{)567,892}$

41. $55,892 \div 64$. Round quotient to the nearest ten.

42. $67,000 \div 43$. Round quotient to the nearest hundred.

43. $225,954 \div 415$. Round quotient to the nearest hundred.

44. $535,843 \div 478$. Round quotient to the nearest hundred.

C *Exercises 45–48. The revenue department of a state had the following collection data for the first 3 weeks of April.*

Taxes Collected

Number of Returns	Total Taxes Paid
Week 1—4563	$24,986,988
Week 2—3981	$19,315,812
Week 3—11,765	$48,660,040

45. Find the taxes paid per return during week 1.

46. Find the taxes paid per return during week 2.

47. Find the taxes paid per return during week 3. Round to the nearest hundred dollars.

48. Find the taxes paid per return during the 3 weeks. Round to the nearest hundred dollars.

49. A forestry survey finds that 1890 trees are ready to harvest on a 14-acre plot. On the average, how many trees are ready to harvest per acre?

50. Green Tract Lumber Company replants 5865 seedling fir trees on a 15-acre plot of logged-over land. What is the average number of seedlings planted per acre?

51. Ms. Munos buys 45 radios to sell in her department store. She pays $1260 for the radios. Ms. Munos reorders an additional 72 radios. What will she pay for the reordered radios if she gets the same price per radio as the original order?

52. Burkhardt Floral orders 25 dozen red roses at the wholesale market. The roses cost $300. The following week they order 34 dozen of the roses. What do they pay for the 34 dozen roses if they pay the same price per dozen as in the original order?

53. In 2009, Bills Gates of Microsoft was the richest person in the United States, with an estimated net worth of $57 billion. How much would you have to spend per day in order to spend all of Bill Gate's $57 billion in 90 years, ignoring leap years? Round to the nearest hundred thousand.

54. How much money would you have to spend per day, ignoring leap years, in order to spend Bill Gates's $57 billion in 50 years? In 20 years? Round to the nearest hundred dollars.

Exercises 55–57. Use the 2009 estimated population and the area of the country as given.

Estimated Population in 2009

Country	Estimated Population	Area, in Square Kilometers
China	1,335,962,000	9,596,960
Italy	60,090,000	301,230
United States	305,967,000	9,629,091

55. What was the population density (people per square kilometer, that is, the number of people divided by the number of square kilometers) of China, rounded to the nearest whole person?

56. What was the population density (people per square kilometer, that is, the number of people divided by the number of square kilometers) of Italy, rounded to the nearest whole person?

57. What was the population density (people per square kilometer, that is, the number of people divided by the number of square kilometers) of the United States, rounded to the nearest whole person?

Exercises 58–59. It is estimated that there are about 72 million dogs and 82 million cats owned in the United States.

58. The American Veterinary Medical Association (AMVA) estimates that dog owners spent about $19,800,000,000 in veterinary fees for their dogs in the last year. What is the average cost per dog?

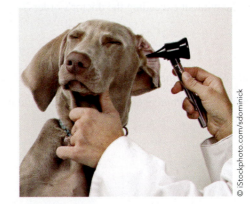

59. The American Veterinary Medical Association (AMVA) estimates that cat owners spent about $6,642,000,000 in veterinary fees for their cats in the last year. What is the average cost per cat?

Exercises 60–62. The 2000 Census population and the number of House of Representative seats in the United States and two states are given below.

Population and House Representation

	Population	Number of House Seats
United States	272,171,813	435
California	33,145,121	53
Montana	882,779	2

60. How many people does each House member represent in the United States?

61. How many people does each representative from California represent?

62. How many people does each representative from Montana represent?

63. In 2008, the estimated population of California was 36,756,666. The gross state product (GSP) was about $1,850,000,000,000. What was the state product per person, rounded to the nearest hundred dollars?

64. In 2008, the estimated population of Kansas was 2,850,000 and the total personal income tax for the state was about $11,205,000,000. What was the per capita income tax, rounded to the nearest ten dollars?

Exercises 65–66. A bag of white cheddar corn cakes contains 14 servings, a total of 630 calories and 1820 mg of sodium.

65. How many calories are there per serving?

66. How many milligrams of sodium are there per serving?

67. Juan is advised by his doctor not to exceed 2700 mg of aspirin per day for his arthritis pain. If he takes capsules containing 325 mg of aspirin, how many capsules can he take without exceeding the doctor's orders?

Exercises 68–69 refer to the chapter application. See Table 1.1, page 3.

68. If the earnings of *Titanic* were halved, where would it appear on the list?

69. If the average ticket price was $8, estimate how many tickets were sold to *Spider-Man*?

70. Jerry Rice of the San Francisco 49ers holds the Super Bowl record for most pass receptions. In the 1989 game, he had 11 receptions for a total of 215 yards. What was the average yardage per reception, rounded to the nearest whole yard?

71. In 2008, the Super Bowl champion Pittsburg Steelers had a roster of 83 players and a total payroll of $119,176,821. Calculate the average salary for the Steelers, rounded to the nearest thousand dollars.

STATE YOUR UNDERSTANDING

72. Explain to an 8-year-old child that $45 \div 9 = 5$.

73. Explain the concept of remainder.

74. Define and give an example of a quotient.

CHALLENGE

75. The Belgium Bulb Company has 171,000 tulip bulbs to market. Eight bulbs are put in a package when shipping to the United States and sold for $3 per package. Twelve bulbs are put in a package when shipping to France and sold for $5 per package. In which country will the Belgium Bulb Company get the greatest gross return? What is the difference in gross receipts?

Exercises 76–77. Complete the problems by writing in the correct digit wherever you see a letter.

76.

$$\begin{array}{r} 5AB2 \\ 3\overline{)1653C} \end{array}$$

77. $\begin{array}{r} 21B \\ A3\overline{)4CC1} \end{array}$

78. Divide 23,000,000 and 140,000,000 by 10, 100, 1000, 10,000, and 100,000. What do you observe? Can you devise a rule for dividing by 10, 100, 1000, 10,000, and 100,000?

HOW & WHY

OBJECTIVE

Solve an equation of the form $ax = b$ or $\dfrac{x}{a} = b$, where x, a, and b are whole numbers.

In Section 1.2, the equations involved the inverse operations addition and subtraction. Multiplication and division are also inverse operations. We can use this idea to solve equations containing those operations. For example, if 4 is multiplied by 2, $4 \cdot 2 = 8$, the product is 8. If the product is divided by 2, $8 \div 2$, the result is 4, the original number. In the same manner, if 12 is divided by 3, $12 \div 3 = 4$, the quotient is 4. If the quotient is multiplied by 3, $4 \cdot 3 = 12$, the original number. We use this idea to solve equations in which the variable is either multiplied or divided by a number.

When a variable is multiplied or divided by a number, the multiplication symbols ($\cdot$ or $\times$) and the division symbol ($\div$) normally are not written. We write $3x$ for 3 times x and $\dfrac{x}{3}$ for x divided by 3.

Consider the following:

$$5x = 30$$
$$\frac{5x}{5} = \frac{30}{5} \qquad \text{Division will eliminate multiplication.}$$
$$x = 6$$

If x in the original equation is replaced by 6, we have

$$5x = 30$$
$$5 \cdot 6 = 30$$
$$30 = 30 \qquad \text{A true statement.}$$

Therefore, the solution is $x = 6$.

Now consider when the variable is divided by a number:

$$\frac{x}{7} = 21$$
$$7 \cdot \frac{x}{7} = 7 \cdot 21 \qquad \text{Multiplication will eliminate division.}$$
$$x = 147$$

If x in the original equation is replaced by 147, we have

$$\frac{147}{7} = 21$$
$$21 = 21 \qquad \text{A true statement.}$$

Therefore, the solution is $x = 147$.

> **To solve an equation using multiplication or division**
>
> 1. Divide both sides by the same number to isolate the variable, or
> 2. Multiply both sides by the same number to isolate the variable.
> 3. Check the solution by substituting it for the variable in the original equation.

EXAMPLES A–E

DIRECTIONS: Solve and check.

STRATEGY: Isolate the variable by multiplying or dividing both sides of the equation by the same number. Check the solution by substituting it for the variable in the original equation.

A. $3x = 24$

$3x = 24$

$\dfrac{3x}{3} = \dfrac{24}{3}$ Isolate the variable by dividing both sides of the equation by 3.

$x = 8$ Simplify.

CHECK: $3x = 24$

$3(8) = 24$ Substitute 8 for x in the original equation.

$24 = 24$ The statement is true.

The solution is $x = 8$.

B. $\dfrac{x}{4} = 9$

$\dfrac{x}{4} = 9$

$4 \cdot \dfrac{x}{4} = 4 \cdot 9$ Isolate the variable by multiplying both sides by 4.

$x = 36$ Simplify.

CHECK: $\dfrac{x}{4} = 9$

$\dfrac{36}{4} = 9$ Substitute 36 for x in the original equation.

$9 = 9$ The statement is true.

The solution is $x = 36$.

C. $\dfrac{c}{7} = 12$

$\dfrac{c}{7} = 12$

$7 \cdot \dfrac{c}{7} = 7 \cdot 12$ Isolate the variable by multiplying both sides of the equation by 7.

$c = 84$ Simplify.

CHECK: $\dfrac{c}{7} = 12$

$\dfrac{84}{7} = 12$ Substitute 84 for c in the original equation.

$12 = 12$ The statement is true.

The solution is $c = 84$.

D. $9y = 117$

$9y = 117$

$\dfrac{9y}{9} = \dfrac{117}{9}$ Isolate the variable by dividing both sides of the equation by 9.

$y = 13$ Simplify.

WARM-UP

A. $6y = 18$

WARM-UP

B. $\dfrac{a}{5} = 10$

WARM-UP

C. $\dfrac{b}{3} = 33.$

WARM-UP

D. $8t = 96$

ANSWERS TO WARM-UPS A–D

A. $y = 3$ **B.** $a = 50$ **C.** $b = 99$

D. $t = 12$

CHECK:
$$9y = 117$$
$$9(13) = 117 \quad \text{Substitute 13 for } y \text{ in the original equation.}$$
$$117 = 117 \quad \text{The statement is true.}$$

The solution is $y = 13$.

WARM-UP

E. What is the length (ℓ) of a second lot in the subdivision if the width (w) is 90 feet and the area (A) is 10,350 square feet? Use the formula $A = \ell w$.

E. What is the width (w) of a rectangular lot in a subdivision if the length (ℓ) is 125 feet and the area (A) is 9375 square feet? Use the formula $A = \ell w$.

STRATEGY: To find the width of the lot, substitute the area, $A = 9375$, and the length, $\ell = 125$, into the formula and solve.

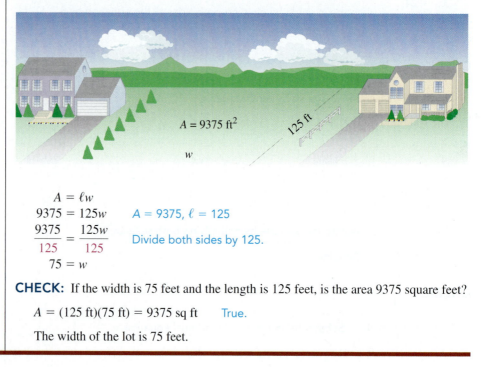

$A = 9375$ ft^2

125 ft

w

$$A = \ell w$$
$$9375 = 125w \quad \text{A = 9375, } \ell = 125$$
$$\frac{9375}{125} = \frac{125w}{125} \quad \text{Divide both sides by 125.}$$
$$75 = w$$

CHECK: If the width is 75 feet and the length is 125 feet, is the area 9375 square feet?

$$A = (125 \text{ ft})(75 \text{ ft}) = 9375 \text{ sq ft} \quad \text{True.}$$

The width of the lot is 75 feet.

ANSWER TO WARM-UP E   **E.** The length of the lot is 115 feet.

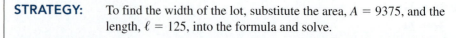

EXERCISES

■ **OBJECTIVE** Solve an equation of the form $ax = b$ or $\dfrac{x}{a} = b$, where x, a, and b are whole numbers. (See page 56.)

Solve and check.

1. $3x = 15$

2. $\dfrac{z}{4} = 5$

3. $\dfrac{c}{3} = 6$

4. $8x = 32$

5. $13x = 52$

6. $\dfrac{y}{4} = 14$

7. $\dfrac{b}{2} = 23$

8. $15a = 135$

9. $12x = 144$

10. $\dfrac{x}{14} = 12$

11. $\dfrac{y}{13} = 24$

12. $23c = 184$

13. $27x = 648$

14. $\dfrac{a}{32} = 1216$

15. $\dfrac{b}{12} = 2034$

16. $57z = 2451$

17. $1098 = 18x$

18. $616 = 11y$

19. $34 = \dfrac{w}{23}$

20. $64 = \dfrac{c}{33}$

21. Find the width of a rectangular garden plot that has a length of 35 feet and an area of 595 square feet. Use the formula $A = \ell w$.

22. Find the length of a room that has an area of 391 square feet and a width of 17 feet.

23. Crab sells at the dock for $2 per pound. A fisherman sells his catch and receives $4680. How many pounds of crab does he sell?

24. Felicia earns $7 an hour. Last week she earned $231. How many hours did she work last week?

25. If the wholesale cost of 18 stereo sets is $5580, what is the wholesale cost of one set? Use the formula $C = np$, where C is the total cost, n is the number of units purchased, and p is the price per unit.

26. Using the formula in Exercise 25, if the wholesale cost of 24 personal computers is $18,864, what is the wholesale cost of one computer?

27. The average daily low temperature in Toronto in July is twice the average high temperature in January. Write an equation that describes this relationship. Be sure to define all variables in your equation. If the average daily low temperature in July is 60°F, what is the average daily high temperature in January?

28. Car manufacturers recommend that the fuel filter in a car be replaced when the mileage is ten times the recommended mileage for an oil change. Write an equation that describes this relationship. Be sure to define all variables in your equation. If a fuel filter should be replaced every 30,000 miles, how often should the oil be changed?

Whole-Number Exponents and Powers of 10

OBJECTIVES

1. Find the value of an expression written in exponential form.
2. Multiply or divide a whole number by a power of 10.

VOCABULARY

A **base** is a number used as a repeated factor. An **exponent** indicates the number of times the base is used as a factor and is always written as a superscript to the base. In 2^3, 2 is the base and 3 is the exponent.

The **value** of 2^3 is 8.

An exponent of 2 is often read **"squared"** and an exponent of 3 is often read **"cubed."**

A **power of 10** is the value obtained when 10 is written with an exponent.

HOW & WHY

■ **OBJECTIVE 1** Find the value of an expression written in exponential form.

Exponents show repeated multiplication. Whole-number *exponents* greater than 1 are used to write repeated multiplications in shorter form. For example,

5^4 means $5 \cdot 5 \cdot 5 \cdot 5$

and since $5 \cdot 5 \cdot 5 \cdot 5 = 625$ we write $5^4 = 625$. The number 625 is sometimes called the "fourth power of five" or "the *value* of 5^4."

$$\text{EXPONENT}$$
$$\downarrow$$
$$\text{BASE} \rightarrow 5^4 = 625 \leftarrow \text{VALUE}$$

Similarly, the value of 7^6 is

$7^6 = 7 \cdot 7 \cdot 7 \cdot 7 \cdot 7 \cdot 7 = 117,649$

The base, the repeated factor, is 7. The exponent, which indicates the number of times the base is used as a factor, is 6.

The exponent 1 is a special case.

In general, $x^1 = x$. So $2^1 = 2$, $13^1 = 13$, $7^1 = 7$, and $(413)^1 = 413$.

We can see a reason for the meaning of $6^1 (6^1 = 6)$ by studying the following pattern.

$6^4 = 6 \cdot 6 \cdot 6 \cdot 6$
$6^3 = 6 \cdot 6 \cdot 6$
$6^2 = 6 \cdot 6$
$6^1 = 6$

Exponential Property of One

If 1 is used as an exponent, the value is equal to the base.

$b^1 = b$

To find the value of an expression with a natural number exponent

1. If the exponent is 1, the value is the same as the base.
2. If the exponent is greater than 1, use the base number as a factor as many times as shown by the exponent. Multiply.

Exponents give us a second way to write an area measurement. Using exponents, we can write 74 square inches as 74 in^2. The symbol 74 in^2. is still read "seventy-four square inches." Also, 65 square feet is written as 65 ft^2.

EXAMPLES A–F

DIRECTIONS: Find the value.

STRATEGY: Identify the exponent. If it is 1, the value is the base number. If it is greater than 1, use it to tell how many times the base is used as a factor and then multiply.

A. Find the value of 11^3.   Use 11 as a factor three times.

$11^3 = 11 \cdot 11 \cdot 11$
$ = 1331$

The value is 1331.

B. Simplify: 29^1

$29^1 = 29$ If the exponent is 1, the value is the base number.

The value is 29.

C. Find the value of 10^7.

$10^7 = (10)(10)(10)(10)(10)(10)(10)$
$ = 10,000,000$ Ten million.

     Note that the value has seven zeros.

The value is 10,000,000.

D. Evaluate: 6^5

$6^5 = 6(6)(6)(6)(6) = 7776$

The value is 7776.

CALCULATOR EXAMPLE:

E. Find the value of 11^6.

Calculators usually have an exponent key marked $\boxed{y^x}$ or $\boxed{\wedge}$. If your calculator doesn't have such a key, you can multiply the repeated factors.

The value is 1,771,561.

F. Mabel needs 125 square yards of carpet to carpet her house wall to wall. Write the amount of carpet needed using an exponent to show the measure.

125 square yards = 125 yd^2 Square yards is written yd^2.

Mabel needs 125 yd^2 of carpet.

WARM-UP

A. Find the value of 16^3.

WARM-UP

B. Simplify: 72^1

WARM-UP

C. Find the value of 10^6.

WARM-UP

D. Evaluate: 7^4

WARM-UP

E. Find the value of 5^9.

WARM-UP

F. Herman's vegetable garden covers a total of 700 square feet. Write the area of his garden using an exponent to show the measure.

HOW & WHY

■ **OBJECTIVE 2** Multiply or divide a whole number by a power of 10.

It is particularly easy to multiply or divide a whole number by a power of 10. Consider the following and their products when multiplied by 10.

$4 \times 10 = 40$ $9 \times 10 = 90$ $17 \times 10 = 170$ $88 \times 10 = 880$

The place value of every digit becomes 10 times larger when the number is multiplied by 10. So to multiply by 10, we need to merely write a zero to the right of the whole

ANSWERS TO WARM-UPS A–F

A. 4096 **B.** 72 **C.** 1,000,000
D. 2401 **E.** 1,953,125 **F.** The area of Herman's garden is 700 ft^2.

number. If a whole number is multiplied by 10 more than once, a zero is written to the right for each 10. So,

$42 \times 10^3 = 42{,}000$ Three zeros are written on the right, one for each 10.

Because division is the inverse of multiplication, dividing by 10 will eliminate the last zero on the right of a whole number. So,

$650{,}000 \div 10 = 65{,}000$ Eliminate the final zero on the right.

If we divide by 10 more than once, one zero is eliminated for each 10. So,

$650{,}000 \div 10^4 = 65$ Eliminate four zeros on the right.

> **To multiply a whole number by a power of 10**
>
> 1. Identify the exponent of 10.
> 2. Write as many zeros to the right of the whole number as the exponent of 10.

> **To divide a whole number by a power of 10**
>
> 1. Identify the exponent of 10.
> 2. Eliminate the same number of zeros to the right of the whole number as the exponent of 10.

Using powers of 10, we have a third way of writing a whole number in expanded form.

$8257 = 8000 + 200 + 50 + 7$, or

$\quad = 8 \text{ thousands} + 2 \text{ hundreds} + 5 \text{ tens} + 7 \text{ ones}$, or

$\quad = (8 \times 10^3) + (2 \times 10^2) + (5 \times 10^1) + (7 \times 1)$

EXAMPLES G–K

DIRECTIONS: Multiply or divide.

STRATEGY: Identify the exponent of 10. For multiplication, write the same number of zeros to the right of the whole number as the exponent of 10. For division, eliminate the same number of zeros on the right of the whole number as the exponent of 10.

WARM-UP
G. Multiply: 9821×10^4

G. Multiply: $45{,}786 \times 10^5$

$45{,}786 \times 10^5 = 4{,}578{,}600{,}000$ The exponent of 10 is 5. To multiply, write 5 zeros to the right of the whole number.

The product is 4,578,600,000.

WARM-UP
H. Simplify: 21×10^7

H. Simplify: 532×10^5

$532 \times 10^5 = 53{,}200{,}000$

The product is 53,200,000.

ANSWERS TO WARM-UPS G–H

G. 98,210,000
H. 210,000,000

I. Divide: $\dfrac{134{,}000}{10^2}$

$\dfrac{134{,}000}{10^2} = 1340$ The exponent of 10 is 2.
To divide, eliminate 2 zeros on the right of the whole number.

The quotient is 1340.

J. Simplify: $606{,}820{,}000 \div 10^4$

$606{,}820{,}000 \div 10^4 = 60{,}682$ Eliminate 4 zeros on the right.

The quotient is 60,682.

K. A recent fund-raising campaign raised an average of $315 per donor. How much was raised if there were 100,000 (10^5) donors?

STRATEGY: To find the total raised, multiply the average donation by the number of donors.

$315 \times 10^5 = 31{,}500{,}000$

The exponent is 5. To multiply, write 5 zeros to the right of the whole number.

The campaign raised $31,500,000.

EXERCISES 1.5

■ OBJECTIVE 1 Find the value of an expression written in exponential form. (See page 60.)

A *Write in exponential form.*

1. 16(16)(16)(16)(16)(16)

2. $43 \times 43 \times 43 \times 43 \times 43 \times 43 \times 43 \times 43$

Find the value.

3. 7^2

4. 6^2

5. 2^3

6. 3^3

7. 15^1

8. 20^1

9. In $13^4 = 28{,}561$, 13 is the _____, 4 is the _____, and 28,561 is the _____.

10. In $4^7 = 16{,}384$, 16,384 is the _____, 4 is the _____, and 7 is the _____.

B *Find the value.*

11. 8^3

12. 3^6

13. 32^2

14. 15^3

15. 10^3

16. 10^6

17. 9^3

18. 7^4

19. 3^8

20. 5^6

■ OBJECTIVE 2 Multiply or divide a whole number by a power of 10. (See page 61.)

A *Multiply or divide.*

21. 88×10^2

22. 29×10^3

23. 21×10^5

24. 69×10^4

25. $2300 \div 10^2$ **26.** $49{,}000 \div 10^2$ **27.** $780{,}000 \div 10^3$ **28.** $346{,}000 \div 10^1$

29. To multiply a number by a power of 10, write as many zeros to the right of the number as the _____ of 10.

30. To divide a number by a power of 10, eliminate as many _____ on the right of the number as the exponent of 10.

B *Multiply or divide.*

31. 202×10^5 **32.** 2120×10^3 **33.** $7{,}270{,}000 \div 10^3$

34. $92{,}000{,}000 \div 10^3$ **35.** 6734×10^4 **36.** 5466×10^5

37. $\dfrac{528{,}000}{10^3}$ **38.** $\dfrac{310{,}000}{10^2}$ **39.** 9610×10^7

40. 732×10^5 **41.** $450{,}000{,}000 \div 10^5$ **42.** $8{,}953{,}000{,}000 \div 10^6$

C

43. Write in exponent form:
10(10)(10)(10)(10)(10)(10)(10)(10)(10)(10)

44. Write in exponent form:
4(4)(4)(4)(4)(4)(4)(4)(4)(4)(4)(4)(4)(4)(4)(4)(4)(4)

Find the value.

45. 11^5 **46.** 23^4 **47.** 14^5 **48.** 31^4

Multiply or divide.

49. $47{,}160 \times 10^9$ **50.** 630×10^{13} **51.** $\dfrac{680{,}000{,}000}{10^7}$ **52.** $\dfrac{3{,}120{,}000{,}000{,}000{,}000}{10^8}$

53. Salvador bought a lot for his new house in a suburban subdivision. The lot consists of 10,800 square feet. Write the size of Salvador's lot using an exponent to express the measure.

54. To make her niece's wedding gown, Marlene needs 25 square meters of material. Write the amount of material she needs using an exponent to express the measure.

55. The operating budget for skate parks in a Midwestern city is approximately 30×10^4 dollars. Write this amount in place value form.

56. In mid-2009, the government warned that ten financial institutions faced a 746×10^8 captial shortfall. Write this amount in place value form.

57. The distance from Earth to the nearest star outside our solar system (Alpha Centauri) is approximately 255×10^{11} miles. Write this distance in place value form.

58. A high roller in Atlantic City places nine consecutive bets at the "Twenty-one" table. The first bet is $4 and each succeeding bet is four times the one before. How much does she wager on the ninth bet? Express the answer as a power of 4 and as a whole number.

59. The distance that light travels in a year is called a light-year. This distance is almost 6 trillion miles. Write the place value name for this number. Write the number as 6 times a power of 10.

60. The average distance from Earth to the sun is approximately 93 million miles. Write the place value name for this distance and write it as 93 times a power of 10.

Exercises 61–63. A number of bacteria is quadrupling (increasing fourfold) in size evey hour. There are four bacteria at the start of the experiment.

61. How many bacteria are there after 3 hours? Write your answer in exponential form and also in place value notation.

62. How many bacteria are there after 9 hours? Write your answer in exponential form and also in place value notation.

63. How many hours will it take for the number of bacteria to exceed 1000?

64. The distance around Earth at the equator is about 25×10^3 miles. Write this number in place value notation.

65. The surface area of the Pacific Ocean is about 642×10^5 square miles. Write this number in place value notation.

66. The area of the three largest deserts in the world is given in the following table.

The Three Largest Deserts in the World

Desert	Sahara (N. Africa)	Arabian (Arabian Peninsula)	Gobi (Mongolia/China)
Approximate Area in Square Miles	3,320,000	900,000	450,000

Write each area as the product of a whole number and a power of 10.

67. Disk space in a personal computer is commonly measured in gigabytes, which are billions of bytes. Write the number of bytes in 1 gigabyte as a power of 10.

68. A googol is 1 followed by 100 zeros. Legend has it that the name for this number was chosen by a 9-year-old nephew of mathematician Edward Kasner. Write a googol as a power of 10.

Exercises 69–72 refer to the chapter application. See Table 1.1, page 3.

69. Round the earnings of *Shrek 2* to the nearest million dollars and then express this amount as a product of a number and a power of 10.

70. Round the earnings of *Star Wars: Episode IV* to the nearest hundred million dollars and then express this amount as a product of a number and a power of 10.

71. What is the smallest power of 60 that is larger than the gross earnings of *Titanic*?

72. What is the smallest power of 5 that is larger than the gross earnings of *E.T. The Extra-Terrestrial*?

STATE YOUR UNDERSTANDING

73. Explain what is meant by 4^{10}.

74. Explain how to multiply a whole number by a power of 10. Give at least two examples.

CHALLENGE

75. Mitchell's grandparents deposit $4 on his first birthday and quadruple that amount on each succeeding birthday until he is 10. What amount did Mitchell's grandparents deposit on his 10th birthday? What is the total amount they have deposited in the account?

76. Find the sum of the cubes of the digits.

77. Find the difference between the sum of the cubes of 5, 10, 12, and 24 and the sum of the fifth powers of 2, 4, 6, and 7.

78. The sun is estimated to weigh 2 octillion tons. Write the place value for this number. Write it as 2 times a power of 10.

SECTION

1.6 Order of Operations

OBJECTIVE

Perform any combination of operations on whole numbers.

VOCABULARY

Parentheses ( ) and **brackets [ ]** are used in mathematics as **grouping** symbols. These symbols indicate that the operations inside are to be performed first. Other grouping symbols that are often used are **braces { }** and the **fraction bar** —.

HOW & WHY

■ **OBJECTIVE 1** Perform any combination of operations on whole numbers.

Without a rule it is possible to interpret $6 + 4 \cdot 11$ in two ways:

$$6 + 4 \cdot 11 = 6 + 44$$
$$= 50$$

or

$$6 + 4 \cdot 11 = 10 \cdot 11$$
$$= 110$$

In order to decide which answer to use, we agree to use a standard set of rules. Among these is the rule that we multiply before adding. So

$$6 + 4 \cdot 11 = 50$$

The order in which the operations are performed is important because the order often determines the answer. Therefore, there is an established *order of operations.* This established order was agreed upon many years ago and is programmed into most of today's calculators and computers.

ORDER OF OPERATIONS

To evaluate an expression with more than one operation

1. Parentheses—Do the operations within grouping symbols first (parentheses, fraction bar, etc.), in the order given in steps 2, 3, and 4.
2. Exponents—Do the operations indicated by exponents.
3. Multiply and Divide—Do multiplication and division as they appear from left to right.
4. Add and Subtract—Do addition and subtraction as they appear from left to right.

So we see that

$$15 - 8 \div 2 = 15 - 4 \qquad \text{Divide first.}$$
$$= 11 \qquad \text{Subtract.}$$
$$(18 - 11)(5) = 7(5) \qquad \text{Subtract in parentheses first.}$$
$$= 35 \qquad \text{Multiply.}$$
$$84 \div 21 \cdot 4 = 4 \cdot 4 \qquad \text{Neither multiplication nor division takes preference over}$$
$$= 16 \qquad \text{the other, so do them from left to right.}$$

As you can see, the rules for the order of operations are fairly complicated and it is important that you learn them all. A standard memory trick is to use the first letters to make an easy-to-remember phrase.

Parentheses

Exponents

Multiplication/**D**ivision

Addition/**S**ubtraction

Consider the phrase **P**lease **E**xcuse **M**y **D**ear **A**unt **S**ally. Note that the first letters of the words in this phrase are exactly the same (and in the same order) as the first letters for the order of operations.

Many students use "Please excuse my dear Aunt Sally" to help them remember the order for operations. Why not give it a try?

CAUTION

Brackets [ ] and braces { } are often used differently in calculators. Consult your calculator manual.

Exercises involving all of the operations are shown in the examples.

DIRECTIONS: Simplify.

STRATEGY: The operations are done in this order: operations in parentheses first, exponents next, then multiplication and division, and finally, addition and subtraction.

WARM-UP
A. Simplify:
$9 \cdot 4 + 7 \cdot 9$

A. Simplify: $6 \cdot 3 + 9 \cdot 6$

$$
\begin{aligned}
6 \cdot 3 + 9 \cdot 6 &= 18 + 54 \qquad \text{Multiply first.} \\
&= 72 \qquad\qquad \text{Add.}
\end{aligned}
$$

The value is 72.

WARM-UP
B. Simplify:
$6 \cdot 11 - 16 \div 4 + 8 \cdot 2$

B. Simplify: $33 - 9 \div 3 + 7 \cdot 4$

$$
\begin{aligned}
33 - 9 \div 3 + 7 \cdot 4 &= 33 - 3 + 28 \qquad \text{Divide and multiply.} \\
&= 30 + 28 \qquad\quad \text{Subtract.} \\
&= 58 \qquad\qquad\quad \text{Add.}
\end{aligned}
$$

WARM-UP
C. Simplify:
$54 \div 6 + 15 - 2(9 - 6)$

C. Simplify: $4 \cdot 8 + 22 - 5(7 + 2)$

$$
\begin{aligned}
4 \cdot 8 + 22 - 5(7 + 2) &= 4 \cdot 8 + 22 - 5(9) \qquad \text{Add in parentheses first.} \\
&= 32 + 22 - 45 \qquad\quad \text{Multiply.} \\
&= 54 - 45 \qquad\qquad\;\; \text{Add.} \\
&= 9 \qquad\qquad\qquad\;\;\; \text{Subtract.}
\end{aligned}
$$

WARM-UP
D. Simplify:
$6 \cdot 4^2 - 3 \cdot 5^2 + 47 - 8 \cdot 4$

D. Simplify: $2 \cdot 3^4 - 5 \cdot 2^2 + 31$

$$
\begin{aligned}
2 \cdot 3^4 - 5 \cdot 2^2 + 31 &= 2 \cdot 81 - 5 \cdot 4 + 31 \qquad \text{Do exponents first.} \\
&= 162 - 20 + 31 \qquad\quad\;\; \text{Multiply.} \\
&= 142 + 31 \qquad\qquad\;\; \text{Subtract.} \\
&= 173 \qquad\qquad\qquad\; \text{Add.}
\end{aligned}
$$

WARM-UP
E. $(4^3 - 110 \div 2)^2 + 55$

E. Simplify: $(5^2 - 6 \cdot 2)^2 + 7^2$

STRATEGY: First do the operations in the parentheses following the order of operations.

$$
\begin{aligned}
(5^2 - 6 \cdot 2)^2 + 7^2 &= (25 - 6 \cdot 2)^2 + 7^2 \qquad \text{Do the exponent in parentheses first.} \\
&= (25 - 12)^2 + 7^2 \qquad\quad \text{Multiply.} \\
&= (13)^2 + 7^2 \qquad\qquad\;\; \text{Subtract.}
\end{aligned}
$$

Now that the operations inside the parentheses are complete, continue using the order of operations.

$$
\begin{aligned}
&= 169 + 49 \qquad \text{Do the exponents.} \\
&= 218 \qquad\qquad \text{Add.}
\end{aligned}
$$

CALCULATOR EXAMPLE:

F. Simplify: $2469 + 281 \cdot 11 - 3041$

Enter the numbers and operations as they appear from left to right.

The answer is 2519.

WARM-UP
F. Simplify:
$4236 + 17 \cdot 584 \div 8$

G. Food for the Poor prepares two types of food baskets for distribution to the victims of the 2004 tsunami in Southeast Asia. The family pack contains 15 pounds of rice, and the elderly pack contains 8 pounds of rice. How many pounds of rice are needed for 325 family packs and 120 elderly packs?

STRATEGY: To find the number of pounds of rice needed for the packs, multiply the number of packs by the number of pounds per pack. Then add the two amounts.

$325(15) + 120(8) = 4875 + 960$ Multiply.

$= 5835$ Add.

Food for the Poor needs 5835 pounds of rice.

WARM-UP

G. The Fruit-of-the Month Club prepares two types of boxes for shipment. Box A contains 12 apples and Box B contains 16 apples. How many apples are needed for 125 orders of Box A and 225 orders of Box B?

ANSWER TO WARM-UP G

G. The orders require 5100 apples.

EXERCISES 1.6

OBJECTIVE   Perform any combination of operations on whole numbers. (See page 66.)

A   *Simplify.*

1. $3 \cdot 9 + 12$

2. $21 + 7 \cdot 5$

3. $46 - 5 \cdot 9$

4. $32 \cdot 5 - 12$

5. $42 + 48 \div 6$

6. $54 \div 6 - 7$

7. $53 - (23 + 2)$

8. $(41 - 15) - 11$

9. $84 \div 7 \times 4$

10. $72 \div 8 \times 3$

11. $44 + 5 \cdot 4 - 12$

12. $55 - 5 \cdot 5 + 5$

13. $2^3 - 5 + 3^3$

14. $6^2 \div 9 + 4^2$

15. $5 \cdot 7 + 3 \cdot 6$

16. $7 \cdot 7 - 6 \cdot 5$

B

17. $4^2 - 5 \cdot 2 + 7 \cdot 6$

18. $7^2 + 30 \div 5 + 2 \cdot 8$

19. $72 \div 8 + 12 - 5$

20. $34 \cdot 6 \div 17 + 12 - 13$

21. $(32 + 25) - (63 - 14)$

22. $(63 - 28) - (20 + 7)$

23. $64 \div 8 \cdot 2^3 + 6 \cdot 7$

24. $90 \div 15 \cdot 5^2 + 4 \cdot 9$

25. $75 \div 15 \cdot 7$

26. $120 \div 15 \cdot 6$

27. $88 - 3(45 - 37) + 42 - 35$

28. $46(37 - 33) \div 8 - 15 + 31$

29. $3^5 + 3^4$

30. $3^5 - 2^4 + 7^2$

31. $102 \cdot 3^3 - 72 \div 6 + 15$

32. $14 \cdot 3^4 - 364 \div 7 + 28$

C

33. $50 - 12 \div 6 - 36 \div 6 + 3$

34. $80 - 24 \div 4 + 30 \div 6 + 4$

Exercises 35–38. The graph shows the count of ducks, by species, at a lake in northern Idaho, as taken by a chapter of Ducks Unlimited.

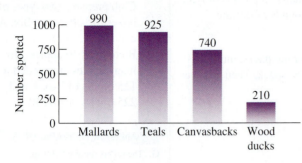

© Chrislofoto/Shutterstock.com

35. How many more mallards and canvasbacks were counted than teals and wood ducks?

36. If twice as many canvasbacks had been counted, how many more canvasbacks would there have been than teals?

37. If four times the number of wood ducks had been counted, how many more wood ducks and mallards would there have been than teals and canvasbacks?

38. If twice the number of teals had been counted, how many more teals and wood ducks would there have been compared with mallards and canvasbacks?

Simplify.

39. $11(2^3 \cdot 3 - 20) \div 4 + 33$

40. $15(4^3 \cdot 5 - 61) \div 7 - 155$

41. $9(4^2 \cdot 3 - 22) \div 9 + 28$

42. $14(3^3 \cdot 6 - 35) \div 14 - 115$

43. On the first of the month, Eltana's Custodial Service has an inventory of 95 packages of 36-count paper towels at $12 each, 47 cans of powdered cleanser at $4 each, and 58 disposable mops at $14 each. At the end of the month, there remain 22 packages of towels, 23 cans of cleanser, and 17 mops. What is the cost of the supplies used for the month?

44. For the year's second quarter, Pamela's Appliance Store stocked dishwashers that cost $335 each and barbecue grills at $117 each. She sells the dishwashers for $409 each and the grills for $131 each. If she sells 36 dishwashers and 74 barbecues, how much net income does Pamela realize?

45. A long-haul trucker is paid $29 for every 100 miles driven and $17 per stop. If he averages 2400 miles and 35 stops per week, what is his average weekly income? Assuming he takes 2 weeks off each year for vacation, what is his average yearly income?

Exercises 46–47. The labels for bread, cereal with milk, orange juice, and jam show their nutrition facts.

Cereal

Nutrition Facts

Serving Size 1/2 cup (58g)
Servings per Container about 12

Amount Per Serving	Cereal	Cereal with 1/2 cup Skim milk
Calories	200	240
Calories from Fat	10	10
	% Daily Values**	
Total Fat 1g*	2%	2%
Saturated Fat 3g	0%	0%
Cholesterol 0mg	0%	0%
Sodium 350mg	15%	17%
Potassium 160mg	5%	10%
Total Carbohydrate 47g	16%	18%
Dietary Fiber 5g	21%	21%
Sugars 7g		
Other Carbohydrates 35g		
Protein 6g		

Orange Juice

Nutrition Facts

Serving Size 8 fl oz (240 mL)
Servings Per Container 8

Amount Per Serving	
Calories 110	Calories from Fat 0
	% Daily Value*
Total Fat 0 g	0%
Sodium 0mg	0%
Potassium 450mg	13%
Total Carbohydrate 26g	9%
Sugars 22g	
Protein 2g	

Vitamin C 120%	•	Calcium 2%
Thiamin 10%	•	Niacin 4%
Vitamin B6 6%	•	Folate 15%

Not a significant source of saturated fat, cholesterol, dietary fiber, vitamin A and iron

*Percent Daily Values are based on a 2,000 calorie diet.

Jam

Nutrition Facts

Serving Size 1 Tbsp. (20g)
Servings per Container about 25

Amount Per Serving	
Calories 50	
	% Daily Value*
Total Fat 0g	0%
Sodium 15mg	1%
Total Carbohydrate 13g	4%
Sugars 13g	
Protein 0g	

*Percent Daily Values are based on a 2,000 calorie diet.

Bread

Nutrition Facts

Serving Size 1 slice (49g)
Servings Per Container 14

Amount Per Serving	
Calories 120	Calories from Fat 10
	% Daily Value*
Total Fat 1 g	2%
Saturated Fat 0g	0%
Cholesterol 0mg	0%
Sodium 360mg	15%
Total Carbohydrate 24g	8%
Dietary Fiber 1g	5%
Sugars 1g	
Protein 4g	

Vitamin A 0%	•	Vitamin C 0%
Thiamin 10%	•	Riboflavin 8%
Calcium 2%	•	Iron 10%
Niacin 8%	•	Folic Acid 10%

*Percent Daily Values are based on a 2,000 calorie diet. Your daily values may be higher or lower depending on your calorie needs.

46. How many milligrams (mg) of sodium are consumed if Marla has 3 servings of orange juice, 2 servings of cereal with one cup of skim milk, and 2 slices of bread with jam for breakfast? Milk contains 62 mg of sodium per one-half cup serving.

47. How many calories does Marla consume when she eats the breakfast listed in Exercise 46?

48. Ruth orders clothes from an Appleseed's catalog. She orders two turtleneck tops for $35 each, two pairs of stretch pants for $45 each, one suede jacket for $130, and three Italian leather belts for $25 each. The shipping and handling charge for her order is $15. What is the total charge for Ruth's order?

49. Sally orders birthday presents for her parents from a catalog. She orders each parent a 9-band AM/FM radio at $30 each. She orders her dad three double decks of playing cards at $3 each and a gold-clad double eagle coin for $20. She orders her mother a deluxe touch panel telephone for $21 and four rainbow garden flower bulb sets for $18 each. The shipping and handling charges for the order are $8. What is the total charge for Sally's order?

50. Ron makes the following purchases at his local auto supply store: four spark plugs at $1 each, a can of tire sealant for $4, two heavy duty shock absorbers at $10 each, a case (12 quarts) of motor oil at $1 per quart, and a gallon of antifreeze for $5. The sales tax for his purchases is $3. Ron has a store credit of $25. How much does he owe the store for this transaction?

51. Clay and Connie have a $100 gift certificate to the Olive Garden. One night they order an appetizer for $6, veal parmesan for $17, and fettuccine alfredo for $13. They each have a glass of wine, at $5 each. They end the dinner with desserts for $6 each and coffee for $2 each. How much do they have left on their gift certificate after the meal?

Exercises 52–53. Ryan makes $1800 a month after taxes. Her rent is $725, utilities are $140, and her phone is $78. She spends $250 on food, $95 on car insurance, and $180 on gas.

52. How much money does Ryan have left each month after paying her expenses?

53. Ryan decides to save $150 per month to build up her emergency fund, which already contains $1650. How many months will Ryan need to save in order for her to have two months' salary in her emergency fund?

Exercises 54–55. Derrick stops at the Koffee People Drive-Thru every morning on his way to work and buys a large espresso for $4.

54. Assuming Derrick works 48 weeks per year, how much does he spend in a year on espresso?

55. As a way to save money, Derrick decided to make his espresso at home. He buys an espresso machine for $120, and spends $22 per month on espresso beans from Costco. How much money will he save in a year?

Exercises 56–58 refer to the chapter application. See Table 1.1, page 3.

56. Calculate the sum of the gross earnings for *Shrek 2* and *Spider-Man* and divide by 2.

57. Calculate the sum of the gross earnings for *Shrek 2* and *Spider-Man* divided by 2.

58. Use the order of operations to explain why the answers to Exercise 56 and Exercise 57 are not the same

STATE YOUR UNDERSTANDING

59. Which of the following is correct? Explain.

$$20 - 10 \div 2 = 10 \div 2 \qquad \text{or} \qquad 20 - 10 \div 2 = 20 - 5$$
$$= 5 \qquad\qquad\qquad = 15$$

60. Explain how to simplify $2(1 + 36 \div 3^2) - 3$ using the order of operations.

CHALLENGE

61. Simplify: $(6 \cdot 3 - 8)^2 - 50 + 2 \cdot 3^2 + 2(9 - 5)^3$

62. USA Media buys three first-run movies: #1 for $185, #2 for $143, and #3 for $198. During the first 2 months, #1 is rented 10 times at the weekend rate of $5 and 26 times at the weekday rate of $3; #2 is rented 12 times at the weekend rate and 30 times at the weekday rate; and #3 is rented 8 times at the weekend rate and 18 times at the weekday rate. How much money must still be raised to pay for the cost of the three movies?

63. It is estimated that hot water heaters need to be big enough to accommodate the water usage for an entire hour. To figure what size heater you need, first identify the single hour of the day in which water usage is highest. Next, identify the types of water usage during this hour. The estimates of water usage for various activities are shown in the table. Now calculate the total number of gallons of water used in your hour. Your water heater must have this capacity.

Water Usage

Activity	Gallons Used
Shower	20
Bath	20
Washing hair	4
Shaving	2
Washing hands/face	4
Dishwasher	14

HOW & WHY

Recall that we have solved equations involving only one operation. Let's look at some equations that involve two operations.

To solve $x - 5 = 7$, we add 5 to both sides of the equation. To solve $4x = 16$, we divide both sides of the equation by 4. The following equation requires both steps.

$$4x - 5 = 7$$

$4x - 5 + 5 = 7 + 5$ First, eliminate the subtraction by adding 5 to both sides.

$4x = 12$ Simplify both sides.

$\dfrac{4x}{4} = \dfrac{12}{4}$ Eliminate the multiplication. Divide both sides by 4.

$x = 3$ Simplify.

CHECK: $4x - 5 = 7$

$4(3) - 5 = 7$ Substitute 3 for x.

$12 - 5 = 7$ Multiply.

$7 = 7$ Subtract.

Thus, if x is replaced by 3 in the original equation, the statement is true. So the solution is $x = 3$. Now solve

$$5x + 11 = 46$$

$5x + 11 - 11 = 46 - 11$ First, eliminate the addition by subtracting 11 from both

$5x = 35$ sides. Simplify.

$\dfrac{5x}{5} = \dfrac{35}{5}$ Eliminate the multiplication by dividing both sides by 5.

$x = 7$ Simplify.

CHECK: $5x + 11 = 46$

$5(7) + 11 = 46$ Substitute 7 for x in the original equation.

$35 + 11 = 46$ Multiply.

$46 = 46$ Add.

Thus, if x is replaced by 7 in the original equation, the statement is true. So the solution is $x = 7$.

Note that in each of the previous examples the operations are eliminated in the opposite order in which they are performed. That is, addition and subtraction were eliminated first and then the multiplication and division.

To solve an equation of the form $ax + b = c$, $ax - b = c$, $\dfrac{x}{a} + b = c$, or $\dfrac{x}{a} - b = c$

1. Eliminate the addition or subtraction by subtracting or adding the same number to both sides.
2. Eliminate the multiplication or division by dividing or multiplying by the same number on both sides.
3. Check the solution by substituting it in the original equation.

EXAMPLES A–E

DIRECTIONS: Solve and check.

STRATEGY: Isolate the variable by first adding or subtracting the same number from both sides. Second, multiply or divide both sides by the same number.

A.

$$3x - 23 = 7$$

$$3x - 23 = 7$$

$$3x - 23 + 23 = 7 + 23 \qquad \text{Add 23 to both sides to eliminate the subtraction.}$$

$$3x = 30 \qquad \text{Simplify.}$$

$$\frac{3x}{3} = \frac{30}{3} \qquad \text{Divide both sides by 3 to eliminate the multiplication.}$$

$$x = 10 \qquad \text{Simplify.}$$

CHECK:

$$3x - 23 = 7$$

$$3(10) - 23 = 7 \qquad \text{Substitute 10 for } x \text{ in the original equation.}$$

$$30 - 23 = 7 \qquad \text{Simplify.}$$

$$7 = 7 \qquad \text{The statement is true.}$$

The solution is $x = 10$.

B.

$$\frac{y}{9} + 3 = 13$$

$$\frac{y}{9} + 3 = 13$$

$$\frac{y}{9} + 3 - 3 = 13 - 3 \qquad \text{Subtract 3 from both sides to eliminate the addition.}$$

$$\frac{y}{9} = 10 \qquad \text{Simplify.}$$

$$9\left(\frac{y}{9}\right) = 9(10) \qquad \text{Multiply both sides by 9 to eliminate the division.}$$

$$y = 90$$

CHECK:

$$\frac{y}{9} + 3 = 13$$

$$\frac{90}{9} + 3 = 13 \qquad \text{Substitute 90 for } y \text{ in the original equation.}$$

$$10 + 3 = 13 \qquad \text{Simplify.}$$

$$13 = 13 \qquad \text{The statement is true.}$$

The solution is $y = 90$.

C.

$$\frac{z}{4} - 8 = 14$$

$$\frac{z}{4} - 8 = 14$$

$$\frac{z}{4} - 8 + 8 = 14 + 8 \qquad \text{Add 8 to both sides to eliminate the subtraction.}$$

$$\frac{z}{4} = 22 \qquad \text{Simplify.}$$

$$4\left(\frac{z}{4}\right) = 4(22) \qquad \text{Multiply both sides by 4 to eliminate the division.}$$

$$z = 88 \qquad \text{Simplify.}$$

CHECK: $\dfrac{z}{4} - 8 = 14$

$\dfrac{88}{4} - 8 = 14$ Substitute 88 for z in the original equation.

$22 - 8 = 14$ Simplify.

$14 = 14$ The statement is true.

The solution is $z = 88$.

WARM-UP

D. $8c + 11 = 43$

D.

$6b + 14 = 26$

$6b + 14 = 26$

$6b + 14 - 14 = 26 - 14$ Subtract 14 from both sides to eliminate the addition.

$6b = 12$ Simplify.

$\dfrac{6b}{6} = \dfrac{12}{6}$ Divide both sides by 6 to eliminate the multiplication.

$b = 2$ Simplify.

CHECK: $6b + 14 = 26$

$6(2) + 14 = 26$ Substitute 2 for b in the original equation.

$12 + 14 = 26$ Simplify.

$26 = 26$ The statement is true.

The solution is $b = 2$.

WARM-UP

E. Find the monthly payment on an original loan of $1155 if the balance after 14 payments is $385.

E. The formula for the balance of a loan (D) is $D + NP = B$, where P represents the monthly payment, N represents the number of payments, and B represents the amount of money borrowed. Find the number of payments that have been made on an original loan of $875 with a current balance of $425 if the payment is $25 per month.

STRATEGY: Substitute the given values in the formula and solve.

$D + NP = B$

$425 + N(25) = 875$ Substitute 425 for D, 25 for P and 875 for B.

$425 - 425 + 25N = 875 - 425$ Subtract 425 from each side.

$25N = 450$

$\dfrac{25N}{25} = \dfrac{450}{25}$ Divide both sides by 25.

$N = 18$

CHECK: If 18 payments have been made, is the balance $425?

$\begin{array}{r} \$875 \\ -450 \\ \hline \$425 \end{array}$ 18 payments of $25 is $450.

True.

Eighteen payments have been made.

ANSWERS TO WARM-UPS D–E

D. $c = 4$

E. The monthly payment is $55.

EXERCISES

Solve equations of the form $ax + b = c$, $ax - b = c$, $\frac{x}{a} + b = c$, and $\frac{x}{a} - b = c$, in which x, a, b, and c are whole numbers. (See page 74.)

Solve and check.

1. $4x - 16 = 12$

2. $\frac{a}{4} + 9 = 13$

3. $\frac{y}{3} - 7 = 4$

4. $36 = 5x + 6$

5. $45 = 6x + 9$

6. $\frac{a}{9} + 5 = 10$

7. $\frac{c}{8} + 23 = 27$

8. $12x - 10 = 38$

9. $11x + 32 = 54$

10. $7y + 53 = 123$

11. $15c - 63 = 117$

12. $6 = \frac{w}{3} - 33$

13. $81 = \frac{a}{14} + 67$

14. $\frac{x}{21} + 92 = 115$

15. $673 = 45b - 272$

16. $804 = 43c + 30$

17. Fast-Tix charges $43 per ticket for a rock concert plus an $8 service charge. How many tickets did Remy buy if he was charged $309? Use the formula $C = PN + S$, where C is the total cost, P is the price per ticket, N is the number of tickets purchased, and S is the service charge.

18. Ticket Master charges Jose $253 for nine tickets to the Festival of Jazz. If the service charge is $10, what is the price per ticket? Use the formula in Exercise 17.

19. Rana is paid $40 per day plus $8 per artificial flower arrangement she designs and completes. How many arrangements did she complete if she earned $88 for the day? Use the formula $S = B + PN$, where S is the total salary earned, B is the base pay for the day, P is the pay per unit, and N is the number of units completed.

20. Rana's sister works at a drapery firm where the pay is $50 per day plus $12 per unit completed. How many units did she complete if she earned $122 for the day?

Exercises 21–24. Several different health spa plans are shown in the table.

Health Spa Plans

Spa	Monthly Fee	Charge per Visit
B-Fit	None	$8
Join-Us	$24	$6
Gym Rats	$32	$4

21. Jessica has $72 budgeted for spas each month. Write an equation for the number of visits she would get from B-Fit. Let v be the number of visits per month. Let C represent Jessica's monthly spa budget. Find the number of visits that Jessica can purchase from B-Fit each month.

22. Jessica has $72 budgeted for spas each month. Write an equation for the number of visits she would get from Join-Us. Let v be the number of visits per month. Let C represent Jessica's monthly spa budget. Find the number of visits that Jessica can purchase from Join-Us each month.

23. Jessica has $72 budgeted for spas each month. Write an equation for the number of visits she would get from Gym Rats. Let v be the number of visits per month. Let C represent Jessica's monthly spa budget. Find the number of visits that Jessica can purchase from Gym Rats each month.

24. Using the results of Exercises 21–23, which company will give Jessica the most visits for her $72?

OBJECTIVES

1. Find the average of a set of whole numbers.
2. Find the median of a set of whole numbers.
3. Find the mode of a set of whole numbers.

VOCABULARY

The **average**, or **mean**, of a set of numbers is the sum of the set of numbers divided by the total number of numbers in the set.

The **median** of a set of numbers, ordered from smallest to largest, is either the middle number of the set or the average of the two middle numbers in the set.

The **mode** of a set of numbers is the number or numbers that appear the most often in the set.

HOW & WHY

■ **OBJECTIVE 1** Find the average of a set of whole numbers.

The *average* or *mean* of a set of numbers is used in statistics. It is one of the ways to find the middle of a set of numbers (like the average of a set of test grades). Mathematicians call the average or mean a "measure of central tendency." The average of a set of numbers is found by adding the numbers in the set and then dividing that sum by the number of numbers in the set. For example, to find the average of 11, 21, and 28:

$11 + 21 + 28 = 60$ Find the sum of the numbers in the set.
$60 \div 3 = 20$ Divide the sum by the number of numbers, 3.

The average is 20.
    The "central" number or average does not need to be one of the members of the set. The average, 20, is not a member of the set.

> **To find the average of a set of whole numbers**
>
> 1. Add the numbers.
> 2. Divide the sum by the number of numbers in the set.

EXAMPLES A–E

DIRECTIONS: Find the average.

STRATEGY: Add the numbers in the set. Divide the sum by the number of numbers in the set.

WARM-UP
A. Find the average of 251, 92, and 449.

A. Find the average of 212, 189, and 253.

$212 + 189 + 253 = 654$ Add the numbers in the group.
$654 \div 3 = 218$ Divide the sum by the number of numbers.

The average is 218.

ANSWER TO WARM-UP A
A. 264

B. Find the average of 23, 57, 352, and 224.

$23 + 57 + 352 + 224 = 656$ Add the numbers in the group.
$656 \div 4 = 164$ Divide the sum by the number of numbers.

The average is 164.

WARM-UP
B. Find the average of 12, 61, 49, 82, and 91.

CALCULATOR EXAMPLE:

C. Find the average of 509, 217, 188, and 658.

STRATEGY: Enter the sum, in parentheses, and divide by 4.

$(509 + 217 + 188 + 658) \div 4$

The average is 393.

WARM-UP
C. Find the average of 19, 29, 92, 77, and 58.

D. The average of 42, 27, 15, and ? is 32. Find the missing number.

STRATEGY: Because the average of the four number is 32, we know that the sum of the four numbers is 4(32), or 128. To find the missing number, subtract the sum of the three given numbers from 128.

$$128 - (42 + 27 + 15) = 128 - (84)$$
$$= 44$$

So the missing number is 44.

WARM-UP
D. The average of 24, 36, 12, 93, and ? is 50. Find the missing number.

E. In order to help Pete lose weight the dietician has him record his caloric intake for a week. He records the following: Monday, 3120; Tuesday, 1885; Wednesday, 1600; Thursday, 2466; Friday, 1434; Saturday, 1955; and Sunday, 2016. What is Pete's average caloric intake per day?

STRATEGY: Add the calories for each day and then divide by 7, the number of days.

```
3120        2068
1885     7)14476
1600        14
2466         4
1434         0
1955        47
2016        42
14476       56
            56
             0
```

Pete's average caloric intake is 2068 calories per day.

WARM-UP
E. The Alpenrose Dairy ships the following number of gallons of milk to local groceries: Monday, 1045; Tuesday, 1325; Wednesday, 2005; Thursday, 1810; and Friday, 2165. What is the average number of gallons shipped each day?

HOW & WHY

OBJECTIVE 2   Find the median of a set of whole numbers.

Another "measure of central tendency" is called the *median*. The median of a set of numbers is a number such that half the numbers in the set are smaller than the median and half the numbers are larger than the median. If the set is ordered from smallest to largest and there is an odd number of numbers in the set, the median is literally the middle number. Consider the set of numbers

ANSWERS TO WARM-UPS B–E

B. 59   **C.** 55   **D.** 85
E. Alpenrose Dairy shipped an average of 1670 gallons of milk each day.

3, 35, 41, 46, 50

The median is the middle number, 41.

If the set is ordered from smallest to largest and there is an even number of numbers in the set, the median is the average (mean) of the two middle numbers. Consider the set of numbers

4, 6, 9, 13, 15, 30

The median is the average of the middle numbers, 9 and 13.

$(9 + 13) \div 2 = 11$

The median is 11.

The median is used when a set of numbers has one or two numbers that are much larger or smaller than the others. For instance, a basketball player scores the following points per game over nine games; 32, 25, 32, 5, 33, 23, 33, 7, and 35. The average of the set is 25, which is near the low end of the set. The median is the middle number, which in this case is 32, and may give us a better center point.

To find the median of a set of numbers

1. List the numbers in order from smallest to largest.
2. If there is an odd number of numbers in the set, the median is the middle number.
3. If there is an even number of numbers in the set, the median is the average (mean) of the two middle numbers.

EXAMPLES F–G

DIRECTIONS: Find the median of the set of whole numbers.

STRATEGY: List the numbers from smallest to largest. If there is an odd number of numbers in the set, choose the middle number. If there is an even number of numbers in the set, find the average of the two middle numbers.

WARM-UP
F. Find the median of 103, 77, 113, 94, and 55.

F. Find the median of 37, 25, 46, 39, 22, 64, and 80.

22, 25, 37, 39, 46, 64, 80   List the numbers from smallest to largest.

The median is 39.   Because there is an odd number of numbers in the set, the median is the middle number.

WARM-UP
G. Find the median of 147, 160, 111, 85, 137, and 287.

G. Find the median of 88, 56, 74, 40, 29, 123, 81, and 9.

9, 29, 40, 56, 74, 81, 88, 123   List the numbers from smallest to largest.

$\dfrac{56 + 74}{2} = \dfrac{130}{2}$ Because there is an even number of numbers in the set, the median is the average of the two middle numbers, 56 and 74.

$= 65$

The median is 65.

ANSWERS TO WARM-UPS F–G

F. 94   G. 142.

HOW & WHY

■ **OBJECTIVE 3** Find the mode of a set of whole numbers.

A third "measure of central tendency" is the *mode*. The mode is the number that occurs most often in the set. For example, consider

45, 67, 88, 88, 92, 100.

The number that occurs most often is 88, so 88 is the mode.

    Each set of numbers has exactly one average and exactly one median. However a set of numbers can have more than one mode or no mode at all. See Examples I and J.

> **To find the mode of a set of numbers**
>
> 1. Find the number or numbers that occur most often.
> 2. If all the numbers occur the same number of times, there is no mode.

EXAMPLES H–K

DIRECTIONS: Find the mode of the set of numbers.

STRATEGY: Find the number or numbers that occur most often. If all the numbers occur the same number of times, there is no mode.

H. Find the mode of 44, 44, 77, 53, 44, 54, and 45.

    The mode is 44.   The number 44 appears three times. No other number appears three times.

I. Find the mode of 22, 56, 72, 22, 81, 72, 93, and 105.

    The modes are 22 and 72.   Both 22 and 72 occur twice and the other numbers appear just once.

J. Find the mode of 13, 35, 16, 35, 16, and 13.

    There is no mode.   All the numbers occur the same number of times, 2. So no number appears most often.

K. During the annual Salmon Fishing Derby, 46 fish are entered. The weights of the fish are recorded as shown:

Number of Fish	Weight per Fish
3	6 lb
2	8 lb
9	11 lb
12	14 lb
8	18 lb
9	23 lb
2	27 lb
1	30 lb

What are the average, median, and mode of the weights of the fish entered in the derby?

STRATEGY: To find the average, first find the total weight of all 46 fish.

$$3(6) = 18$$
$$2(8) = 16$$
$$9(11) = 99$$
$$12(14) = 168$$
$$8(18) = 144$$
$$9(23) = 207$$
$$2(27) = 54$$
$$1(30) = \underline{30}$$
$$736$$

Multiply 3 times 6 because there are 3 fish that weigh 6 lb, for a total of 18 lb, and so on.

Now divide the total weight by the number of salmon, 46.

$$736 \div 46 = 16$$

The average weight per fish is 16 lb.

Because both the 23rd and 24th fish weigh 14 lb, the median weight of the fish is 14 lb.

Because there are 46 fish, we need to find the average of the weights of the 23rd and the 24th fish.

The mode of the weight of the fish is 14 lb.

14 lb occurs most often in the list.

EXERCISES 1.7

■ **OBJECTIVE 1** Find the average of a set of whole numbers. (See page 78.)

A *Find the average.*

1. 5, 11

2. 9, 17

3. 10, 22

4. 21, 31

5. 7, 13, 14, 18

6. 11, 15, 19

7. 11, 13, 13, 15

8. 9, 9, 17, 17

9. 10, 8, 5, 5

10. 20, 15, 3, 2

11. 9, 11, 6, 8, 11

12. 15, 7, 3, 31, 4

Find the missing number to make the average correct.

13. The average of 11, 14, 15, and ? is 14.

14. The average of 12, 17, 21, and ? is 17.

B *Find the average.*

15. 22, 26, 40, 48

16. 22, 43, 48, 67

17. 23, 33, 43, 53

18. 17, 21, 29, 53, 35

19. 14, 17, 25, 34, 50, 82

20. 93, 144, 221, 138

21. 111, 131, 113, 333

22. 472, 247, 563, 446

23. 100, 151, 228, 145

24. 45, 144, 252, 291

25. 82, 95, 101, 153, 281, 110

26. 149, 82, 105, 91, 262, 217

Find the missing number.

27. The average of 39, 86, 57, 79, and ? is 64.

28. The average of 32, 40, 57, 106, 44, and ? is 58.

A *Find the median.*

29. 10, 58, 61

30. 44, 53, 67

31. 33, 34, 77

32. 21, 26, 56, 87

33. 82, 17, 64, 104

34. 77, 9, 57, 93

B

35. 7, 55, 49, 131, 108

36. 2, 3, 507, 61, 129

37. 97, 101, 123, 129, 133, 145

38. 17, 42, 18, 18, 51, 48, 67

39. 39, 77, 95, 103, 41, 123

40. 175, 309, 174, 342, 243, 189, 233, 94

■ **OBJECTIVE 3** Find the mode of a set of whole numbers. (See page 81.)

A *Find the mode.*

41. 17, 17, 38, 65, 72

42. 9, 12, 62, 62, 68

43. 1, 1, 2, 2, 3, 3, 3, 6, 7, 7

44. 9, 27, 44, 67, 27, 20

45. 23, 45, 19, 36, 22, 46, 89

46. 17, 32, 63, 17, 18, 34, 12

B

47. 27, 43, 43, 33, 54, 43, 33, 26

48. 64, 42, 70, 64, 42, 70, 79, 42

49. 44, 55, 55, 44, 55, 44, 180

50. 22, 25, 29, 22, 26, 29, 29, 22

51. 19, 23, 14, 14, 19, 23

52. 35, 27, 88, 55, 55, 35, 88, 27

C

53. On a Saturday in January 2010, the eight winning teams in the NBA scored the following number of points: 105, 106, 106, 99, 98, 94, 107, and 101. Find the average and the median number of points scored by the winning teams.

54. The following number of fish were counted at the Bonneville fish ladder during one day in May: coho, 0; shad, 247; steelhead, 73; and chinook, 2504. Find the average and median number of fish.

55. After two rounds of the Champions Tour in Hawaii, the top 10 golfers had the following scores: 128, 131, 132, 132, 134, 134, 134, 135, 135, and 135. Find the average and the median scores after two rounds.

56. A bowler has the following scores for nine games: 260, 200, 210, 300, 195, 211, 271, 200, and 205. What are the average, median, and mode scores per game?

57. A consumer magazine tests 24 makes of cars for gas mileage. The results are shown in the table.

New Car Gas Mileage

Number of Makes	Gas Mileage Based on 200 Miles
2	19 mpg
4	22 mpg
2	24 mpg
4	30 mpg
5	34 mpg
4	40 mpg
3	48 mpg

What are the average, median, and mode of the gas mileage of the cars?

58. Stan's Skate Board Shop had a sale on shoes. The number of pairs and the price are given in the table.

Shoes on Sale

Pairs by Styles	Price per Pair
3 pair "Carbon"	$50
4 pair "Swindle"	$40
5 pair "The Wire"	$55
7 pair "Halen"	$30
6 pair "Skeet"	$60
5 pair "Vulcan"	$45

What are the average, median, and mode prices per pair of shoes?

59. Twenty wrestlers are weighed in on the first day of practice.

Weights of Wrestlers

Number of Players	Weight
1	99 lb
3	110 lb
2	115 lb
5	124 lb
3	130 lb
2	155 lb
2	167 lb
1	197 lb
1	210 lb

What is the average weight of the wrestlers? What is the median weight of the wrestlers?

60. Eighty-two ladies at the Rock Creek Country Club ladies championship tournament recorded the following scores:

Ladies Championship Scores

Number of Golfers	Score
1	66
3	68
7	69
8	70
12	71
16	72
15	74
10	76
6	78
3	82
1	85

What are the average, median, and mode scores for the tournament?

61. A West Coast city is expanding its mass transit system. It is building a 15-mile east–west light rail line for $780 million and an 11-mile north–south light rail line for $850 million. What is the average cost per mile, to the nearest million dollars, of the new lines?

62. A city in Missouri has a budget of $73,497,400 and serves a population of 35,800. A second city in Missouri has a budget of $59,925,000 and serves a population of 79,900. What is the average cost per resident in each city? What is the average combined cost per resident in the cities, rounded to the nearest hundred dollars?

63. The following table lists the prices of three top-rated coffee makers, as tested by the Good Housekeeping Institute in 2009.

Brand	DeLonghi	Mr. Coffee	Krups KM1000
Price	$80	$40	$90

What is the average price of the three models of coffee makers? How many cost less than the average price?

64. The following table lists the prices of five (of the ten) customer favorite mp3 players, as reported by one Internet guide.

Brand	Apple iPod touch 8 GB	Apple iPod classic 80 GB	Creative Zen 4 GB	Zune 80 GB	SanDisk 1 GB
Price	$299	$247	$136	$199	$49

What is the average price (rounded to the nearest dollar) of the five mp3 players? How many of them are more expensive than the average cost?

Exercises 65–68. According to the U.S. Census Bureau, the population of Nevada in 2000 was 1,998,257 and the population estimate in 2008 was 2,600,167.

65. What was the average population of Nevada over the 8-year period?

66. What was the average increase per year over the 8-year period? (Round to the nearest whole person per year.)

67. Use your answer to Exercise 66 to complete the table.

Population by Year in Nevada

Year	Population	Year	Population
2000	1,998,257	2005	
2001		2006	
2002		2007	
2003		2008	
2004			

68. Is it true that the actual population of Nevada is given by the table in Exercise 67?

69. The attendance at Disneyland was 14,500,000 in 2005, 14,400,000 in 2006, 14,900,000 in 2007, and 14,700,000 in 2008. What was the average yearly attendance, to the nearest hundred thousand, for the 4-year period?

70. The following table gives the states with the greatest number of hazardous waste sites in 2004, according to the Environmental Protection Agency.

New Jersey	California	Pennsylvania	New York	Michigan
114	94	94	85	65

What is the average number of hazardous waste sites for the top five states? Round to the nearest whole number.

Exercises 71–74. The assets of five of the top U.S. commercial banks in a month of 2008 are given in the table.

Assets of the Top Five U.S. Commercial Banks

Bank	J.P. Morgan Chase Bank NA	Bank of America, NA	Citibank, NA	Wachovia Bank, NA	Wells Fargo Bank, NA
Assets (in billions)	$1,746	$1,472	$1,231	$635	$539

71. What were the average assets for the top three banks? Round to the nearest billion dollars.

72. What were the average assets for the top four banks? Round to the nearest billion dollars.

73. What were the average assets for the top five banks? Round to the nearest billion dollars.

74. Compare your answers for Exercises 71, 72, and 73. Are they increasing or decreasing? Explain.

Exercises 75–78. The following table lists the percentage of Internet users in major continents in 2009. (Source: Internet World Stats internetworldstats.com)

Continent	Percentage of Population Who Are Internet Users
Africa	6
Asia	17
Europe	49
North America	74
South America	34

75. What is the mean percentage of Internet users?

76. How many of the continents are above the mean?

77. What is the median percentage of Internet usage?

78. What is the mean of the continent with the highest percentage of Internet users and the continent with the lowest percentage?.

79. On Jupiter Island in Florida, the median home price is about $4 million. State in words what this means.

80. In 2009, the city of Boston, Massachusetts, had a population of 585,862 and a median age of 33. (Source: zoomprospector.com) Approximately how many people in the city of Boston were over 33 years old?

81. In 1950, the median age of men in the United States at their first marriage was 23, whereas in 2002 the median age was 27. Explain in words what this means.

Exercises 82–85 refer to the chapter application. See Table 1.1, page 3.

82. Calculate the average earnings of the top three movies Round to the nearest dollar.

83. Calculate the average earnings of the top five movies. Round to the nearest dollar.

84. Calculate the median gross earnings of the top five movies.

85. Round the gross earnings for all ten top-grossing movies to the nearest million dollars and then calculate the average earnings of all ten movies.

STATE YOUR UNDERSTANDING

86. Explain what is meant by the average of two or more numbers.

87. Explain how to find the average (mean) of 2, 4, 5, 5, and 9. What does the average of a set of numbers tell you about the set?

CHALLENGE

88. A patron of the arts estimates that the average donation to a fund-raising drive will be $72. She will donate $150 for each dollar by which she misses the average. The 150 donors made the contributions listed in the table. How much does the patron donate to the fund drive?

Contributions to the Arts

Number of Donors	Donation
5	$153
13	$125
24	$110
30	$100
30	$ 75
24	$ 50
14	$ 25
10	$ 17

89. Divide 35, 68, 120, 44, 56, 75, 82, 170, and 92 by 2 and 5. Which ones are divisible by 2 (the division has no remainder)? Which ones are divisible by 5? See if you can find simple rules for looking at a number and telling whether or not it is divisible by 2 and/or 5.

90. Using the new car ads in the newspaper, find four advertised prices for the same model of a car. What is the average price, to the nearest 10 dollars?

1.8 Drawing and Interpreting Graphs

OBJECTIVES

1. Read data from bar, pictorial, and line graphs.
2. Construct a bar or line graph.

VOCABULARY

Graphs are used to illustrate sets of numerical information.

A **bar graph** uses solid lines or heavy bars of fixed length to represent numbers from a set. Bar graphs contain two **scales**, a vertical scale and a horizontal scale. The **vertical scale** represents one set of values and the **horizontal scale** represents a second set of values. These depend on the information to be presented. The bar graph in Figure 1.7 illustrates four types of cars (first set of values) and the number of each type of car sold (second set of values).

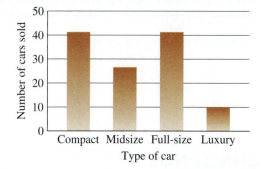

Figure 1.7

A **line graph** uses lines connecting points to represent numbers from a set. A line graph has a vertical and a horizontal scale, like a bar graph. A line graph showing the percentage of women in real estate is shown in Figure 1.8.

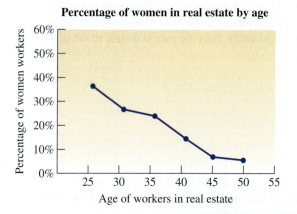

Figure 1.8

A **pictograph** uses symbols or simple drawings to represent numbers from a set. The pictograph in Figure 1.9 shows the distribution of mathematics students at a community college.

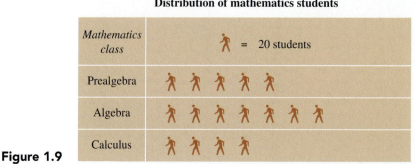

Figure 1.9

HOW & WHY

■ **OBJECTIVE 1** Read data from bar, pictorial, and line graphs.

A graph or chart is a picture used for presenting data for the purpose of comparison. To "read a graph" means to find values from the graph.

Examine the bar graph in Figure 1.10.

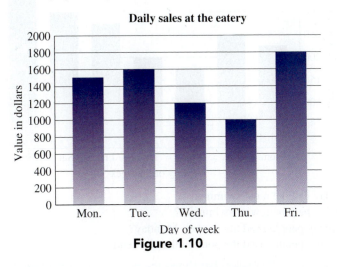

Figure 1.10

The vertical scale shows dollar values and is divided into units of $200. The horizontal scale shows days of the week. From the graph we see that

1. Friday had the greatest sales (highest bar), with $1800 in sales.
2. Thursday had the least sales (lowest bar), with sales of $1000.
3. Monday's sales appear to be $1500 (the bar falls between the scale divisions).
4. Friday had $600 more in sales than those for Wednesday.
5. The total sales for the week were $7100.

Some advantages of displaying data with a graph:

1. Each person can easily find the data most useful to him or her.
2. The visual display is easy for most people to read.
3. Some questions can be answered by a quick look at the graph. For example, "What day does the Eatery need the least staff?"

EXAMPLES A–B

DIRECTIONS: Answer the questions associated with the graph.

STRATEGY: Examine the graph to determine the values that are related.

WARM-UP

A. The percent of people in a western state who did not have health insurance in 2004 is shown in the bar graph.

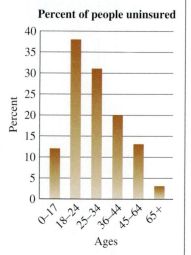

Percent of people uninsured

1. Which age group had the most uninsured people?
2. Which age group had the fewest uninsured people?
3. What percent of the 0–17 age group was uninsured?
4. What percent of the people in the 25–34 age group was uninsured?

WARM-UP

B. The number of birds spotted during a recent expedition of the Huntsville Bird Society is shown in the pictorial graph.

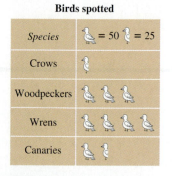

Birds spotted

1. Which species of bird was spotted most often?
2. How many woodpeckers and wrens were spotted?
3. How many more canaries were spotted than crows?

ANSWERS TO WARM-UPS A–B

A. 1. 18–24 **2.** 65+ **3.** 12 percent
  4. 31 percent
B. 1. Wrens **2.** 350 **3.** 50

A. The graph shows the number of people who used the Harmon Pool during a 1-week period.

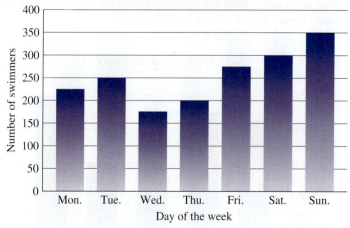

Harmon Pool head count

1. What day had the most swimmers?
2. What day had the fewest swimmers?
3. How many people used the pool on Monday?
4. How many people used the pool on the weekend?

1. Sunday *The tallest bar shows the greatest number of swimmers.*
2. Wednesday *The shortest bar shows the fewest number of swimmers.*
3. 225 *Read the vertical scale at the top of the bar for Monday. The value is estimated because the top of the bar is not on a scale line.*
4. 650 *Add the number of swimmers for Saturday and Sunday.*

B. The total sales from hot dogs, soda, T-shirts, and buttons during an air show are shown in the pictorial graph.

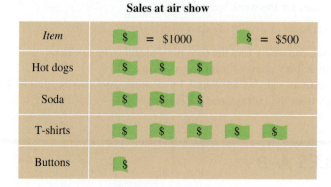

Sales at air show

1. What item has the largest dollar sales?
2. What were the total sales from hot dogs and buttons.
3. How many more dollars were realized from the sale of T-shirts than from buttons?

1. T-shirts *T-shirts have the greatest number of symbols.*
2. Hot dogs: $3000
   Buttons: +$ 500
   Total: $3500
3. T-shirts: $5000
   Buttons: −$ 500 *Subtract the sales of buttons from the sales of T-shirts.*
   Difference: $4500

The sales from T-shirts were $4500 more than from buttons.

HOW & WHY

■ OBJECTIVE 2 Construct a bar or line graph.

Let us construct a bar graph to show variations in the heating bill for the Morales Family. The data are shown in Table 1.4.

To draw and label the bar graph for these data, we show the cost on the vertical scale and the months on the horizontal scale. This is a logical display because we will most likely be asked to find the highest and lowest monthly heating costs and a vertical display of numbers is easier to read than a horizontal display of numbers. This is the typical way bar graphs are displayed. Be sure to write the labels on the vertical and horizontal scales as soon as you have chosen how the data will be displayed. Now title the graph so that the reader will recognize the data it contains.

The next step is to construct the two scales of the graph. Because each monthly total is divisible by 25, we choose multiples of 25 for the vertical scale. We could have chosen one unit for the vertical scale, but the bars would be very long and the graph would take up a lot of space. If we had chosen a larger scale, for instance 100, then the graph might be too compact and we would need to find fractional values on the scale. It is easier to draw the graph if we use a scale that divides each unit of data. The months are displayed on the horizontal scale. Be sure to draw the bars with uniform width, because each of them represents the cost for one month. A vertical display of between 5 and 12 units is typical. The vertical display should start with zero. See Figure 1.11.

TABLE 1.4 Heating Bills for the Morales Family

Month	Cost, in Dollars
January	200
February	175
March	150
April	100
May	75
June	25

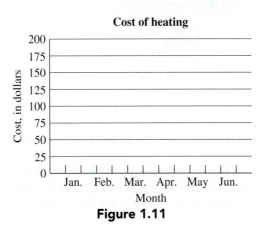

Figure 1.11

We stop the vertical scale at 200 because the maximum heating cost to be displayed is $200. The next step is to draw the bars. Start by finding the cost for January. The cost was $200 for January, so we draw the bar for January until the height of 200 is reached. This is the top of the bar. Now draw the solid bar for January. See Figure 1.12.

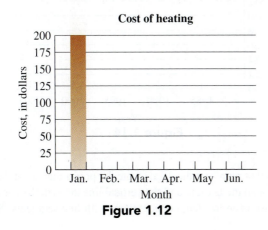

Figure 1.12

Complete the graph by drawing the bars for the other months. See Figure 1.13.

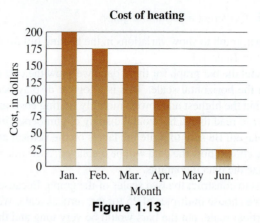

Cost of heating

Figure 1.13

A line graph is similar to a bar graph in that it has vertical and horizontal scales. The data are represented by points rather than bars, and the points are connected by line segments. We use a line graph to display the data in Table 1.5.

TABLE 1.5 Property Taxes

Year	Tax Rate (per $1000)
1985	$12.00
1990	$15.00
1995	$14.00
2000	$16.00
2005	$20.00

The vertical scale represents the tax rate, and each unit represents $2. This requires using a half space for the $15.00 rate.

Another possibility is to use a vertical scale in which each unit represents $1, but this would require 20 units on the vertical scale and would make the graph much taller. We opt to save space by using $2 units on the vertical scale. See Figure 1.14.

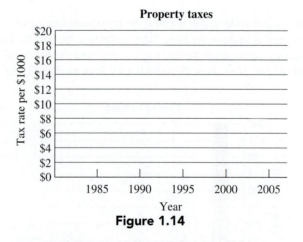

Property taxes

Figure 1.14

To find the points that represent the data, locate the points that are at the intersection of the horizontal line through the tax rate and the vertical line through the corresponding year. Once all the points have been located, connect them with line segments. See Figure 1.15.

Property taxes

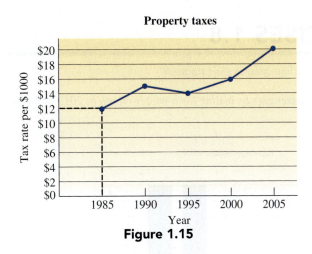

Figure 1.15

From the graph we can conclude the following:

1. Only during one 5-year period (1990–1995) did the tax rate decline.
2. The largest increase in the tax rate took place from 2000 to 2005.
3. The tax rate has increased $8 per thousand from 1985 to 2005.

EXAMPLE C

DIRECTIONS: Construct a bar graph.

STRATEGY: List the related values in pairs and draw two scales to show the pairs of values.

C. The number of phone calls recorded during the week Mary was on vacation: Monday, 12; Tuesday, 9; Wednesday, 6; Thursday, 10; Friday, 8; Saturday, 15; Sunday, 4.

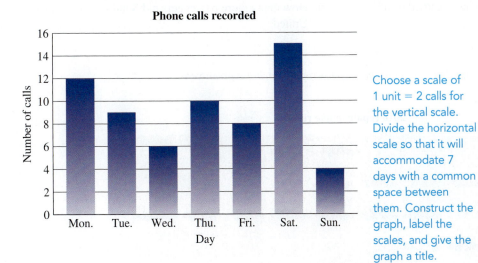

Choose a scale of 1 unit = 2 calls for the vertical scale. Divide the horizontal scale so that it will accommodate 7 days with a common space between them. Construct the graph, label the scales, and give the graph a title.

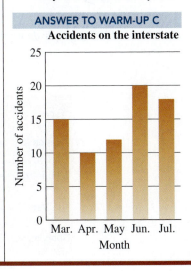

EXERCISES 1.8

OBJECTIVE 1 Read data from bar, pictorial, and line graphs. (See page 89.)

A *Exercises 1–6. The graph shows the number of passengers by airline for a winter month in 2010 at the Portland International Airport.*

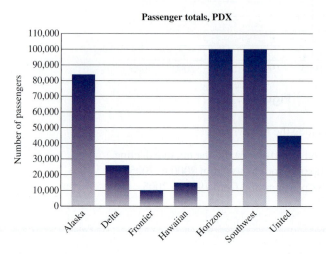

Passenger totals, PDX

1. Which airline had the least number of passengers?

2. Which airline had the greatest number of passengers?

3. How many passengers did Hawaiian have?

4. How many passengers did Delta and United together have?

5. Estimate the total number of passengers carried by all these airlines.

6. How many more passengers did Southwest have than United?

Exercises 7–12. The graph shows the number of cars in the slop for repair during a given year.

2010 repair intake record

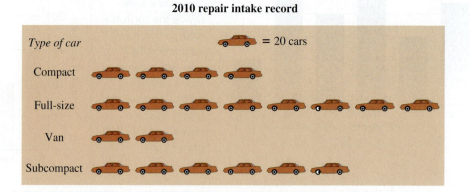

7. How many vans are in the shop for repair during the year?

8. How many compacts and subcompacts are in for repair during the year?

9. What type of vehicle has the most cars in for repair?

10. Are more subcompacts or compacts in for repair during the year?

11. How many vehicles are in for repair during the year?

12. If the average repair cost for compacts is $210, what is the gross income on compact repairs for the year?

B *Exercises 13–18. The graph shows the recent number of television sets in selected countries.*

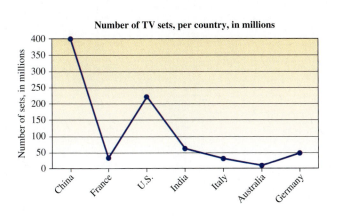

Number of TV sets, per country, in millions

13. What country had the fewest TV sets?

14. What country had the most TV sets?

15. How many TV sets did India have?

16. To the nearest ten million, what was the average number of sets in these countries?

17. Estimate the difference in the number of TV sets in India and Italy?

18. Estimate the approximate total number of TV sets in all the countries except China. Round to the nearest ten million. How many more sets is this than the number of sets in China?

Exercises 19–24. The graph shows the amounts paid for raw materials at Southern Corporation during a production period.

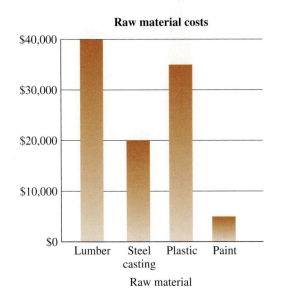

Raw material costs

19. What is the total paid for paint and lumber?

20. What is the total paid for raw materials?

21. How much less is paid for steel casting than for plastic?

22. How much more is paid for plastic than for paint?

23. If Southern Corporation decides to double its production during the next period, what will it pay for steel casting?

24. If Southern Corporation decides to double its production during the next period, how much more will it pay for lumber and steel casting than plastic and paint?

A *In Exercises 25–28, draw bar graphs to display the data. Be sure to title the graph and label the axes and scales.*

25. Distribution of grades in a history class: A, 10; B, 12; C, 25; D, 6; F, 4.

26. The distribution of monthly income: rent, $550; automobile, $325; taxes, $250; clothes, $100; food, $350; miscellaneous, $200.

27. Career preferences as expressed by a senior class: business, 100; law, 30; medicine, 60; science, 80; engineering, 55; renewable energy, 60; armed services, 20.

28. In 2009, according to statistics from *Forbes*, the most valuable sports franchises were Manchester United (soccer), $1800 million; Dallas Cowboys (football), $1600 million; Washington Redskins (football), $1500 million; New England Patriots (football), $1320 million; New York Yankees (baseball), $1300 million; and Real Madrid (soccer), $1200 million. Use a scale of 1 unit = $400 million on the vertical axis.

In Exercises 29–32, draw line graphs to display the data. Be sure to title the graph and label the axes and scales.

29. Daily sales at the local men's store: Monday, $1500; Tuesday, $2500; Wednesday, $1500; Thursday, $3500; Friday, $4000; Saturday, $6000; Sunday, $4500.

30. The gallons of water used each quarter of the year by a small city in New Mexico:

Jan.–Mar. 20,000,000
Apr.–Jun. 30,000,000
Jul.–Sept. 45,000,000
Oct.–Dec. 25,000,000

31. Profits from a recent church bazaar: bingo, $750; craft sales, $1500; quilt raffle, $450; bake sale, $600; refreshments, $900.

32. Jobs in the electronics industry in a western state are shown in the table.

Year	Number of Jobs
2000	15,000
2001	21,000
2002	18,000
2003	16,000
2004	15,000
2005	12,000

B

33. Draw a bar graph to display the median cost of a three-bedroom house in Austin, Texas.

Year	Cost
1990	$90,000
1995	$125,000
2000	$150,000
2005	$185,000
2009	$193,000

34. Draw a line graph to display the oil production from a local well over a 5-year period.

Year	Barrels Produced
2006	15,000
2007	22,500
2008	35,000
2009	32,500
2010	40,000

35. The age at which a person is eligible for full Social Security benefits is increasing. The table gives year of birth and full retirement age, according to the Social Security Administration. Make a line graph for this data.

Year of Birth	Full Retirement Age
1935	65
1940	65 yr 6 mo
1945	66
1950	66
1955	66 yr 2 mo
1960	67

36. The following table gives the cost of a hamburger meal in various European cities, according to the Worldwide Cost of Living Survey 2009. Draw a bar graph for the data.

City	Amsterdam	Athens	Dublin	Prague	Warsaw
Cost	$7.88	$6.66	$9.16	$4.91	$3.86

37. Draw a bar graph to display the median price of existing single-family housing in the Miami, Florida, area, which was $138,700 in 2002, $139,200 in 2004, $371,200 in 2006, and $285,100 in 2008, according to data from the National Association of Realtors.

38. The table gives the number of British troops deployed in Iraq for the first 6 years of the Iraqi war (Source: www.Mod.uk). Make a line graph for the data.

Year	2003	2004	2005	2006	2007	2008
Numbers of troops	18,000	8600	8500	7200	5500	4100

C *Exercises 39–42 refer to the following bar graph, which shows the population (and estimated population) of the three largest urban areas in the year 2000.*

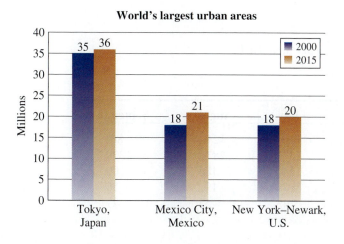

World's largest urban areas

39. Which city had the largest population in the year 2000?

40. In the year 2015, is Tokyo expected to have a larger population than Mexico City and New York–Newark combined?

41. Which urban area is expected to grow the most during the 15-year period?

42. How many million people lived in the three largest urban areas in the year 2000?

Exercises 43–44. The table gives the number of visits to the national park system for the last century. (Source: National Park Service.)

Year	Number of Visits
1920	1,058,455
1930	3,246,656
1940	16,755,251
1950	33,252,589
1960	79,229,000
1970	172,004,600
1980	220,463,211
1990	255,581,467
2000	285,891,275
2010 (projected)	284,000,000

43. Which ten-year period had the greatest amount of growth in the number of visits?

44. Make a bar graph for the data.

Exercises 45–46. The table lists visitor information at Lizard Lake State Park.

Visitors at Lizard Lake State Park

	May	June	July	August	September
Overnight camping	231	378	1104	1219	861
Picnics	57	265	2371	2873	1329
Boat rental	29	147	147	183	109
Hiking/climbing	48	178	178	192	56
Horse rental	22	43	43	58	27

45. Draw a line graph to display the data on overnight camping for the 5 months.

46. Draw a bar graph to display the data on hiking/climbing for the 5 months.

Exercise 47 refers to the chapter application. See Table 1.1, page 3.

47. Make a bar graph that shows the gross earnings of the top ten movies. Use figures rounded to the nearest million dollars.

© Photos 12/Alamy

STATE YOUR UNDERSTANDING

48. Explain the advantages of each type of graph. Which is preferable? Why?

CHALLENGE

49. The figures for U.S. casualties in four declared wars of the twentieth century are World War I, 321,000; World War II, 1,076,000; Korean War, 158,000; Vietnam War, 211,000. Draw a bar graph and a line graph to illustrate the information. Which of your graphs do you think does the best job of displaying the data?

KEY CONCEPTS

SECTION 1.1 Whole Numbers and Tables: Writing, Rounding, and Inequalities

Definitions and Concepts	Examples

The whole numbers are 0, 1, 2, 3, and so on.

238   two hundred thirty-eight
6,198,349   six million, one hundred ninety-eight thousand, three hundred forty-nine

One whole number is smaller than another if it is to the left on the number line.

$3 < 6$

One whole number is larger than another if it is to the right on the number line.

$14 > 2$

To round a whole number:
- Round to the larger number if the digit to the right is 5 or more.
- Round to the smaller number if the digit to the right is 4 or less.

$6,745 \approx 7,000$ (nearest thousand)

$6,745 \approx 6,700$ (nearest hundred)

Tables are a method of organizing information or data in rows and columns.

Enrollment by Gender at River CC

	Males	Females
English	52	67
Math	71	64
Science	69	75
History	63	59

There are 71 males taking math and 75 females taking science.

SECTION 1.2 Adding and Subtracting Whole Numbers

Definitions and Concepts	Examples

To add whole numbers, write the numbers in columns so the place values are aligned. Add each column starting with the ones. Carry as necessary.

addend + addend = sum

$$\begin{array}{r} \overset{1}{3}72 \\ +594 \\ \hline 966 \end{array} \qquad \begin{array}{r} \overset{11}{3}6 \\ +785 \\ \hline 821 \end{array}$$

To subtract whole numbers, write the numbers in columns so the place values are aligned. Subtract, starting with the ones column. Borrow if necessary.

The answer to a subtraction problem is called the *difference.*

$$\begin{array}{r} 4597 \\ -\ 362 \\ \hline 4235 \end{array} \qquad \begin{array}{r} {}^{2\ 14\ 4\ 12} \\ 3452 \\ -\ 735 \\ \hline 2717 \end{array}$$

The perimeter of a polygon is the distance around the outside.

To calculate the perimeter, add the lengths of the sides.

4 ft   Perimeter
12 ft   10 ft   7 ft
$= 12 + 4 + 10 + 7$
$= 33$
$P = 33$ ft

SECTION 1.3 Multiplying Whole Numbers

Definitions and Concepts	Examples

To multiply whole numbers, multiply the first factor by each digit in the second factor, keeping alignment. Add the partial products.

$(factor)(factor) = product$

The area of a rectangle is the space inside it: $A = \ell \cdot w$

$$
\begin{array}{r}
482 \\
\times\ 12 \\
\hline
964 \\
482 \\
\hline
5784
\end{array}
$$

2×482
1×482

4 ft

6 ft

$$
\begin{aligned}
\text{Area} &= \ell \cdot w \\
&= 6(4) \\
&= 24
\end{aligned}
$$

The area is 24 square feet.

SECTION 1.4 Dividing Whole Numbers

Definitions and Concepts	Examples

To divide whole numbers, use long division as shown.

$$
divisor \overline{)dividend} \quad \text{quotient}
$$

$$
\begin{array}{r}
205 \\
32\overline{)6578} \\
64 \\
\hline
17 \\
0 \\
\hline
178 \\
160 \\
\hline
18
\end{array}
$$

$32(2) = 64$
Subtract and bring down the 7
$32(0) = 0$
Subtract and bring down the 8
$32(5) = 160$
Subtract. The remainder is 18

SECTION 1.5 Whole-Number Exponents and Powers of 10

Definitions and Concepts	Examples

An exponent indicates how many times a number is used as a factor.

$base^{exponent} = value$

A power of 10 is the value of 10 with some exponent.

$2^3 = (2)(2)(2) = 8$

$10^4 = (10)(10)(10)(10) = 10,000$

SECTION 1.6 Order of Operations

Definitions and Concepts	Examples
The order of operations is Parentheses Exponents Multiplication and Division Addition and Subtraction	$2(7 + 4) - 36 \div 3^2 + 5$ $2(11) - 36 \div 3^2 + 5$ $2(11) - 36 \div 9 + 5$ $22 - 4 + 5$ $18 + 5$ 23

SECTION 1.7 Average, Median, and Mode

Definitions and Concepts	Examples
To find the average of a set of numbers, • Add the numbers. • Divide by the number of numbers.	$4, 13, 26, 51, 51$ $4 + 13 + 26 + 51 + 51 = 145$ $145 \div 5 = 29$ The average is 29.
To find the median of a set of numbers, • List the numbers in order from smallest to largest. • If there is an odd number of numbers in the set, the median is the middle number. • If there is an even number of numbers in the set, the median is the average (mean) of the middle two.	The median is 26.
To find the mode of a set of numbers, • Find the number or numbers that occur most often. • If all the numbers occur the same number of times, there is no mode.	The mode is 51

SECTION 1.8 Drawing and Interpreting Graphs

Definitions and Concepts	Examples

A line graph uses a line to connect data points.

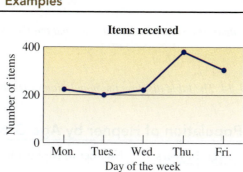

A bar graph uses bars to represent data values.

A pictograph uses pictures to represent data values.

Board-feet of timber produced

Umatilla	🌲 🌲 🌲
Wasco	🌲 🌲 🌲 🌲
Tillamook	🌲 🌲

🌲 = 1,000,000 board-feet

  REVIEW EXERCISES

SECTION 1.1

Write the word name for each of these numbers.

1. 607,321

2. 9,070,800

Write the place value name for each of these numbers.

3. Sixty-two thousand, three hundred thirty-seven

4. Five million, four hundred forty-four thousand, nineteen

Insert > or < between the numbers to make a true statement.

5. 347 351

6. 76 69

7. 809 811

Round to the nearest ten, hundred, thousand, and ten thousand.

8. 79,437

9. 183,659

Exercises 10–14. The table displays the population of Hepner by age group.

Population of Hepner by Age Group

Age, in years	Number of Residents	Age, in years	Number of Residents
Under 15	472	36–50	1098
15–25	398	51–70	602
26–35	612	Over 70	89

10. Which age group is the largest?

11. How many more residents are in the under-15 group than in the 15–25 group?

12. What is the total population in the 26–50 age group?

13. How many more people are there in the under-15 age group as opposed to the over-70 age group?

14. What is the population of Hepner?

SECTION 1.2

Add.

15. 336
  72
+509

16. 3834
  510
+ 519

17. 34
455
17
377
+881

18. 6,891
12,055
51,932
4,772
+ 492

Subtract.

19. 943
−722

20. 803
−738

21. 8315
−6983

22. 246,892
−149,558

23. Find the perimeter of the polygon.

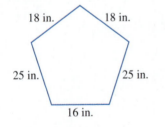

18 in.   18 in.

25 in.   25 in.

16 in.

24. Estimate the sum by rounding each addend to the nearest ten thousand and to the nearest thousand.

$$\begin{array}{r} 34{,}683 \\ 5{,}278 \\ 11{,}498 \\ 678 \\ + 56{,}723 \\ \hline \end{array}$$

25. Estimate the difference by rounding each number to the nearest thousand and to the nearest hundred.

$$\begin{array}{r} 7534 \\ - 4267 \\ \hline \end{array}$$

SECTION 1.3

Multiply.

26.
$$\begin{array}{r} 73 \\ \times 28 \\ \hline \end{array}$$

27.
$$\begin{array}{r} 406 \\ \times\ 29 \\ \hline \end{array}$$

28.
$$\begin{array}{r} 407 \\ \times\ 68 \\ \hline \end{array}$$

29. (449)(171)

30. The local Girl Scout Troop sold 54 cases of cookies during the recent sale. Each case contained 24 boxes of cookies. If the cookies sold for $4 a box, how much did they gross from the sale?

31. Estimate the product by front rounding the factors.

$$\begin{array}{r} 5{,}810 \\ \times\ 462 \\ \hline \end{array}$$

32. Find the area of the rectangle.

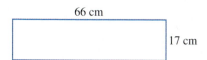

66 cm

17 cm

SECTION 1.4

Divide.

33. $14\overline{)210}$

34. $18\overline{)576}$

35. $176\overline{)15{,}141}$

36. $274{,}486 \div 74$

37. $65\overline{)345{,}892}$ (Round to the nearest hundred.)

38. The Candy Basket packs boxes containing 32 pieces of assorted chocolates. How many boxes can be made from 4675 chocolates? Will any pieces of candy be left over? If so how many?

SECTION 1.5

Find the value.

39. 11^3

40. 4^5

Multiply or divide.

41. 23×10^3

42. $78{,}000{,}000 \div 10^5$

43. 712×10^6

44. $35{,}600{,}000 \div 10^4$

45. In 2005, President Bush's plan to privatize part of Social Security required borrowing approximately 34×10^8. Write this amount in place value form.

SECTION 1.6

Simplify.

46. $40 - 24 + 8$ **47.** $6 \cdot 10 + 5$ **48.** $18 \div 2 - 3 \cdot 2$

49. $94 \div 47 + 47 - 16 \cdot 2 + 6$

50. $35 - (25 - 17) + (12 - 10)^2 + 5 \cdot 2$

SECTION 1.7

Find the average, median, and mode.

51. 41, 64, 23, 70, 87   **52.** 93, 110, 216, 317

53. 63, 74, 53, 63, 37, 82   **54.** 1086, 4008, 3136, 8312, 8312, 1474

55. The six children of Jack and Mary Barker had annual incomes of \$54,500, \$45,674, \$87,420, \$110,675, \$63,785, and \$163,782. Find the average salary of the six siblings to the nearest hundred dollars.

SECTION 1.8

Exercises 56–57. The graph displays the temperature readings for a 24-hour period.

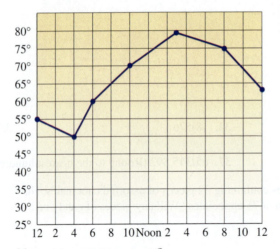

56. At what times do the highest and lowest temperatures occur?

57. What is the difference in temperature from 6 A.M. and 8 P.M.?

58. The grade distribution in an algebra class is displayed in the table.

Algebra Grades

Grade	Number of Students
A	5
B	10
C	12
D	4
F	2

Construct a bar graph to display the data.

TRUE/FALSE CONCEPT REVIEW

Check your understanding of the language of basic mathematics. Tell whether each of the following statements is true (always true) or false (not always true). For each statement you judge to be false, revise it to make a statement that is true.

Answers

1. All whole numbers can be written using nine digits.

1. _____

2. In the number 8425, the digit 4 represents 400.

2. _____

3. The word *and* is not used when writing the word names of whole numbers.

3. _____

4. The symbols $7 < 23$ can be read "seven is greater than twenty-three."

4. _____

5. $2567 < 2566$

5. _____

6. To the nearest thousand, 7398 rounds to 7000.

6. _____

7. It is possible for the rounded value of a number to be equal to the original number.

7. _____

8. The expanded form of a whole number shows the plus signs that are usually not written.

8. _____

9. The sum of 80 and 7 is 807.

9. _____

10. The process of "carrying" when doing an addition problem with pencil and paper is based on the place value of the numbers.

10. _____

11. A line graph has at least two scales.

11. _____

12. The product of 15 and 3 is 18.

12. _____

13. It is possible to subtract 47 from 65 without "borrowing."

13. _____

14. The number 8 is a factor of 72.

14. _____

15. The multiplication sign is sometimes omitted when writing a multiplication problem.

15. _____

16. Whenever a number is multiplied by zero, the value remains unchanged.

16. _____

17. There is more than one method for doing division problems.

17. _____

18. In $104 \div 4 = 26$, the quotient is 104.

18. _____

19. If a division exercise has a remainder, then we know that there is no whole number quotient.

19. _____

20. When zero is divided by any whole number from 25 to 91, the result is 0.

20. _____

21. The result of zero divided by zero can be either 1 or 0.

21. _____

22. The value of 7^2 is 14.

22. _____

23. The value of 222^1 is 222.

23. _____

24. One billion is a power of 10.

24. _____

25. The product of 450×10^3 is equal to 45,000.

25. _____

26. The quotient of 9000 and 10 is 900.

26. _____

27. In the order of operations, exponents are always evaluated before addition.

27. _____

28. In the order of operations, multiplication is always evaluated before subtraction.

28. _____

29. The value of $2^3 + 2^3$ is the same as the value of 2^4.

29. _____

30. The average of three different numbers is smaller than the largest of the three numbers.

30. _____

31. The word *mean* sometimes has the same meaning as *average*.

31. _____

32. A table is a method of displaying data in an array using a horizontal and vertical arrangement to distinguish the type of data.

32. _____

33. The median of 34, 54, 14, 44, 67, 81, and 90 is 44.

33. _____

34. The distance around a polygon is called the perimeter.

34. _____

TEST

1. Divide: $72\overline{)15{,}264}$

1. _____

2. Subtract: $9615 - 6349$

2. _____

3. Simplify: $55 \div 5 + 6 \cdot 4 - 7$

3. _____

4. Multiply: $37(428)$

4. _____

5. Insert $<$ or $>$ to make the statement true: 368   371

5. _____

6. Multiply: 55×10^6

6. _____

7. Multiply: $608(392)$

7. _____

8. Write the place value name for seven hundred thirty thousand sixty-one.

8. _____

9. Find the average of 3456, 812, 4002, 562, and 1123.

9. _____

10. Multiply: $65(5733)$. Round the product to the nearest hundred.

10. _____

11. Round 38,524 to the nearest thousand.

11. _____

12. Estimate the sum of 95,914, 31,348, 68,699, and 30,341 by rounding each number to the nearest ten thousand.

12. _____

13. Find the value of 9^3.

13. _____

14. Add: $84 + 745 + 56 + 7802$

14. _____

15. Find the perimeter of the rectangle.

15. _____

42 ft

18 ft

16. Estimate the product by front rounding: 752(38)

16. _____

17. Subtract:
$$7053$$
$$- 895$$

17. _____

18. Write the word name for 4005.

18. _____

19. Simplify: $65 + 3^3 - 66 \div 11$

19. _____

20. Add:
$$42{,}888$$
$$67{,}911$$
$$93{,}467$$
$$23{,}567$$
$$+ \ 31{,}823$$

20. _____

21. Divide: $7{,}730{,}000{,}000 \div 10^6$

21. _____

22. Round 675,937,558 to the nearest million.

22. _____

23. Divide: $75{,}432 \div 65$

23. _____

24. Simplify: $95 - 8^2 + 48 \div 4$

24. _____

25. Find the area of a rectangle that measures 23 cm by 15 cm.

25. _____

26. Simplify: $(5 \cdot 3)^2 + (4^2)^2 + 11 \cdot 3$

26. _____

27. Find the average, median, and mode of 795, 576, 691, 795, 416, and 909.

27. _____

28. A secretary can type an average of 75 words per minute. If there are approximately 700 words per page, how long will it take the secretary to type 18 pages?

28. _____

29. Nine people share in a Power Ball lottery jackpot. If the jackpot is worth $124,758,000, how much will each person receive? If each person's share is to be distributed evenly over a 20-year period, how much will each person receive per year?

29. _____

Exercises 30–32. The graph shows the home sales for a month at a local real estate office.

Monthly home sales

30. What price range had the greatest sales?

30. _____

31. What was the total number of homes sold in the top two price ranges?

31. _____

32. How many more homes were sold in the lowest range than in the highest range?

32. _____

Exercises 33–35. The table shows the number of employees by division and shift for Beaver Horseradish.

Employees by Division: Beaver Horseradish

Division	Day Shift	Night Shift
A	215	175
B	365	120
C	95	50

33. Which division has the greatest number of employees?

33. _____

34. How many more employees are in the day shift in division A as compared to the day shift in division C?

34. _____

35. How many employees are in the three divisions?

35. _____

CLASS ACTIVITY 1

To determine the intensity of cardiovascular training, a person needs to know his or her maximal heart rate (MHR) and resting heart rate (RHR). The MHR may be approximated by subtracting your age from 220. The RHR can be measured by sitting quietly for about 15 minutes, taking your pulse for 15 seconds, and multiplying by 4. The difference between the MHR and RHR is the heart rate reserve (HRR), the number of beats available to go from rest to all-out effort. To determine a target training rate, divide the HRR by 2 and add the results to the RHR.

1. Jessie is 24 years old and has a RHR of 68 beats per minute. Determine her cardiovascular target training rate.

2. June is 52 years old. She measures her resting heart rate as 18 beats in 15 seconds. Determine her cardiovascular target training rate.

3. Determine the cardiovascular target training rate for every member of your group.

4. As you age, does your target training rate increase or decrease? Why?

CLASS ACTIVITY 2

The following table gives nutritional information for various beverages and snacks.

Beverage	Serving Size	Calories	Sugar (grams)	Caffeine (mg)
Coca-Cola Classic	8 oz	97	27	23
Diet Coke	8 oz	1	0	31
Powerade	8 oz	50	14	0
Full Throttle	8 oz	111	29	72
Black coffee	8 oz	2	0	95
M&Ms	1 cup	1023	132	29

SOURCE: thecoca-colacompany.com and nutritiondata.com

1. At the Minute Mart, you can get a supersize (32 oz) soft drink. Compare the calories, sugar, and caffeine in one supersize Coke to one cup of M&Ms.

2. Genevieve has decided that she is going to replace her three servings a day of Coca-Cola Classic with three servings of Diet Coke.
   a. Calculate how her intake of calories, sugar, and caffeine will change if she makes the switch for a month.
   b. What are the benefits of such a change? What are the drawbacks?

3. Josh drinks one bottle of Powerade (32 oz) every time he works out. If he works out three times per week, calculate Josh's intake of calories and sugar for an entire year from the Powerade.

4. Lucas works the midnight shift at FedEx four nights per week and drinks one 12-oz can of Full Throttle before his shift to keep him awake.
   a. Calculate his intake of calories, sugar, and caffeine for a month (4 weeks) from the Full Throttle.
   b. If Lucas switched to 12 oz of black coffee each day for a month instead of the Full Throttle, how would his intake of calories, sugar, and caffeine change?
   c. What are the benefits of such a change? What are the drawbacks?

GROUP PROJECT (1–2 WEEKS)

All tables, graphs, and charts should be clearly labeled and computer-generated if possible. Written responses should be typed and checked for spelling and grammar.

1. Go to the library and find the population and area for each state in the United States. Organize your information by geographic region. Record your information in a table.

2. Calculate the total population and the total area for each region. Calculate the population density (number of people per square mile, rounded to the nearest whole person) for each region, and put this and the other regional totals in a regional summary table. Then make three separate graphs, one for regional population, one for regional area, and the third for regional population density.

3. Calculate the average population per state for each region, rounding as necessary. Put this information in a bar graph. What does this information tell you about the regions? How is it different from the population density of the region?

4. How did your group decide on the makeup of the regions? Explain your reasoning.

5. Are your results what you expected? Explain. What surprised you?

GOOD ADVICE FOR *Studying*

PLANNING MAKES PERFECT

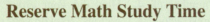
© Image Source/Jupiter Images

Reserve Math Study Time

- Aim for 1 to 2 hours every day.
- Consider location—home, library, campus tutoring center.
- Schedule some study time as close to the end of class as possible.
- Find a study buddy and/or form a study group.

Use Basic Learning Theory Techniques

- Most forgetting happens directly after first exposure. Combat this by starting your homework as soon after class as possible.
- Learning is more effective with frequent repetition. This is why 1 to 2 hours each day is preferable to 7 hours on one day.
- Talking about what you are learning helps your brain remember. This is why study buddies and study groups are so effective and valuable.

Form an Effective Study Group

- Meet regularly (maybe once or twice per week).
- Schedule extra review sessions before tests.
- Stay on task—do not allow the group to spend time complaining.
- Encourage and support each other.

Protect Your Physical Health

- Eat breakfast every day, including a high-quality protein.
- Allow for adequate sleep (experts recommend 7 to 8 hours daily).
- Exercise regularly.
- Schedule time for friends and family.
- Handle stress in a positive manner—try meditation, yoga, progressive relaxation.

© Lordface/Dreamstime.com

Primes and Multiples

APPLICATION

Mathematicians have always been fascinated by numbers—their structure and their uses. All the other chapters in this textbook explain ways to use numbers. This chapter explores the structure of counting numbers. The structure of the counting numbers—their basic building blocks—is similar in nature to the concept of all molecules being made up of atoms of the basic elements. As a molecule of water is formed from two atoms of hydrogen and one atom of oxygen, H_2O, so the number 12 is made up of two factors of 2 and one factor of 3, that is, $12 = 2 \cdot 2 \cdot 3$.

Throughout the history of humanity, people have studied different sets of numbers. Some groups have even attributed magical powers to certain numbers ("lucky seven") or groups of numbers because of some special properties. We investigate here the notion of a "magic square." Magic squares are arrangements of the numbers 1–4, or 1–9, or 1–16, or 1–25, or 1–any number squared. These numbers are written in square arrays that are 2×2, or 3×3, or 4×4, and so on. The numbers are placed so that the sum of each row and of each column and of each diagonal is the same. Figure 2.1 shows an example of a 3×3 magic square, which uses the numbers from 1 to 9 (3^2).

2	9	4
7	5	3
6	1	8

Figure 2.1

Row 1	$2 + 9 + 4 = 15$	Column 1	$2 + 7 + 6 = 15$
Row 2	$7 + 5 + 3 = 15$	Column 2	$9 + 5 + 1 = 15$
Row 3	$6 + 1 + 8 = 15$	Column 3	$4 + 3 + 8 = 15$
Diagonal	$2 + 5 + 8 = 15$	Diagonal	$4 + 5 + 6 = 15$

In medieval times, some people wore magic squares and used them as talismans. The talisman wearers considered them powerful enough to provide protection from evil spirits.

Benjamin Franklin was a big fan of magic squares. As clerk to the Pennsylvania Assembly, he admitted to creating them when he was bored with the proceedings. He created more and more complex squares as time passed. He even experimented with magic circles. Figure 2.2 shows an 8×8 square, using the numbers from 1 to 64 (8^2), that Franklin created. This square has several interesting features.

52	61	4	13	20	29	36	45
14	3	62	51	46	35	30	19
53	60	5	12	21	28	37	44
11	6	59	54	43	38	27	22
55	58	7	10	23	26	39	42
9	8	57	56	41	40	25	24
50	63	2	15	18	31	34	47
16	1	64	49	48	33	32	17

Figure 2.2

GROUP DISCUSSION

1. What is the sum of each row and each column?
2. What is the sum of the four corners? Of the four middle squares?
3. Starting at the lower left corner, 16, move diagonally up three times, move right once, move down and right diagonally to the lower right corner, 17. What is the sum of this path?
4. Start at 50 and trace the path "parallel" to the one in Exercise 3. What is the sum of this path?
5. Find at least six other paths of eight numbers through the square that have the same sum.
6. What is the sum of the first four numbers in each row? In each column? Of the last four numbers in each row or column? What is the sum of these half rows and half columns?

SECTION

2.1 Divisibility Tests

OBJECTIVES

1. Determine whether a natural number is divisible by 2, 3, or 5.
2. Determine whether a natural number is divisible by 6, 9, or 10.

VOCABULARY

A whole number is **divisible** by another whole number if the quotient of these numbers is a natural number and the remainder is 0. The second number is said to be a **divisor** of the first. Thus, 7 is a **divisor** of 42, because $42 \div 7 = 6$.

We also say 42 is **divisible** by 7.

The **even digits** are 0, 2, 4, 6, and 8.

The **odd digits** are 1, 3, 5, 7, and 9.

HOW & WHY

■ **OBJECTIVE 1** Determine whether a natural number is divisible by 2, 3, or 5.

To ask if a number is divisible by 3 is to ask if the division of the number by 3 comes out even (has no remainder). We can answer this question by doing the division and checking to see that there is no remainder. Or, we can bypass the division by using divisibility tests. For many numbers we can answer questions about divisibility mentally. Table 2.1 provides clues for some of these tests.

TABLE 2.1 **Patterns of Multiplying**

Some Natural Numbers	Multiply by 2	Multiply by 3	Multiply by 5	Multiply by 9
1	2	3	5	9
2	4	6	10	18
3	6	9	15	27
4	8	12	20	36
5	10	15	25	45
6	12	18	30	54
10	20	30	50	90
17	34	51	85	153
26	52	78	130	234
45	90	135	225	405
71	142	213	355	639

In the second column (Multiply by 2), the ones digit of each number is an even digit, that is either 0, 2, 4, 6, or 8. Because the ones place is even, the number itself is also even.

In the third column (Multiply by 3), the sum of the digits of each number in the column is divisible by 3. For example, the sum of the digits of 51 (17 · 3) is 5 + 1, or 6, and 6 is divisible by 3. Likewise, the sum of the digits of 213 is divisible by 3 because 2 + 1 + 3 is 6.

In the fourth column (Multiply by 5), the ones digit of each number is 0 or 5.

> **To test for divisibility of a natural number by 2, 3, or 5**
>
> If the ones-place digit of the number is even (0, 2, 4, 6, or 8), the number is divisible by 2.
>
> If the sum of the digits of the number is divisible by 3, then the number is divisible by 3.
>
> If the ones-place digit of the number is 0 or 5, the number is divisible by 5.

EXAMPLES A–D

DIRECTIONS: Determine whether the natural number is divisible by 2, 3, or 5.

STRATEGY: First check the ones-place digit. If it is even, the number is divisible by 2. If it is 0 or 5, the number is divisible by 5. Next find the sum of the digits. If the sum is divisible by 3, the number is divisible by 3.

A. Is 66 divisible by 2, 3, or 5?

66 is divisible by 2.	The ones-place digit is 6.
66 is divisible by 3.	6 + 6 = 12, which is divisible by 3.
66 is not divisible by 5.	The ones-place digit is neither 0 nor 5.

B. Is 210 divisible by 2, 3, or 5?

210 is divisible by 2.	The ones-place digit is 0.
210 is divisible by 3.	2 + 1 + 0 = 3, which is divisible by 3.
210 is divisible by 5.	The ones-place digit is 0.

C. Is 931 divisible by 2, 3, or 5?

931 is not divisible by 2.	The ones-place digit is not even.
931 is not divisible by 3.	9 + 3 + 1 = 13, which is not divisible by 3.
931 is not divisible by 5.	The ones-place digit is neither 0 nor 5.

WARM-UP
A. Is 63 divisible by 2, 3, or 5?

WARM-UP
B. Is 390 divisible by 2, 3, or 5?

WARM-UP
C. Is 435 divisible by 2, 3, or 5?

ANSWERS TO WARM-UPS A–C

A. 63 is divisible by 3; 63 is not divisible by 2 or 5. **B.** 390 is divisible by 2, 3, and 5. **C.** 435 divisible by 3 and 5; 435 is not divisible by 2.

WARM-UP

D. If Anna has ¥ 626 to divide among the children, will each child receive the same number of yen in whole numbers. Why or why not?

D. Georgio and Anna and their three children are on a trip of Japan. Anna has a total of ¥ 585 to divide among the children. She wants each child to receive the same number of yen in whole numbers. Is this possible? Why or why not?

Yes, each child will receive the same number of yen in whole numbers (¥ 195 each) because 585 is divisible by 3.

HOW & WHY

■ **OBJECTIVE 2** Determine whether a natural number is divisible by 6, 9, or 10.

In Table 2.1, some numbers appear in both the Multiply by 2 column and the Multiply by 3 column. These numbers are divisible by 6. Because $6 = 2 \cdot 3$, every number divisible by 6 must also be divisible by 2 and 3. For example, 78 is divisible by both 2 and 3. Therefore, it is also divisible by 6.

In the Multiply by 9 column, the sum of the digits of each number is divisible by 9. For example, the sum of the digits in 36 is $3 + 6$, or 9, which is divisible by 9. Also, the sum of the digits in 981 is $9 + 8 + 1$, or 18, which is divisible by 9.

Notice that, because 0 is even, all natural numbers ending in 0 are divisible by 2. They are also divisible by 5 because they end in 0. These numbers are also divisible by 10. These numbers appear in both the Multiply by 2 column and the Multiply by 5 column. The numbers 20, 90, and 130 are all divisible by 10.

> ## To test for divisibility of a natural number by 6, 9, or 10
>
> If the number is divisible by both 2 and 3, then the number is divisible by 6.
> If the sum of the digits of the number is divisible by 9, then the number is divisible by 9.
> If the ones-place digit of the number is 0, then the number is divisible by 10.

EXAMPLES E–F

DIRECTIONS: Determine whether a natural number is divisible by 6, 9, or 10.

STRATEGY: First, check whether the number is divisible by both 2 and 3. If so, the number is divisible by 6. Second, find the sum of the digits. If the sum is divisible by 9, then the number is divisible by 9. Finally, check the ones-place digit. If the digit is 0, the number is divisible by 10.

E. Is 960 divisible by 6, 9, or 10?

960 is divisible by 6.	960 is divisible by both 2 and 3.
960 is not divisible by 9.	$9 + 6 + 0 = 15$, which is not divisible by 9.
960 is divisible by 10.	The ones-place digit is 0.

ANSWERS TO WARM-UPS D-E

D. No, each child will not receive the same amount because 626 is not divisible by 3.

E. 4632 is divisible by 6; 4632 is not divisible by 9 or 10.

F. Is 2880 divisible by 6, 9, or 10?

2880 is divisible by 6.   2880 is divisible by both 2 and 3.
2880 is divisible by 9.   2 + 8 + 8 + 0 = 18, which is divisible by 9.
2880 is divisible by 10.   The ones-place digit is 0.

EXERCISES 2.1

■ **OBJECTIVE 1** Determine whether a natural number is divisible by 2, 3, or 5. (See page 118.)

A *Is each number divisible by 2?*

1. 37

2. 56

3. 80

4. 75

5. 48

6. 102

Is each number divisible by 5?

7. 56

8. 23

9. 45

10. 315

11. 551

12. 710

Is each number divisible by 3?

13. 81

14. 90

15. 54

16. 37

17. 78

18. 53

B *Determine whether the natural number is divisible by 2, 3, or 5.*

19. 2190

20. 2670

21. 3998

22. 1668

23. 7535

24. 1845

25. 4175

26. 6360

27. 11,205

28. 11,206

■ **OBJECTIVE 2** Determine whether a natural number is divisible by 6, 9, or 10. (See page 120.)

A *Is each number divisible by 6?*

29. 114

30. 181

31. 254

32. 333

33. 684

34. 452

Is each number divisible by 9?

35. 351

36. 171

37. 933

38. 333

39. 585

40. 765

Is each number divisible by 10?

41. 747

42. 330

43. 920

44. 706

45. 1927

46. 9210

B *Determine whether the natural number is divisible by 6, 9, or 10.*

47. 7470

48. 6894

49. 8352

50. 3780

51. 5555

52. 8888

53. 5700

54. 7880

55. 7290

56. 9156

C

57. Charlie left his coin collection to be divided evenly among his three children. His collection contains 439 coins. Is it possible for each child to receive the same number of coins? Explain.

58. Charlie's coin collection is valued at $126. Is it possible to divide the collection so that each of his three children receives the same dollar amount?

59. Pedro and two friends rent an apartment for $987 per month. Is it possible for each of them to spend the same whole number of dollars on the rent? Explain

60. Helen and four of her friends won the lottery. They received a check for $65,475. Is it possible for them to split the money evenly? Explain.

61. A marching band has 175 members. Can the band march in rows of 3 without any member being left over? Rows of 5? Rows of 10?

62. Karla and her five friends are driving 1362 miles to attend a wedding. Is it possible for each of them to drive the same number of miles?

63. Pete purchased 5 six-packs of beer to serve at his poker party. If 9 men attended the party, could they each have the same number of bottles of beer?

64. The Tigard recreational basketball league has 120 participants. Is it possible to form teams of 5, 6, 9 or 10 players in each team with no one left off a team?

65. Stephan and his friends are raising money that is to be split among 3 charities. So far they have raised $310. What is the minimum additional amount they must raise so each charity receives the same amount of money?

66. A movie theater is to have 250 seats. The manager plans to arrange them in rows with the same number of seats in each row. Is it possible to have rows of 10 seats each? Rows of 15 seats each? Determine the answers by using divisibility tests.

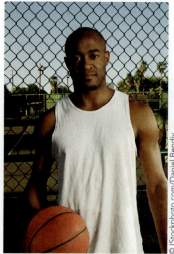

© IStockphoto.com/Daniel Bendjy

67. Mark and Barbara attend the zoo and see a pen with peacocks and water buffalo. On the way out, Barbara remarks that the pen has 30 eyes and 44 feet. How many animals are in the pen?

68. In Exercise 67, how many peacocks and how many water buffalo are in the pen?

69. A small circus has an act with elephants and riders. In all, there are 48 eyes and 68 feet. What is the total number of elephants plus riders?

70. In Exercise 69, how many elephants and how many riders are in the act?

STATE YOUR UNDERSTANDING

71. Explain what it means to say, "This number is divisible by 5." Give an example of a number that is divisible by 5 and one that is not divisible by 5.

72. Explain why a number that is divisible by both 2 and 3 must also be divisible by 6.

73. Explain the difference in the divisibility test for 2 and 3.

74. Write a short statement to explain why every number divisible by 9 is also divisible by 3.

CHALLENGE

75. Is 35,766 divisible by 6?

76. Is 11,370 divisible by 15? Write a divisibility test for 15.

77. Is 11,370 divisible by 30? Write a divisibility test for 30.

78. Is 99,000,111,370 divisible by:
   a. 2?_____ , because the _____

   b. 3?_____ , because the _____

   c. 5?_____ , because the _____

   d. 6?_____ , because the _____

   e. 9?_____ , because the _____

   f. 10?_____ , because the _____

MAINTAIN YOUR SKILLS

79. Round 67,482 to the nearest thousand and nearest ten thousand.

80. Round 5,056,857 to the nearest ten thousand and nearest hundred thousand.

81. Divide: $10,985 \div 845$

82. Divide: $3120 \div 195$

83. Find the perimeter of a square that is 18 cm on a side.

84. Find the area of a square that is 18 cm on a side.

85. Multiply 12 by 1, 2, 3, 4, 5, and 6.

86. Multiply 13 by 1, 2, 3, 4, 5, and 6.

87. Multiply 123 by 1, 2, 3, 4, 5, and 6.

88. Multiply 1231 by 1, 2, 3, 4, 5, and 6.

SECTION

2.2 Multiples

OBJECTIVES

1. List multiples of a whole number.
2. Determine whether a given whole number is a multiple of another whole number.

VOCABULARY

A **multiple** of a whole number is the product of that number and a natural number. For instance,

21 is a multiple of 7 because 7(3) = 21.

77 is a multiple of 7 because 7(11) = 77.

98 is a multiple of 7 because 7(14) = 98.

147 is a multiple of 7 because 7(21) = 147.

HOW & WHY

■ **OBJECTIVE 1** List multiples of a whole number.

To list the multiples of 7, we multiply 7 by each natural number. See Table 2.2.

TABLE 2.2 Multiples of 7

Natural Number	Multiple of 7
1	7
2	14
3	21
4	28
5	35
6	42
…	…
15	105
…	…
47	329
…	…

Table 2.2 can be continued without end. We say that the first multiple of 7 is 7, the second multiple of 7 is 14, the 15th multiple of 7 is 105, the 47th multiple of 7 is 329, and so on. To find a particular multiple of 7, say, the 23rd, we multiply 7 by 23. The 23rd multiple of 7 is 161.

EXAMPLES A–E

DIRECTIONS: List the designated multiples.

STRATEGY: Multiply the natural number by the given value.

A. List the first five multiples of 9.

$1 \cdot 9 = 9$ Multiply 9 by 1, 2, 3, 4, and 5.
$2 \cdot 9 = 18$
$3 \cdot 9 = 27$
$4 \cdot 9 = 36$
$5 \cdot 9 = 45$

The first five multiples of 9 are 9, 18, 27, 36, and 45.

B. List the first five multiples of 16.

$1 \cdot 16 = 16$ Multiply 16 by 1, 2, 3, 4, and 5.
$2 \cdot 16 = 32$
$3 \cdot 16 = 48$
$4 \cdot 16 = 64$
$5 \cdot 16 = 80$

The first five multiples of 16 are 16, 32, 48, 64, and 80.

CALCULATOR EXAMPLE:

C. Find the 7th, 23rd, 28th, and 452nd multiples of 16.

STRATEGY: Use a calculator to multiply 7, 23, 28, and 452 by 16.

The 7th multiple of 16 is 112, the 23rd multiple of 16 is 368, the 28th multiple of 16 is 448, and the 452nd multiple of 16 is 7232.

CALCULATOR EXAMPLE:

D. Find all of the multiples of 9 between 611 and 650.

STRATEGY: Use a calculator (see Example C) to make a quick estimate. Start with, say, the 60th multiple. Multiply 60 times 9.

$60(9) = 540$ The product is too small.

So try 65.

$65(9) = 585$ The product, 585, is still too small.

So try 68.

$68(9) = 612$ This is the first multiple of 9 larger than 611.

The 68th multiple is 612, so keep on multiplying.

$69(9) = 621$ $70(9) = 630$ $71(9) = 639$
$72(9) = 648$ $73(9) = 657$
                       This multiple is larger than 650.

The multiples of 9 between 611 and 650 are 612, 621, 630, 639, and 648.

E. Maria's mathematics teacher assigns homework problems numbered from 1 to 70 that are multiples of 4. Which problems should she work?

E. Jordan's mathematics teacher assigns homework problems numbered from 1 to 60 that are multiples of 6. Which problems should he work?

STRATEGY: Find the multiples of 6 from 1 to 60 by multiplying by 1, 2, 3, 4, and so on, until the product is 60 or larger.

$1(6) = 6$	$2(6) = 12$	$3(6) = 18$	$4(6) = 24$	$5(6) = 30$
$6(6) = 36$	$7(6) = 42$	$8(6) = 48$	$9(6) = 54$	$10(6) = 60$

Jordan should work problems 6, 12, 18, 24, 30, 36, 42, 48, 54, and 60.

HOW & WHY

■ **OBJECTIVE 2** Determine whether a given whole number is a multiple of another whole number.

If one number is a multiple of another number, the first must be divisible by the second. To determine whether 852 is a multiple of 6, we check to see if 852 is divisible by 6; that is, check to see whether 852 is divisible by both 2 and 3.

852   Divisible by 2 because the ones-place digit is even (2).
852   Divisible by 3 because $8 + 5 + 2 = 15$, which is divisible by 3.

So 852 is a multiple of 6.

If there is no divisibility test, use long division. For example, is 299 a multiple of 13? Divide 299 by 13.

$$\begin{array}{r} 23 \\ 13\overline{)299} \\ \underline{26} \\ 39 \\ \underline{39} \\ 0 \end{array}$$

Because $299 \div 13 = 23$ with no remainder, $299 = 13 \times 23$. Therefore, 299 is a multiple of 13.

EXAMPLES F–J

DIRECTIONS: Determine whether a given whole number is a multiple of another whole number.

STRATEGY: Use divisibility tests or long division to determine whether the first number is divisible by the second.

F. Is 873 a multiple of 9?
Use the divisibility test for 9.
$8 + 7 + 3 = 18$ The sum of the digits is divisible by 9.
So 873 is a multiple of 9.

G. Is 2706 a multiple of 6?
Use the divisibility tests for 2 and 3.
2706 is divisible by 2.   The ones-place digit is even (6).
2706 is divisible by 3.   The sum of the digits, $2 + 7 + 0 + 6 = 15$, is divisible by 3.
So 2706 is a multiple of 6.

ANSWERS TO WARM-UPS E–G

E. Maria should work problems 4, 8, 12, 16, 20, 24, 28, 32, 36, 40, 44, 48, 52, 56, 60, 64, and 68.
F. yes   **G.** no

H. Is 2103 a multiple of 19?

Because we have no divisibility test for 19, we use long division.

$$
\begin{array}{r}
110 \\
19)\overline{2103} \\
\underline{19} \\
20 \\
\underline{19} \\
13
\end{array}
$$

20   The remainder is not 0.

So 2103 is a not multiple of 19.

I. Is 6915 a multiple of 15?

Use the divisibility tests for 3 and 5 or use long division.

6915 is divisible by 3.   The sum of the digits, 6 + 9 + 1 + 5 = 21, is divisible by 3.

6915 is divisible by 5.   The ones-place digit is 5.
Long division gives the same result because 6915 ÷ 15 = 461.
So 6915 is a multiple of 15.

J. Is 4 a multiple of 20?

The number 4 is not divisible by 20. The multiples of 20 are 20, 40, 60, and so on.
The smallest multiple of 20 is 20 · 1 = 20.
No, 4 is not a multiple of 20.

EXERCISES 2.2

■ **OBJECTIVE 1** List multiples of a whole number. (See page 124.)

A *List the first five multiples of the whole number.*

1. 5   **2.** 14   **3.** 23

4. 18   **5.** 21   **6.** 24

7. 30   **8.** 41   **9.** 50

10. 45

B

11. 53   **12.** 59   **13.** 67

14. 72   **15.** 85   **16.** 113

17. 155   **18.** 234   **19.** 375

20. 427

A *Is each number a multiple of 6?*

21. 84   **22.** 44   **23.** 90

24. 96   **25.** 95   **26.** 102

Is each number a multiple of 9?

27. 58   **28.** 81   **29.** 89

30. 117   **31.** 324   **32.** 378

B *Is each number a multiple of 7?*

33. 84   **34.** 86   **35.** 91

36. 111   **37.** 126   **38.** 181

Is each number a multiple of 6? of 9? of 15?

39. 432   **40.** 600   **41.** 660

42. 765   **43.** 780   **44.** 768

Is each number a multiple of 17? of 31?

45. 651   **46.** 935   **47.** 1581

48. 3689

C

49. Jean is driving home from work one evening when she comes across a police safety inspection team stopping cars. She knows that the team chooses every fourth car to inspect. She counts and determines that she is 14th in line. Will she be selected to have her car safety-checked?

50. In Exercise 49, would Jean be selected if she were 28th in line?

51. Frank assigns his algebra class problems numbered from 1 to 55 that are multiples of 5. Which problems should the students work?

52. Russ assigns his calculus class problems numbered from 1 to 99 that are multiples of 7. Which problems should the students work?

53. Katy reported that she counted 30 goat feet in a pen at a petting zoo. Explain how you know that she made a mistake.

54. Vance reported that he counted 37 duck feet in a pen. Can he be correct?

55. Thanh and his crew are to arrange 780 flower pots in rows in the public square. They want to have the same number of pots in each row. What arrangements are possible if there must be at least 6 rows? Assume that the number of pots in a row must be larger than the number of rows.

56. Minh and his crew are setting up 450 chairs in a large banquet hall for a lecture. They want to have the same number of chairs in each row. What different arrangements are possible if there must be at least 5 rows? Assume that the number of chairs in a row must be larger than the number of rows.

57. A gear has 20 teeth. The gear is rotated through 240 teeth. Is it in the original position after the rotation?

58. A gear has 24 teeth. The gear is rotated through 212 teeth. Is it in its original position after the rotation?

59. A mutual fund holds 9 million shares of Hewlett-Packard Co. and 27 million shares of Yahoo! Inc. Is one a multiple of the other? If you were reporting a comparison of the two holdings, how might you write the comparison so that it is easily understood?

© Brukoid/Shutterstock.com

60. A mutual fund holds 48 million shares of Microsoft Corp. and 6 million shares of International Business Machines Corp. (IBM). Is one a multiple of the other? If you were reporting a comparison of the two holdings, how might you write the comparison so that it is easily understood?

Exercises 61–64 relate to the chapter application. Pythagoras (ca. 580–500 B.C.) was a Greek mathematician who contributed to number theory and geometry. The Pythagorean theorem states that the sides of a right triangle satisfy the equation $a^2 + b^2 = c^2$, where a, b, and c are the sides.

Any set of three numbers that satisfies the equation is called a Pythagorean triple. A famous Pythagorean triple is 3, 4, 5.

61. Verify that 3, 4, 5 is a Pythagorean triple.

62. The branch of mathematics called "number theory" informs us that all multiples of a Pythagorean triple are also Pythagorean triples. Find three different Pythagorean triples that are multiples of 3, 4, and 5, and verify that they are also Pythagorean triples.

63. Another Pythagorean triple includes $a = 5$ and $b = 12$. What is the third number in this triple?

64. Find three triples that are multiples of the triple 5, 12, 13 and verify that they are Pythagorean triples.

STATE YOUR UNDERSTANDING

65. Write a short statement to explain why *every* multiple of 12 is also a multiple of 6.

66. One of the factors of 135 is 9 because $9 \cdot 15 = 135$. The number 9 is also a divisor of 135 because $135 \div 9 = 15$. Every factor of a number is also a divisor of that number. Explain in your own words why you think we have two different words, *factor* and *divisor,* for such numbers.

67. Suppose you are asked to list all of the multiples of 7 from 126 to 175. Describe a method you could use to be sure all the multiples are listed.

CHALLENGE

68. Is 5920 a multiple of 32?

69. Is 9743 a multiple of 87?

70. Find the largest number less than 6000 that is a multiple of 6, 9, and 17.

71. How many multiples of 3 are there between 1000 and 5000?

MAINTAIN YOUR SKILLS

List all of the numbers from 1 to 30 that divide the given number evenly (0 remainder).

72. 12

73. 14

74. 15

75. 18

76. 23

77. 30

Find the smallest number whose square is greater than the given number.

78. 56

79. 87

80. 345

81. 500

82. Is every number divisible by both 3 and 6 also divisible by 18? Explain.

83. Is every number divisible by both 2 and 8 also divisible by 16? Explain.

Divisors and Factors

VOCABULARY

The product of two numbers is a multiple of each. The two numbers are called **factors**. Thus, 8 and 5 are factors of 40 because $8 \cdot 5 = 40$.

Recall that $12^2 = 12 \cdot 12 = 144$, so the **square** of 12 is 144. The number 144 is called a **perfect square**.

When two or more numbers are multiplied, each number is a factor of the product. If a number is a factor of a second number, it is also a **divisor** of the second number.

OBJECTIVES

1. Write a counting number as the product of two factors in all possible ways.
2. List all of the factors (divisors) of a counting number.

HOW & WHY

■ **OBJECTIVE 1** Write a counting number as the product of two factors in all possible ways.

Finding the factors of a number can be visualized using blocks, pennies, or other small objects. To illustrate, let's examine the factors of 12. Arrange 12 blocks or squares in a rectangle. The rectangle will have an area of 12 square units. The length and width of the rectangle are factors of 12. For instance, Figure 2.3 shows a rectangle with 3 rows of 4 squares.

Figure 2.3

Because $3 \times 4 = 12$, 3 and 4 are factors of 12. Figure 2.4 shows the same 12 squares arranged into a different rectangle.

Figure 2.4

Figure 2.4 shows that $2 \times 6 = 12$, so 2 and 6 are also factors of 12.

Figure 2.5

Figure 2.5 shows a third arrangement: $1 \times 12 = 12$, so 1 and 12 are factors of 12. Figure 2.6 shows that the results of using rows of 5, 7, 8, 9, 10, or 11 do not form rectangles.

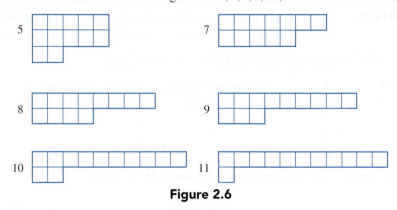

Figure 2.6

We conclude that the *only* pairs of factors of 12 are $1 \cdot 12$, $2 \cdot 6$, and $3 \cdot 4$.

To write a larger number, say, 250, as the product of two factors in all possible ways, we could again draw rectangles, or divide 250 by every number smaller than 250. Either method takes too long. The following steps save time.

1. List all the counting numbers from 1 to the first number whose square is larger than 250. Because $15 \times 15 = 225$ and $16 \times 16 = 256$, we stop at 16.

1	6	11	16	We can stop at 16 because 250 divided by
2	7	12		any number larger than 16 gives a quotient
3	8	13		that is less than 16. But all the possible fac-
4	9	14		tors less than 16 are already in the chart.
5	10	15		

2. Divide each of the listed numbers into 250. List the factors of the numbers that divide evenly. Otherwise, cross out the number.

$1 \cdot 250$	6̶	1̶1̶	16	When you find a number that is *not* a
$2 \cdot 125$	7̶	1̶2̶		factor, you can also eliminate all the
3̶	8̶	1̶3̶		multiples of that number. For example,
4̶	9̶	1̶4̶		because 3 is not a factor of 250, we can
$5 \cdot 50$	$10 \cdot 25$	1̶5̶		also eliminate 6, 9, 12, and 15.

These steps give us a list of all the two-factor products. Hence, 250 written as a product in all possible ways is

$$1 \cdot 250 \qquad 2 \cdot 125 \qquad 5 \cdot 50 \qquad 10 \cdot 25$$

To use the square method to write a counting number as the product of two factors in all possible ways

1. List all the counting numbers from 1 to the first number whose square is larger than the given number.
2. For each number on the list, test whether the number is a divisor of the given number.
3. If the number is not a divisor, cross it and all of its multiples off the list.
4. If the number is a divisor, write the indicated product of the two factors. The first factor is the tested number; the second factor is the quotient.

EXAMPLES A–C

DIRECTIONS: Write the counting number as the product of two factors in all possible ways.

STRATEGY: Use the square method. Begin by testing all of the counting numbers from 1 to the first number whose square is larger than the given number.

WARM-UP

A. Write 84 as the product of two factors in all possible ways.

A. Write 98 as the product of two factors in all possible ways.

$1 \cdot 98$	6̶	We can stop at 10 because $10^2 = 100$, which
$2 \cdot 49$	$7 \cdot 14$	is larger than 98.
3̶	8̶	
4̶	9̶	
5̶	1̶0̶	

The pairs of factors whose product is 98 are $1 \cdot 98$, $2 \cdot 49$, and $7 \cdot 14$.

B. Write 216 as the product of two factors in all possible ways.

1 · 216	6 · 36	~~14~~	We can stop at 15 because $15^2 = 225$,
2 · 108	~~7~~	12 · 18	which is larger than 216.
3 · 72	8 · 27	~~13~~	
4 · 54	9 · 24	~~14~~	
~~5~~	~~10~~	~~15~~	

The pairs of factors whose product is 216 are 1 · 216, 2 · 108, 3 · 72, 4 · 54, 6 · 36, 8 · 27, 9 · 24, and 12 · 18.

C. A television station has 130 minutes of late-night programming to fill. In what ways can the time be scheduled if each program must last a whole number of minutes, if each schedule must include programs all the same length, and each program must be 10 minutes or longer in length?

STRATEGY: List the pairs of factors of 130.

1 · 130
[1 program that is 130 minutes long or 130 programs that are 1 minute long (1-minute programs are not allowed)]
2 · 65
[2 programs that are 65 minutes long or 65 programs that are 2 minutes long (2-minute programs are not allowed)]
~~3~~
~~4~~
5 · 26
[5 programs that are 26 minutes long or 26 programs that are 5 minutes long (5-minute programs are not allowed)]
~~6~~
~~7~~
~~8~~
~~9~~
10 · 13
[10 programs that are 13 minutes long or 13 programs that are 10 minutes long]
~~11~~
~~12~~   Stop here because $12^2 = 144$, which is larger than 130.

So the station can schedule 1 program of 130 minutes, 2 programs of 65 minutes, 5 programs of 26 minutes, 10 programs of 13 minutes, or 13 programs of 10 minutes

WARM-UP
B. Write 150 as the product of two factors in all possible ways.

WARM-UP
C. A television station has 180 minutes of programming to fill. In what ways can the time be scheduled if each program must last a whole number of minutes and if each schedule must include programs all the same length? In addition, no program can be less than 15 minutes.

HOW & WHY

■ **OBJECTIVE 2** List all of the factors (divisors) of a counting number.

The square method to find pairs of factors also gives us a list of *all* factors or divisors of a given whole number. To make a list of all factors of 250, in order, we can use the chart for all pairs of factors of 250.

↓
1 · 250
2 · 125
5 · 50
10 · 25
└───↑

Reading in the direction of the arrows, we see that the ordered list of all factors of 250 is 1, 2, 5, 10, 25, 50, 125, and 250.

ANSWERS TO WARM-UPS B–C

B. 1 · 150, 2 · 75, 3 · 50, 5 · 30, 6 · 25, and 10 · 15

C. The time can be filled with 1 program of 180 minutes, 2 programs of 90 minutes, 3 programs of 60 minutes, 4 programs of 45 minutes, 5 programs of 36 minutes, 6 programs of 30 minutes, 9 programs of 20 minutes, 12 programs of 15 minutes.

EXAMPLES D–E

DIRECTIONS: List, in order, all factors of a given whole number.

STRATEGY: Use the square method to find all the pairs of factors. List the factors in order by reading down the left column of factors and up the right column.

WARM-UP
D. List all the factors of 170.

D. List all the factors of 342.

$1 \cdot 342$ 7 $\cancel{13}$ We can stop at 19 because $19^2 = 361$,
$2 \cdot 171$ 8 $\cancel{14}$ which is larger than 342.
$3 \cdot 114$ $9 \cdot 38$ $\cancel{15}$
$\cancel{4}$ $\cancel{10}$ $\cancel{16}$
$\cancel{5}$ $\cancel{11}$ $\cancel{17}$
$6 \cdot 57$ $\cancel{12}$ $18 \cdot 19$

$\downarrow$
$1 \cdot 342$ List the pairs and list the factors following the arrows.
$2 \cdot 171$
$3 \cdot 114$
$6 \cdot 57$
$9 \cdot 38$
$18 \cdot 19$

In order, all the factors of 168 are 1, 2, 3, 6, 9, 18, 19, 38, 57, 114, 171, and 342.

WARM-UP
E. List all the factors of 53.

E. List all the factors of 41.

$1 \cdot 41$ $\cancel{4}$ $\cancel{7}$ Stop at 7 because $7^2 = 49$.
2 5
3 6

All the factors of 41 are 1 and 41.

ANSWERS TO WARM-UPS D–E   **D.** 1, 2, 5, 10, 17, 34, 85, and 170   **E.** 1 and 53

EXERCISES 2.3

■ **OBJECTIVE 1** Write a counting number as the product of two factors in all possible ways. (See page 131.)

A *Write the whole number as the product of two factors in all possible ways.*

1. 16   **2.** 20   **3.** 31

4. 29   **5.** 33   **6.** 44

7. 46	8. 48	9. 49
10. 67	11. 72	12. 75
13. 88	14. 90	

B

15. 95	16. 98	17. 100
18. 104	19. 105	20. 108
21. 112	22. 117	23. 124
24. 128	25. 441	26. 445
27. 459	28. 465	

■ **OBJECTIVE 2** List all of the factors (divisors) of the counting number. (See page 133.)

A *List all of the factors (divisors) of the whole number.*

29. 15	30. 18	31. 19
32. 37	33. 39	34. 34
35. 46	36. 48	37. 52
38. 57	39. 64	40. 68

B

41. 72	42. 76	43. 78
44. 98	45. 112	46. 116
47. 134	48. 136	49. 162
50. 192		

C *Write the counting number as a product of two factors in all possible ways.*

51. 444	52. 550	53. 652
54. 672	55. 680	56. 852

57. In what ways can a television station schedule 120 minutes of time if each program must last a whole number of minutes, and each schedule must include programs of the same length, and no program can be less than 15 minutes in lenght?

58. A bag of Cheetos contains 14 oz. How many combinations of people and whole ounces of Cheetos can be made from the bag?

59. A card game requires that all the cards be divided among the players (there must be at least 2 players). A standard deck of cards contains 52 cards. How many players can play and each receive the same number of cards?

60. Child care experts recommend that child care facilities have 1 adult for every 3 or 4 infants. The Bee Fore School Day Care has 24 infants. If they staff according to the low end of the recommendation, how many adults do they need? If they staff according to the high end of the recommendation, how many adults do they need?

61. Child care experts recommend that child care facilities have 1 caregiver for 7 to 10 preschoolers. KinderCare has 65 preschoolers. If they have 5 caregivers, do they meet the recommendations?

62. The Hillsboro High School marching band has 64 members. List all the rectangular configurations possible for the band if no row can contain fewer than 6 members.

63. The director of the Forefront Marching Band arranged the band members in rows of 4 but had one person left over. He then arranged the band in rows of 5, but still had one person left over. What do you know about the number of members in the band? What is the smallest possible number of band members?

64. Juan is planting 42 tulip bulbs. He is planting them in the shape of a rectangle. How many different ways can he do this?

© Kim Doucette/Shutterstock.com

Exercises 65–68 relate to the chapter application.

*Numbers fascinated the ancient Greeks, in part because every Greek letter has a number associated with it. In particular, each Greek name had its own number and everyone was very interested in the properties of the number for his or her name. An exceptional individual was one whose number was a **perfect number**. **A perfect number** is a number that is the sum of all its divisors, excluding the number itself. Historical note: The Greeks did not consider a number to be a divisor of itself, although today we do.*

65. List all the divisors of 6. Find the sum of the divisors that are less than 6. Is 6 a perfect number?

66. Find another perfect number less than 20.

67. Find a perfect number between 20 and 30.

68. The third perfect number is 496. Verify that it is a perfect number.

STATE YOUR UNDERSTANDING

69. Explain the difference between a factor and a divisor of a number.

70. Describe how multiples, factors, and divisors are related to each other.

CHALLENGE

71. Find the largest factor of 2973 that is less than 2973.

72. Find three numbers that have exactly 5 different factors.

MAINTAIN YOUR SKILLS

Multiply.

73. 48(63)

74. 68(404)

75. 92(381)

76. 407(702)

Divide.

77. $78\overline{)2574}$

78. $82\overline{)24,682}$

79. $306\overline{)8265}$

80. $306\overline{)92,110}$

81. How many speakers can be wired from a spool of wire containing 1000 feet if each speaker requires 24 feet of wire? How much wire is left?

82. A consumer magazine tested 15 brands of tires to determine the number of miles they could travel before the tread would be worn away. The results are shown in the graph. What is the average mileage of the 15 brands?

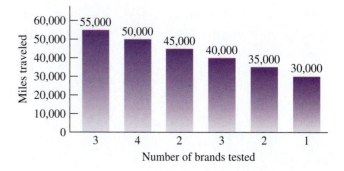

SECTION 2.4 Primes and Composites

VOCABULARY

A **prime number** is a whole number greater than 1 with exactly two factors (divisors). The two factors are the number 1 and the number itself.

A **composite number** is a whole number greater than 1 with more than two factors (divisors).

HOW & WHY

■ **OBJECTIVE**   Determine whether a whole number is prime or composite.

The whole numbers zero (0) and one (1) are neither prime nor composite. The number 2 is the first prime number ($2 = 1 \cdot 2$), because 1 and 2 are the only factors of 2. The number 3 ($3 = 3 \cdot 1$) is also prime because 1 and 3 are its only factors. The number 4 is a composite number ($4 = 1 \cdot 4$ and $4 = 2 \cdot 2$) because 4 has more than two factors.

To determine whether a number is prime or composite, list its factors or divisors in a chart like those in Section 2.3. Then count the number of factors. For instance, the chart for 247 is

1 · 247	7	13 · 19
2	8	14
3	9	15
4	10	16
5	11	
6	12	

We would have to test up to 16 in order to find all the factors of 247, as $16^2 = 256$ and $256 > 247$. Because we do not need *all* the factors, we stop at 13 because we know that 247 has at least four factors.

Therefore, 247 is a composite number.

The chart for 311 is

1 · 311	6	11	16
2	7	12	17
3	8	13	18
4	9	14	
5	10	15	

Stop here since $18 \cdot 18 = 324$ and $324 > 311$.

The number 311 is a prime number because it has exactly two factors, 1 and itself.

All prime numbers up to any given number may be found by a method called the Sieve of Eratosthenes. Eratosthenes (born ca. 230 B.C.) is remembered for both the prime Sieve and his method of measuring the circumference of Earth. The accuracy of his measurement, compared with modern methods, is within 50 miles, or six-tenths of 1%.

To use the famous Sieve to find all the primes from 2 to 35, list the numbers from 2 to 35.

②	3	4	5	6	7	8	9
10	11	12	13	14	15	16	17
18	19	20	21	22	23	24	25
26	27	28	29	30	31	32	33
34	35						

The number 2 is prime, but all other multiples of 2 are not prime. They are crossed off.

The next number is 3, which is prime.

②	③	~~4~~	5	~~6~~	7	8	9
~~10~~	11	~~12~~	13	~~14~~	~~15~~	~~16~~	17
~~18~~	19	~~20~~	~~21~~	~~22~~	23	~~24~~	25
~~26~~	~~27~~	~~28~~	29	~~30~~	31	~~32~~	~~33~~
~~34~~	~~35~~						

All remaining multiples of 3 are not prime, so they are crossed off; that is 9, 15, 21, etc.

The number 4 has already been eliminated. The next number, 5, is prime.

②	③	~~4~~	⑤	~~6~~	⑦	8	9
~~10~~	⑪	~~12~~	⑬	~~14~~	~~15~~	~~16~~	⑰
~~18~~	⑲	~~20~~	~~21~~	~~22~~	㉓	~~24~~	~~25~~
~~26~~	~~27~~	~~28~~	㉙	~~30~~	㉛	~~32~~	~~33~~
~~34~~	~~35~~						

All remaining multiples of 5 are not prime, so they are crossed off.

The multiples of the remaining numbers, except themselves, have been crossed off. We need to test divisors only up to the first number whose square is larger than 30 ($6 \cdot 6 = 36$).

The prime numbers less than 35 are 2, 3, 5, 7, 11, 13, 17, 19, 23, 29, and 31.

Looking at the Sieve, we see that we can shorten the factor chart by omitting all numbers except those that are prime. For example, is 427 prime or composite?

$1 \cdot 427$
~~2~~
~~3~~
~~5~~
$7 \cdot 61$ Stop here, since we do not need all the factors.

The number 427 is composite, because it has *at least* four factors, 1, 7, 61, and 427. We know that a number is prime if no smaller prime number divides it evenly.

Keep the divisibility tests for 2, 3, and 5 in mind since they are prime numbers

> ## To determine whether a number is prime or composite
>
> Test every prime number whose square is less than the number.
>
> a. If the number has exactly two divisors (factors), the number 1 and itself, it is prime.
> b. If the number has more than two divisors (factors), it is composite.
> c. Remember: 0 and 1 are neither prime nor composite.

EXAMPLES A–F

DIRECTIONS: Determine whether the number is prime or composite.

STRATEGY: Test all possible prime factors of the number. If there are exactly two factors, the number 1 and itself, the number is prime.

A. Is 101 prime or composite?

Test 1 and the prime numbers from 2 to 7. We stop at 7 because $11^2 > 101$.

$1 \cdot 101$ ~~7~~   The numbers 2, 3, and 5 can be crossed out using the
~~2~~   divisibility tests. Eliminate 7 by division.
~~3~~
~~5~~

The number 101 is a prime number.

WARM-UP

B. Is 119 prime or composite?

B. Is 209 prime or composite?

Test 1 and the prime numbers from 2 to 13, since $17^2 > 209$.

1 · 209	7̸	
2̸	11 · 19	Stop testing at 11, because we have at least four
3̸	13	factors.
5̸		

The number 209 is a composite number.

WARM-UP

C. Is 277 prime or composite?

C. Is 253 prime or composite?

Test 1 and the prime numbers from 2 to 13. Stop the list at 13 because $17^2 > 253$.

1 · 253	7̸	
2̸	11 · 23	Stop testing at 11, as we have at least four factors.
3̸	13	
5̸		

So 253 is a composite number.

WARM-UP

D. Is 749 prime or composite?

D. Is 467 prime or composite?

Test 1 and the prime numbers from 2 to 19, since $23^2 > 467$.

1 · 467	1̸1̸
2̸	1̸3̸
5̸	1̸7̸
7̸	1̸9̸

So 467 is a prime number.

WARM-UP

E. Is 301,755 prime or composite?

E. Is 124,653 prime or composite?

124,653 is divisible by 3   The sum of the digits $1 + 2 + 4 + 6 + 5 + 3 = 21$, which is divisible by 3.

We have at least three factors, 1, 3, and 124,653.
So 124,653 is a composite number.

WARM-UP

F. In the same contest Mikey was asked, "Is the number 234,425 prime or composite?" What should he have answered?

F. Christine won a math contest at her school. One of the questions in the contest was "Is 234,423 prime or composite?" What should Christine have answered?

234,423 is not divisible by 2.   The ones-place digit is 3 (not even).
234,423 is divisible by 3. $2 + 3 + 4 + 4 + 2 + 3 = 18$, which is divisible by 3.

We have at least three factors, 1, 3, and 234,423.
Christine should have answered "composite."

ANSWERS TO WARM-UPS B–F

B. composite **C.** prime
D. composite **E.** composite
F. composite

EXERCISES 2.4

■ **OBJECTIVE**   Determine whether a whole number is prime or composite. (See page 138.)

A *Tell whether the number is prime or composite.*

1. 6   **2.** 7   **3.** 8   **4.** 11

5. 15   **6.** 16   **7.** 17   **8.** 21

9. 22　　　　　**10.** 23　　　　　**11.** 29　　　　　**12.** 33

13. 38　　　　　**14.** 39　　　　　**15.** 41　　　　　**16.** 43

17. 46　　　　　**18.** 50　　　　　**19.** 53　　　　　**20.** 57

21. 58　　　　　**22.** 59

B

23. 52　　　　　**24.** 61　　　　　**25.** 60　　　　　**26.** 67

27. 70　　　　　**28.** 77　　　　　**29.** 79　　　　　**30.** 82

31. 83　　　　　**32.** 89　　　　　**33.** 91　　　　　**34.** 93

35. 97　　　　　**36.** 110　　　　　**37.** 127　　　　　**38.** 133

39. 146　　　　　**40.** 157　　　　　**41.** 183　　　　　**42.** 187

43. 213　　　　　**44.** 231　　　　　**45.** 233　　　　　**46.** 321

C

47. 383　　　　　**48.** 389　　　　　**49.** 391　　　　　**50.** 437

51. 581　　　　　**52.** 587　　　　　**53.** 1323　　　　　**54.** 1337

55. The year 2011 is a prime number. What is the next year that is a prime number?

56. How many prime numbers are there between 1 and 100? How many between 100 and 200?

57. Explain how one can tell just by looking that 12,345,678 is not a prime number.

58. Is the year of your birth a prime or composite number?

59. Three and 7 are both prime and so are 37 and 73. Is it true that any combination of two prime digits is also prime? Either give a justification, or find an example that does not work.

60. Find a pair of primes, one with one digit and one with two digits, so that both three-digit numbers built from the two primes are prime.

61. The *Exxon Valdez* ran aground off the shore of Alaska on March 24, 1989, spilling 11 million gallons of oil. Is 1989 a prime number?

© tkachuk/Shutterstock.com

62. According to the U.S. Census Bureau the estimated population of the four major regions of the United States for 2008 is given in the following table:

Population of the United States in 2008

Northeast	54,924,779
Midwest	66,561,448
South	111,718,539
West	70,854,948

Three of the four populations are clearly not prime numbers. Identify them and tell how you know they are not prime. (Optional: Is the fourth number prime?)

Exercises 63–66 relate to the chapter application.

Many mathematicians have been fascinated by prime numbers. Computer searches for new prime numbers are common. A French mathematician named Marin Mersenne (1588–1648) identified a special form of prime numbers that bears his name. A Mersenne prime has the form $M_p = 2^p - 1$, where p is a prime number. The smallest Mersenne prime number is 3 because when p = 2, the smallest prime, $M_2 = 2^2 - 1 = 4 - 1 = 3$.

63. Find the value of the next three Mersenne primes by substituting the prime numbers 3, 5, and 7 for p.

64. Mersenne was originally hoping that all numbers with his special form were prime. He discovered that this is false with M_{11}. Calculate M_{11} by substituting 11 for p and show that M_{11} is not prime.

65. Every Mersenne prime has a companion perfect number. See Exercise 55 in Section 2.3. The perfect number is expressed by $P = M_p \times (2^{p-1})$. Find the perfect number that is companion to M_3.

66. What perfect number is companion to M_5?

STATE YOUR UNDERSTANDING

67. Explain the difference between prime numbers and composite numbers. Give an example of each.

68. True or false: The product of two composite numbers is composite. Explain your reasoning.

69. Is it possible for a composite number to have exactly four factors? If so, give two examples. If not, tell why not.

CHALLENGE

70. Is 23,341 a prime or composite number?

71. Is 37,789 a prime or composite number?

72. A sphenic number is a number that is a product of three unequal prime numbers. The smallest sphenic number is $2 \cdot 3 \cdot 5 = 30$. Is 4199 a sphenic number?

53. A gear with 36 teeth is engaged with another [that] has 24 teeth. How many turns of the first gear [are nec]essary in order for the two gears to return to t[heir origi]nal positions?

Exercises 55–56. All the planets revolve around t[he sun ...] *for Jupiter to complete its orbit, and about 30 ye[ars ...]*

55. If both Mars and Jupiter are visible together i[n ...] how long will it be before they are both visibl[e ...] a second time?

57. John runs 6 miles a day. John's friend runs 1[...] each day. What is the least number of days e[ach must] run so they covered the exact same distance?

STATE YOUR UNDERS[TANDING]

59. Explain how to find the LCM of 20, 24, and [...]

CHALLENGE

61. Find the LCM of 144, 240, and 360.

63. Find the LCM of 60, 210, 315 and 350.

64. Here is a classic Hindu puzzle from the sev[enth cen]tury. A galloping horse frightens a woman [carrying a] basket of eggs. She drops the basket and br[eaks the] eggs. Concerned passersby ask how many e[ggs she] lost, but she can't remember. She does rem[ember that] there was one egg left over when she count[ed by 2s,] 2 eggs left over when she counted by 3s, 3 [eggs] left over when she counted by 4s, and 4 egg[s left over] when she counted by 5s. How many eggs d[id she lose?]

MAINTAIN YOUR SKILLS

73. Because the last digit of the number 120 is 0, 120 is divisible by 2. Is the quotient of 120 and 2 divisible by 2? Continue dividing by 2 until the quotient is not divisible by 2. What is the final quotient?

74. Because the last digit of the number 416 is 6, 416 is divisible by 2. Is the quotient of 416 and 2 divisible by 2? Continue dividing by 2 until the quotient is not divisible by 2. What is the final quotient?

75. Because the last digit of the number 1040 is 0, 1040 is divisible by 2. Is the quotient of 1040 and 2 divisible by 2? Continue dividing by 2 until the quotient is not divisible by 2. What is the final quotient?

76. Because the sum of the digits of 1701 is 9, 1701 is divisible by 3. Is the quotient of 1701 and 3 divisible by 3? Continue dividing by 3 until the quotient is not divisible by 3. What is the final quotient?

77. Because the sum of the digits of 1029 is 12, 1029 is divisible by 3. Is the quotient of 1029 and 3 divisible by 3? Continue dividing by 3 until the quotient is not divisible by 3. What is the final quotient?

78. Because the last digit of the number 3500 is 0, 3500 is divisible by 5. Is the quotient of 3500 and 5 divisible by 5? Continue dividing by 5 until the quotient is not divisible by 5. What is the final quotient?

79. Because the last digit of the number 2880 is 0, 2880 is divisible by 5. Is the quotient of 2880 and 5 divisible by 5? Continue dividing by 5 until the quotient is not divisible by 5. What is the final quotient?

80. The number 1029 is divisible by 7. Is the quotient of 1029 and 7 divisible by 7? Continue dividing by 7 until the quotient is not divisible by 7. What is the final quotient?

81. The number 1859 is divisible by 13. Is the quotient of 1859 and 13 divisible by 13? Continue dividing by 13 until the quotient is not divisible by 13. What is the final quotient?

82. Because the last digit of the number 17,408 is 8, 17,408 is divisible by 2. Is the quotient of 17,408 and 2 divisible by 2? Continue dividing by 2 until the quotient is not divisible by 2. What is the final quotient?

Prime Factorization

VOCABULARY

The **prime factorization** of a counting number is the indicated product of prime numbers. There are two ways of asking the same question.

1. "What is the prime factorization of this number?"
2. "Write this number in prime-factored form."

$51 = 3 \cdot 17$ and $66 = 2 \cdot 3 \cdot 11$ are prime factorizations.

$30 = 3 \cdot 10$ is not a prime factorization, because 10 is not a prime number.

Recall that exponents show repeated factors. This can save space in writing.

$2 \cdot 2 \cdot 2 = 2^3$ and $3 \cdot 3 \cdot 3 \cdot 3 \cdot 7 \cdot 7 \cdot 7 = 3^4 \cdot 7^3$

OBJECTIVES

1. Find the prime factorization of a counting number by repeated division.
2. Find the prime factorization of a counting number using the Tree Method.

■ **OBJECTIVE 1** Find the least common multiple
Method. (See page 150.)

■ **OBJECTIVE 2** Find the least common multiple
Method. (See page 152.)

A *Find the LCM of each group of whole numbers usi*

1. 4, 18   **2.** 6, 12

5. 3, 30   **6.** 4, 30

9. 5, 9   **10.** 9, 24

13. 4, 10, 20   **14.** 4, 6, 8

17. 2, 6, 10   **18.** 8, 9, 12

B

21. 18, 24   **22.** 20, 30

25. 18, 20   **26.** 14, 20

29. 20, 24   **30.** 28, 42

33. 9, 12, 15   **34.** 12, 15, 18

37. 12, 16, 24   **38.** 20, 30, 40

C

41. 10, 14, 21, 35   **42.** 8, 14,

44. 12, 32, 48, 96   **45.** 35, 5(

Find the least common denominator for each set of fr

47. $\dfrac{2}{3}, \dfrac{3}{4}, \dfrac{5}{8}$ **48.** $\dfrac{5}{6}, \dfrac{4}{5}, \dfrac{1}{9}$

51. Mary is a paramedic who works 4 days and has 1
off. Her husband, Frank, works 5 days and has 1 d
off. On May 1, they were both off. What is the ne>
their days off will coincide?

SECTION 2.4 Primes and Composites

Definitions and Concepts	Examples
A prime number is a counting number with exactly two factors, itself and 1.	2, 3, 5, 7, 11, 13, 17, and 19 are the primes less than 20.
A composite number is a counting number with more than two factors.	4, 6, 8, 9, 10, 12, 14, 15, 16, and 18 are the composite numbers less than 20.

To determine if a number is prime,
• Systematically search for factor pairs up to the number whose square is larger than the number.
• If none are found then the number is prime.

Is 53 prime?
$1 \cdot 53$ 6̸
2   7̸
3   8̸
4̸   Stop because $8^2 = 64 > 53$.
5̸
53 is a prime number.

SECTION 2.5 Prime Factorization

Definitions and Concepts	Examples
The prime factorization of a number is the number written as the product of primes.	$40 = 2^3 \cdot 5$

To find the prime factorization of a number,
• Divide the number and each succeeding quotient by a prime number until the quotient is 1.
• Write the indicated product of all the primes.

$2\overline{)60}$
$2\overline{)30}$
$3\overline{)15}$
$5\underline{)\ 5}$
$\ \ \ \ 1$

$60 = 2^2 \cdot 3 \cdot 5$

SECTION 2.6 Least Common Multiple

Definitions and Concepts	Examples
The least common multiple (LCM) of two or more numbers is the smallest natural number that is a multiple of each of the numbers.	The LCM of 6 and 8 is 24.

To find the least common multiple,
• Write each number in prime-factored form.
• The LCM is the product of the highest power of each prime factor.

Find the LCM of 20 and 150.
$20 = 2^2 \quad\ \cdot 5$
$150 = 2 \cdot 3 \cdot 5^2$
$\text{LCM} = 2^2 \cdot 3 \cdot 5^2 = 300$

REVIEW EXERCISES

SECTION 2.1

1. Which of these numbers is divisible by 2?
9, 44, 50, 478, 563

2. Which of these numbers is divisible by 3?
6, 36, 63, 636, 663

3. Which of these numbers is divisible by 5?
15, 51, 255, 525, 552

4. Which of these numbers is divisible by 6?
36, 65, 144, 714

5. Which of these numbers is divisible by 9?
6, 36, 63, 636, 663

6. Which of these numbers is divisible by 10?
50, 55, 505, 550, 555

7. Which of these numbers is divisible by both 2 and 3?
444, 555, 666, 777, 888, 999

8. Which of these numbers is divisible by both 2 and 9?
450, 550, 660, 770, 880, 990

9. Which of these numbers is divisible by both 3 and 5?
445, 545, 645, 745, 845

10. Which of these numbers is divisible by both 3 and 10?
440, 550, 660, 770, 880

Determine whether the number is divisible by 6, 7, or 9.

11. 567

12. 576

13. 693

14. 765

Determine whether the number is divisible by 4, 5, or 10.

15. 840

16. 575

Determine whether the number is divisible by 6 or 15.

17. 705

18. 975

19. Tiger Woods finished the 4-day tournament with a stroke total of 268. Is it possible for him to have shot the same score in each of the rounds? Explain.

20. Bobbie and her five partners incurred expenses of $3060 in their satellite dish installation enterprise. Can the expenses be divided evenly in whole dollars among them? Explain.

SECTION 2.2

List the first five multiples of the whole number.

21. 7

22. 63

23. 73

24. 64

25. 85

26. 97

27. 131

28. 142

29. 211

30. 252

Is each number a multiple of 6? of 9? of 15?

31. 84

32. 135

33. 210

34. 310

35. 315

36. 540

Is each number a multiple of 12? of 16?

37. 288

38. 560

39. Shelly bought shares of Intel at $23 each. The stock cost $2691 before brokerage fees. Did Shelly buy a whole number of shares?

40. In Exercise 39, if Shelly paid $5678 for the shares, did she buy a whole number of shares?

SECTION 2.3

Write the whole number as the product of two factors in all possible ways.

41. 15

42. 28

43. 42

44. 136

45. 236

46. 325

47. 354

48. 341

49. 343

List all of the factors (divisors) of the whole number.

50. 33

51. 48

52. 57

53. 60

54. 78

55. 97

56. 99

57. 108

58. 110

59. In a recent year, the average single-family detached residence consumed $1590 in energy costs. Find the possible costs of energy consumption (whole number amounts) per person. Assume that each household has from 1 to 10 people.

60. A rectangular floor area requires 240 one-square-foot tiles. List all the possible whole-number dimensions that this floor could measure that would require all the titles. Assume that the floor must be at least 3 feet wide.

SECTION 2.4

Tell whether the number is prime or composite.

61. 17

62. 30

63. 43

64. 53

65. 57

66. 61

67. 78

68. 89

69. 93

70. 111

71. 117

72. 91

73. 337

74. 339

75. 391

76. 359

77. 713

78. 853

79. The year 2017 is a prime number. What is the next year that is a prime number?

80. What was the last year before 1997 that was a prime number?

SECTION 2.5

Write the prime factorization of each number.

81. 32

82. 36

83. 38

84. 42

85. 75

86. 96

87. 222

88. 276

89. 252

90. 256

91. 290

92. 297

93. 340

94. 342

95. 373

96. 264

97. 265

98. 266

99. Write the prime factorization of the next several years.

100. Write the prime factorization of the years your parents were born.

SECTION 2.6

Find the LCM of each group of numbers.

101. 6, 8

102. 8, 10

103. 18, 40

104. 20, 36

105. 36, 44

106. 24, 72

107. 25, 35

108. 28, 35

109. 40, 45

110. 3, 9, 18

111. 6, 8, 12

112. 8, 10, 12

113. 9, 24, 36

114. 18, 24, 36

115. 15, 50, 60

116. 32, 40, 60

117. 40, 60, 105

118. 55, 66, 90

119. Find two numbers that have an LCM of 98.

120. Find two numbers that have an LCM of 102.

TRUE/FALSE CONCEPT REVIEW

Check your understanding of the language of basic mathematics. Tell whether each of the following statements is true (always true) or false (not always true). For each statement you judge to be false, revise it to make a statement that is true.

Answers

1. Every multiple of 3 ends with the digit 3.

   1. _____

2. Every multiple of 10 ends with the digit 0.

   2. _____

3. Every multiple of 11 is divisible by 11.

   3. _____

4. Every multiple of 5 is the product of 5 and some natural number.

   4. _____

5. Every natural number, except the number 1, has at least two different factors.

   5. _____

6. Every factor of 300 is also a divisor of 300.

   6. _____

7. Every multiple of 300 is also a factor of 300.

   7. _____

8. The square of 25 is 50.

   8. _____

9. Every natural number ending in 6 is divisible by 6.

   9. _____

10. Every natural number ending in 6 is divisible by 2.

    10. _____

11. Every natural number ending in 9 is divisible by 3.

    11. _____

12. The number 3192 is divisible by 7.

    12. _____

13. The number 77,773 is divisible by 3.

    13. _____

14. The number 123,321,231 is divisible by 9.

    14. _____

15. The number 111,111,115 is divisible by 5.

    15. _____

16. All prime numbers are odd.

    16. _____

17. All numbers that end with the digit 8 are composite.

    17. _____

18. Every prime number has exactly two multiples.

    18. _____

19. It is possible for a composite number to have exactly seven factors.

    19. _____

20. All of the prime factors of a composite number are smaller than the number.

    20. _____

21. The least common multiple (LCM) of three different numbers is the product of the three numbers.

21. _____

22. The product of two prime numbers is also the LCM of the two numbers.

22. _____

23. The largest divisor of the least common multiple (LCM) of three numbers is the largest of the three numbers.

23. _____

24. It is possible for the LCM of three natural numbers to be one of the three numbers.

24. _____

TEST

Answers

1. Is 411,234 divisible by 6?

1. _____

2. List all of the factors (divisors) of 112.

2. _____

3. Is 8617 divisible by 7?

3. _____

4. Is 2,030,000 divisible by 3?

4. _____

5. What is the LCM of 12 and 32?

5. _____

6. Write 75 as the product of two factors in as many ways as possible.

6. _____

7. Write the prime factorization of 280.

7. _____

8. Find the LCM of 18, 42, and 84.

8. _____

9. Is 15,075 a multiple of 15?

9. _____

10. Write all multiples of 13 between 200 and 250.

10. _____

11. Is 200 a multiple of 400?

11. _____

12. Is 109 a prime or a composite number?

12. _____

13. Is 111 a prime or a composite number?

13. _____

14. Write the prime factorization of 605.

14. _____

15. What is the LCM of 18, 21, and 56?

15. _____

16. What is the smallest prime number?

16. _____

17. What is the largest composite number that is less than 300?

17. _____

18. What is the smallest natural number that 6, 18, 24, and 30 will divide evenly?

18. _____

19. Can two different numbers have the same prime factorization? Explain.

19. _____

20. List two sets of three different numbers whose LCM is 42.

20. _____

CLASS ACTIVITY 1

1. Have each person in the group choose a whole number, making sure each person chooses a different whole number. Have each person with an even number divide by 2. Have each person with an odd number multiply by 3 and add 1. Record the results. Repeat the same process with the new number—that is, divide each even number by 2 and multiply each odd number by 3 and add 1. Do this a third time. Continue until you know you have gone as far as possible. Now compare your final result with the other member of the group. You should all have ended with the same number, regardless of the number you started with.
Mathematicians believe that this will always happen, yet they have been unable to prove that it is true for all whole numbers. This is called the Syracuse conjecture.

2. Mathematicians have established that every prime number greater than 3 is either one more than or one less than a multiple of 6.

   a. Use the Sieve of Eratosthenes to identify all the prime numbers less than 100. Divide up the list of these primes and verify that each of them is one away from a multiple of 6.

   b. Here is a step-by-step confirmation of this fact. As a group, supply the reasons for each step of the confirmation.

Consider the portion of the following number line. The first point, a, is at an even number.

a b c d e f g h i j k l m n o p q r s t u v

Statement

   a. If a is even, then so are c, e, g, i, k, m, o, q, s, and u.
   b. If a is a multiple of 3, then so are d, g, j, m, p, s, and v.
   c. a, g, m, and s, are multiples of 6.
   d. None of the points in steps a, b, or c can be prime.
   e. The only possible places for prime number to occur are at b, f, h, l, n, r, and t.
   f. All primes are one unit from a multiple of 6.

Reason

   a.
   b.
   c.
   d.
   e.
   f.

CLASS ACTIVITY 2

The Goldbach conjecture states that every even number greater than 4 can be written as the sum of two prime numbers. Mathematicians believe this to be true, but as yet no one has proved that it is true for all even numbers. Complete the table.

Do you believe the Goldbach conjecture?

Even Number	Sum of Primes	Even Number	Sum of Primes
6	3 + 3	26	
8	3 + 5	28	
10		30	
12		32	
14		34	
16		36	
18		38	
20		40	
22		42	
24		44	

GROUP PROJECT (2–3 WEEKS)

In the early 1200s, the Italian mathematician Leonardo de Pisa, also known as Fibonacci, was interested in a set of numbers that today bear his name. The set begins {1, 1, 2, 3, 5, . . .} and continues infinitely using the rule that the next number in the set is the sum of the previous two.

1. Mathematicians often name numbers of a set according to the order in which they occur. For the Fibonacci numbers, it is customary to designate F_n as the nth element of the set. So $F_3 = 2$ and $F_4 = 3$. Calculate all the Fibonacci numbers less than 500. Make a table that lists them in order along with their name.

2. The Fibonacci numbers are strongly related to items in nature that spiral, such as the cells on pineapples or the petals on pinecones. The number of spirals in normal specimens is always a Fibonacci number, although the number changes from species to species. Find two whole pineapples and two pinecones from different kinds of trees. Look at each item from the top and count the spirals that move clockwise. Then count the spirals that move counterclockwise. Record your results in the following table. Write a paragraph that summarizes your findings.

Item	Number of Clockwise Spirals	Number of Counterclockwise Spirals
Pineapple 1		
Pineapple 2		
Pinecone 1		
Pinecone 2		

3. Just as each number has a unique prime factorization, each number can also be written as the sum of Fibonacci numbers. Always start with the largest Fibonacci number less than the given number, then go to the largest Fibonacci number less than the difference, and continue until the whole number is accounted for. For example,

$$46 = 34 + 12$$
$$= 34 + 8 + 4$$
$$= 34 + 8 + 3 + 1$$

Make a table that gives each number from 1 to 50 as the sum of Fibonacci numbers.

4. The game of Fibonacci Nim is an easy game in which players take turns removing chips from a central pot. You may start with any number of chips. Each player must take at least one chip out of the pile but not more than twice the number just removed by his opponent. (The first player may not remove all the chips on the first move.) The player who takes the last chip(s) wins. Start with 20 chips and play Nim five times. What do you notice about the number of chips remaining just before the game was won?

5. A winning strategy is to take the smallest Fibonacci number in the sum of Fibonacci numbers that makes the number of chips remaining. For example, if 46 chips are remaining, then one chip should be taken, because $46 = 34 + 8 + 3 + 1$. Play Nim five more times starting with 20 chips and using this strategy. Did it work? If both players use the strategy, which player will win? Write a paragraph summarizing how to win at Nim.

GOOD ADVICE FOR *Studying*

NEW HABITS FROM OLD

© Thinkstock/Comstock/Jupiter Images

General Strategy for Studying Math

- Read the assigned section before class.
- Go to class and take notes.
- Review the text and your notes before starting the exercises.
- Work through the exercises.
- Make a note of any questions you have.
- End the session by summarizing the techniques you have learned.

How to Read a Math Text

- Begin by reading the objectives and new vocabulary.
- Read with a pencil/highlighter and mark up the text with notes to yourself.
- After reading an example, try the corresponding warm-up exercise.
- If you don't remember an earlier concept, go back and review before continuing.

Keep a Positive Attitude

- Begin each study session by expecting that you will be able to learn the skills.
- Do not spend more than 15 minutes on any one problem. If you can't solve it, mark it to ask about later and go on.
- Follow up on all your questions with your instructor or in the tutoring center or with a study buddy.
- Always end a study session with a problem that you have successfully solved—even if it means going backward. Your brain will remember that you were successful.
- Do not allow negative self-talk. Remind yourself that you have a new chance to be successful and that you are capable of learning this material.

Fractions and Mixed Numbers

APPLICATION

Shane is building a cedar deck in back of his house. His plans include stairs at one end of the deck and planter boxes on one side. See Figure 3.1.

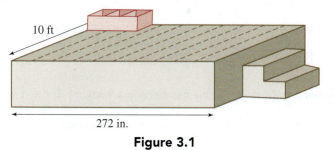

10 ft

272 in.

Figure 3.1

Shane would like to keep the sawing to a minimum, as he has only a hand-held circular saw, so he decides that he will design the deck so that he can use standard lengths of lumber. When he goes to Home Depot to purchase the lumber, the first thing he learns is that the standard sizes of lumber do not coincide with the actual size of the boards. For instance, a 10 ft 2-by-4 would seem to imply that its dimensions are 10 ft long, 4 in. wide, and 2 in. deep. However, the actual dimensions are 10 ft long, 3½ in. wide, and 1½ in. deep. (The reduction in dimensions happens when the rough cut lumber is "surfaced.") This means Shane will have to recalculate what he needs. He had been planning to use 2-by-4s for the deck boards. Now that he knows they are only 3½ in. wide instead of 4 in. wide, he will need more of them than he originally planned. Table 3.1 shows the actual sizes of all the lumber Shane needs for the deck.

TABLE 3.1 **Lumber Sizes**

Standard Size (in.)	Actual Size (in.)
4 × 4 posts	3½ × 3½
2 × 12 joists	1½ × 11¼
2 × 4 deck boards	1½ × 3½
2 × 6 stair treads	1½ × 5½
1 × 10 trim	¾ × 9¼

GROUP DISCUSSION

Use a tape measure to determine the dimensions of the room you are in using inches. Make a drawing of all four walls, including doors and windows. Measure as accurately as you can (specifying fractions of an inch). When you have finished, compare your measurements to those of another group. Are they identical? Why or why not? How can you determine which group has the most accurate measurements?

3.1 Proper and Improper Fractions; Mixed Numbers

OBJECTIVES

1. Write a fraction to describe the parts of a unit.
2. Select proper or improper fractions from a list of fractions.
3. Change improper fractions to mixed numbers.
4. Change mixed numbers to improper fractions.

VOCABULARY

A **fraction** $\left(\text{for example, } \dfrac{3}{7} \right)$ is a name for a number. The upper numeral (3) is the **numerator**. The lower numeral (7) is the **denominator**.

A **proper fraction** is one in which the numerator is less than the denominator $\left(\text{for example, } \dfrac{5}{11} \right)$.

An **improper fraction** is one in which the numerator is not less than the denominator $\left(\text{for example, } \dfrac{12}{5} \text{ or } \dfrac{13}{13} \right)$.

A **mixed number** is the sum of a whole number and a fraction $\left(3 + \dfrac{2}{3} \right)$ with the addition sign omitted $\left(3\dfrac{2}{3} \right)$. The fraction part is usually a proper fraction.

HOW & WHY

■ **OBJECTIVE 1** Write a fraction to describe the parts of a unit.

A unit (here we use a rectangle) may be divided into smaller parts of equal size in order to picture a fraction. The rectangle in Figure 3.2 is divided into seven parts, and six of the parts are shaded. The fraction $\dfrac{6}{7}$ represents the shaded part. The denominator (7) tells the number of parts in the unit. The numerator (6) tells the number of shaded parts. The fraction $\dfrac{1}{7}$ represents the part that is not shaded.

Figure 3.2

Because fractions are another way of writing a division problem and because division by zero is not defined, the denominator can never be zero. There will always be at least one part in a unit.

The unit many also be shown on a ruler. The fraction $\frac{6}{10}$ represents the distance from 0 to the arrowhead in Figure 3.3.

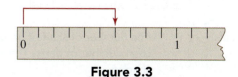

Figure 3.3

To write a fraction to describe the parts of a unit

Write the fraction:

$$\frac{numerator}{denominator} = \frac{number\ of\ shaded\ parts}{total\ number\ of\ parts\ in\ one\ unit}$$

To write a fraction from a ruler or a number line

Write the fraction:

$$\frac{numerator}{denominator} = \frac{number\ of\ spaces\ between\ zero\ and\ end\ of\ arrow}{number\ of\ spaces\ between\ zero\ and\ one}$$

EXAMPLES A–E

DIRECTIONS: Write the fraction represented by the figure.

STRATEGY: First count the number of parts that are shaded or marked. This number is the numerator. Now count the total number of parts in the unit. This number is the denominator.

A. Write the fraction represented by

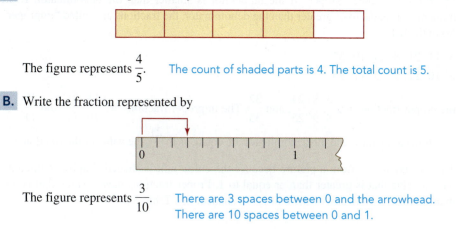

The figure represents $\frac{4}{5}$. The count of shaded parts is 4. The total count is 5.

B. Write the fraction represented by

The figure represents $\frac{3}{10}$. There are 3 spaces between 0 and the arrowhead. There are 10 spaces between 0 and 1.

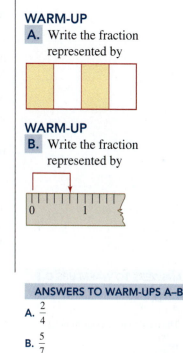

WARM-UP

C. Write the fraction represented by

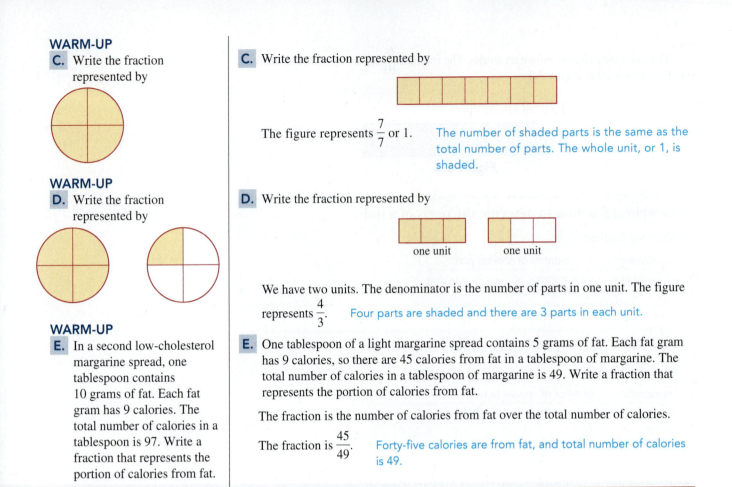

C. Write the fraction represented by

The figure represents $\dfrac{7}{7}$ or 1.   The number of shaded parts is the same as the total number of parts. The whole unit, or 1, is shaded.

WARM-UP

D. Write the fraction represented by

D. Write the fraction represented by

one unit   one unit

We have two units. The denominator is the number of parts in one unit. The figure represents $\dfrac{4}{3}$.   Four parts are shaded and there are 3 parts in each unit.

WARM-UP

E. In a second low-cholesterol margarine spread, one tablespoon contains 10 grams of fat. Each fat gram has 9 calories. The total number of calories in a tablespoon is 97. Write a fraction that represents the portion of calories from fat.

E. One tablespoon of a light margarine spread contains 5 grams of fat. Each fat gram has 9 calories, so there are 45 calories from fat in a tablespoon of margarine. The total number of calories in a tablespoon of margarine is 49. Write a fraction that represents the portion of calories from fat.

The fraction is the number of calories from fat over the total number of calories.

The fraction is $\dfrac{45}{49}$.   Forty-five calories are from fat, and total number of calories is 49.

HOW & WHY

■ **OBJECTIVE 2** Select proper or improper fractions from a list of fractions.

Fractions are called "proper" if the numerator is smaller than the denominator. If the numerator is equal to or greater than the denominator, the fractions are called "improper." So in the list

$$\frac{3}{8}, \frac{11}{10}, \frac{21}{21}, \frac{8}{9}, \frac{24}{25}, \frac{19}{11}, \frac{32}{35}$$

the proper fractions are $\dfrac{3}{8}, \dfrac{8}{9}, \dfrac{24}{25}$, and $\dfrac{32}{35}$. The improper fractions are $\dfrac{11}{10}, \dfrac{21}{21}$, and $\dfrac{19}{11}$.

If the numerator and the denominator are equal, as in $\dfrac{21}{21}$, the value of the fraction is 1.

This is easy to see from a picture because the entire unit is shaded. Improper fractions have a value that is greater than or equal to 1. Proper fractions have a value that is less than 1 because some part of the unit is not shaded. See Table 3.2.

TABLE 3.2 Regions That Show Proper and Improper Fractions

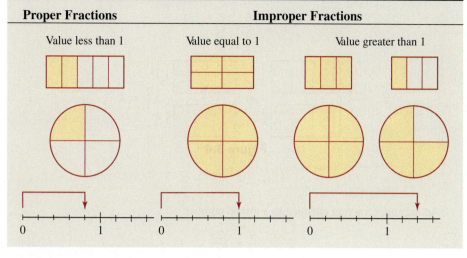

Proper Fractions	Improper Fractions	
Value less than 1	Value equal to 1	Value greater than 1

To determine if a fraction is proper or improper

1. Compare the size of the numerator and the denominator.
2. If the numerator is smaller, the fraction is proper. Otherwise the fraction is improper.

EXAMPLE F

DIRECTIONS: Identify the proper and improper fractions in the list.

STRATEGY: Compare the numerator and the denominator. If the numerator is smaller, the fraction is proper. If not, the fraction is improper.

F. Identify the proper and improper fractions: $\dfrac{5}{7}, \dfrac{9}{10}, \dfrac{14}{14}, \dfrac{16}{13}, \dfrac{18}{18}, \dfrac{27}{25}, \dfrac{29}{31}, \dfrac{37}{35}$

The proper fractions are

$\dfrac{5}{7}, \dfrac{9}{10}, \dfrac{29}{31}$ The numerators are smaller than the denominators.

The improper fractions are

$\dfrac{14}{14}, \dfrac{16}{13}, \dfrac{18}{18}, \dfrac{27}{25}, \dfrac{37}{35}$ The numerators are not smaller than the denominators.

HOW & WHY

■ **OBJECTIVE 3** Change improper fractions to mixed numbers.

An improper fraction is equal to a whole number or to a mixed number. A mixed number is the sum of a whole number and a fraction. Figures 3.4 and 3.5 show the conversions.

An improper fraction changed to a whole number:

$$\frac{6}{3} = 2$$

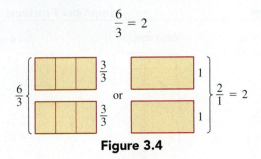

Figure 3.4

An improper fraction changed to a mixed number:

$$\frac{9}{7} = 1\frac{2}{7}$$

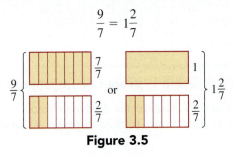

Figure 3.5

The shortcut for changing an improper fraction to a mixed number is to divide.

$$\frac{15}{3} = 15 \div 3 = 5$$

$$\frac{17}{5} = 17 \div 5 = 3\frac{2}{5} \qquad \begin{array}{r} 3 \\ 5\overline{)17} \\ \underline{15} \\ 2 \end{array}$$

To change an improper fraction to a mixed number

1. Divide the numerator by the denominator.
2. If there is a remainder, write the whole number and then write the fraction: $\dfrac{\text{remainder}}{\text{divisor}}$.

CAUTION

Do not confuse the process of changing an improper fraction to a mixed number with "simplifying." Simplifying fractions is a different procedure. See Section 3.2.

EXAMPLES G–I

DIRECTIONS: Change the improper fraction to a mixed number.

STRATEGY: Divide the numerator by the denominator to find the whole number. If there is a remainder, write it over the denominator to form the fraction part.

G. Change $\frac{11}{3}$ to a mixed number.

$$\frac{11}{3} = 11 \div 3$$

$$\begin{array}{r} 3 \\ 3\overline{)11} \\ \underline{9} \\ 2 \end{array}$$

So $\frac{11}{3} = 3\frac{2}{3}$

H. Change $\frac{145}{5}$ to a mixed number.

$$\frac{145}{5} = 145 \div 5$$

$$\begin{array}{r} 29 \\ 5\overline{)145} \\ \underline{10} \\ 45 \\ \underline{45} \end{array}$$

So $\frac{145}{5} = 29.$ Because there is no remainder, the fraction is equal to a whole number.

CALCULATOR EXAMPLE:

I. Change $\frac{348}{7}$ to a mixed number.

If your calculator has a key for fractions, refer to the manual to see how you can use it to change fractions to mixed numbers. If your calculator does not have a fraction key, divide 348 by 7. The quotient $\frac{348}{7} \approx 49.714$, so the whole-number part is 49.

Now subtract to find the remainder: $348 - 7(49) = 5.$

$$\begin{array}{r} 49 \\ 7\overline{)348} \\ \underline{343} \\ 5 \end{array}$$

$348 \div 7 = 49 \text{ R } 5$ so $\frac{348}{7} = 49\frac{5}{7}$

WARM-UP

G. Change $\frac{17}{4}$ to a mixed number.

WARM-UP

H. Change $\frac{117}{3}$ to a mixed number.

WARM-UP

I. Change $\frac{411}{23}$ to a mixed number.

HOW & WHY

■ **OBJECTIVE 4**   Change mixed numbers to improper fractions.

Despite the value judgment attached to the name "improper," in many cases improper fractions are a more convenient and useful form than mixed numbers. Thus, it is important to be able to convert mixed numbers to improper fractions.

Every mixed number can be changed to an improper fraction. See Figure 3.6.

$$1\frac{3}{7} = \frac{10}{7}$$

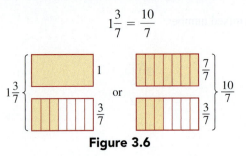

Figure 3.6

The shortcut uses multiplication and addition:

$$1\frac{3}{7} = \frac{1(7) + 3}{7} = \frac{7 + 3}{7} = \frac{10}{7} \quad \text{and} \quad 4\frac{5}{7} = \frac{4(7) + 5}{7} = \frac{33}{7}$$

To change a mixed number to an improper fraction

1. Multiply the denominator times the whole number.
2. Add the numerator to the product from step 1.
3. Write the sum from step 2 over the denominator.

To change a whole number to an improper fraction

Write the whole number over 1.

EXAMPLES J–L

DIRECTIONS: Change each mixed number to an improper fraction.

STRATEGY: Multiply the whole number times the denominator. Add the numerator. Write the sum over the denominator.

WARM-UP

J. Change $5\frac{2}{9}$ to an improper fraction.

J. Change $3\frac{4}{9}$ to an improper fraction.

$$3\frac{4}{9} = \frac{3(9) + 4}{9}$$

Multiply the whole number times the denominator and add the numerator. Write the sum over the denominator.

$$= \frac{31}{9}$$

ANSWER TO WARM-UP J

J. $\frac{47}{9}$

K. Change $6\dfrac{6}{11}$ to an improper fraction.

$6\dfrac{6}{11} = \dfrac{6(11) + 6}{11}$ Multiply the whole number times the denominator and add the numerator. Write the sum over the denominator.

$\qquad = \dfrac{72}{11}$

L. Change 31 to an improper fraction.

$\dfrac{31}{1}$ Write the whole number over 1.

WARM-UP

K. Change $1\dfrac{14}{15}$ to an improper fraction.

WARM-UP

L. Change 19 to an improper fraction.

ANSWERS TO WARM-UPS K–L

K. $\dfrac{29}{15}$ **L.** $\dfrac{19}{1}$

EXERCISES 3.1

■ **OBJECTIVE 1** Write a fraction to describe the parts of a unit. (See page 168.)

A *Write the fraction represented by the shaded part of the figure.*

1.

2.

3.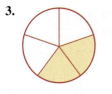

4.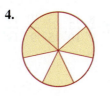

5.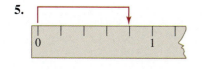

6.

B

7.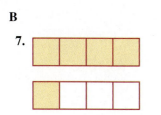

8.

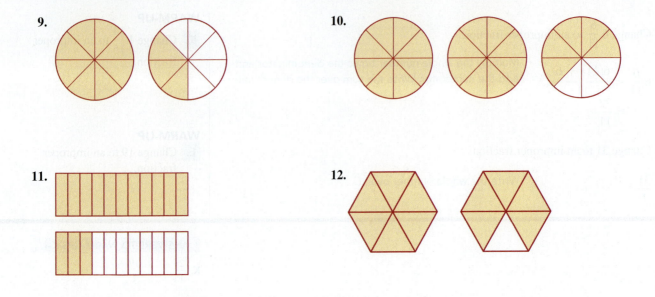

9.

10.

11.

12.

■ **OBJECTIVE 2** Select proper or improper fractions from a list of fractions. (See page 170.)

A *Identify the proper and improper fractions from the list.*

13. $\dfrac{2}{13}, \dfrac{4}{13}, \dfrac{5}{13}, \dfrac{7}{13}, \dfrac{10}{13}, \dfrac{13}{13}, \dfrac{15}{13}, \dfrac{17}{13}$

14. $\dfrac{5}{6}, \dfrac{8}{7}, \dfrac{14}{15}, \dfrac{16}{18}, \dfrac{17}{17}, \dfrac{23}{25}, \dfrac{28}{26}$

15. $\dfrac{17}{16}, \dfrac{18}{17}, \dfrac{29}{19}, \dfrac{30}{21}, \dfrac{23}{23}$

16. $\dfrac{7}{12}, \dfrac{11}{12}, \dfrac{13}{12}, \dfrac{12}{13}, \dfrac{7}{13}$

17. $\dfrac{8}{9}, \dfrac{12}{10}, \dfrac{13}{15}, \dfrac{10}{9}, \dfrac{19}{19}, \dfrac{6}{10}, \dfrac{16}{15}$

18. $\dfrac{7}{4}, \dfrac{10}{11}, \dfrac{13}{13}, \dfrac{20}{19}, \dfrac{3}{5}$

B

19. $\dfrac{3}{3}, \dfrac{6}{5}, \dfrac{8}{8}, \dfrac{11}{12}, \dfrac{15}{14}, \dfrac{19}{19}, \dfrac{21}{20}, \dfrac{24}{23}, \dfrac{29}{30}, \dfrac{32}{31}$

20. $\dfrac{5}{6}, \dfrac{6}{7}, \dfrac{7}{8}, \dfrac{8}{9}, \dfrac{10}{10}, \dfrac{11}{12}, \dfrac{12}{13}, \dfrac{13}{13}, \dfrac{15}{15}$

■ **OBJECTIVE 3** Change improper fractions to mixed numbers. (See page 171.)

A *Change to a mixed number*

21. $\dfrac{23}{6}$

22. $\dfrac{31}{7}$

23. $\dfrac{89}{7}$

24. $\dfrac{99}{7}$

25. $\dfrac{123}{11}$

26. $\dfrac{134}{9}$

B

27. $\dfrac{89}{13}$

28. $\dfrac{98}{13}$

29. $\dfrac{413}{27}$

30. $\dfrac{500}{17}$

31. $\dfrac{331}{15}$

32. $\dfrac{421}{15}$

■ **OBJECTIVE 4** Change mixed numbers to improper fractions. (See page 174.)

A *Change to an improper fraction.*

33. $5\dfrac{5}{7}$

34. $6\dfrac{7}{8}$

35. 15

36. 21

37. $8\dfrac{4}{9}$

38. $6\dfrac{5}{6}$

B

39. $43\dfrac{2}{5}$

40. $37\dfrac{3}{4}$

41. $42\dfrac{5}{7}$

42. $42\dfrac{3}{8}$

43. $61\dfrac{5}{8}$

44. $46\dfrac{6}{7}$

C *Exercises 45–46. Fill in the boxes so the statement is true. Explain your answer.*

45. The fraction $\dfrac{0}{\square}$ is a proper fraction.

46. $41\dfrac{\square}{7} = \dfrac{292}{7}$

47. Find the error(s) in the statement: $4\dfrac{2}{3} = \dfrac{8}{12}$. Correct the statement.

48. Find the error(s) in the statement: $\dfrac{45}{7} = 4\dfrac{5}{7}$. Correct the statement.

Write the fraction represented by the figure.

49.

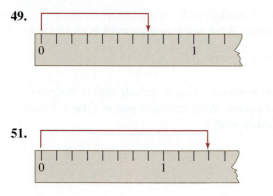

50.

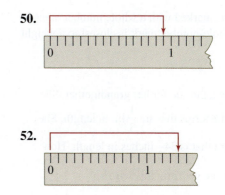

51.

52.

53. Draw a rectangular unit divided into equal parts that shows $\frac{5}{11}$.

54. Draw rectangular units divided into equal parts that shows $\frac{7}{4}$.

55. In a new housing development of 35 homes, 23 of the homes have shake roofs. What fraction represents the part of the homes in the development that have shake roofs.

56. In a recent poll of 90 people, 61 listed themselves as Cacucasian. What fraction represents the people polled who are Caucasian?

57. The U.S. Postal Service defines nonstandard mail as anything that is longer than $11\frac{1}{2}$ in., taller than $6\frac{1}{8}$ in., and/or thicker than $\frac{1}{4}$ in. Which of these dimensions can be changed to improper fractions and which cannot? Rewrite the appropriate dimensions as improper fractions.

58. The Adams family budgets $1155 for food and housing. They spend $772 per month for housing. What fractional part of their food and housing budget is spent for food?

59. What fraction of a full tank of gas is indicated by the gas gauge?

60. What fraction of a full tank of gas is indicated by the gas gauge?

61. A weight scale is marked with a whole number at each pound. What whole-number mark is closest to a weight of $\frac{85}{12}$ lb?

62. A ruler is marked with a whole number at each centimeter. What whole-number mark is closest to a length of $\frac{97}{10}$ cm?

63. Jenna is creating a mosaic for her grandmother. She needs copper wire strips that are $\frac{1}{3}$ in. in length. She has a copper wire that is $34\frac{2}{3}$ inches in length. How many strips can she get from this wire?

64. Cedale worked a 5 hour 43 minute shift at Walgreens one afternoon. What fractional part of a day (24 hours) did Cedale work?

65. Thangh was given the task of moving 510 hay bales from the field and storing them in the barn. On the first day he moved 153 bales. What fraction of the hay bales are still in the field?

66. Anna finished her comprehensive exam in 2 hours and 17 minutes. What fractional part of an hour is 17 minutes? Express the time it took her to finish the exam as a mixed number of hours.

67. The figure shows a measuring cup that contains oil for a zucchini bread recipe. The oil is what fractional part of a cup?

Exercises 68–70 relate to the chapter application. See page 167.

68. A joist labeled "2 × 12" is actually $1\frac{1}{2}$ × $11\frac{1}{4}$ in. Convert these measurements to improper fractions.

69. According to Shane's plans for his deck, each joist must be 272 in. long. However, lumber is sold by the foot, not the inch. So Shane must convert 272 in. into feet. Because he knows there are 12 in. in a foot, he knows that 272 in. $= \dfrac{272}{12}$ ft. Is this a proper or improper fraction? How do you know?

70. Change the length of a joist, $\dfrac{272}{12}$ ft, to a mixed number.

STATE YOUR UNDERSTANDING

71. Explain how to change $\dfrac{34}{5}$ to a mixed number. Explain how to change $7\dfrac{3}{8}$ to an improper fraction.

72. Tell why mixed numbers and improper fractions are both useful. Give examples of the use of each.

73. Explain why a proper fraction cannot be changed into a mixed number.

74. Write the whole number 23 as an improper fraction with (a) the numerator 253 and (b) the denominator 253.

75. Write the whole number 16 as an improper fraction with (a) the numerator 144 and (b) the denominator 144.

76. The Swete Tuth candy company packs 30 pieces of candy in its special Mother's Day box. Write, as a mixed number, the number of special boxes that can be made from 67,892 pieces of candy. The company then packs 25 of the special boxes in a carton for shipping. Write, as a mixed number, the number of cartons that can be filled. If it costs Swete Tuth $45 per carton to ship the candy, what is the shipping cost for the number of full cartons that can be shipped?

77. Jose has $21\frac{3}{4}$ yd of rope to use in a day care class. If the rope is cut into $\frac{1}{4}$-yd pieces, how many pieces will there be? If there are 15 children in the class, how many pieces of rope will each child get? How many pieces will be left over?

78. Using the following figure, write a fraction to name each of the division marks. Now divide each section in half and write new names for each mark. Divide the new sections in half and again write new fraction names for the marks. You now have three names for each of the original marks. What conclusion can you make about these names?

79. Write $\dfrac{45678}{37}$ as a mixed number with the aid of a calculator. Use the calculator to find the whole-number part and the numerator of the fraction part of the mixed number.

80. Divide: $102 \div 3$

81. Divide: $72 \div 4$

82. Divide: $390 \div 6$

83. Divide: $390 \div 26$

84. Is 413 prime or composite?

85. Is 419 prime or composite?

86. Find the prime factorization of 275.

87. Find the prime factorization of 414.

88. Wendy averages 45 miles per gallon of gasoline in her Prius. On a recent trip she traveled 1035 miles. How many gallons of gas did she use?

89. Bonnie traveled 208 miles, 198 miles, 214 miles, and 192 miles on four tanks of gas on her vintage motorcycle. If her tank holds 4 gallons of gas, what was her average number on miles per tank? To the nearest whole number, what was her average number of miles per gallon?

Simplifying Fractions

VOCABULARY

Equivalent fractions are fractions that are the different names for the same number.

Simplifying a fraction is the process of renaming it by using a smaller numerator and denominator.

A fraction is **completely simplified** when its numerator and denominator have no common factors other than 1. For instance, $\dfrac{12}{18} = \dfrac{2}{3}$.

HOW & WHY

■ **OBJECTIVE** Simplify a fraction.

Fractions are *equivalent* if they represent the same quantity. When we compare the two units in Figure 3.7, we see that each is divided into four parts. The shaded part on the left is named $\dfrac{2}{4}$, whereas the shaded part on the right is labeled $\dfrac{1}{2}$. It is clear that the two are the same size, and therefore we say $\dfrac{2}{4} = \dfrac{1}{2}$.

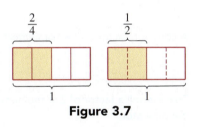

Figure 3.7

The arithmetical way of showing that $\dfrac{2}{4} = \dfrac{1}{2}$ is to eliminate the common factors by dividing:

$$\frac{2}{4} = \frac{2 \div 2}{4 \div 2} = \frac{1}{2}$$

The division can also be shown by eliminating the common factors.

$$\frac{2}{4} = \frac{1 \cdot \cancel{2}}{2 \cdot \cancel{2}} = \frac{1}{2}$$

This method works for all fractions. To simplify $\dfrac{35}{40}$:

$$\frac{35}{40} = \frac{\cancel{5} \cdot 7}{\cancel{5} \cdot 8} = \frac{7}{8} \qquad \text{Eliminate the common factors.}$$

or

$$\frac{35}{40} = \frac{35 \div 5}{40 \div 5} = \frac{7}{8} \qquad \text{Divide out the common factors.}$$

When all common factors have been eliminated (divided out), the fraction is completely simplified.

$$\frac{12}{48} = \frac{6}{24} = \frac{3}{12} = \frac{1}{4}$$ Completely simplified.

If the common factors are not discovered easily, they can be found by writing the numerator and denominator in prime-factored form. See Example E.

> **To simplify a fraction completely**
>
> Eliminate all common factors, other than 1, in the numerator and the denominator.

EXAMPLES A–H

DIRECTIONS: Simplify completely.

STRATEGY: Eliminate the common factors in the numerator and the denominator.

A. Simplify: $\dfrac{30}{42}$

$$\frac{30}{42} = \frac{5 \cdot \overset{1}{\cancel{6}}}{7 \cdot \underset{1}{\cancel{6}}} = \frac{5}{7}$$ The common factor is 6. Eliminate the common factor by dividing.

B. Simplify: $\dfrac{16}{24}$

$$\frac{16}{24} = \frac{\overset{1}{\cancel{2}} \cdot 8}{\underset{1}{\cancel{2}} \cdot 12} = \frac{8}{12}$$ A common factor of 2 is eliminated.

$$= \frac{\overset{1}{\cancel{2}} \cdot 4}{\underset{1}{\cancel{2}} \cdot 6} = \frac{4}{6}$$ There is still a factor of 2 in the numerator and the denominator.

$$= \frac{\overset{1}{\cancel{2}} \cdot 2}{\underset{1}{\cancel{2}} \cdot 3} = \frac{2}{3}$$ Again, a common factor of 2 is eliminated by division.

or

$$\frac{16}{24} = \frac{2 \cdot \overset{1}{\cancel{8}}}{3 \cdot \underset{1}{\cancel{8}}} = \frac{2}{3}$$ Rather than divide by 2 three times, divide by 8 once, and the fraction is simplified completely.

C. Simplify: $\dfrac{90}{72}$

$$\frac{90}{72} = \frac{90 \div 18}{72 \div 18} = \frac{5}{4}$$ Divide both numerator and denominator by 18.

or

$$\frac{90}{72} = \frac{5}{4}$$ Divide 90 and 72 by 18 mentally.

WARM-UP
A. Simplify: $\dfrac{66}{77}$

WARM-UP
B. Simplify: $\dfrac{18}{45}$

WARM-UP
C. Simplify: $\dfrac{90}{35}$

ANSWERS TO WARM-UPS A–C

A. $\dfrac{6}{7}$ B. $\dfrac{2}{5}$ C. $\dfrac{18}{7}$

D. Simplify: $\dfrac{100}{600}$

$$\dfrac{100}{600} = \dfrac{100 \div 100}{600 \div 100} = \dfrac{1}{6}$$ Divide both numerator and denominator by 100.

or

$$\dfrac{100}{600} = \dfrac{1}{6}$$ Divide by 100 mentally using the shortcut for dividing by powers of 10: $100 = 10^2$.

E. Simplify: $\dfrac{300}{630}$

STRATEGY: Because the numbers are large, write them in prime-factored form.

$$\dfrac{300}{630} = \dfrac{2 \cdot 2 \cdot 3 \cdot 5 \cdot 5}{2 \cdot 3 \cdot 3 \cdot 5 \cdot 7}$$ Eliminate the common factors.

$$= \dfrac{2 \cdot 5}{3 \cdot 7}$$

$$= \dfrac{10}{21}$$ Multiply.

F. Simplify: $\dfrac{18}{35}$

$$\dfrac{18}{35} = \dfrac{2 \cdot 3 \cdot 3}{5 \cdot 7} = \dfrac{18}{35}$$ There are no common factors. The fraction is already completely simplified.

CALCULATOR EXAMPLE:

G. Simplify: $\dfrac{493}{551}$

$$\dfrac{493}{551} = \dfrac{17}{19}$$ Use the fraction key on your calculator.

H. Mitchell mows lawns to earn money. One week, he has 18 lawns to mow. He finished 10 of them by Wednesday. What fraction of the lawns has he mowed? Simplify the fraction completely.

STRATEGY: Form the fraction: $\dfrac{\text{number of lawns mowed}}{\text{total number of lawns}}$

$$\dfrac{10}{18}$$ Ten lawns out of 18 are mowed.

$$\dfrac{10}{18} = \dfrac{2 \cdot 5}{2 \cdot 9} = \dfrac{5}{9}$$ Simplify.

So $\dfrac{5}{9}$ of the lawns are mowed.

EXERCISES 3.2

A   *Simplify completely.*

1. $\dfrac{8}{12}$

2. $\dfrac{9}{21}$

3. $\dfrac{12}{18}$

4. $\dfrac{15}{18}$

5. $\dfrac{18}{20}$

6. $\dfrac{16}{18}$

7. $\dfrac{45}{60}$

8. $\dfrac{40}{70}$

9. $\dfrac{32}{80}$

10. $\dfrac{21}{28}$

11. $\dfrac{9}{30}$

12. $\dfrac{18}{22}$

13. $\dfrac{35}{40}$

14. $\dfrac{36}{40}$

15. $\dfrac{72}{40}$

16. $\dfrac{66}{44}$

17. $\dfrac{42}{77}$

18. $\dfrac{28}{36}$

19. $\dfrac{65}{91}$

20. $\dfrac{25}{45}$

21. $\dfrac{108}{12}$

22. $\dfrac{198}{22}$

B

23. $\dfrac{93}{62}$

24. $\dfrac{60}{35}$

25. $\dfrac{14}{42}$

26. $\dfrac{88}{40}$

27. $\dfrac{56}{80}$

28. $\dfrac{20}{36}$

29. $\dfrac{27}{36}$

30. $\dfrac{14}{98}$

31. $\dfrac{35}{48}$

32. $\dfrac{27}{38}$

33. $\dfrac{50}{75}$

34. $\dfrac{30}{75}$

35. $\dfrac{80}{75}$

36. $\dfrac{60}{75}$

37. $\dfrac{300}{900}$

38. $\dfrac{700}{800}$

39. $\dfrac{75}{80}$

40. $\dfrac{65}{80}$

41. $\dfrac{72}{96}$

42. $\dfrac{36}{39}$

43. $\dfrac{55}{11}$

44. $\dfrac{96}{16}$

45. $\dfrac{85}{105}$

46. $\dfrac{56}{140}$

47. $\dfrac{72}{140}$

48. $\dfrac{72}{100}$

49. $\dfrac{99}{132}$

50. $\dfrac{84}{120}$

C

51. Which of these fractions is completely simplified?

    $\dfrac{72}{124}$, $\dfrac{16}{81}$, $\dfrac{35}{84}$, $\dfrac{93}{132}$

52. Which of these fractions is completely simplified?

    $\dfrac{26}{91}$, $\dfrac{44}{76}$, $\dfrac{102}{119}$, $\dfrac{56}{145}$

Simplify completely.

53. $\dfrac{84}{144}$

54. $\dfrac{114}{174}$

55. $\dfrac{196}{210}$

56. $\dfrac{268}{402}$

57. $\dfrac{546}{910}$

58. $\dfrac{840}{1060}$

Exercises 59–69. Show answers in simplest form.

59. In clothing, petite sizes are designed for women who are 5′4″ or less. What fractional part of a foot is 4″? Express the height 5′4″ as a mixed number of feet.

60. In the 2008–2009 basketball season, Steve Nash of the Phoenix Suns attempted 246 three-point shots and made 108 of them. What fraction of the three-point attempts did he make? What fraction did he miss?

61. In the 2008–2009 hockey season, Jimmy Howard, a goalie with the Detroit Red Wings defended 28 shots on goal and made 24 saves. What fraction of the total number of shots did he save? How many points did the opposing teams score on Howard?

62. In 2009, the top three countries that the United Stated imported goods from were China ($110 billion), Canada ($86 billion), and Mexico ($66 billion). What fraction of the total amount of imports of the top three countries is the amount imported from China?

63. Maria performs a tune-up on her automobile. She finds that two of the six spark plugs are fouled. What fraction represents the number of fouled plugs?

64. The float on a tank registers 12 feet. If the tank is full when it registers 28 feet, what fraction of the tank is full?

65. Gyrid makes a payment of $140 on her tuition of $540 at the local community college. What fraction of her bill does she still owe?

66. The Bonneville Dam fish counter shows 4320 salmon passing over the dam. Of this number, 220 are cohos. What fraction of the salmon are cohos?

67. A local food bank was able to meet the needs of 64 clients last week but had to turn away another 16 clients. What fraction of the clients was served by the food bank?

68. The energy used to produce 1 lb of virgin rubber is 15,800 BTUs. Producing 1 lb of recycled rubber requires only 4600 BTUs. What fraction of the BTUs needed to produce 1 lb of virgin rubber is used to produce 1 lb of recycled rubber?

69. One hundred forty elk are tallied at the Florence Refuge. Thirty-five of the elk are bulls. What fraction of the elk are cows?

© IPK Photography/Shutterstock.com

Exercises 70–71 relate to the chapter application.

70. The joists on Shane's deck are each $\frac{272}{12}$ ft long. Write this as a simplified fraction. Write as a simplified mixed number.

71. Shane is considering making his deck 270 in. long instead of 272 in. Write a simplified fraction that is the number of feet in 270 in. Convert this to a mixed number.

STATE YOUR UNDERSTANDING

72. Draw a picture that illustrates that $\frac{9}{12} = \frac{6}{8} = \frac{3}{4}$.

73. Explain how to simplify $\frac{525}{1125}$.

74. Explain why $\frac{12}{16}$ and $\frac{15}{20}$ are equivalent fractions.

CHALLENGE

75. Are these four fractions equivalent? Justify your answer.

$$\frac{495}{1188} \quad \frac{665}{1596} \quad \frac{1095}{2628} \quad \frac{890}{2136}$$

76. Are these four fractions equivalent? Justify your answer.

$$\frac{1170}{2925} \quad \frac{864}{2160} \quad \frac{672}{1440} \quad \frac{1134}{2430}$$

Multiply.

77. 6(42)

78. 9(88)

79. 11(99)

80. 7(333)

81. 8(444)

Divide.

82. 270 ÷ 18

83. 240 ÷ 15

84. 270 ÷ 45

85. 336 ÷ 21

86. 798 ÷ 42

Multiplying and Dividing Fractions

SECTION

3.3

VOCABULARY

A **product** is the answer to a multiplication problem. If two fractions have a product of 1, each fraction is called the **reciprocal** of the other. For example, $\frac{2}{3}$ is the reciprocal of $\frac{3}{2}$. A **quotient** is the answer to a division exercise. The quotient of 33 and 3 is 11.

OBJECTIVES

1. Multiply fractions.
2. Find the reciprocal of a number.
3. Divide fractions.

HOW & WHY

■ **OBJECTIVE 1** Multiply fractions.

The word *of* often indicates multiplication. For example, what is $\frac{1}{2}$ of $\frac{1}{3}$? In other words, $\frac{1}{2} \cdot \frac{1}{3} = ?$

See Figure 3.8. The rectangle is divided into three parts. One part, $\frac{1}{3}$, is shaded yellow. To find $\frac{1}{2}$ of the yellow shaded region, divide each of the thirds into two parts (halves). Figure 3.9 shows the rectangle divided into six parts. So, $\frac{1}{2}$ of the shaded third is $\frac{1}{6}$ of the rectangle, which is shaded blue.

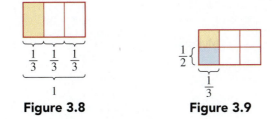

Figure 3.8   **Figure 3.9**

$$\frac{1}{2} \text{ of } \frac{1}{3} = \frac{1}{2} \cdot \frac{1}{3} = \frac{1}{6} = \frac{\text{number of parts shaded blue}}{\text{total number of parts}}$$

What is $\frac{1}{4}$ of $\frac{3}{4}$? $\left(\frac{1}{4} \cdot \frac{3}{4} = ? \right)$ In Figure 3.10 the rectangle has been divided into four parts, and $\frac{3}{4}$ is represented by the parts that are shaded yellow. To find $\frac{1}{4}$ of the yellow shaded regions, divide each of the fourths into four parts. The rectangle is now divided into 16 parts, and $\frac{1}{4}$ of each of the three original yellow regions is shaded blue. The blue shaded region represents $\frac{3}{16}$. See Figure 3.11.

$$\frac{1}{4} \text{ of } \frac{3}{4} = \frac{1}{4} \cdot \frac{3}{4} = \frac{3}{16}$$

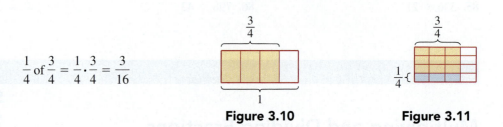

Figure 3.10   **Figure 3.11**

We have seen that $\frac{1}{2} \cdot \frac{1}{3} = \frac{1}{6}$ and that $\frac{1}{4} \cdot \frac{3}{4} = \frac{3}{16}$. The mathematical method is to multiply the numerators and multiply the denominators.

$$\frac{14}{45} \cdot \frac{18}{35} = \frac{252}{1575}$$

$$= \frac{252 \div 9}{1575 \div 9} \quad \text{Simplify by dividing both the numerator and the denominator by 9.}$$

$$= \frac{28}{175}$$

$$= \frac{28 \div 7}{175 \div 7} \quad \text{Simplify by dividing both the numerator and the denominator by 7.}$$

$$= \frac{4}{25} \quad \text{The fraction is completely simplified.}$$

Multiplying two or more fractions like those just given can often be done more quickly by simplifying before multiplying.

$$\frac{14}{45} \cdot \frac{18}{35} = \frac{\overset{2}{\cancel{14}}}{45} \cdot \frac{18}{\underset{5}{\cancel{35}}} \quad \text{Simplify by dividing 14 and 35 by 7. Write the quotients above (or below) the original factors.}$$

$$= \frac{\overset{2}{\cancel{14}}}{\underset{5}{\cancel{45}}} \cdot \frac{\overset{2}{\cancel{18}}}{\underset{5}{\cancel{35}}} \quad \text{Divide 18 and 45 by 9.}$$

$$= \frac{4}{25} \quad \text{Multiply.}$$

The next example shows all of the simplifying done in one step.

$$\frac{20}{36}\cdot\frac{18}{25}=\frac{\overset{\overset{1}{\cancel{5}}}{\cancel{20}}}{\underset{\underset{1}{9}}{\cancel{36}}}\cdot\frac{\overset{2}{\cancel{18}}}{\underset{5}{\cancel{25}}}$$

Divide 20 and 36 by 4, then divide 5 and 25 by 5.

Next divide 18 and 9 by 9.

$$=\frac{2}{5}$$

Then multiply.

If the numbers are large, find the prime factorization of each numerator and denominator. See Example E.

> ### To multiply fractions
>
> 1. Simplify.
> 2. Write the product of the numerators over the product of the denominators.

> ### CAUTION
>
> Simplifying before doing the operation works only for multiplication because it is based on multiplying by 1. It does **not** work for addition, subtraction, or division.

EXAMPLES A–G

DIRECTIONS: Multiply. Simplify completely.

STRATEGY: Simplify and then multiply.

A. Multiply and simplify: $\dfrac{2}{3}\cdot\dfrac{4}{7}\cdot\dfrac{1}{5}$

$$\frac{2}{3}\cdot\frac{4}{7}\cdot\frac{1}{5}=\frac{8}{105}$$

There are no common factors. Multiply the numerators and multiply the denominators.

B. Multiply and simplify: $\dfrac{7}{3}\cdot 2$

STRATEGY: First change 2 to an improper fraction.

$$\frac{7}{3}\cdot 2=\frac{7}{3}\cdot\frac{2}{1}$$

$2=\dfrac{2}{1}.$

$$=\frac{14}{3}\quad\text{or}\quad 4\frac{2}{3}$$

Write the product of the numerators over the product of the denominators.

C. Multiply and simplify: $\dfrac{5}{8}\cdot\dfrac{4}{7}$

$$\frac{5}{8}\cdot\frac{4}{7}=\frac{5}{\underset{2}{\cancel{8}}}\cdot\frac{\overset{1}{\cancel{4}}}{7}$$

Eliminate the common factor of 4 in 8 and 4.

$$=\frac{5}{14}$$

Multiply.

WARM-UP

D. Multiply and simplify:

$$\frac{8}{16} \cdot \frac{12}{9} \cdot \frac{15}{18}$$

D. Multiply and simplify: $\dfrac{7}{9} \cdot \dfrac{18}{5} \cdot \dfrac{10}{21}$

$$\frac{7}{9} \cdot \frac{18}{5} \cdot \frac{10}{21} = \frac{\overset{1}{\cancel{7}}}{\underset{1}{\cancel{9}}} \cdot \frac{\overset{2}{\cancel{18}}}{\underset{1}{\cancel{5}}} \cdot \frac{\overset{2}{\cancel{10}}}{\underset{3}{\cancel{21}}}$$
Eliminate the common factors of 9, 5, and 7.

$$= \frac{4}{3}$$
Multiply.

WARM-UP

E. Multiply and simplify:

$$\frac{60}{70} \cdot \frac{42}{66}$$

E. Multiply and simplify: $\dfrac{24}{28} \cdot \dfrac{35}{50}$

STRATEGY: Finding the prime factorization makes the common factors easier to detect.

$$\frac{24}{28} \cdot \frac{35}{50} = \frac{2 \cdot 2 \cdot 2 \cdot 3}{2 \cdot 2 \cdot 7} \cdot \frac{5 \cdot 7}{2 \cdot 5 \cdot 5}$$

$$= \frac{\cancel{2} \cdot \cancel{2} \cdot 2 \cdot 3}{\cancel{2} \cdot \cancel{2} \cdot \cancel{7}} \cdot \frac{\cancel{5} \cdot \cancel{7}}{2 \cdot \cancel{5} \cdot 5}$$
Eliminate the common factors.

$$= \frac{3}{5}$$
Multiply.

CALCULATOR EXAMPLE:

WARM-UP

F. Multiply and simplify:

$$\frac{18}{21} \cdot \frac{28}{30}$$

F. Multiply and simplify: $\dfrac{18}{80} \cdot \dfrac{60}{72}$
Use the fraction key. The calculator will automatically simplify the product.

$$\frac{18}{80} \cdot \frac{60}{72} = \frac{3}{16}$$

WARM-UP

G. During one year $\dfrac{7}{9}$ of all the taxes collected were sales taxes. Of the sales taxes collected, $\dfrac{5}{7}$ were paid by individuals. What fraction of all the taxes paid were sales taxes paid by individuals?

G. In one state $\dfrac{3}{8}$ of taxes collected were from income taxes. Of the income taxes collected, $\dfrac{1}{3}$ was paid by corporations. What fraction of all the taxes collected were income taxes paid by corporations?

$$\frac{\cancel{3}}{8} \cdot \frac{1}{\cancel{3}} = \frac{1}{8}$$

So $\dfrac{1}{8}$ of all the taxes collected were income taxes paid by corporations.

HOW & WHY

■ **OBJECTIVE 2** Find the reciprocal of a number.

Finding the reciprocal of a fraction is often called "inverting" the fraction. For instance, the reciprocal of $\dfrac{4}{7}$ is $\dfrac{7}{4}$. We check by showing that the product of the fraction and its reciprocal is 1:

$$\frac{4}{7} \cdot \frac{7}{4} = \frac{28}{28} = 1$$

> **To find the reciprocal of a fraction**
>
> Interchange the numerator and denominator.

> **CAUTION**
>
> The number zero (0) does not have a reciprocal.

EXAMPLES H–I

DIRECTIONS: Find the reciprocal.

STRATEGY: Interchange the numerator and the denominator; that is, "invert" the fraction.

H. Find the reciprocal of $\dfrac{6}{11}$.

$\dfrac{11}{6}$ Exchange the numerator and denominator.

CHECK: $\dfrac{6}{11} \cdot \dfrac{11}{6} = \dfrac{66}{66} = 1$

The reciprocal of $\dfrac{6}{11}$ is $\dfrac{11}{6}$, or $1\dfrac{5}{6}$.

I. Find the reciprocal of $1\dfrac{4}{9}$.

STRATEGY: First write $1\dfrac{4}{9}$ as an improper fraction.

$1\dfrac{4}{9} = \dfrac{13}{9}$

$\dfrac{9}{13}$ Invert the fraction.

CHECK: $\dfrac{13}{9} \cdot \dfrac{9}{13} = \dfrac{117}{117} = 1$

The reciprocal of $1\dfrac{4}{9}$ is $\dfrac{9}{13}$.

WARM-UP
H. Find the reciprocal of $\dfrac{14}{5}$.

WARM-UP
I. Find the reciprocal of $3\dfrac{4}{7}$.

HOW & WHY

■ **OBJECTIVE 3**   Divide fractions.

In Chapter 1, you learned that division is the inverse of multiplication; that is, the quotient of a division problem is the number that is multiplied times the divisor (second number) to get the dividend. For example, the quotient of $21 \div 3$ is 7. When 7 is multiplied by the divisor, 3, the product is 21, the dividend.

ANSWERS TO WARM-UPS H–I

H. $\dfrac{5}{14}$ **I.** $\dfrac{7}{25}$

Another way of thinking of division is to ask "How many groups of a given size are contained in a number?"

	Think	Answer
$21 \div 3$	How many 3s in 21?	7
$\dfrac{4}{5} \div \dfrac{1}{10}$	How many $\dfrac{1}{10}$s in $\dfrac{4}{5}$?	See Figure 3.12

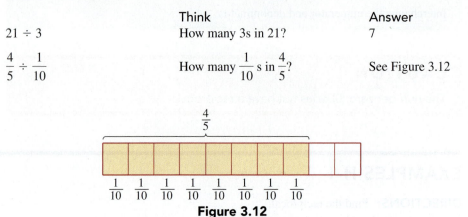

Figure 3.12

Figure 3.12 shows that there are eight $\dfrac{1}{10}$s in $\dfrac{4}{5}$. Therefore, we write $\dfrac{4}{5} \div \dfrac{1}{10} = 8$. Check by multiplying. Because $8 \cdot \dfrac{1}{10} = \dfrac{8}{10} = \dfrac{4}{5}$, we know that the quotient is correct.

The quotient can also be found by multiplying $\dfrac{4}{5}$ by the reciprocal of $\dfrac{1}{10}$. This is the customary shortcut.

$$\frac{4}{5} \div \frac{1}{10} = \frac{4}{5} \cdot \frac{10}{1} = \frac{40}{5} = 8$$

To divide fractions

Multiply the first fraction by the reciprocal of the divisor; that is, invert the divisor and multiply.

CAUTION

Do not simplify the fractions before changing the division to multiplication; that is, invert before simplifying.

EXAMPLES J–N

DIRECTIONS: Divide. Simplify completely.

STRATEGY: Multiply by the reciprocal of the divisor.

J. Divide: $\dfrac{9}{13} \div \dfrac{10}{13}$

$$\frac{9}{13} \div \frac{10}{13} = \frac{9}{\cancel{13}} \cdot \frac{\cancel{13}}{10} \qquad \text{\color{blue}Multiply by the reciprocal of the divisor.}$$

$$= \frac{9}{10}$$

K. Divide: $\dfrac{9}{4} \div \dfrac{3}{7}$

$$\dfrac{9}{4} \div \dfrac{3}{7} = \dfrac{9}{4} \cdot \dfrac{7}{\overset{3}{\underset{1}{\cancel{3}}}}$$ Invert the divisor and multiply.

$$= \dfrac{21}{4} \text{ or } 5\dfrac{1}{4}$$

L. Divide: $\dfrac{1}{12} \div \dfrac{3}{5}$

$$\dfrac{1}{12} \div \dfrac{3}{5} = \dfrac{1}{12} \cdot \dfrac{5}{3}$$ Invert the divisor and multiply.

$$= \dfrac{5}{36}$$

CALCULATOR EXAMPLE: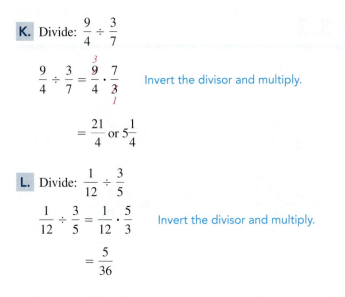

M. Divide and simplify: $\dfrac{21}{32} \div \dfrac{42}{80}$

STRATEGY: Use the fraction key. The calculator will automatically invert the divisor and simplify the quotient.

$$\dfrac{21}{32} \div \dfrac{42}{80} = \dfrac{5}{4}$$

N. The distance a nut moves one turn on a bolt is $\dfrac{3}{16}$ inch. How many turns will it take to move the nut $\dfrac{3}{4}$ inch?

STRATEGY: To find the number of turns required to move the nut $\dfrac{3}{4}$ inch, divide $\dfrac{3}{4}$ by the distance the nut moves in one turn.

$$\dfrac{3}{4} \div \dfrac{3}{16} = \dfrac{\overset{1}{\cancel{3}}}{\underset{1}{\cancel{4}}} \cdot \dfrac{\overset{4}{\cancel{16}}}{\underset{1}{\cancel{3}}} = 4$$ Invert the divisor and multiply.

It will take 4 turns to move the nut $\dfrac{3}{4}$ inch.

WARM-UP

K. Divide: $\dfrac{5}{3} \div \dfrac{10}{7}$

WARM-UP

L. Divide: $\dfrac{5}{11} \div \dfrac{2}{3}$

WARM-UP

M. Divide and simplify:

$$\dfrac{54}{70} \div \dfrac{36}{50}$$

WARM-UP

N. Suppose in Example N that the distance the nut moves on the bolt with one turn is $\dfrac{3}{32}$ inch. How many turns will it take to move the nut $\dfrac{3}{4}$ inch?

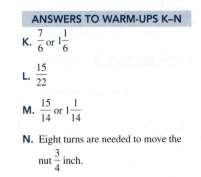

ANSWERS TO WARM-UPS K–N

K. $\dfrac{7}{6}$ or $1\dfrac{1}{6}$

L. $\dfrac{15}{22}$

M. $\dfrac{15}{14}$ or $1\dfrac{1}{14}$

N. Eight turns are needed to move the nut $\dfrac{3}{4}$ inch.

EXERCISES 3.3

■ **OBJECTIVE 1** Multiply fractions. (See page 187.)

A *Multiply. Simplify completely.*

1. $\dfrac{1}{5} \cdot \dfrac{2}{5}$

2. $\dfrac{5}{8} \cdot \dfrac{7}{9}$

3. $\dfrac{3}{2} \cdot \dfrac{5}{14}$

4. $\dfrac{3}{7} \cdot \dfrac{6}{5}$

5. $\dfrac{6}{11} \cdot \dfrac{5}{9}$

6. $\dfrac{4}{9} \cdot \dfrac{3}{8}$

7. $\dfrac{4}{9} \cdot \dfrac{9}{16}$

8. $\dfrac{5}{2} \cdot \dfrac{2}{15}$

9. $14 \cdot \dfrac{8}{35}$

10. $12 \cdot \dfrac{3}{14}$

B

11. $\dfrac{3}{16} \cdot \dfrac{4}{9} \cdot \dfrac{5}{8}$

12. $\dfrac{2}{15} \cdot \dfrac{5}{6} \cdot \dfrac{3}{4}$

13. $8 \cdot \dfrac{1}{5} \cdot \dfrac{15}{16}$

14. $\dfrac{16}{5} \cdot 7 \cdot \dfrac{3}{28}$

15. $\dfrac{11}{3} \cdot \dfrac{3}{4} \cdot \dfrac{4}{11}$

16. $\dfrac{21}{16} \cdot 2 \cdot \dfrac{8}{7}$

17. $\dfrac{32}{55} \cdot \dfrac{44}{51} \cdot \dfrac{34}{16}$

18. $\dfrac{32}{39} \cdot \dfrac{24}{96} \cdot \dfrac{52}{72}$

19. $\dfrac{7}{6} \cdot \dfrac{8}{23} \cdot \dfrac{5}{8} \cdot \dfrac{0}{9}$

20. $\dfrac{7}{3} \cdot \dfrac{5}{0} \cdot \dfrac{20}{21} \cdot \dfrac{8}{3}$

■ **OBJECTIVE 2** Find the reciprocal of a number. (See page 190.)

A *Find the reciprocal.*

21. $\dfrac{8}{13}$

22. $\dfrac{7}{5}$

23. 21

24. 0

25. $\dfrac{2}{0}$

B

26. $3\dfrac{5}{9}$

27. $5\dfrac{1}{12}$

28. 19

29. $\dfrac{1}{17}$

30. 25

■ **OBJECTIVE 3** Divide fractions. (See page 191.)

A *Divide. Simplify completely.*

31. $\dfrac{7}{20} \div \dfrac{14}{15}$

32. $\dfrac{8}{13} \div \dfrac{2}{13}$

33. $\dfrac{4}{7} \div \dfrac{3}{8}$

34. $\dfrac{5}{7} \div \dfrac{9}{8}$

35. $\dfrac{8}{25} \div \dfrac{16}{25}$

36. $\dfrac{5}{6} \div \dfrac{5}{3}$

37. $\dfrac{7}{44} \div \dfrac{14}{33}$

38. $\dfrac{5}{12} \div \dfrac{25}{6}$

39. $\dfrac{5}{8} \div \dfrac{15}{32}$

40. $\dfrac{3}{4} \div \dfrac{9}{16}$

B

41. $\dfrac{21}{40} \div \dfrac{9}{28}$

42. $\dfrac{10}{55} \div \dfrac{10}{11}$

43. $\dfrac{28}{30} \div \dfrac{14}{15}$

44. $\dfrac{9}{12} \div \dfrac{12}{7}$

45. $\dfrac{6}{5} \div \dfrac{9}{25}$

46. $\dfrac{9}{5} \div \dfrac{56}{60}$

47. $\dfrac{90}{55} \div \dfrac{9}{5}$

48. $\dfrac{28}{35} \div \dfrac{7}{5}$

49. $\dfrac{63}{80} \div \dfrac{9}{10}$

50. $\dfrac{45}{105} \div \dfrac{3}{7}$

C *Multiply. Simplify completely.*

51. $\dfrac{56}{72} \cdot \dfrac{35}{80} \cdot \dfrac{6}{14}$

52. $\dfrac{21}{14} \cdot \dfrac{15}{42} \cdot \dfrac{18}{15}$

53. $\dfrac{124}{160} \cdot \dfrac{40}{62} \cdot \dfrac{44}{11}$

54. $\dfrac{32}{675} \cdot \dfrac{250}{16} \cdot \dfrac{135}{50}$

Divide. Simplify completely.

55. $\dfrac{22}{140} \div \dfrac{55}{56}$

56. $\dfrac{8}{108} \div \dfrac{16}{81}$

57. $\dfrac{16}{81} \div \dfrac{8}{108}$

58. $\dfrac{72}{75} \div \dfrac{32}{50}$

Fill in the boxes with a single number so the statement is true. Explain your answer.

59. $\dfrac{3}{4} \cdot \dfrac{\square}{6} = \dfrac{7}{8}$

60. $\dfrac{4}{\square} \cdot \dfrac{11}{20} = \dfrac{11}{50}$

61. Find the error in the statement: $\dfrac{3}{8} \cdot \dfrac{4}{5} = \dfrac{15}{32}$. Correct the statement. Explain how you would avoid this error.

62. Find the error in the statement: $\dfrac{3}{8} \div \dfrac{4}{5} = \dfrac{32}{15}$. Correct the statement. Explain how you would avoid this error.

63. For families with four children, there are 16 possible combinations of gender and birth order. Half of these have a boy as firstborn. How many of the possible combinations have a boy as firstborn? How many have a girl as firstborn?

64. For families with four children, there are 16 possible combinations of gender and birth order. One-fourth of these have three girls and one boy. How many of the possible combinations have three girls?

65. For families with four children, there are 16 possible combinations of gender and birth order. One-fourth of the possible combinations have three girls and one boy, and one-fourth of these (3 girl–1 boy combinations) have a firstborn boy followed by three girls. How many of the possible combinations have a firstborn boy followed by three girls?

66. In the United States in 2008, there were 80,970 people on the waitlist for a kidney transplant. About one-fifth of those people received transplants. How many people received kidney transplants that year?

67. Proctor & Gamble sells Tide detergent in a 100-oz size. How many ounces of detergent are left when the container is $\frac{3}{5}$ full?

68. The National Zoo has T-shirts that regularly sell for $27. The shirts are marked $\frac{1}{3}$ off during a Labor Day sale. What is the sale price of a T-shirt?

69. Underinflation of car tires can waste up to $\frac{1}{20}$ of a car's fuel due to an increase in "rolling resistance." If Becky uses 820 gallons of gas in a year, how many gallons can she save with proper tire inflation?

70. Underinflation of truck tires can waste up to $\frac{1}{15}$ of a truck's fuel due an increase in "rolling resistance." If Melvin uses 500 gallons of gas while unaware that the tires are low, how many gallons might be wasted?

71. In the United States, 6 out of 100 people diagnosed with lung cancer survive 5 years. About 200,000 new cases of lung cancer are diagnosed every year. How many of these patients are expected to survive 5 years?

72. The website iheard.com lists 110 country radio stations. Of these, about 3 out of every 10 have the highest rating (5 stars). How many country radio stations on the iheard list have the highest rating?

73. According to statistics gathered by the United Nations, 73 of 100 people in Mexico have access to safe drinking water. The population of Mexico in 2008 was estimated at 109,955,000. How many Mexicans did not have access to safe drinking water that year?

STATE YOUR UNDERSTANDING

74. Explain how to find the product of $\frac{35}{24}$ and $\frac{40}{14}$.

75. Explain how to find the quotient of $\frac{35}{24}$ and $\frac{40}{14}$.

76. Evalynne's supervisor tells her that her salary is to be divided by one-half. Should she quit her job? Explain.

CHALLENGE

77. Simplify. $\left(\dfrac{81}{75} \cdot \dfrac{96}{99} \cdot \dfrac{55}{125} \right) \div \dfrac{128}{250}$

78. The In-n-Out Grocery has a standard workweek of 40 hours. Jan works $\dfrac{3}{4}$ of a standard week, Jose works $\dfrac{5}{8}$ of a standard week, Aria works $\dfrac{9}{8}$ of a standard week, and Bill works $\dfrac{6}{5}$ of a standard week. How many hours did each employee work? In-n-Out pays an average salary of $11 an hour. What is the total week's payroll for these four employees?

MAINTAIN YOUR SKILLS

Change each improper fraction to a mixed number.

79. $\dfrac{22}{5}$

80. $\dfrac{18}{7}$

81. $\dfrac{45}{13}$

82. $\dfrac{29}{15}$

Change each mixed number to an improper fraction.

83. $3\dfrac{3}{8}$

84. $3\dfrac{2}{3}$

85. $7\dfrac{4}{9}$

86. $2\dfrac{5}{8}$

87. Kevin needs a load of gravel for a drainage field. His truck can safely haul a load of $1\dfrac{3}{4}$ tons (3500 lb).

If the gravel is sold in 222-lb scoops, how many scoops can Kevin haul?

Multiplying and Dividing Mixed Numbers

SECTION 3.4

HOW & WHY

OBJECTIVES

1. Multiply mixed numbers.
2. Divide mixed numbers.

■ **OBJECTIVE 1** Multiply mixed numbers.

In Section 3.1, we changed mixed numbers to improper fractions. For instance,

$$4\dfrac{2}{9} = \dfrac{4(9) + 2}{9} = \dfrac{38}{9}$$

To multiply mixed numbers we change them to improper fractions and then multiply.

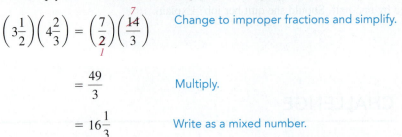

$$\left(3\frac{1}{2}\right)\left(4\frac{2}{3}\right) = \left(\frac{7}{\underset{1}{2}}\right)\left(\frac{\overset{7}{14}}{3}\right) \qquad \text{Change to improper fractions and simplify.}$$

$$= \frac{49}{3} \qquad \text{Multiply.}$$

$$= 16\frac{1}{3} \qquad \text{Write as a mixed number.}$$

Products may be left as improper fractions or as mixed numbers; either is acceptable. In this text, we write mixed numbers. In algebra, improper fractions are often preferred.

> **To multiply mixed numbers and/or whole numbers**
>
> 1. Change to improper fractions.
> 2. Simplify and multiply.

EXAMPLES A–D

DIRECTIONS: Multiply. Write as a mixed number.

STRATEGY: Change the mixed numbers and whole numbers to improper fractions. Simplify and multiply. Write the answer as a mixed number.

WARM-UP

A. Multiply: $\dfrac{3}{5}\left(1\dfrac{1}{2}\right)$

A. Multiply: $\dfrac{3}{7}\left(4\dfrac{2}{5}\right)$

$$\frac{3}{7}\left(4\frac{2}{5}\right) = \frac{3}{7} \cdot \frac{22}{5} \qquad 4\frac{2}{5} = \frac{22}{5}$$

$$= \frac{66}{35} \qquad \text{Multiply.}$$

$$= 1\frac{31}{35} \qquad \text{Write as a mixed number.}$$

WARM-UP

B. Multiply: $\left(2\dfrac{1}{3}\right)\left(4\dfrac{4}{5}\right)$

B. Multiply: $\left(2\dfrac{1}{3}\right)\left(2\dfrac{4}{7}\right)$

$$\left(2\frac{1}{3}\right)\left(2\frac{4}{7}\right) = \frac{7}{3} \cdot \frac{18}{7}$$

$$= \frac{\overset{1}{7}}{\underset{1}{3}} \cdot \frac{\overset{6}{18}}{\underset{1}{7}} \qquad \text{Simplify.}$$

$$= 6 \qquad \text{Multiply and write as a whole number.}$$

ANSWERS TO WARM-UPS A–B

A. $\dfrac{9}{10}$

B. $11\dfrac{1}{5}$

C. Multiply: $6\left(\dfrac{3}{4}\right)\left(2\dfrac{1}{3}\right)$

WARM-UP

C. Multiply: $4\left(3\dfrac{3}{5}\right)\left(\dfrac{5}{9}\right)$

CAUTION

$6\left(\dfrac{3}{4}\right)$ means $6\cdot\dfrac{3}{4}$, but $6\dfrac{3}{4}$ means $6+\dfrac{3}{4}$.

$$6\left(\dfrac{3}{4}\right)\left(2\dfrac{1}{3}\right) = \dfrac{\overset{3}{\cancel{6}}}{1}\cdot\dfrac{\overset{1}{\cancel{3}}}{\underset{2}{\cancel{4}}}\cdot\dfrac{7}{\underset{1}{\cancel{3}}}$$

$$= \dfrac{21}{2} = 10\dfrac{1}{2} \qquad \text{Multiply and write as a mixed number.}$$

CALCULATOR EXAMPLE:

D. Multiply: $3\dfrac{1}{9}\left(5\dfrac{1}{4}\right)$

$$3\dfrac{1}{9}\left(5\dfrac{1}{4}\right) = 16\dfrac{1}{3}$$

On a calculator with fraction keys, it is not necessary to change the mixed numbers to improper fractions. The calculator is programmed to operate with simple fractions or with mixed numbers.

WARM-UP

D. Multiply: $6\dfrac{2}{5}\left(4\dfrac{2}{3}\right)$

HOW & WHY

■ **OBJECTIVE 2** Divide mixed numbers.

Division of mixed numbers is also done by changing to improper fractions first.

$$\left(6\dfrac{1}{6}\right)\div\left(2\dfrac{1}{2}\right) = \left(\dfrac{37}{6}\right)\div\left(\dfrac{5}{2}\right) \qquad \text{Change to improper fractions.}$$

$$= \left(\dfrac{37}{\underset{3}{\cancel{6}}}\right)\left(\dfrac{\overset{1}{\cancel{2}}}{5}\right) \qquad \text{Multiply by the reciprocal of the divisor and simplify.}$$

$$= \dfrac{37}{15} \qquad \text{Multiply.}$$

$$= 2\dfrac{7}{15} \qquad \text{Write as a mixed number.}$$

To divide mixed numbers and/or whole numbers

1. Change to improper fractions.
2. Divide.

EXAMPLES E–I

DIRECTIONS: Divide. Write as a mixed number.

STRATEGY: Change the mixed numbers and whole numbers to improper fractions. Divide and simplify completely. Write the answer as a mixed number.

WARM-UP
E. Divide: $3\dfrac{1}{6} \div 2\dfrac{5}{12}$

E. Divide: $3\dfrac{1}{3} \div 2\dfrac{4}{9}$

$$3\dfrac{1}{3} \div 2\dfrac{4}{9} = \dfrac{10}{3} \div \dfrac{22}{9} \qquad 3\dfrac{1}{3} = \dfrac{10}{3} \text{ and } 2\dfrac{4}{9} = \dfrac{22}{9}$$

$$= \dfrac{\overset{5}{\cancel{10}}}{\underset{1}{\cancel{3}}} \cdot \dfrac{\overset{3}{\cancel{9}}}{\underset{11}{\cancel{22}}} \qquad \text{Invert the divisor and multiply.}$$

$$= \dfrac{15}{11} = 1\dfrac{4}{11}$$

WARM-UP
F. Divide: $14\dfrac{2}{3} \div 15\dfrac{5}{7}$

F. Divide: $10\dfrac{5}{16} \div 16\dfrac{1}{2}$

$$10\dfrac{5}{16} \div 16\dfrac{1}{2} = \dfrac{165}{16} \div \dfrac{33}{2}$$

$$= \dfrac{\overset{5}{\cancel{165}}}{\underset{8}{\cancel{16}}} \cdot \dfrac{\overset{1}{\cancel{2}}}{\underset{1}{\cancel{33}}} \qquad \text{Invert the divisor and multiply.}$$

$$= \dfrac{5}{8}$$

WARM-UP
G. Divide: $6\dfrac{3}{5} \div 6$

G. Divide: $8 \div 1\dfrac{1}{5}$

$$8 \div 1\dfrac{1}{5} = \dfrac{8}{1} \div \dfrac{6}{5}$$

$$= \dfrac{\overset{4}{\cancel{8}}}{1} \cdot \dfrac{5}{\underset{3}{\cancel{6}}} \qquad \text{Invert the divisor and multiply.}$$

$$= \dfrac{20}{3} = 6\dfrac{2}{3}$$

CALCULATOR EXAMPLE:

WARM-UP
H. Divide: $12\dfrac{5}{6} \div 10\dfrac{1}{12}$

H. Divide: $25\dfrac{1}{4} \div 30\dfrac{3}{4}$

$$25\dfrac{1}{4} \div 30\dfrac{3}{4} = \dfrac{101}{123}$$

On a calculator with fraction keys, it is not necessary to change the mixed numbers to improper fractions. The calculator is programmed to operate with simple fractions or with mixed numbers.

ANSWERS TO WARM-UPS E–H

E. $1\dfrac{9}{29}$ **F.** $\dfrac{14}{15}$

G. $1\dfrac{1}{10}$ **H.** $1\dfrac{3}{11}$

I. The Portland Iron Works produces steel beams that are $12\frac{3}{10}$ in. thick. What is the height in feet of a stack of 15 beams?

$$15\left(12\frac{3}{10}\right) = \frac{15}{1} \cdot \frac{123}{10}$$

$$= \frac{\overset{3}{\cancel{15}}}{1} \cdot \frac{123}{\underset{2}{\cancel{10}}}$$

$$= \frac{369}{2} = 184\frac{1}{2}$$

The height of the stack is $184\frac{1}{2}$ in. To find the height in feet, divide by 12 since there are 12 in. in 1 ft.

$$184\frac{1}{2} \div 12 = \frac{369}{2} \div \frac{12}{1}$$

$$= \frac{\overset{123}{\cancel{369}}}{2} \cdot \frac{1}{\underset{4}{\cancel{12}}}$$

$$= \frac{123}{8} = 15\frac{3}{8}$$

The stack of steel beams is $15\frac{3}{8}$ ft high.

WARM-UP

I. The Portland Iron Works producers another steel beam that is $18\frac{3}{4}$ in. thick. What is the height in feet of a stack of 14 beans?

ANSWER TO WARM-UP I

I. The height is $21\frac{7}{8}$ ft.

EXERCISES 3.4

■ OBJECTIVE 1 Multiply mixed numbers. (See page 197.)

A *Multiply. Simplify completely and write as a mixed number if possible.*

1. $\left(\frac{2}{5}\right)\left(1\frac{2}{5}\right)$

2. $\left(\frac{2}{7}\right)\left(2\frac{5}{7}\right)$

3. $\left(2\frac{3}{8}\right)\left(3\frac{1}{4}\right)$

4. $\left(2\frac{3}{5}\right)\left(1\frac{1}{4}\right)$

5. $2\left(3\frac{2}{3}\right)$

6. $3\left(2\frac{1}{9}\right)$

7. $\left(3\frac{3}{5}\right)\left(1\frac{1}{4}\right)(5)$

8. $\left(\frac{3}{5}\right)\left(2\frac{1}{3}\right)(3)$

9. $\left(7\frac{1}{2}\right)\left(3\frac{1}{5}\right)$

10. $\left(4\frac{1}{6}\right)\left(3\frac{3}{5}\right)$

11. $\left(4\frac{3}{8}\right)\left(\frac{16}{25}\right)$

12. $\left(4\frac{4}{9}\right)\left(\frac{12}{25}\right)$

B

13. $5\left(3\dfrac{3}{4}\right)$

14. $7\left(5\dfrac{1}{8}\right)$

15. $\left(7\dfrac{3}{4}\right)\left(\dfrac{2}{3}\right)(0)$

16. $\left(3\dfrac{5}{9}\right)\left(6\dfrac{5}{7}\right)(0)$

17. $2\left(4\dfrac{3}{8}\right)\left(3\dfrac{1}{5}\right)$

18. $3\left(2\dfrac{1}{4}\right)\left(2\dfrac{1}{3}\right)$

19. $\left(2\dfrac{2}{3}\right)\left(\dfrac{11}{12}\right)\left(6\dfrac{3}{4}\right)$

20. $\left(3\dfrac{3}{7}\right)\left(\dfrac{7}{9}\right)\left(3\dfrac{1}{2}\right)$

21. $\left(1\dfrac{1}{5}\right)\left(4\dfrac{1}{3}\right)\left(6\dfrac{1}{6}\right)$

22. $\left(12\dfrac{1}{4}\right)\left(1\dfrac{1}{7}\right)\left(2\dfrac{1}{3}\right)$

23. $(14)\left(6\dfrac{1}{2}\right)\left(1\dfrac{2}{13}\right)$

24. $14\left(3\dfrac{11}{15}\right)\left(2\dfrac{1}{7}\right)$

■ **OBJECTIVE 2** Divide mixed numbers. (See page 199.)

A   *Divide. Simplify completely and write as a mixed number if possible.*

25. $6 \div 2\dfrac{1}{3}$

26. $5 \div 1\dfrac{1}{5}$

27. $3\dfrac{5}{9} \div 2\dfrac{1}{18}$

28. $4\dfrac{2}{3} \div 8\dfrac{2}{3}$

29. $5\dfrac{2}{3} \div 1\dfrac{1}{9}$

30. $4\dfrac{1}{4} \div 3\dfrac{1}{16}$

31. $2\dfrac{1}{4} \div 1\dfrac{5}{8}$

32. $3\dfrac{5}{9} \div 4\dfrac{2}{3}$

33. $3\dfrac{1}{3} \div 2\dfrac{1}{2}$

34. $6\dfrac{1}{2} \div 2\dfrac{1}{5}$

35. $4\dfrac{4}{15} \div 6\dfrac{2}{3}$

36. $6\dfrac{1}{4} \div 7\dfrac{1}{2}$

B

37. $\dfrac{5}{6} \div 4\dfrac{1}{3}$

38. $\dfrac{7}{3} \div 1\dfrac{4}{9}$

39. $8\dfrac{2}{3} \div 6$

40. $5\dfrac{5}{8} \div 6$

41. $4\dfrac{5}{6} \div \dfrac{1}{3}$

42. $7\dfrac{1}{3} \div 1\dfrac{1}{9}$

43. $8\dfrac{1}{10} \div \dfrac{3}{5}$

44. $7\dfrac{1}{6} \div \dfrac{7}{3}$

45. $23\dfrac{1}{4} \div 1\dfrac{1}{3}$

46. $26\dfrac{7}{8} \div 3\dfrac{3}{4}$

47. $8\dfrac{3}{4} \div 3\dfrac{4}{7}$

48. $22\dfrac{2}{3} \div 6\dfrac{6}{7}$

C   *Multiply. Simplify completely and write as a mixed number if possible.*

49. $12\left(3\dfrac{3}{8}\right)\left(2\dfrac{4}{9}\right)\left(3\dfrac{2}{11}\right)$

50. $3\left(12\dfrac{1}{4}\right)\left(1\dfrac{1}{7}\right)\left(2\dfrac{1}{3}\right)$

Divide. Simplify completely and write as a mixed number if possible.

51. $\left(1\dfrac{2}{3}\right) \div \left(3\dfrac{3}{4}\right) \div \left(1\dfrac{3}{7}\right)$

52. $\left(2\dfrac{1}{2}\right) \div \left(4\dfrac{1}{3}\right) \div \left(1\dfrac{1}{4}\right)$

53. Find the error in the statement: $1\dfrac{2}{3} \cdot 1\dfrac{1}{2} = 1\dfrac{1}{3}$.
Correct the statement. Explain how you would avoid this error.

54. Find the error in the statement: $6\dfrac{2}{9} \div 2\dfrac{2}{3} = 3\dfrac{1}{3}$.
Correct the statement. Explain how you would avoid this error.

55. A 6-ft-by-8-ft readymade storage shed has interior dimensions of $7\frac{2}{3}$ ft wide by $5\frac{1}{2}$ ft deep. How many square feet are in the interior of the shed?

56. Jose plans to mulch his garden plot that measures $8\frac{3}{4}$ ft by $6\frac{1}{2}$ ft. A bag of mulch will cover 25 square feet of ground. Can Jose cover his garden using two bags of mulch?

57. A jewelry store advertises two diamond rings. One ring is $\frac{1}{2}$ carat total weight for $700. Another ring is $1\frac{1}{2}$ carats total weight for $3000. What is the price per carat of each of the two rings?

58. Joe has an 8-ft board that he wants to cut into $20\frac{1}{2}$-in. lengths to make shelves. How many shelves can Joe get from the board?

59. The iron content in a water sample at Lake Hieda is eight parts per million. The iron content in Swan Lake is $2\frac{3}{4}$ times greater than the content in Lake Hieda. What is the iron content in Swan Lake in parts per million?

60. The water pressure during a bad neighborhood grass fire is reduced to $\frac{5}{9}$ its original pressure at the hydrant. What is the reduced pressure if the original pressure was $70\frac{1}{5}$ pounds per square inch?

© iStockphoto.com/George Peters

61. Krogers advertised extra-lean ground beef on sale for $7 a pack. If each pack weighs $3\frac{1}{2}$ lb, what is the price per pound of the ground beef?

62. The American Heart Association estimates that a little more than $\frac{1}{4}$ of all Americans have some form of hypertension (high blood pressure), which puts them at greater risk for heart attacks and strokes. The population of Alabama was estimated to be 4,988,000 in 2010. About how many Alabamians would be expected to have hypertension in 2010?

63. The amount of CO_2 a car emits is directly related to the amount of gas it uses. Cars give off 20 lb of CO_2 for every gallon of gas used. Cars in 2005 were averaging 21 mpg and would emit 2000 lb of CO_2 in 2100 miles. In 2009, President Obama ordered that all cars average 35 mpg by 2016. A car averaging 35 mpg will emit $\frac{2}{5}$ less CO_2 in the same distance. How many fewer pounds of CO_2 does the more-efficient car emit in the 2100 miles?

64. Nutritionists recommend that not more than $\frac{3}{10}$ of your daily intake in calories should come from fat. If each gram of fat is 10 calories, what is the recommended upper limit on fat for a diet of 2400 calories?

Exercises 65–67 relate to the chapter application.

65. Shane read in his *How to Build Decks* book that stairs with 10-in. treads (widths) are easy to build using either 2 × 4s or 2 × 6s. Either size will make treads with a slight overhang at the front, which is recommended. If Shane wants the actual tread to be $10\frac{1}{2}$ in., how many 2 × 4s per tread will he need? See Table 3.1 on page 167 for the actual size of a 2 × 4.

66. If Shane wants the actual tread to be 11 in., how many 2 × 6s per tread will he need?

67. Shane will be using 2 × 4 deck boards. How many will he need for a 272-in. long deck?

STATE YOUR UNDERSTANDING

68. Explain how to simplify $5\frac{1}{4} \div 1\frac{7}{8}$.

69. When a number is multiplied by $1\frac{1}{2}$, the result is larger than the original number. But when you divide by $1\frac{1}{2}$ the result is smaller. Explain why.

70. Why is it helpful to change mixed numbers to improper fractions before multiplying or dividing?

CHALLENGE

71. Multiply and simplify completely:

$$\left(\frac{2}{15}\right)\left(1\frac{6}{7}\right)\left(2\frac{2}{49}\right)\left(16\frac{2}{5}\right)\left(8\frac{3}{4}\right)\left(5\frac{1}{4}\right)\left(3\frac{3}{13}\right)$$

72. The Celtic Candy Company has two packs of mints that they sell in discount stores. One pack contains $1\frac{1}{4}$ lb of mints and the other contains $3\frac{1}{3}$ lb of mints.

If the smaller pack sells for $2 and the larger pack for $5, which size should they use to get the most income from 3000 lb of mints? How much more is the income?

MAINTAIN YOUR SKILLS

Change to a mixed number.

73. $\dfrac{77}{9}$

74. $\dfrac{553}{15}$

Change to an improper fraction.

75. $3\dfrac{6}{7}$

76. $4\dfrac{11}{12}$

77. $14\dfrac{7}{9}$

Multiply.

78. $7200\left(\dfrac{1}{60}\right)\left(\dfrac{1}{60}\right)$

79. $60\left(\dfrac{1}{40}\right)\left(\dfrac{3}{20}\right)\left(\dfrac{4800}{1}\right)$

80. $25\left(\dfrac{1}{12}\right)\left(\dfrac{1}{12}\right)\left(\dfrac{15}{1}\right)$

81. To be eligible for a drawing at the Mideast Film Festival, your ticket stub number must be a multiple of 6. If Jean's ticket number is 456,726, is she eligible for the drawing?

82. Sales for the Fireyear Company totaled $954,000 last year. During the first 6 months of last year, monthly sales were $72,400, $68,200, $85,000, $89,500, $92,700, and $87,200. What were the average monthly sales for the rest of the year?

HOW & WHY

OBJECTIVE

Solve an equation of the form $\dfrac{ax}{b} = \dfrac{c}{d}$, where a, b, c, and d are whole numbers.

We have previously solved equations in which variables (letters) are either multiplied or divided by whole numbers. We performed the inverse operations to solve for the variable. To eliminate multiplication, we divide by the number being multiplied. To eliminate division, we multiply by the number that is the divisor. Now we solve equations in which variables are multiplied by fractions. Recall from Chapter 1 that if a number is multiplied by a variable, there is usually no multiplication sign between them. For example, $2x$ is understood to mean 2 times x, and $\dfrac{2}{3}x$ means $\dfrac{2}{3}$ times x. However, we usually do not write $\dfrac{2}{3}x$. Instead, we write this as $\dfrac{2x}{3}$. We can do this because

$$\frac{2}{3}x = \frac{2}{3} \cdot x = \frac{2}{3} \cdot \frac{x}{1} = \frac{2x}{3}$$

While we usually write $\dfrac{2x}{3}$ instead of $\dfrac{2}{3}x$, for convenience we may use either of these forms. Recall that $\dfrac{2x}{3}$ means the *product* of 2 and x *divided* by 3.

> **To solve an equation of the form** $\dfrac{ax}{b} = \dfrac{c}{d}$
>
> 1. Multiply both sides of the equation by b to eliminate the division on the left side.
> 2. Divide both sides by a to isolate the variable.

EXAMPLES A–C

DIRECTIONS: Solve

STRATEGY: Multiply both sides by the denominator of the fraction containing the variable. Solve as before.

WARM-UP

A. Solve: $\dfrac{6x}{7} = 12$

A. Solve: $\dfrac{3x}{4} = 2$

$4\left(\dfrac{3x}{4}\right) = 4(2)$ — To eliminate the division, multiply both sides by 4.

$3x = 8$ — Simplify.

$\dfrac{3x}{3} = \dfrac{8}{3}$ — To eliminate the multiplication, divide both sides by 3.

$x = \dfrac{8}{3}$

CHECK: $\dfrac{3}{4} \cdot \dfrac{8}{3} = 2$ — Substitute $\dfrac{8}{3}$ for x in the original equation. Recall that $\dfrac{3x}{4} = \dfrac{3}{4}x$.

$2 = 2$

The solution is $x = \dfrac{8}{3}$, or $x = 2\dfrac{2}{3}$.

ANSWER TO WARM-UP A

A. $x = 14$

B. Solve: $\dfrac{2}{3} = \dfrac{4x}{5}$

$5\left(\dfrac{2}{3}\right) = 5\left(\dfrac{4x}{5}\right)$ To eliminate the division by 5, multiply both sides by 5.

$\dfrac{10}{3} = 4x$

$\dfrac{10}{3(4)} = \dfrac{4x}{4}$ To eliminate the multiplication by 4, divide both sides by 4.

$\dfrac{10}{12} = x$

The solution is $x = \dfrac{5}{6}$.   The check is left for the student.

C. Commuting to work accounts for one-third of the mileage put on a private car. If Nancy averages 510 miles per month in commuting to work, how many miles does she put on her car in a month?

STRATEGY: First write the English statement of the equation.

One-third times the total number of miles = miles of commute

$\dfrac{1}{3}x = 510$ Let x represent the number of total miles. Miles of commute is 510.

$3\left(\dfrac{1}{3}x\right) = 3(510)$ Multiply both sides by 3 to eliminate the division.

$x = 1530$

CHECK: Is one-third of 1530 equal to 510?

$\dfrac{1}{3}(1530) = 510$ Yes

Nancy puts 1530 miles on her car per month.

WARM-UP

B. Solve: $\dfrac{5x}{8} = \dfrac{3}{2}$

WARM-UP

C. It requires about two-fifths the amount of energy to make "new" paper from recycled paper as from trees. If the amount of energy needed to make a given amount of paper from recycled paper is equivalent to 1500 BTUs, how much energy is needed to make the same amount from trees?

ANSWERS TO WARM-UPS B–C

B. $x = \dfrac{12}{5}$ or $x = 2\dfrac{2}{5}$

C. It would take 3750 BTUs.

EXERCISES

Solve.

1. $\dfrac{2x}{3} = \dfrac{1}{2}$

2. $\dfrac{2x}{3} = \dfrac{2}{5}$

3. $\dfrac{3y}{5} = \dfrac{2}{3}$

4. $\dfrac{3y}{4} = \dfrac{5}{8}$

5. $\dfrac{4z}{5} = \dfrac{1}{4}$

6. $\dfrac{5z}{4} = \dfrac{5}{6}$

7. $\dfrac{17}{9} = \dfrac{8x}{9}$

8. $\dfrac{29}{10} = \dfrac{9x}{5}$

9. $\dfrac{7a}{4} = \dfrac{5}{2}$

10. $\dfrac{15b}{4} = \dfrac{13}{5}$

11. $\dfrac{47}{8} = \dfrac{47b}{12}$

12. $\dfrac{13}{3} = \dfrac{52w}{9}$

13. $\dfrac{35z}{6} = \dfrac{35}{12}$

14. $\dfrac{5b}{18} = \dfrac{2}{9}$

15. $\dfrac{15a}{2} = \dfrac{11}{4}$

16. $\dfrac{119x}{12} = \dfrac{119}{8}$

17. Vince walks $\dfrac{2}{3}$ of the distance from his home to school. If he walks $\dfrac{1}{2}$ mi, what is the distance from his home to school?

18. Mona cut a board into seven pieces of equal length. If each piece is $1\dfrac{4}{7}$ ft long, what was the length of the board?

19. In a southern city, $3\dfrac{1}{2}$ times more pounds of glass are recycled than tin. How many pounds of tin are recycled when 630 lb of glass are recycled?

20. Washing machines use about $\dfrac{7}{50}$ of all the water consumed in the home. If Matthew uses 280 gallons of water per month to operate his washing machine, how many gallons of water does he use in a month?

Building Fractions; Listing in Order; Inequalities

OBJECTIVES

1. Rename fractions by multiplying by 1 in the form $\frac{a}{a}$.
2. Build a fraction by finding the missing numerator.
3. List a group of fractions from smallest to largest.

VOCABULARY

Recall that **equivalent fractions** are fractions that are different names for the same number. For instance, $\frac{4}{8}$ and $\frac{11}{22}$ are equivalent because both represent one-half $\left(\frac{1}{2}\right)$ of a unit. Two or more fractions have a **common denominator** when they have the same denominator.

HOW & WHY

■ **OBJECTIVE 1** Rename fractions by multiplying by 1 in the form $\frac{a}{a}$.

The process of renaming fractions is often referred to as "building fractions." Building a fraction means renaming the fraction by multiplying both numerator and denominator by a common factor. This process is often necessary when we add and subtract fractions. We "build fractions" to a common denominator so they can be compared, added, or subtracted. Building fractions is the opposite of "simplifying" fractions.

Simplifying a Fraction	Building a Fraction
$\frac{8}{10} = \frac{4 \cdot 2}{5 \cdot 2} = \frac{4}{5}$	$\frac{4}{5} = \frac{4 \cdot 2}{5 \cdot 2} = \frac{8}{10}$

Visually, in Figure 3.13 we have

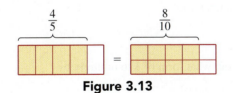

$$\frac{4}{5} \qquad \frac{8}{10}$$

Figure 3.13

Table 3.3 shows five fractions built to equivalent fractions.

TABLE 3.3 Equivalent Fractions

	Multiply numerator and denominator by					
	2	3	4	6	10	15
$\frac{2}{5}$ =	$\frac{4}{10}$ =	$\frac{6}{15}$ =	$\frac{8}{20}$ =	$\frac{12}{30}$ =	$\frac{20}{50}$ =	$\frac{30}{75}$
$\frac{1}{2}$ =	$\frac{2}{4}$ =	$\frac{3}{6}$ =	$\frac{4}{8}$ =	$\frac{6}{12}$ =	$\frac{10}{20}$ =	$\frac{15}{30}$
$\frac{5}{9}$ =	$\frac{10}{18}$ =	$\frac{15}{27}$ =	$\frac{20}{36}$ =	$\frac{30}{54}$ =	$\frac{50}{90}$ =	$\frac{75}{135}$
$\frac{7}{8}$ =	$\frac{14}{16}$ =	$\frac{21}{24}$ =	$\frac{28}{32}$ =	$\frac{42}{48}$ =	$\frac{70}{80}$ =	$\frac{105}{120}$
$\frac{7}{4}$ =	$\frac{14}{8}$ =	$\frac{21}{12}$ =	$\frac{28}{16}$ =	$\frac{42}{24}$ =	$\frac{70}{40}$ =	$\frac{105}{60}$

> **To rename a fraction**
>
> Multiply both the numerator and the denominator of the fraction by a common factor; that is, multiply the fraction by 1 in the form $\dfrac{a}{a}$, $a \neq 1$ and $a \neq 0$.

EXAMPLES A–B

DIRECTIONS: Rename the fraction.

STRATEGY: Multiply the fraction by 1 in the form $\dfrac{a}{a}$.

WARM-UP

A. Rename $\dfrac{5}{12}$, using $\dfrac{8}{8}$ for 1.

A. Rename $\dfrac{7}{10}$, using $\dfrac{5}{5}$ for 1.

$$\frac{7}{10} \cdot \frac{5}{5} = \frac{35}{50}$$

The new fraction, $\dfrac{35}{50}$, is equivalent to $\dfrac{7}{10}$.

WARM-UP

B. Write three fractions equivalent to $\dfrac{7}{8}$, using $\dfrac{4}{4}, \dfrac{7}{7}$, and $\dfrac{10}{10}$.

B. Write three fractions equivalent to $\dfrac{4}{11}$, using $\dfrac{3}{3}, \dfrac{6}{6}$, and $\dfrac{8}{8}$.

$$\frac{4}{11} \cdot \frac{3}{3} = \frac{12}{33}$$

$$\frac{4}{11} \cdot \frac{6}{6} = \frac{24}{66}$$

$$\frac{4}{11} \cdot \frac{8}{8} = \frac{32}{88}$$

HOW & WHY

■ OBJECTIVE 2 Build a fraction by finding the missing numerator.

To find the missing numerator in

$$\frac{5}{7} = \frac{?}{42}$$

divide 42 by 7 to find out what form of 1 to multiply by.

$42 \div 7 = 6.$

The correct factor (multiplier) is $\dfrac{6}{6}$. So

$$\frac{5}{7} = \frac{5}{7} \cdot \frac{6}{6} = \frac{30}{42}$$

The shortcut is to write 42, the target denominator. Then multiply the original numerator, 5, by 6, the quotient of the target denominator and the original denominator.

$$\frac{5}{7} = \frac{5 \cdot 6}{42} = \frac{30}{42}$$

The fractions $\dfrac{5}{7}$ and $\dfrac{30}{42}$ are equivalent. Either fraction can be used in place of the other.

> **To find the missing numerator when building fractions**
>
> 1. Divide the target denominator by the original denominator.
> 2. Multiply this quotient by the original numerator.

EXAMPLES C–E

DIRECTIONS: Find the missing numerator.

STRATEGY: Multiply the given numerator by the same factor used in the denominator.

C. Find the missing numerator: $\dfrac{3}{5} = \dfrac{?}{80}$

$80 \div 5 = 16$ Divide the denominators.

$\dfrac{3}{5} = \dfrac{3(16)}{80} = \dfrac{48}{80}$ Multiply the quotient, 16, by the given numerator.

D. Find the missing numerator: $\dfrac{5}{12} = \dfrac{?}{60}$

$60 \div 12 = 5$ Divide the denominators.

$\dfrac{5}{12} = \dfrac{5(5)}{60} = \dfrac{25}{60}$ Multiply the quotient, 5, by the given numerator.

E. Find the missing numerator: $\dfrac{8}{5} = \dfrac{?}{140}$

$140 \div 5 = 28$

$\dfrac{8}{5} = \dfrac{8(28)}{140} = \dfrac{224}{140}$

WARM-UP

C. Find the missing numerator:
$\dfrac{3}{4} = \dfrac{?}{80}$

WARM-UP

D. Find the missing numerator:
$\dfrac{7}{13} = \dfrac{?}{65}$

WARM-UP

E. Find the missing numerator:
$\dfrac{19}{15} = \dfrac{?}{135}$

HOW & WHY

■ **OBJECTIVE 3** List a group of fractions from smallest to largest.

If two fractions have the same denominator, the one with the smaller numerator has the smaller value. Figure 3.14 shows that $\dfrac{2}{5}$ is smaller than $\dfrac{4}{5}$; that is, $\dfrac{2}{5} < \dfrac{4}{5}$.

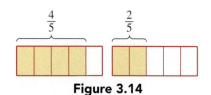

Figure 3.14

$\dfrac{2}{5} < \dfrac{4}{5}$ means "two-fifths is less than four-fifths."

$\dfrac{4}{5} > \dfrac{2}{5}$ means "four-fifths is greater than two-fifths."

If fractions to be compared do not have a common denominator, then one or more can be renamed so that all have a common denominator. The preferred common denominator is the least common multiple (LCM) of all the denominators.

To list $\dfrac{3}{8}, \dfrac{5}{16}, \dfrac{1}{2}$, and $\dfrac{9}{16}$ from smallest to largest, we write each with a common denominator. Then we compare the numerators. The LCM of the denominators is 16. We build each fraction so that is has a denominator of 16.

$\dfrac{3}{8} = \dfrac{6}{16}$ $\dfrac{5}{16} = \dfrac{5}{16}$ $\dfrac{1}{2} = \dfrac{8}{16}$ $\dfrac{9}{16} = \dfrac{9}{16}$ Each fraction now has a denominator of 16.

We arrange the fractions with denominator 16 in order from smallest to largest according the values of the numerators.

ANSWERS TO WARM-UPS C–E

C. 60   **D.** 35   **E.** 171

$$\frac{5}{16} < \frac{6}{16} < \frac{8}{16} < \frac{9}{16}$$ The fractions are listed in order from the smallest to the largest with common denominator 16.

Next, replace each fraction by the original, so

$$\frac{5}{16} < \frac{3}{8} < \frac{1}{2} < \frac{9}{16}$$ The original fractions are listed in order from smallest to largest.

> **To list fractions from smallest to largest**
> 1. Build the fractions so that they have a common denominator. Use the LCM of the denominators.
> 2. List the fractions with common denominators so the numerators range from smallest to largest.
> 3. Simplify.

EXAMPLES F–K

DIRECTIONS: Tell which fraction is larger.

STRATEGY: Write the fractions with a common denominator. The fraction with the larger numerator is the larger.

WARM-UP
F. Which is larger $\frac{5}{12}$ or $\frac{7}{16}$?

F. Which is larger $\frac{5}{9}$ or $\frac{4}{7}$?

STRATEGY: The LCM of 9 and 7 is 63. Build each fraction so it has 63 for a denominator.

$$\frac{5}{9} = \frac{35}{63} \quad \text{and} \quad \frac{4}{7} = \frac{36}{63}$$

$$\frac{36}{63} > \frac{35}{63}$$ Compare using the numerators, 36 > 35.

$$\frac{4}{7} > \frac{5}{9}$$

So $\frac{4}{7}$ is larger.

DIRECTIONS: List the group of fractions from smallest to largest.

STRATEGY: Build each of the fractions to a common denominator. List the fractions from smallest to largest by the value of the numerator. Simplify.

WARM-UP
G. List from smallest to largest:
$\frac{2}{3}, \frac{5}{7}$, and $\frac{9}{14}$

G. List from smallest to largest: $\frac{3}{8}, \frac{7}{20}$, and $\frac{2}{5}$

$$\frac{3}{8} = \frac{15}{40} \qquad \frac{7}{20} = \frac{14}{40} \qquad \frac{2}{5} = \frac{16}{40}$$ The LCM of 8, 20, and 5 is 40. Build the fractions to the denominator 40.

$$\frac{14}{40}, \frac{15}{40}, \frac{16}{40}$$ List the fractions in the order of the numerators: 14 <15 <16

The list is $\frac{7}{20}, \frac{3}{8}$, and $\frac{2}{5}$. Simplify.

ANSWERS TO WARM-UPS F–G

F. $\frac{7}{16}$ **G.** $\frac{9}{14}, \frac{2}{3}, \frac{5}{7}$

H. List from smallest to largest: $2\frac{3}{8}$, $2\frac{2}{5}$, and $2\frac{3}{10}$.

STRATEGY: Each mixed number has the same whole-number part. Compare the fraction parts.

$2\frac{3}{8} = 2\frac{15}{40}$ $2\frac{2}{5} = 2\frac{16}{40}$ $2\frac{3}{10} = 2\frac{12}{40}$ The LCM of 8, 5, and 10 is 40.

$2\frac{12}{40}, 2\frac{15}{40}, 2\frac{16}{40}$ List the numbers in the order of the numerators from smallest to largest.

The list is $2\frac{3}{10}$, $2\frac{3}{8}$, and $2\frac{2}{5}$.   Simplify.

I. The Acme Hardware Store sells bolts with diameters of $\frac{5}{16}$, $\frac{3}{8}$, $\frac{1}{2}$, $\frac{5}{8}$, $\frac{1}{4}$, and $\frac{7}{16}$ in. List the diameters from smallest to largest.

$\frac{5}{16} = \frac{5}{16}$ $\frac{3}{8} = \frac{6}{16}$ $\frac{1}{2} = \frac{8}{16}$ Write each diameter using the common denominator 16.

$\frac{5}{8} = \frac{10}{16}$ $\frac{1}{4} = \frac{4}{16}$ $\frac{7}{16} = \frac{7}{16}$

$\frac{4}{16}, \frac{5}{16}, \frac{6}{16}, \frac{7}{16}, \frac{8}{16}, \frac{10}{16}$ List the diameters in order of the numerators from smallest to largest.

From smallest to largest, the diameters are

$\frac{1}{4}, \frac{5}{16}, \frac{3}{8}, \frac{7}{16}, \frac{1}{2}$, and $\frac{5}{8}$ in.   Simplify.

DIRECTIONS: Tell whether the statement is true or false.

STRATEGY: Build both fractions to a common denominator and compare the numerators.

J. True or false: $\frac{4}{5} > \frac{7}{9}$?

$\frac{4}{5} = \frac{36}{45}$ $\frac{7}{9} = \frac{35}{45}$ The common denominator is 45.

The statement is true.   Compare the numerators.

K. True or false: $\frac{16}{27} < \frac{19}{36}$?

$\frac{16}{27} = \frac{64}{108}$ $\frac{19}{36} = \frac{57}{108}$ The common denominator is 108.

The statement is false.   Compare the numerators.

EXERCISES 3.5

■ **OBJECTIVE 1** Rename fractions by multiplying by 1 in the form $\frac{a}{a}$. (See page 209.)

A *Write four fractions equivalent to each of the given fractions by multiplying by $\frac{2}{2}, \frac{3}{3}, \frac{4}{4}$, and $\frac{5}{5}$.*

1. $\frac{2}{3}$

2. $\frac{3}{5}$

3. $\frac{5}{6}$

4. $\frac{5}{8}$

5. $\frac{4}{9}$

6. $\frac{7}{10}$

B

7. $\frac{11}{13}$

8. $\frac{11}{15}$

9. $\frac{14}{5}$

10. $\frac{6}{5}$

■ **OBJECTIVE 2** Build a fraction by finding the missing numerator. (See page 210.)

A *Find the missing numerator.*

11. $\frac{1}{2} = \frac{?}{16}$

12. $\frac{3}{4} = \frac{?}{28}$

13. $\frac{3}{7} = \frac{?}{42}$

14. $\frac{7}{12} = \frac{?}{96}$

15. $\frac{4}{5} = \frac{?}{35}$

16. $\frac{7}{8} = \frac{?}{32}$

17. $\frac{2}{17} = \frac{?}{34}$

18. $\frac{5}{6} = \frac{?}{24}$

19. $\frac{2}{15} = \frac{?}{90}$

20. $\frac{5}{9} = \frac{?}{45}$

B

21. $\frac{?}{12} = \frac{1}{3}$

22. $\frac{?}{88} = \frac{7}{22}$

23. $\frac{22}{7} = \frac{?}{21}$

24. $\frac{11}{5} = \frac{?}{100}$

25. $\frac{?}{400} = \frac{3}{8}$

26. $\frac{2}{3} = \frac{?}{108}$

27. $\frac{?}{72} = \frac{5}{18}$

28. $\frac{?}{126} = \frac{5}{6}$

29. $\frac{13}{18} = \frac{?}{144}$

30. $\frac{13}{21} = \frac{?}{105}$

■ **OBJECTIVE 3** List a group of fractions from smallest to largest. (See page 211.)

A *List the fractions from smallest to largest.*

31. $\frac{3}{23}, \frac{5}{23}, \frac{2}{23}$

32. $\frac{6}{17}, \frac{5}{17}, \frac{3}{17}$

33. $\frac{5}{8}, \frac{3}{4}, \frac{1}{2}$

34. $\frac{1}{2}, \frac{3}{5}, \frac{7}{10}$

35. $\frac{1}{4}, \frac{2}{5}, \frac{3}{10}$

36. $\frac{2}{3}, \frac{8}{15}, \frac{3}{5}$

Are the following statements true or false?

37. $\frac{4}{11} < \frac{3}{11}$

38. $\frac{7}{9} > \frac{2}{9}$

39. $\frac{11}{16} > \frac{7}{8}$

40. $\frac{9}{16} > \frac{5}{8}$

41. $\frac{17}{30} < \frac{11}{15}$

42. $\frac{11}{10} < \frac{8}{9}$

B *List the fractions from smallest to largest.*

43. $\dfrac{5}{9}, \dfrac{11}{18}, \dfrac{7}{12}$

44. $\dfrac{3}{8}, \dfrac{3}{10}, \dfrac{2}{5}$

45. $\dfrac{13}{15}, \dfrac{4}{15}, \dfrac{5}{6}, \dfrac{9}{10}$

46. $\dfrac{3}{5}, \dfrac{11}{20}, \dfrac{7}{10}, \dfrac{5}{8}$

47. $\dfrac{11}{24}, \dfrac{17}{36}, \dfrac{35}{72}$

48. $\dfrac{8}{25}, \dfrac{31}{50}, \dfrac{59}{100}$

49. $\dfrac{13}{28}, \dfrac{17}{35}, \dfrac{3}{7}$

50. $\dfrac{3}{16}, \dfrac{5}{32}, \dfrac{1}{8}$

51. $1\dfrac{9}{16}, 1\dfrac{13}{20}, 1\dfrac{5}{8}$

52. $5\dfrac{14}{15}, 5\dfrac{19}{20}, 5\dfrac{11}{12}$

Are the following statements true or false?

53. $\dfrac{17}{60} < \dfrac{4}{15}$

54. $\dfrac{8}{25} < \dfrac{59}{100}$

55. $\dfrac{11}{12} > \dfrac{7}{8}$

56. $\dfrac{19}{40} > \dfrac{31}{60}$

57. $\dfrac{5}{32} < \dfrac{7}{40}$

58. $\dfrac{11}{27} > \dfrac{29}{36}$

C

59. Find the LCM of the denominators of $\dfrac{1}{2}, \dfrac{2}{3}, \dfrac{1}{6}$, and $\dfrac{5}{8}$. Build the four fractions so that each has the LCM as the denominator.

60. Find the LCM of the denominators of $\dfrac{1}{4}, \dfrac{4}{13}$, and $\dfrac{5}{26}$. Build the three fractions so that each has the LCM as the denominator.

61. The night nurse at Malcolm X Community Hospital finds bottles containing codeine tablets out of the usual order. The bottles contain tablets having the following strengths of codeine: $\dfrac{1}{8}, \dfrac{3}{32}, \dfrac{5}{16}, \dfrac{3}{8}, \dfrac{9}{16}, \dfrac{1}{2}$, and $\dfrac{1}{4}$ grain, respectively. Arrange the bottles in order of the strength of codeine from the smallest to the largest.

62. Joe, an apprentice, is given the task of sorting a bin of bolts according to their diameters. The bolts have the following diameters: $\dfrac{11}{16}, \dfrac{7}{8}, 1\dfrac{1}{16}, \dfrac{3}{4}, 1\dfrac{1}{8}$, and $1\dfrac{3}{32}$ in. How should he list the diameters from the smallest to the largest?

63. According to various surveys, 42 out of 50 Americans believe in heaven, whereas 2 out of 3 Americans believe in an afterlife. Do more Americans believe in heaven or an afterlife?

64. According to the General Services Administration, the federal government owns about $\dfrac{1}{30}$ of the acreage in Alaska and about $\dfrac{6}{125}$ of the acreage of Kentucky. Which state has the larger portion of federally owned land?

65. According to various surveys, 9 out of 20 Americans believe in ghosts and 12 out of 50 Americans believe in black magic. Do more Americans believe in ghosts or black magic?

66. Islam is the second largest religion in the world, with more than a billion people practicing worldwide. Indonesia is the country with the largest number of Muslims, $\frac{22}{25}$ of its population. In Kuwait, $\frac{17}{20}$ of the population practices Islam. Which country has a larger portion of their total population practicing Islam?

67. Three chemistry students weigh a container of a chemical. Mary records the weight as $3\frac{1}{8}$ lb. George reads the weight as $3\frac{3}{16}$ lb. Chang reads the weight as $3\frac{1}{4}$ lb. Whose measurement is heaviest?

68. Three rulers are marked in inches. On the first ruler, the spaces are divided into tenths. On the second, they are divided into sixteenths, and on the third, they are divided into eighths. All are used to measure a line on a scale drawing. The nearest mark on the first ruler is $5\frac{7}{10}$, the nearest mark on the second is $5\frac{11}{16}$, and the nearest mark on the third is $5\frac{6}{8}$. Which is the largest (longest) measurement?

69. Larry, Moe, and Curly bought a Subway franchise. Larry contributed $\frac{1}{4}$ of the costs, Moe contributed $\frac{7}{12}$ of the costs, and Curly contributed the remaining $\frac{1}{6}$. Which man owns the largest part of the business and which man owns the smallest?

STATE YOUR UNDERSTANDING

70. Explain why it is easier to compare the sizes of two fractions if they have common denominators.

71. What is the difference between simplifying fractions and building fractions?

CHALLENGE

72. List $\frac{12}{25}$, $\frac{14}{29}$, $\frac{29}{60}$, $\frac{35}{71}$, $\frac{39}{81}$, and $\frac{43}{98}$ from smallest to largest.

73. Build $\frac{5}{7}$ so that it has denominators 70, 91, 161, 784, and 4067.

74. Fernando and Filipe are hired to sell tickets for the holiday raffle. Fernando sells $\frac{14}{17}$ of his quota of 765 tickets. Filipe sells $\frac{19}{23}$ of his quota of 759 tickets. Who sells more of his quota? Who sells more tickets?

Find the LCM of the denominators of each set of fractions.

75. $\dfrac{3}{7}, \dfrac{2}{3}$

76. $\dfrac{1}{8}, \dfrac{3}{4}$

77. $\dfrac{5}{8}, \dfrac{3}{10}$

78. $\dfrac{3}{4}, \dfrac{5}{14}$

79. $\dfrac{11}{96}, \dfrac{35}{72}$

80. $\dfrac{1}{20}, \dfrac{5}{28}, \dfrac{3}{5}$

81. Find the prime factorization of 96.

82. Find the prime factorization of 120.

83. The Forest Service rents a two-engine plane at $625 per hour and a single-engine plane at $365 per hour to drop fire retardant. During a forest fire, the two-engine plane was used for 4 hr and the single-engine plane was used for 2 hr. What was the cost of using the two planes?

84. A patient in the local hospice is taking one capsule containing 150 mg of pain medication every 6 hours. Starting Monday, the doctor is increasing the dosage to 300 mg every 4 hours. How many 150-mg capsules should the hospice attendant order for the next week (7 days)?

Adding Fractions

VOCABULARY

Like fractions are fractions with common denominators. **Unlike fractions** are fractions with different denominators.

OBJECTIVES

1. Add like fractions.
2. Add unlike fractions.

HOW & WHY

■ **OBJECTIVE 1** Add like fractions.

What is the sum of $\dfrac{1}{5}$ and $\dfrac{2}{5}$? The denominators tell us the number of parts in the unit. The numerators tell us how many of these parts are shaded. By adding the numerators we find the total number of shaded parts. The common denominator keeps track of the size of the parts. See Figure 3.15.

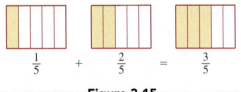

$$\frac{1}{5} \quad + \quad \frac{2}{5} \quad = \quad \frac{3}{5}$$

Figure 3.15

To add like fractions

1. Add the numerators.
2. Write the sum over the common denominator.
3. Simplify.

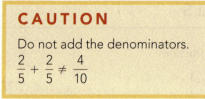

CAUTION

Do not add the denominators.

$$\frac{2}{5} + \frac{2}{5} \neq \frac{4}{10}$$

EXAMPLES A–D

DIRECTIONS: Add and simplify.

STRATEGY: Add the numerators and write the sum over the common denominator. Simplify.

WARM-UP

A. Add: $\frac{3}{11} + \frac{5}{11}$

A. Add: $\frac{4}{9} + \frac{1}{9}$

$$\frac{4}{9} + \frac{1}{9} = \frac{5}{9} \qquad \text{Add the numerators. Keep the common denominator.}$$

WARM-UP

B. Add: $\frac{3}{7} + \frac{2}{7} + \frac{4}{7}$

B. Add: $\frac{3}{4} + \frac{3}{4} + \frac{1}{4}$

$$\frac{3}{4} + \frac{3}{4} + \frac{1}{4} = \frac{7}{4} \qquad \text{Add.}$$

$$= 1\frac{3}{4} \qquad \text{Write as a mixed number.}$$

WARM-UP

C. Add: $\frac{1}{14} + \frac{3}{14} + \frac{3}{14}$

C. Add: $\frac{1}{13} + \frac{5}{13} + \frac{7}{13}$

$$\frac{1}{13} + \frac{5}{13} + \frac{7}{13} = \frac{13}{13} \qquad \text{Add.}$$

$$= 1 \qquad \text{Simplify.}$$

WARM-UP

D. According to a report from Leed Community College, $\frac{3}{14}$ of their total revenues were from alumni contributions and $\frac{9}{14}$ were from tuition. What portion of the total revenues were from these two sources?

D. According to the annual report from Radd Community College, $\frac{11}{20}$ of their total revenues were from state taxes and $\frac{7}{20}$ were from tuition and fees. What portion of the total revenues were from these two sources?

$$\frac{11}{20} + \frac{7}{20} = \frac{18}{20} \qquad \text{Add.}$$

$$= \frac{9}{10} \qquad \text{Simplify.}$$

Taxes and tuition and fees were $\frac{9}{10}$ of the revenue for the school.

ANSWERS TO WARM-UPS A–D

A. $\frac{8}{11}$ **B.** $\frac{9}{7}$ or $1\frac{2}{7}$ **C.** $\frac{1}{2}$

D. Contributions and tuition were $\frac{6}{7}$ of the revenue for the school.

HOW & WHY

■ **OBJECTIVE 2** Add unlike fractions.

The sum $\frac{1}{2} + \frac{1}{5}$ cannot be found in this form. A look at Figure 3.16 shows that the parts are not the same size.

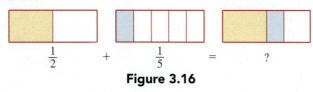

$$\frac{1}{2} \qquad + \qquad \frac{1}{5} \qquad = \qquad ?$$

Figure 3.16

To add, rename $\frac{1}{2}$ and $\frac{1}{5}$ as like fractions. The LCM (least common multiple) of the two denominators serves as the least common denominator. The LCM of 2 and 5 is 10. Renaming the fractions, we write

$$\frac{1}{2} = \left(\frac{1}{2}\right)\left(\frac{5}{5}\right) = \frac{5}{10} \qquad \text{and} \qquad \frac{1}{5} = \left(\frac{1}{5}\right)\left(\frac{2}{2}\right) = \frac{2}{10}$$

Figure 3.17 shows the regions with equal size parts.

$$\frac{1}{2} \qquad + \qquad \frac{1}{5} \qquad = \qquad ?$$

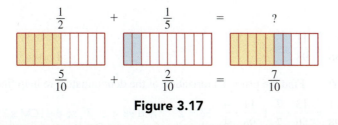

$$\frac{5}{10} \qquad + \qquad \frac{2}{10} \qquad = \qquad \frac{7}{10}$$

Figure 3.17

> ### To add unlike fractions
>
> 1. Build the fractions so that they have a common denominator.
> 2. Add and simplify.

EXAMPLES E–J

DIRECTIONS: Add and simplify.

STRATEGY: Build each of the fractions to a common denominator, add, and simplify.

E. Add: $\frac{1}{8} + \frac{3}{4}$

$\frac{1}{8} + \frac{3}{4} = \frac{1}{8} + \frac{3}{4} \cdot \frac{2}{2}$ The LCM of 8 and 4 is 8. Build the fractions.

$\phantom{\frac{1}{8} + \frac{3}{4}} = \frac{1}{8} + \frac{6}{8}$

$\phantom{\frac{1}{8} + \frac{3}{4}} = \frac{7}{8}$ Add.

WARM-UP

F. Add: $\dfrac{5}{12} + \dfrac{2}{9}$

F. Add: $\dfrac{3}{14} + \dfrac{8}{35}$

$\dfrac{3}{14} + \dfrac{8}{35} = \dfrac{3}{14} \cdot \dfrac{5}{5} + \dfrac{8}{35} \cdot \dfrac{2}{2}$ The LCM of 14 and 35 is 70.

$= \dfrac{15}{70} + \dfrac{16}{70}$

$= \dfrac{31}{70}$ Add.

WARM-UP

G. Add: $\dfrac{5}{8} + \dfrac{7}{24}$

G. Add: $\dfrac{5}{6} + \dfrac{1}{10}$

$\dfrac{5}{6} + \dfrac{1}{10} = \dfrac{5}{6} \cdot \dfrac{5}{5} + \dfrac{1}{10} \cdot \dfrac{3}{3}$ The LCM of 6 and 10 is 30.

$= \dfrac{25}{30} + \dfrac{3}{30}$

$= \dfrac{28}{30}$ Add.

$= \dfrac{14}{15}$ Simplify.

WARM-UP

H. Add: $\dfrac{13}{45} + \dfrac{28}{75}$

H. Add: $\dfrac{13}{56} + \dfrac{11}{98}$

STRATEGY: Find the prime factorization of the denominators to help find the LCM.

$\dfrac{13}{56} + \dfrac{11}{98} = \dfrac{13}{56} \cdot \dfrac{7}{7} + \dfrac{11}{98} \cdot \dfrac{4}{4}$ $56 = 2^3 \cdot 7$ and $98 = 2 \cdot 7^2$, so the LCM is $2^3 \cdot 7^2 = 392$.

$= \dfrac{13(7)}{392} + \dfrac{11(4)}{392}$

$= \dfrac{91}{392} + \dfrac{44}{392}$

$= \dfrac{135}{392}$ Add.

CALCULATOR EXAMPLE:

WARM-UP

I. Add: $\dfrac{9}{40} + \dfrac{5}{56}$

I. Add: $\dfrac{51}{80} + \dfrac{25}{96}$

$\dfrac{51}{80} + \dfrac{25}{96} = \dfrac{431}{480}$

On a calculator with fraction keys, it is not necessary to find the common denominator. The calculator is programmed to add and simplify.

ANSWERS TO WARM-UPS F–I

F. $\dfrac{23}{36}$

G. $\dfrac{11}{12}$

H. $\dfrac{149}{225}$

I. $\dfrac{11}{35}$

J. Sheila is assembling a composting bin for her lawn and garden debris. She needs a bolt that will reach through a $\frac{1}{32}$-in.-thick washer, a $\frac{3}{16}$-in.-thick plastic bushing, a $\frac{3}{4}$-in.-thick piece of steel tubing, a second $\frac{1}{32}$-in.-thick washer, and a $\frac{1}{4}$-in.-thick nut. How long a bolt does she need?

STRATEGY: Add the thicknesses of each part to find the total length needed.

$$\frac{1}{32} + \frac{3}{16} + \frac{3}{4} + \frac{1}{32} + \frac{1}{4}$$ The LCM of 32, 16, and 4 is 32.

$$\frac{1}{32} + \frac{6}{32} + \frac{24}{32} + \frac{1}{32} + \frac{8}{32}$$ Build the fractions to a common denominator.

$$= \frac{40}{32}$$ Add.

$$= \frac{5}{4} = 1\frac{1}{4}$$ Simplify and write as a mixed number.

The bolt must be $1\frac{1}{4}$ in. long.

WARM-UP

J. A nail must be long enough to reach through three thicknesses of wood and penetrate a fourth piece $\frac{1}{4}$ in. If the first piece of wood is $\frac{5}{16}$ in. thick, the second is $\frac{3}{8}$ in. thick, and the third is $\frac{9}{16}$ in. thick, how long must the nail be?

ANSWER TO WARM-UP J

J. The nail must be at least $1\frac{1}{2}$ in. long.

EXERCISES 3.6

■ **OBJECTIVE 1** Add like fractions. (See page 217.)

A *Add. Simplify completely.*

1. $\frac{4}{17} + \frac{5}{17}$

2. $\frac{5}{12} + \frac{2}{12}$

3. $\frac{3}{10} + \frac{4}{10} + \frac{1}{10}$

4. $\frac{1}{8} + \frac{2}{8} + \frac{3}{8}$

5. $\frac{3}{11} + \frac{8}{11}$

6. $\frac{6}{7} + \frac{8}{7}$

7. $\frac{4}{15} + \frac{7}{15} + \frac{11}{15}$

8. $\frac{4}{12} + \frac{3}{12} + \frac{8}{12}$

9. $\frac{7}{19} + \frac{6}{19} + \frac{3}{19}$

10. $\frac{5}{11} + \frac{2}{11} + \frac{1}{11}$

11. $\frac{5}{12} + \frac{5}{12} + \frac{5}{12}$

12. $\frac{9}{16} + \frac{7}{16} + \frac{4}{16}$

B

13. $\frac{3}{20} + \frac{5}{20} + \frac{7}{20}$

14. $\frac{7}{32} + \frac{8}{32} + \frac{5}{32}$

15. $\frac{3}{18} + \frac{5}{18} + \frac{4}{18}$

16. $\frac{3}{16} + \frac{2}{16} + \frac{5}{16}$

17. $\frac{7}{30} + \frac{11}{30} + \frac{3}{30}$

18. $\frac{4}{25} + \frac{8}{25} + \frac{3}{25}$

19. $\frac{2}{35} + \frac{8}{35} + \frac{4}{35}$

20. $\frac{7}{18} + \frac{2}{18} + \frac{7}{18}$

A *Add. Simplify completely.*

21. $\dfrac{1}{6} + \dfrac{3}{8}$

22. $\dfrac{3}{7} + \dfrac{2}{5}$

23. $\dfrac{1}{14} + \dfrac{3}{7}$

24. $\dfrac{7}{15} + \dfrac{1}{3}$

25. $\dfrac{7}{16} + \dfrac{3}{8}$

26. $\dfrac{4}{9} + \dfrac{5}{18}$

27. $\dfrac{1}{3} + \dfrac{1}{6} + \dfrac{1}{18}$

28. $\dfrac{1}{4} + \dfrac{2}{5} + \dfrac{3}{20}$

29. $\dfrac{3}{7} + \dfrac{1}{14} + \dfrac{1}{2}$

30. $\dfrac{5}{8} + \dfrac{1}{2} + \dfrac{3}{4}$

B

31. $\dfrac{5}{18} + \dfrac{11}{24}$

32. $\dfrac{1}{18} + \dfrac{1}{24}$

33. $\dfrac{3}{5} + \dfrac{9}{10} + \dfrac{7}{20}$

34. $\dfrac{7}{8} + \dfrac{7}{12} + \dfrac{1}{6}$

35. $\dfrac{2}{3} + \dfrac{3}{4} + \dfrac{1}{2} + \dfrac{5}{12}$

36. $\dfrac{1}{2} + \dfrac{3}{10} + \dfrac{3}{5} + \dfrac{1}{4}$

37. $\dfrac{2}{3} + \dfrac{5}{12} + \dfrac{5}{9} + \dfrac{5}{18}$

38. $\dfrac{15}{36} + \dfrac{1}{6} + \dfrac{5}{12} + \dfrac{1}{18}$

39. $\dfrac{5}{7} + \dfrac{2}{3} + \dfrac{1}{63} + \dfrac{1}{9}$

40. $\dfrac{7}{10} + \dfrac{4}{5} + \dfrac{1}{15} + \dfrac{31}{35}$

41. $\dfrac{3}{5} + \dfrac{2}{25} + \dfrac{4}{75} + \dfrac{2}{3}$

42. $\dfrac{4}{5} + \dfrac{62}{75} + \dfrac{2}{15} + \dfrac{9}{25}$

43. $\dfrac{3}{5} + \dfrac{7}{8} + \dfrac{1}{4} + \dfrac{7}{10}$

44. $\dfrac{11}{15} + \dfrac{7}{12} + \dfrac{9}{10} + \dfrac{17}{20}$

C

45. $\dfrac{2}{45} + \dfrac{4}{9} + \dfrac{1}{15}$

46. $\dfrac{3}{8} + \dfrac{7}{24} + \dfrac{1}{12}$

47. $\dfrac{11}{40} + \dfrac{13}{25} + \dfrac{19}{50}$

48. $\dfrac{25}{72} + \dfrac{19}{144} + \dfrac{1}{12}$

49. $\dfrac{25}{36} + \dfrac{19}{48}$

50. $\dfrac{64}{95} + \dfrac{51}{114}$

51. According to the National Solid Wastes Management Associates, in the United States $\frac{1}{20}$ of old tires are exported, $\frac{3}{50}$ of them are recycled, $\frac{1}{10}$ of them are burned for energy, and the rest are illegally dumped or sent to a land-fill. What part of old tires in the United States are not dumped or sent to landfills? Is this more or less than $\frac{1}{4}$ of all old tires?

52. North America accounts for more than $\frac{9}{32}$ of the total petroleum consumption worldwide. Central and South America together account for another $\frac{1}{16}$ of the total petroleum consumption. Do the Americas consume more or less than half of the petroleum consumed worldwide?

53. According to the Environmental Protection Agency, of all people in this country using pesticides, about $\frac{1}{200}$ of them are in commerce or government, $\frac{19}{100}$ are in agriculture, and the rest are home and garden users. Are the home and garden users more or less than $\frac{3}{4}$ of all the pesticide users?

54. The Republic of Panama has four major ethnic groups. The largest is Mestizo. About $\frac{7}{50}$ of the population is West Indian, $\frac{1}{10}$ of the population is white, and $\frac{6}{100}$ of the population is Amerindian. What portion of Panama's population is not Mestizo?

55. Find the error in the statement: $\frac{1}{5} + \frac{2}{5} = \frac{3}{10}$. Correct the statement. Explain how you would avoid this error.

56. Find the error in the statement: $\frac{1}{2} + \frac{4}{7} = \frac{5}{14}$. Correct the statement. Explain how you would avoid this error.

57. Chef Ramon prepares a punch for a wedding party. The punch calls for $\frac{1}{4}$ gallon of lemon juice, $\frac{3}{4}$ gallon of raspberry juice, $\frac{1}{2}$ gallon of cranberry juice, $\frac{1}{4}$ gallon of lime juice, $\frac{5}{4}$ gallons of 7-Up, and $\frac{3}{4}$ gallon of vodka. How many gallons of punch does the recipe make?

58. A physical therapist advises Belinda to swim $\frac{1}{4}$ mi on Monday and increase this distance by $\frac{1}{16}$ mi each day from Tuesday through Friday. What is the total number of miles she advises?

59. Jonnie is assembling a rocking horse for his granddaughter. He needs a bolt to reach through a $\frac{7}{8}$-in. piece of steel tubing, a $\frac{1}{16}$-in. bushing, a $\frac{1}{2}$-in. piece of tubing, a $\frac{1}{8}$-in.-thick washer, and a $\frac{1}{4}$-in.-thick nut. How long a bolt does he need?

60. What is the perimeter of this triangle?

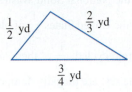

61. Find the length of this pin.

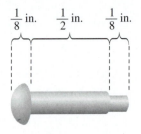

$\frac{1}{8}$ in. $\frac{1}{2}$ in. $\frac{1}{8}$ in.

62. The Sandoz family spends $\frac{2}{15}$ of their income on rent, $\frac{1}{4}$ on food, $\frac{1}{20}$ on clothes, $\frac{1}{10}$ on transportation, and $\frac{5}{24}$ on taxes. What fraction of their income is spent on these costs?

63. Find the length of the rod in the figure. Assume that the grooves and teeth are uniform in length.

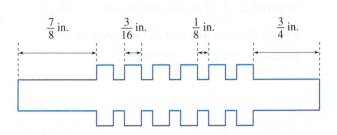

$\frac{7}{8}$ in. $\frac{3}{16}$ in. $\frac{1}{8}$ in. $\frac{3}{4}$ in.

64. Elena walked $\frac{3}{8}$ mi from her house to the bus stop, then $\frac{1}{10}$ mi from where the bus let her off to the library. For lunch, she walked $\frac{1}{5}$ mi to a coffee shop. Elena then returned to the library, finished her research, and caught the bus home. How many miles did she walk on the entire trip?

Exercise 65 relates to the chapter application. See page 167.

65. One weekend Shane got his friend Mike to help him with the deck. Together they installed $\frac{1}{3}$ of the deck boards. The next weekend Shane's sister Carrie helped him and they installed $\frac{1}{2}$ of the deck boards. How much of the deck is installed after the two weekends?

66. Explain how to find the sum of $\dfrac{5}{12}$ and $\dfrac{3}{20}$.

67. Why is it important to write fractions with a common denominator before adding?

CHALLENGE

68. Find the sum of $\dfrac{107}{372}$ and $\dfrac{41}{558}$.

69. Find the sum of $\dfrac{67}{124} + \dfrac{27}{868}$.

70. Janet left $\dfrac{1}{7}, \dfrac{3}{14}$, and $\dfrac{1}{6}$ of her estate to Bob, Greta, and Joe Guerra. She also left $\dfrac{1}{8}, \dfrac{5}{16}$, and $\dfrac{1}{9}$ of the estate to Pele, Rhonda, and Shauna Contreras. Which family received the greater share of the estate?

71. Jim is advised by his doctor to limit his fat intake. For breakfast, his fat intake is a bagel, $\dfrac{3}{4}$ g; a banana, $\dfrac{13}{16}$ g; cereal, $\dfrac{17}{10}$ g; milk, $\dfrac{9}{8}$ g; jelly, 0 g; and coffee, 0 g. Rounded to the nearest whole number, how many grams of fat does Jim consume at breakfast? If each gram of fat represents 9 calories and the total calories for breakfast is 330, what fraction represents the number of calories from fat? Use the rounded whole number of grams of fat.

MAINTAIN YOUR SKILLS

Add.

72. $2 + 8 + \dfrac{1}{4} + \dfrac{1}{6}$

73. $5 + 9 + \dfrac{5}{16} + \dfrac{5}{24}$

74. $1 + 7 + 10 + \dfrac{1}{6} + \dfrac{7}{10} + \dfrac{1}{5}$

75. $3 + 11 + 12 + \dfrac{2}{9} + \dfrac{2}{3} + \dfrac{7}{12}$

76. $3 + \dfrac{3}{8} + 2 + \dfrac{1}{8} + 1 + \dfrac{3}{16}$

77. $4 + \dfrac{1}{12} + 1 + \dfrac{3}{4} + 3 + \dfrac{1}{8}$

Perform the indicated operations.

78. $(32 - 27)8 - 5(17 - 10)$

79. $(13 - 2^3)^3 - 7(11)$

80. Mrs. McCallister bought 1000 shares of Lunar stock in each of the last 5 years. She paid $35 per share the first year, $38 the second, $42 the third, $48 the fourth, and $42 the last year. What was the average she price paid for the stock?

81. In a metal benchwork class that has 36 students, each student is allowed $11\dfrac{5}{8}$ in. of wire solder. How many inches of wire must the instructor provide for the class?

Adding Mixed Numbers

HOW & WHY

■ **OBJECTIVE**   Add mixed numbers.

What is the sum of $3\frac{1}{6}$ and $5\frac{1}{4}$? Pictorially we can show the sum by drawing rectangles such as those in Figure 3.18.

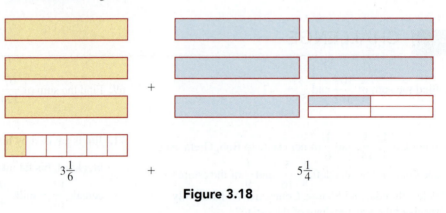

$3\frac{1}{6}$ + $5\frac{1}{4}$

Figure 3.18

It is easy to see that the sum contains eight whole units. The sum of the fraction parts requires finding a common denominator. The LCM of 6 and 4 is 12. Figure 3.19 shows the divided rectangles.

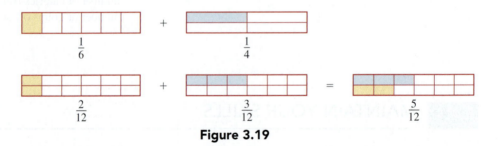

Figure 3.19

So the sum is $8\frac{5}{12}$.

Mixed numbers can be added horizontally or in columns. The sum of $3\frac{1}{6}$ and $5\frac{1}{4}$ is shown both ways.

$$\left(3 + \frac{1}{6}\right) + \left(5 + \frac{1}{4}\right) = (3 + 5) + \left(\frac{1}{6} + \frac{1}{4}\right)$$

$$= 8 + \left(\frac{2}{12} + \frac{3}{12}\right)$$

$$= 8\frac{5}{12}$$

When we write the sum vertically, the grouping of the whole numbers and the fractions takes place naturally.

$$3\frac{1}{6} = 3\frac{2}{12}$$
$$+5\frac{1}{4} = 5\frac{3}{12}$$
$$8\frac{5}{12}$$

Sometimes the sum of the fractions is greater than 1. In this case, change the fraction sum to a mixed number and add it to the whole-number part.

$$12\frac{3}{4} = 12\frac{15}{20}$$ Write the fractions with a common denominator.

$$+10\frac{7}{10} = 10\frac{14}{20}$$

$$= 22\frac{29}{20}$$ Add the whole numbers and the fractions.

$$= 22 + 1\frac{9}{20}$$ Rewrite the improper fraction as a mixed number.

$$= 23\frac{9}{20}$$ Add the mixed number to the whole number.

> **To add mixed numbers**
>
> 1. Add the whole numbers.
> 2. Add the fractions. If the sum of the fractions is more than 1, change the fraction to a mixed number and add again.
> 3. Simplify.

EXAMPLES A–E

DIRECTIONS: Add. Write as a mixed number.

STRATEGY: Add the whole numbers and add the fractions. If the sum of the fractions is an improper fraction, rewrite it as a mixed number and add to the sum of the whole numbers. Simplify.

A. Add: $3\frac{3}{4} + 2\frac{5}{12}$

STRATEGY: Write the mixed numbers in a column to group the whole numbers and to group the fractions.

$$3\frac{3}{4} = 3\frac{9}{12}$$ Build the fractions to the common denominator 12.

$$+2\frac{5}{12} = 2\frac{5}{12}$$

$$5\frac{14}{12}$$ Add.

$$5\frac{14}{12} = 5 + 1\frac{2}{12}$$ Write the improper fraction as a mixed number.

$$= 6\frac{2}{12}$$ Add the whole numbers.

$$= 6\frac{1}{6}$$ Simplify.

WARM-UP

A. Add: $4\frac{1}{2} + 9\frac{5}{16}$

ANSWER TO WARM-UP A

A. $13\frac{13}{16}$

WARM-UP

B. Add: $4\dfrac{2}{5} + 11\dfrac{1}{6} + 8\dfrac{7}{10}$

B. Add: $13\dfrac{3}{5} + 12\dfrac{2}{15} + 8\dfrac{3}{10}$

$$13\dfrac{3}{5} = 13\dfrac{18}{30}$$ The LCM of 5, 15, and 10 is 30.

$$12\dfrac{2}{15} = 12\dfrac{4}{30}$$

$$+8\dfrac{3}{10} = 8\dfrac{9}{30}$$

$$= 33\dfrac{31}{30}$$ Add.

$$33\dfrac{31}{30} = 33 + 1\dfrac{1}{30}$$ Write the improper fraction as a mixed number.

$$= 34\dfrac{1}{30}$$ Add.

WARM-UP

C. Add: $44\dfrac{7}{23} + 18$

C. Add: $36 + 7\dfrac{11}{13}$

$$36 + 7\dfrac{11}{13} = 43\dfrac{11}{13}$$ Add the whole numbers.

WARM-UP

D. Add: $51\dfrac{7}{8} + 36\dfrac{5}{12}$

CALCULATOR EXAMPLE:

D. Add: $31\dfrac{3}{8} + 62\dfrac{13}{15}$

$$31\dfrac{3}{8} + 62\dfrac{13}{15} = \dfrac{11{,}309}{120}$$

$$= 94\dfrac{29}{120}$$

On a calculator with fraction keys, it is not necessary to find the common denominator. The calculator is programmed to add and simplify. Some calculators may not change the improper sum to a mixed number. See your calculator manual.

WARM-UP

E. The same report also records that the city recycles $10\dfrac{1}{3}$ tons of paper, $3\dfrac{3}{8}$ tons of aluminum, and $4\dfrac{5}{12}$ tons of glass each month. How many tons of material are recycled each month?

E. A report by Environmental Hazards Management lists the following amounts of hazardous material that a city of 100,000 discharges into city drains each month: $3\dfrac{3}{4}$ tons of toilet bowl cleaner, $13\dfrac{3}{4}$ tons of liquid household cleaners, and $3\dfrac{2}{5}$ tons of motor oil. How many tons of these materials are discharged each month?

STRATEGY: Find the sum of the number of tons of hazardous material.

$$3\dfrac{3}{4} = 3\dfrac{15}{20}$$ The LCM of 4 and 5 is 20.

$$13\dfrac{3}{4} = 13\dfrac{15}{20}$$

$$+ 3\dfrac{2}{5} = 3\dfrac{8}{20}$$

$$= 19\dfrac{38}{20}$$ Add.

$$19\dfrac{38}{20} = 19 + 1\dfrac{18}{20} = 20\dfrac{9}{10}$$ Change the improper fraction to a mixed number and simplify.

The residents discharge $20\dfrac{9}{10}$ tons of hazardous material each month.

ANSWERS TO WARM-UPS B–E

B. $24\dfrac{4}{15}$ **C.** $62\dfrac{7}{23}$ **D.** $88\dfrac{7}{24}$

E. The city recycles $18\dfrac{1}{8}$ tons of material per month.

EXERCISES 3.7

■ **OBJECTIVE**   Add mixed numbers. (See page 226.)

A   *Write the results as mixed numbers where possible.*

1. $9\dfrac{1}{9}$

    $+3\dfrac{7}{9}$

2. $6\dfrac{1}{3}$

    $+2\dfrac{1}{3}$

3. $3\dfrac{4}{5}$

    $+7\dfrac{4}{5}$

4. $2\dfrac{5}{8}$

    $+3\dfrac{7}{8}$

5. $5\dfrac{7}{15}$

    $+6\dfrac{2}{3}$

6. $2\dfrac{5}{6}$

    $+7\dfrac{2}{3}$

7. $9\dfrac{9}{14}$

    $+3\dfrac{2}{7}$

8. $8\dfrac{5}{12}$

    $+7\dfrac{1}{2}$

9. $4\dfrac{4}{7}$

    $+2\dfrac{11}{14}$

10. $9\dfrac{5}{18}$

     $+7\dfrac{5}{6}$

11. $2\dfrac{8}{15}$

     $+7\dfrac{3}{5}$

12. $4\dfrac{7}{15}$

     $+6\dfrac{2}{3}$

13. $10\dfrac{3}{10}$

     $+11\dfrac{17}{20}$

14. $8\dfrac{5}{9}$

     $+2\dfrac{13}{27}$

15. $2\dfrac{2}{3}$

     $3\dfrac{2}{3}$

     $+5\dfrac{1}{6}$

16. $5\dfrac{2}{5}$

     $8\dfrac{7}{10}$

     $+1\dfrac{1}{2}$

17. $7\dfrac{1}{5} + 6\dfrac{3}{20} + 8 + 4\dfrac{1}{4}$

18. $5\dfrac{2}{5} + 4 + 3\dfrac{2}{3} + 9\dfrac{7}{15}$

B

19. $11\dfrac{2}{3}$

     $+6\dfrac{11}{18}$

20. $12\dfrac{4}{15}$

     $+3\dfrac{5}{6}$

21. $7\dfrac{3}{7}$

     $5\dfrac{9}{14}$

     $+8\dfrac{5}{28}$

22. $7\dfrac{1}{8}$

     $3\dfrac{5}{12}$

     $+8\dfrac{5}{6}$

23. $11\dfrac{7}{20}$

     $+9\dfrac{7}{30}$

24. $15\dfrac{7}{18}$

     $+21\dfrac{6}{27}$

25. $16\dfrac{3}{10} + 11\dfrac{7}{15} + 7\dfrac{1}{6}$

26. $18\dfrac{3}{4} + 17\dfrac{7}{8} + 23\dfrac{1}{6}$

27. $119\dfrac{11}{12}$

$+217\dfrac{7}{18}$
___

28. $219\dfrac{3}{10}$

$+308\dfrac{5}{6}$
___

29. $59\dfrac{4}{11} + 32\dfrac{5}{22} + 15\dfrac{5}{6}$

30. $26\dfrac{5}{8} + 33 + 41\dfrac{5}{6}$

31. $15\dfrac{3}{4} + 18\dfrac{2}{3} + 21\dfrac{1}{2}$

32. $39\dfrac{11}{12} + 49\dfrac{7}{8} + 59\dfrac{3}{4}$

33. $72 + 15\dfrac{7}{24} + 23\dfrac{11}{36}$

34. $42 + 11\dfrac{7}{20} + 6\dfrac{17}{30}$

35. $14\dfrac{9}{10} + 15\dfrac{7}{15} + 11$

36. $12\dfrac{11}{12} + 22\dfrac{5}{8} + 8$

37. $11\dfrac{1}{6}$

$12\dfrac{3}{10}$

$13\dfrac{1}{12}$

$+14\dfrac{1}{20}$
___

38. $11\dfrac{5}{18}$

$21\dfrac{2}{9}$

$31\dfrac{1}{3}$

$+41\dfrac{1}{6}$
___

C

39. $6\dfrac{1}{8} + 2\dfrac{3}{4} + \dfrac{5}{12} + 9\dfrac{1}{3}$

40. $12\dfrac{3}{10} + 5\dfrac{2}{5} + 19\dfrac{2}{3} + \dfrac{11}{15}$

41. $12\dfrac{3}{5} + 7\dfrac{1}{8} + 29\dfrac{3}{4} + 14\dfrac{9}{10}$

42. $18\dfrac{5}{21} + 7\dfrac{5}{7} + 18\dfrac{1}{3} + 20$

43. Nancy painted a portrait of her daughters. The canvas measures $18\dfrac{1}{2}$ by $24\dfrac{3}{4}$ in. She plans to frame it with molding that will require an extra inch and a half on each end to make mitered corners, as in the figure. How much molding does Nancy need to frame her portrait?

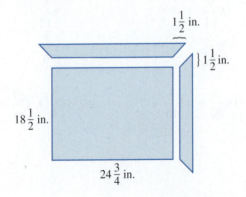

44. Tom is making a bookshelf with six pieces of wood, as shown in the figure. He needs three pieces of wood for the shelves, which are 24 in. long, one piece of wood for the top, which is $25\dfrac{1}{2}$ in. long, and two pieces of wood for the vertical supports, which are $30\dfrac{3}{4}$ inches long. What is the total length of wood that Tom needs?

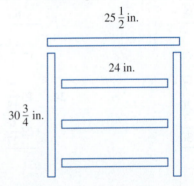

45. Elizabeth is sewing her daughter's wedding dress. The bodice of the dress requires $1\frac{3}{8}$ yd of fabric, the skirt requires $6\frac{1}{2}$ yd, and the jacket requires $2\frac{3}{4}$ yd. How many yards of fabric does she need for the dress and jacket?

46. Juanita worked the following hours at her part-time job during the month of October:

Week	Oct. 1–7	Oct. 8–14	Oct. 15–21	Oct. 22–28	Oct. 29–31
Hours	$25\frac{1}{2}$	$19\frac{2}{3}$	10	$16\frac{5}{6}$	$4\frac{3}{4}$

How many hours did she work during October?

47. After his open heart surgery Jeff went on a diet. His weight loss during an 8-week period is given in the following table. What was Jeff's total weight loss for the 8 weeks?

Week	1	2	3	4	5	6	7	8
Pounds Lost	$3\frac{1}{2}$	$2\frac{1}{4}$	$4\frac{7}{8}$	$3\frac{1}{3}$	$5\frac{3}{4}$	2	$1\frac{3}{8}$	$2\frac{2}{3}$

48. Find the perimeter of a rectangle that has length $22\frac{1}{4}$ ft and width $16\frac{2}{3}$ ft.

49. Find the perimeter of a triangle

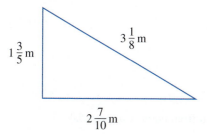

50. What is the overall length of this bolt?

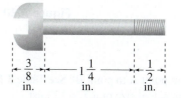

51. One year Dallas, Texas, has the following rainfall amounts: April, $3\frac{9}{20}$ in., May, $5\frac{3}{10}$ in., June, $3\frac{23}{25}$ in. What was the total rainfall for the 3 months?

52. Michael is making bread for a bake sale. His recipe calls for 2 c of rye flour, $3\frac{1}{2}$ c of whole wheat flour, and $1\frac{3}{4}$ c bread flour. What is the total amount of flour the recipe calls for?

53. The State Department of Transportation must resurface parts of seven roads this summer. The distances to be paved are $6\frac{11}{16}$ mi, $8\frac{3}{5}$ mi, $9\frac{3}{4}$ mi, $17\frac{1}{2}$ mi, $5\frac{1}{8}$ mi, $12\frac{4}{5}$ mi, and $\frac{7}{8}$ mi. How many miles of highway are to be resurfaced? If it costs $15,000 to resurface 1 mile, what is the cost of the resurfacing project? Round to the nearest thousand dollars.

Exercises 54–55 relate to the chapter application. See page 167.

The ledger board is a beam that supports the joists on one end. Shane is using 2 × 12s for the joists. Beginning on one end, there is a pair of joists right next to each other that are called the rim joists (for extra strength at the edge of the deck). The rest of the joists are laid out "16 in. on center," meaning that from the center of one joist to the center of the next joist is 16 in. See Figure 3.20.

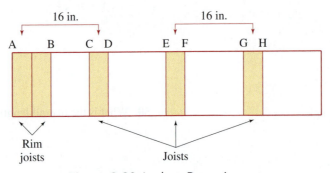

Figure 3.20 Ledger Beam Layout

54. How long is it from point A to point B? See Table 3.1 on page 167 for the actual size of a 2 × 12 joist.

55. How long is it from point A to point D?

STATE YOUR UNDERSTANDING

56. Explain why it is sometimes necessary to rename the sum of two mixed numbers after adding the whole numbers and the fractional parts. Give an example in which this happens.

57. Add $8\frac{4}{5} + 7\frac{3}{8}$ by the procedures of this section. Then change each mixed number to an improper fraction and add. Be sure you get the same result for both. Which method do you prefer? Why?

58. Is $3\left(4\dfrac{1}{4}\right) + 2\dfrac{2}{3} + 5\left(2\dfrac{5}{6}\right) = 3\left(3\dfrac{1}{8}\right) + 7\dfrac{1}{12}$ a true statement?

59. Is $6\left(6\dfrac{4}{9}\right) + 5\left(2\dfrac{5}{6}\right) + 7\left(4\dfrac{1}{3}\right) =$

$6\left(5\dfrac{1}{3}\right) + 8\left(3\dfrac{2}{3}\right) + 7\left(3\dfrac{5}{42}\right)$ a true statement?

60. During the month of January, the rangers at Yellowstone National Park record the following snowfall: week 1, $6\dfrac{9}{10}$ in.; week 2, $8\dfrac{3}{4}$ in.; week 3, $13\dfrac{5}{6}$ in.; and week 4, $9\dfrac{2}{3}$ in. How many inches of snow have fallen during the month? If the average snowfall for the month is $38\dfrac{17}{32}$ in., does this January exceed the average?

MAINTAIN YOUR SKILLS

Subtract.

61. $103 - 77$

62. $318 - 257$

63. $1111 - 889$

64. $3455 - 1689$

Find the missing number.

65. $\dfrac{7}{8} = \dfrac{?}{40}$

66. $\dfrac{5}{18} = \dfrac{?}{90}$

67. Simplify: $\dfrac{1950}{4095}$

68. Simplify: $\dfrac{630}{2730}$

69. Par on the first nine holes of the Ricochet Golf Course is 36. If Millie records scores of 5, 4, 6, 2, 3, 3, 3, 5, and 4 on the nine holes, what is her total score? Is she under or over par for the first nine holes?

70. Dried prunes weigh one-third the weight of fresh prunes. How many pounds of fresh prunes are required to make 124 half-pound packages of dried prunes?

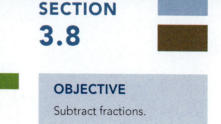

SECTION 3.8

Subtracting Fractions

HOW & WHY

■ **OBJECTIVE** Subtract fractions.

OBJECTIVE

Subtract fractions.

What is the difference of $\dfrac{2}{3}$ and $\dfrac{1}{3}$? In Figure 3.21 we can see that we subtract the numerators and keep the common denominator (subtract the cross-hatched region from the blue region).

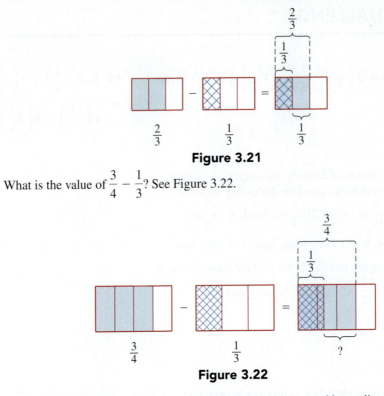

Figure 3.21

What is the value of $\dfrac{3}{4} - \dfrac{1}{3}$? See Figure 3.22.

Figure 3.22

The region shaded blue with the question mark cannot be named immediately, because the original parts are not the same size. If the fractions had a common denominator, we could subtract as in Figure 3.21. Using the common denominator 12, we see in Figure 3.23 that the difference is $\dfrac{5}{12}$.

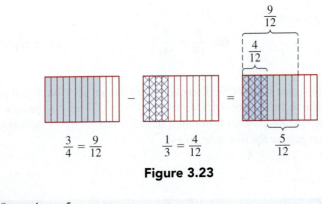

Figure 3.23

$$\frac{3}{4} - \frac{1}{3} = \frac{9}{12} - \frac{4}{12} = \frac{5}{12}$$

The method for subtracting fractions is similar to that for adding fractions.

> **To subtract fractions**
>
> 1 Build each fraction to a common denominator.
> 2 Subtract the numerators and write the difference over the common denominator.
> 3 Simplify.

EXAMPLES A–F

DIRECTIONS: Subtract and simplify.

STRATEGY: Build each fraction to a common denominator.

A. Subtract: $\dfrac{19}{25} - \dfrac{4}{25}$

$\dfrac{19}{25} - \dfrac{4}{25} = \dfrac{15}{25}$ Subtract the numerators.

$= \dfrac{3}{5}$ Simplify.

B. Subtract: $\dfrac{5}{8} - \dfrac{1}{6}$

$\dfrac{5}{8} - \dfrac{1}{6} = \dfrac{5}{8} \cdot \dfrac{3}{3} - \dfrac{1}{6} \cdot \dfrac{4}{4}$ The LCM of 8 and 6 is 24.

$= \dfrac{15}{24} - \dfrac{4}{24}$

$= \dfrac{11}{24}$ Subtract the numerators.

C. Subtract: $\dfrac{13}{15} - \dfrac{5}{12}$

$\dfrac{13}{15} - \dfrac{5}{12} = \dfrac{13}{15} \cdot \dfrac{4}{4} - \dfrac{5}{12} \cdot \dfrac{5}{5}$ The LCM of 15 and 20 is 60.

$= \dfrac{52}{60} - \dfrac{25}{60}$

$= \dfrac{27}{60}$ Subtract the numerators.

$= \dfrac{9}{20}$ Simplify.

D. Subtract: $\dfrac{43}{60} - \dfrac{5}{48}$

$\dfrac{43}{60} - \dfrac{5}{48} = \dfrac{43}{60} \cdot \dfrac{4}{4} - \dfrac{5}{48} \cdot \dfrac{5}{5}$ The LCM of 60 and 48 is 20.

$= \dfrac{172}{240} - \dfrac{25}{240}$

$= \dfrac{147}{240}$ Subtract the numerators.

$= \dfrac{49}{80}$ Simplify.

CALCULATOR EXAMPLE:

E. Subtract: $\dfrac{31}{45} - \dfrac{22}{75}$

$\dfrac{31}{45} - \dfrac{22}{75} = \dfrac{89}{225}$ On a calculator with fraction keys, it is not necessary to find the common denominator. The calculator is programmed to subtract and simplify.

WARM-UP

F. Mike must plane $\frac{3}{32}$ in. from the thickness of a board. If the board is now $\frac{3}{8}$ in. thick, how thick will it be after he has planed it?

ANSWER TO WARM-UP F

F. The board will be $\frac{9}{32}$ in. thick.

F. Lumber mill operators must plan for the shrinkage of "green" (wet) boards when they cut logs. If the shrinkage for a $\frac{5}{8}$-in.-thick board is expected to be $\frac{1}{16}$ in., what will the thickness of the dried board be?

STRATEGY: To find the thickness of the dried board, subtract the shrinkage from the thickness of the green board.

$$\frac{5}{8} - \frac{1}{16} = \frac{10}{16} - \frac{1}{16} \qquad \text{Build } \frac{5}{8} \text{ to a fraction with denominator 16.}$$
$$= \frac{9}{16}$$

The dried board will be $\frac{9}{16}$ in. thick.

EXERCISES 3.8

OBJECTIVE Subtract fractions. (See page 233.)

A *Subtract. Simplify completely.*

1. $\frac{5}{6} - \frac{1}{6}$
2. $\frac{5}{8} - \frac{3}{8}$
3. $\frac{8}{9} - \frac{5}{9}$
4. $\frac{9}{20} - \frac{3}{20}$

5. $\frac{11}{16} - \frac{5}{16}$
6. $\frac{13}{15} - \frac{8}{15}$
7. $\frac{5}{7} - \frac{3}{14}$
8. $\frac{11}{15} - \frac{2}{5}$

9. $\frac{3}{8} - \frac{5}{16}$
10. $\frac{8}{9} - \frac{5}{18}$
11. $\frac{3}{15} - \frac{2}{45}$
12. $\frac{11}{18} - \frac{4}{9}$

13. $\frac{11}{21} - \frac{3}{7}$
14. $\frac{5}{6} - \frac{1}{3}$
15. $\frac{13}{18} - \frac{2}{3}$
16. $\frac{5}{8} - \frac{5}{24}$

17. $\frac{17}{24} - \frac{1}{8}$
18. $\frac{19}{30} - \frac{1}{5}$
19. $\frac{17}{54} - \frac{1}{18}$
20. $\frac{17}{36} - \frac{1}{4}$

B

21. $\frac{7}{8} - \frac{5}{6}$
22. $\frac{4}{5} - \frac{5}{8}$
23. $\frac{11}{15} - \frac{7}{20}$
24. $\frac{7}{15} - \frac{3}{20}$

25. $\frac{7}{9} - \frac{5}{18}$
26. $\frac{2}{3} - \frac{4}{15}$
27. $\frac{9}{16} - \frac{1}{6}$
28. $\frac{5}{6} - \frac{4}{5}$

29. $\frac{13}{20} - \frac{7}{30}$
30. $\frac{4}{7} - \frac{5}{14}$
31. $\frac{11}{12} - \frac{11}{18}$
32. $\frac{7}{10} - \frac{7}{15}$

33. $\frac{7}{10} - \frac{1}{4}$
34. $\frac{8}{15} - \frac{5}{12}$
35. $\frac{7}{10} - \frac{4}{15}$
36. $\frac{13}{15} - \frac{7}{12}$

37. $\frac{17}{48} - \frac{1}{16}$
38. $\frac{21}{32} - \frac{5}{16}$
39. $\frac{13}{16} - \frac{11}{24}$
40. $\frac{17}{18} - \frac{5}{12}$

C

41. $\frac{13}{18} - \frac{7}{12}$
42. $\frac{7}{20} - \frac{1}{8}$
43. $\frac{17}{50} - \frac{13}{40}$
44. $\frac{32}{35} - \frac{17}{20}$

45. $\frac{22}{35} - \frac{3}{40}$
46. $\frac{17}{42} - \frac{8}{63}$

47. According to research from Clairol, Inc., $\frac{4}{5}$ of all Caucasian Americans have brown or black hair (naturally). What fraction of Caucasian Americans have naturally blonde or red hair?

48. On July 1, the reservoir at Bull Run watershed was at $\frac{3}{4}$ capacity. During the month, the reservoir lost $\frac{1}{20}$ of its capacity due to evaporation. What fraction of its capacity does the reservoir hold at the end of July?

49. In May 2009, Willie received a stimulus check from the federal government for $480. He spent $\frac{1}{3}$ of it to pay his bills and put $\frac{1}{4}$ in savings. What fraction of the stimulus does he still have left? How many dollars does he have?

50. A water sample from Lake Tuscumba contains 21 parts per million of phosphate. A sample from Lost Lake contains 2 parts per hundred thousand. Which lake has the greater phosphate content? By how much?

51. In a recent year, 16 of every 25 Americans owned stock in a publicly held company or mutual fund.

What fraction of Americans did not own stock?

52. In a recent year, approximately 138 million people attended a professional ice hockey, basketball, football, or baseball game. If $\frac{14}{25}$ of them attended baseball, $\frac{3}{20}$ of them attended basketball, and $\frac{3}{25}$ of them attended football, what fraction attended ice hockey?

53. In April 2009, Oregon had a workforce of approximately 2,000,000. During the same period approximately 3 out of 25 people were unemployed. what fraction of the work force was employed?

54. The Oregon workforce in April 2009 had approximately $\frac{1}{10}$ employed in manufacturing and $\frac{9}{50}$ in trade, transportation, and utilities. What fraction of the workforce was in other job categories?

55. According to the U.S. Bureau of the Census, in July 2008, $\frac{17}{25}$ of Americans 80 or older were female. What portion of this population was male?

56. Oxygen and carbon are the two most plentiful elements in the human body. On average, $\frac{13}{20}$ of the body is oxygen and $\frac{9}{50}$ of the body is carbon. What part of the body do the remaining elements account for?

57. The diameter at the large end of a tapered pin is $\frac{7}{8}$ in., and at the smaller end, it is $\frac{3}{16}$ in. What is the difference between the diameters?

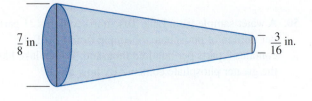

$\frac{7}{8}$ in. $-\frac{3}{16}$ in.

58. Aunt Gertrude left her estate to her four nephews. Two of them each received $\frac{1}{4}$ of the estate, and the third received $\frac{3}{8}$ of the estate. What portion of the estate did the fourth nephew receive?

59. President Barack Obama's proposed federal budget for fiscal year 2010 totaled $3,550,000,000,000. The budget was divided into categories of mandatory spending, discretionary spending, and interest/disaster relief. The mandatory spending was $\frac{14}{25}$ of the total and the interest/disaster relief payments were $\frac{1}{20}$ of the total. What part of the proposed budget was for discretionary spending?

STATE YOUR UNDERSTANDING

60. Explain in writing how you would teach a child to subtract fractions.

61. Explain why $\frac{3}{4} - \frac{1}{2}$ is not equal to $\frac{2}{2}$.

CHALLENGE

Subtract.

62. $\frac{213}{560} - \frac{109}{430}$

63. $\frac{93}{125} - \frac{247}{625}$

64. A donor agrees to donate $1000 for each foot that Skola outdistances Sheila in 13 minutes. Skola walks $\frac{19}{24}$ mi. Sheila walks $\frac{47}{60}$ mi. Does Skola outdistance Sheila? By what fraction of a mile? How much does the donor contribute? (A mile equals 5280 ft.)

65. A landscaper is building a brick border, one brick-length wide (8 in.) around a formal rose garden. The garden is a 10-ft-by-6-ft rectangle. Standard bricks are 8 in. by $3\frac{3}{4}$ in. by $2\frac{1}{4}$ in., and the landscaper is planning to use a $\frac{3}{8}$-in.-wide mortar in the joints. How many whole bricks are needed for the project? Explain your reasoning.

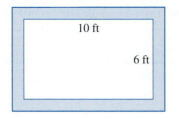

MAINTAIN YOUR SKILLS

Subtract.

66. $56{,}789 - 32{,}671$

67. $21{,}596 - 9396$

Subtract. Write as a mixed number

68. $(11 - 9) + \left(\dfrac{5}{8} - \dfrac{5}{12}\right)$

69. $(22 - 16) + \left(\dfrac{2}{3} - \dfrac{4}{15}\right)$

70. $(92 - 87) + \left(\dfrac{2}{3} - \dfrac{7}{18}\right)$

71. $(34 - 27) + \left(\dfrac{7}{10} - \dfrac{1}{4}\right)$

72. $(54 - 45) + \left(\dfrac{3}{4} - \dfrac{5}{12}\right)$

73. $(63 - 36) + \left(\dfrac{3}{4} - \dfrac{1}{20}\right)$

74. Three bricklayers can each lay 795 bricks per day, on the average. How many bricks can they lay in 5 days?

75. If a retaining wall requires 19,080 bricks, how many days will it take the three bricklayers in Exercise 74 to build the wall?

Subtracting Mixed Numbers

HOW & WHY

■ **OBJECTIVE**   Subtract mixed numbers.

A subtraction problem may be written in horizontal or vertical form. Horizontally:

$$5\frac{7}{11} - 3\frac{5}{11} = (5 - 3) + \left(\frac{7}{11} - \frac{5}{11}\right)$$

Because the denominators are the same, subtract the whole number parts and then subtract the fraction parts.

$$= 2 + \frac{2}{11}$$

$$= 2\frac{2}{11}$$

Vertically:

$$5\frac{7}{11}$$

The process is similar to that for adding mixed numbers.

$$-3\frac{5}{11}$$

$$2\frac{2}{11}$$

It is sometimes necessary to "borrow" from the whole number in order to subtract the fractions. For example,

$$8\frac{2}{5} - 3\frac{3}{4} = ?$$

First, write in columns and build each fraction to the common denominator, 20.

$$8\frac{2}{5} = 8\frac{8}{20}$$

$$-3\frac{3}{4} = 3\frac{15}{20}$$

Because we cannot subtract $\frac{15}{20}$ from $\frac{8}{20}$, we need to "borrow." To do this we rename $8\frac{8}{20}$ by "borrowing" 1 from 8.

$$8\frac{8}{20} = 7 + 1\frac{8}{20}$$

Borrow 1 from 8 and add it to the fraction part.

$$= 7 + \frac{28}{20}$$

Change the mixed number, $1\frac{8}{20}$, to an improper fraction.

$$= 7\frac{28}{20}$$

Write as a mixed number.

CAUTION

Do not write $\frac{18}{20}$. If we "borrow" 1 from 8, we must add 1 $\left(\text{that is, } \frac{20}{20}\right)$ to $\frac{8}{20}$.

The example can now be completed.

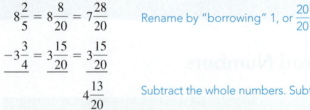

$$8\frac{2}{5} = 8\frac{8}{20} = 7\frac{28}{20}$$

Rename by "borrowing" 1, or $\frac{20}{20}$.

$$-3\frac{3}{4} = 3\frac{15}{20} = 3\frac{15}{20}$$

$$4\frac{13}{20}$$

Subtract the whole numbers. Subtract the fractions.

To subtract mixed numbers

1. Build the fractions so they have a common denominator.
2. Subtract the fractions. If the fractions cannot be subtracted, rename the first mixed number by "borrowing" 1 from the whole-number part to add to the fraction part. Then subtract the fractions.
3. Subtract the whole numbers.
4. Simplify.

EXAMPLES A–G

DIRECTIONS: Subtract. Write as a mixed number.

STRATEGY: Subtract the fractions and subtract the whole numbers. If necessary, borrow. Simplify.

A. Subtract: $21\dfrac{7}{8} - 15\dfrac{5}{12}$

STRATEGY: Write the mixed numbers in columns to group the whole numbers and group the fractions.

$$\begin{aligned}
21\dfrac{7}{8} &= 21\dfrac{21}{24} \qquad \text{\color{blue}Build the fractions to the common denominator, 24.}\\
-15\dfrac{5}{12} &= 15\dfrac{10}{24}\\[4pt]
\hline
&= 6\dfrac{11}{24} \qquad \text{\color{blue}Subtract.}
\end{aligned}$$

B. Subtract: $17\dfrac{2}{3} - 10$

$$17\dfrac{2}{3} - 10 = 7\dfrac{2}{3} \qquad \text{\color{blue}Subtract the whole numbers.}$$

C. Subtract: $21 - 9\dfrac{3}{4}$

STRATEGY: Notice the difference between Examples B and C. Here we must also subtract the fraction. We may think of 21 as $21\dfrac{0}{4}$ in order to get a common denominator for the improper fraction. Or think

$$21 = 20 + 1 = 20 + \dfrac{4}{4} = 20\dfrac{4}{4}$$
$$\begin{aligned}
&\quad -9\dfrac{3}{4}\\
\hline
&\ \ 11\dfrac{1}{4}
\end{aligned}$$

CAUTION

Do not just bring down the fraction $\dfrac{3}{4}$ and then subtract the whole numbers:

$21 - 9\dfrac{3}{4} \neq 12\dfrac{3}{4}$. The fraction must also be subtracted.

WARM-UP

A. Subtract: $22\dfrac{3}{5} - 15\dfrac{2}{9}$

WARM-UP

B. Subtract: $56\dfrac{7}{24} - 34$

WARM-UP

C. Subtract: $47 - 29\dfrac{13}{15}$

ANSWERS TO WARM-UPS A–C

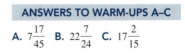

A. $7\dfrac{17}{45}$ **B.** $22\dfrac{7}{24}$ **C.** $17\dfrac{2}{15}$

D. Subtract: $18\dfrac{11}{15} - 8\dfrac{12}{15}$

D. Subtract: $23\dfrac{3}{8} - 9\dfrac{7}{8}$

STRATEGY: Because $\dfrac{7}{8}$ cannot be subtracted from $\dfrac{3}{8}$, we need to borrow.

$$23\dfrac{3}{8} = 22 + 1\dfrac{3}{8} = 22\dfrac{11}{8}$$ Borrow 1 from 23 and change the mixed number to an improper fraction.

$$-9\dfrac{7}{8} \qquad\qquad = \; 9\dfrac{7}{8}$$ Improper fraction.

$$\overline{\qquad\qquad\qquad 13\dfrac{4}{8}}$$ Subtract.

$$= \; 13\dfrac{1}{2}$$ Simplify.

WARM-UP

E. Subtract: $69\dfrac{17}{30} - 53\dfrac{13}{20}$

E. Subtract: $23\dfrac{5}{12} - 11\dfrac{7}{15}$

$$23\dfrac{5}{12} = 23\dfrac{25}{60} = 22 + 1\dfrac{25}{60} = 22\dfrac{85}{60}$$

$$-11\dfrac{7}{15} = 11\dfrac{28}{60} \qquad\qquad = 11\dfrac{28}{60}$$

The LCM of 12 and 15 is 60. Borrow 1 from 23 and change the mixed number to an improper fraction.

$$\overline{\qquad\qquad\qquad\qquad 11\dfrac{57}{60}}$$ Subtract.

$$= 11\dfrac{19}{20}$$ Simplify.

CALCULATOR EXAMPLE:

WARM-UP

F. Subtract: $72\dfrac{7}{9} - 25\dfrac{13}{15}$

F. Subtract: $31\dfrac{1}{3} - 18\dfrac{3}{4}$

$$31\dfrac{1}{3} - 18\dfrac{3}{4} = \dfrac{151}{12}$$

$$= 12\dfrac{7}{12}$$

On a calculator with fraction keys, it is not necessary to find the common denominator. The calculator is programmed to subtract and simplify. Some calculators may not change the improper result to a mixed number.

D. $9\dfrac{14}{15}$

E. $15\dfrac{11}{12}$

F. $46\dfrac{41}{45}$

G. Jane enters a walkathon for cancer research that is 12 miles long. After one and a half hours she has walked $4\frac{3}{8}$ miles. How many miles does she have left to finish the walkathon?

STRATEGY: Subtract the miles she has walked from the distance of the walkathon.

$$12 = 11 + 1\frac{0}{8} = 11\frac{8}{8}$$ Borrow 1 from 12 and rename it as a improper fraction.

$$-4\frac{3}{8} \qquad = \quad 4\frac{3}{8}$$

$$\qquad\qquad = \quad 7\frac{5}{8}$$

Jane has $7\frac{5}{8}$ miles left to finish the walkathon.

EXERCISES 3.9

■ **OBJECTIVE** Subtract mixed numbers. (See page 239.)

A *Subtract. Write the results as mixed numbers where possible.*

1. $12\frac{9}{13}$
   $-5\frac{7}{13}$

2. $31\frac{5}{9}$
   $-20\frac{2}{9}$

3. $147\frac{49}{90}$
   $-135\frac{19}{90}$

4. $205\frac{6}{11}$
   $-199\frac{4}{11}$

5. $10\frac{7}{8}$
   $-5\frac{3}{4}$

6. $19\frac{3}{8}$
   $-8\frac{3}{16}$

7. 12
   $-9\frac{4}{7}$

8. 16
   $-12\frac{1}{3}$

9. $87\frac{5}{12}$
   $-71\frac{2}{15}$

10. $66\frac{5}{6}$
    $-52\frac{9}{10}$

11. $5\frac{1}{4}$
    $-2\frac{3}{4}$

12. $7\frac{3}{8}$
    $-4\frac{5}{8}$

13. $18\frac{5}{12}$
    $-9\frac{7}{12}$

14. $17\frac{3}{8}$
    $-5\frac{5}{8}$

15. $234\frac{7}{9}$
    -62

16. $285\frac{3}{7}$
    -171

17. $11\frac{2}{3} - 5\frac{1}{4}$

18. $8\frac{1}{2} - 6\frac{5}{12}$

B

19. $31\dfrac{2}{15}$

$\quad -22\dfrac{11}{15}$

20. $24\dfrac{7}{12}$

$\quad -15\dfrac{11}{12}$

21. $310\dfrac{23}{24}$

$\quad -254\dfrac{5}{8}$

22. $118\dfrac{7}{12}$

$\quad -93\dfrac{1}{4}$

23. $66\dfrac{7}{15}$

$\quad -51\dfrac{1}{12}$

24. $19\dfrac{9}{16}$

$\quad -13\dfrac{5}{12}$

25. $64\dfrac{3}{10}$

$\quad -41\dfrac{7}{15}$

26. $28\dfrac{1}{3}$

$\quad -15\dfrac{7}{9}$

27. $40\dfrac{1}{6}$

$\quad -24\dfrac{3}{16}$

28. $73\dfrac{9}{20}$

$\quad -55\dfrac{19}{30}$

29. 78

$\quad -14\dfrac{4}{5}$

30. 62

$\quad -16\dfrac{2}{3}$

31. $3\dfrac{17}{32}$

$\quad -1\dfrac{5}{16}$

32. $9\dfrac{3}{10}$

$\quad -\dfrac{17}{20}$

33. $10\dfrac{2}{5}$

$\quad -\dfrac{5}{8}$

34. $4\dfrac{5}{6}$

$\quad -\dfrac{7}{8}$

35. $93\dfrac{7}{8}$

$\quad -19$

36. $99\dfrac{5}{12}$

$\quad -37$

37. $46\dfrac{14}{15}$

$\quad -19\dfrac{27}{40}$

38. $60\dfrac{13}{18}$

$\quad -42\dfrac{37}{45}$

39. $92\dfrac{3}{16} - 17\dfrac{11}{24}$

40. $82\dfrac{4}{15} - 56\dfrac{7}{12}$

C

41. $34\dfrac{2}{39}$

$\quad -17\dfrac{21}{26}$

42. $38\dfrac{5}{24}$

$\quad -21\dfrac{3}{40}$

43. $6\dfrac{5}{21} - 5\dfrac{7}{15}$

44. $6\dfrac{1}{3} - 5\dfrac{5}{6}$

45. Find the error(s) in the statement: $16 - 13\dfrac{1}{4} = 3\dfrac{3}{4}$.
Correct the statement.

46. Find the error(s) in the statement: $5\dfrac{1}{2} - 2\dfrac{3}{4} = 3\dfrac{1}{4}$.
Correct the statement.

47. Han Kwong trims bone and fat from an $8\frac{1}{2}$-lb roast. The meat left weighs $6\frac{1}{8}$ lb. How many pounds does she trim off?

48. Patti has a piece of lumber that measures $8\frac{5}{12}$ ft that is to be used in a spot that calls for a length of $6\frac{1}{2}$ ft. How much of the board must be cut off?

49. Dick harvests $30\frac{3}{4}$ tons of wheat. He sells $18\frac{7}{10}$ tons to the Cartwright Flour Mill. How many tons of wheat does he have left?

50. A $14\frac{3}{4}$-in. casting shrinks $\frac{5}{32}$ in. on cooling. Find the size when the casting is cold.

51. Jason Colwick of Rice won the 2009 NCAA pole vault championship with a vault of 18 ft $8\frac{1}{4}$ in. Second place was won by Scott Roth of Washington, with a vault of 18 ft $2\frac{1}{2}$ in. By how many inches did Jason Colwick win?

52. According to the International Game Fish Association, the largest cubera snapper ever caught weighed $121\frac{5}{8}$ lb and was caught in Louisiana in 1982 by Mike Hebart. The largest red snapper ever caught weighed $50\frac{1}{3}$ lb and was caught in the Gulf of Mexico off Louisiana in 1996 by Doc Kennedy. How much larger was the record cubera snapper than the red snapper?

53. The Boeing 777-200LR has a length of $209\frac{1}{2}$ ft. Its sister plane the 777-300ER has a length of $242\frac{1}{3}$ ft. How much longer is the 777-300ER?

© Vladimir Melnik/Shutterstock.com

54. A McDonnell Douglas DC-9 seats 158 and is $147\frac{5}{6}$ ft long. A Boeing 737 seats 128 to 149 and is $109\frac{7}{12}$ ft long. How much longer is the DC-9?

55. Larry and Greg set out to hike 42 mi in 2 days. At the end of the first day they have covered $22\frac{7}{10}$ mi. How many miles do they have to go?

56. Frank pours $9\frac{1}{10}$ yd of cement for a fountain. Another fountain takes $6\frac{7}{8}$ yd. How much more cement is needed for the larger fountain?

57. Joaquim bought a 48-in. board to make shelves in his closet. One shelf is $25\frac{3}{8}$ in. and the other is $16\frac{3}{4}$ in. How much of the board will be left over?

58. Most flights that originate outside the United States have a total checked baggage limit of only 20 kg for passengers flying coach. Nadia has baggage that weighs $31\frac{3}{5}$ kg. How much extra weight is she charged for?

59. Haja needs to replace a bolt that holds her headboard on the bed frame. She wants to use the washer that is pictured below. What is the largest-diameter bolt she can buy and use the washer?

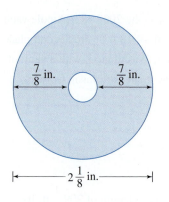

Exercise 60 relates to the chapter application. See page 167.

60. Shane is almost done installing the deck boards on his deck. He has already installed 67 boards. How many inches of his 272-in. deck still need to be covered?

STATE YOUR UNDERSTANDING

61. Explain how to simplify $4\frac{1}{3} - 2\frac{5}{8}$.

62. When you "borrow" 1 to subtract mixed numbers, explain the fraction form it is written in and explain why.

63. Does $4\left(5\dfrac{5}{6}\right) - 3\left(2\dfrac{7}{12}\right) = 6\left(5\dfrac{1}{2}\right) - 5\dfrac{3}{4}$?

64. Does $3\left(7\dfrac{7}{8}\right) - 13\dfrac{1}{10} = 6\left(3\dfrac{3}{16}\right) - 7\left(1\dfrac{8}{35}\right)$?

65. A snail climbs $5\dfrac{1}{3}$ ft in a day and slips back $1\dfrac{2}{3}$ ft at night. What is the snail's net distance in 24 hr? How many days will it take the snail to make a net gain of over 20 ft?

MAINTAIN YOUR SKILLS

Simplify.

66. $97 - 5 \cdot 3 + 12$

67. $8^2 - 4 \cdot 3 + 10$

68. $82 - 4^3 + 10$

69. $14 \cdot 3 \div 7 \cdot 3$

70. $6 \cdot 18 \div 3^2$

71. $5 + 32 \div 4^2$

Multiply.

72. $\dfrac{9}{16} \cdot \dfrac{24}{35} \cdot \dfrac{14}{15}$

73. $\dfrac{1}{2} \cdot \dfrac{5}{8} \cdot \dfrac{3}{5} \cdot 24$

74. Last week when Karla filled the tank of her car with gasoline, the odometer read 89,735 miles. Yesterday when she filled the tank with 18 gallons of gasoline, the odometer read 90,545 miles. What is Karla's mileage, that is, how many miles to the gallon did she get?

75. A pet food canning company packs Feelein Cat Food in cans, each containing $7\dfrac{3}{4}$ oz of cat food. Each empty can weights $1\dfrac{1}{2}$ oz. Twenty-four cans are packed in a case that weighs 10 oz empty. What is the shipping weight of five cases of the cat food?

OBJECTIVE

Solve equations of the form $x + \dfrac{a}{b} = \dfrac{c}{d}$, where a, b, c, and d are whole numbers.

HOW & WHY

We have solved equations in which whole numbers are either added to or subtracted from a variable. Now we solve equations in which fractions or mixed numbers are either added to or subtracted from the variable. We use the same procedure as with whole numbers.

EXAMPLES A–C

DIRECTIONS: Solve.

STRATEGY: Add or subtract the same number from each side of the equation to isolate the variable.

WARM-UP

A. Solve: $x - \dfrac{5}{8} = 3\dfrac{7}{8}$

A. Solve: $x - \dfrac{4}{5} = 3\dfrac{1}{2}$

$x - \dfrac{4}{5} + \dfrac{4}{5} = 3\dfrac{1}{2} + \dfrac{4}{5}$ To eliminate the subtraction, add $\dfrac{4}{5}$ to each side of the equation.

$x = 3\dfrac{1}{2} + \dfrac{4}{5}$ Simplify the left side.

$x = 3\dfrac{5}{10} + \dfrac{8}{10}$ Build each fraction to the common denominator, 10.

$x = 3\dfrac{13}{10}$ Add.

$x = 4\dfrac{3}{10}$ Change the improper fraction to a mixed number and add.

CHECK: $4\dfrac{3}{10} - \dfrac{4}{5} = 3\dfrac{1}{2}$ Substitute $4\dfrac{3}{10}$ for x in the original equation.

$3\dfrac{1}{2} = 3\dfrac{1}{2}$

The solution is $x = 4\dfrac{3}{10}$.

WARM-UP

B. Solve: $6\dfrac{1}{2} = a + 5\dfrac{5}{8}$

B. Solve: $4\dfrac{1}{2} = x + 2\dfrac{2}{3}$

$4\dfrac{1}{2} - 2\dfrac{2}{3} = x + 2\dfrac{2}{3} - 2\dfrac{2}{3}$ To eliminate the addition, subtract $2\dfrac{2}{3}$ from each side of the equation.

$4\dfrac{1}{2} - 2\dfrac{2}{3} = x$

$1\dfrac{5}{6} = x$

CHECK: $4\dfrac{1}{2} = 1\dfrac{5}{6} + 2\dfrac{2}{3}$ Substitute $1\dfrac{5}{6}$ for x in the original equation.

$4\dfrac{1}{2} = 4\dfrac{1}{2}$

The solution is $x = 1\dfrac{5}{6}$.

ANSWERS TO WARM-UPS A–B

A. $x = 4\dfrac{1}{2}$ **B.** $a = \dfrac{7}{8}$

C. On Tuesday, $2\frac{3}{8}$ in. of rain fell on Springfield. This brought the total for the last 5 consecutive days to $14\frac{1}{2}$ in. What was the rainfall for the first 4 days?

STRATEGY: First write the English version of the equation.

$$\left(\begin{array}{c}\text{rain on first}\\ \text{4 days}\end{array}\right) + \left(\begin{array}{c}\text{rain on}\\ \text{Tuesday}\end{array}\right) = \text{total rain}$$

Let x represent the number of inches of rain on the first 4 days.

$x + 2\dfrac{3}{8} = 14\dfrac{1}{2}$ Translate to algebra.

$\underline{-2\dfrac{3}{8} \qquad -2\dfrac{3}{8}}$ Subtract $2\dfrac{3}{8}$ from each side.

$x = 12\dfrac{1}{8}$

Since $12\dfrac{1}{8} + 2\dfrac{3}{8} = 14\dfrac{1}{2}$, $12\dfrac{1}{8}$ in. of rain fell during the first 4 days.

WARM-UP

C. When $3\frac{7}{8}$ mi of new freeway opened last month, it bought the total length to $21\frac{3}{5}$ mi. What was the original length of the freeway?

ANSWER TO WARM-UP C

C. The original freeway was $17\frac{29}{40}$ mi long.

EXERCISES

Solve.

1. $a + \dfrac{1}{8} = \dfrac{5}{8}$

2. $y + \dfrac{3}{8} = \dfrac{7}{8}$

3. $c - \dfrac{3}{16} = \dfrac{7}{16}$

4. $w + \dfrac{5}{12} = \dfrac{11}{12}$

5. $x + \dfrac{2}{9} = \dfrac{3}{8}$

6. $x - \dfrac{7}{8} = \dfrac{3}{4}$

7. $y - \dfrac{5}{7} = \dfrac{8}{9}$

8. $y + \dfrac{5}{9} = \dfrac{9}{10}$

9. $a + \dfrac{9}{8} = \dfrac{12}{5}$

10. $a - \dfrac{5}{4} = \dfrac{3}{8}$

11. $c - 1\dfrac{1}{8} = 2\dfrac{1}{3}$

12. $c + 1\dfrac{1}{8} = 3\dfrac{3}{4}$

13. $x + 6\dfrac{3}{4} = 7\dfrac{7}{9}$

14. $x - 2\dfrac{5}{9} = 2\dfrac{5}{8}$

15. $12 = w + 8\dfrac{5}{6}$

16. $25 = m + 15\dfrac{5}{8}$

17. $a - 13\dfrac{5}{6} = 22\dfrac{11}{18}$

18. $b + 23\dfrac{11}{12} = 34\dfrac{1}{3}$

19. $c + 44\dfrac{13}{21} = 65\dfrac{5}{7}$

20. $x - 27\dfrac{5}{8} = 48\dfrac{2}{3}$

21. A native pine tree grew $1\frac{15}{16}$ ft in the past 10 years to its present height of $45\frac{1}{2}$ ft. What was the height of the tree 10 years ago?

22. Juan brought in $35\frac{3}{4}$ lb of tin to be recycled. This brings his total for the month to $122\frac{1}{4}$ lb. How many pounds had he already brought in this month?

23. Freeda purchased a supply of nails for a construction project. She has used $18\frac{2}{3}$ lb and has $27\frac{1}{3}$ lb left. How many pounds of nails did she buy?

24. For cross-country race practice, Althea has run $10\frac{7}{10}$ mi. She needs to run an additional $13\frac{3}{10}$ mi to meet the goal set by her coach. How many miles does the coach want her to run?

Order of Operations; Average

OBJECTIVES

1. Do any combinations of operations with fractions.
2. Find the average of a group of fractions.

VOCABULARY

Recall that the **mean**, or **average**, of a set of numbers is the sum of the set of numbers divided by the total number of numbers in the set.

HOW & WHY

■ **OBJECTIVE 1** Do any combinations of operations with fractions.

The order of operations for fractions is the same as for whole numbers.

ORDER OF OPERATIONS

To evaluate an expression with more than one operation

1. Parentheses—Do the operations within grouping symbols first (parentheses, fraction bar, etc.) in the order given in steps 2, 3, and 4.
2. Exponents—Do the operations indicated by exponents.
3. Multiply and Divide—Do multiplication and division as they appear from left to right.
4. Add and Subtract—Do addition and subtraction as they appear from left to right.

Table 3.4 summarizes some of the processes that need to be remembered when working with fractions.

TABLE 3.4   Operations with Fractions

Operation	Find the LCM and Build	Change Mixed Numbers to Improper Fractions	Invert Divisor and Multiply	Simplify Answer
Add	Yes	No	No	Yes
Subtract	Yes	No	No	Yes
Multiply	No	Yes	No	Yes
Divide	No	Yes	Yes	Yes

EXAMPLES A–D

DIRECTIONS: Perform the indicated operations.

STRATEGY: Follow the order of operations that are used for whole numbers.

WARM-UP

A. Simplify: $\dfrac{9}{14} - \dfrac{3}{7} \cdot \dfrac{5}{6}$

A. Simplify: $\dfrac{5}{6} - \dfrac{1}{2} \cdot \dfrac{2}{3}$

$\dfrac{5}{6} - \dfrac{1}{2} \cdot \dfrac{2}{3} = \dfrac{5}{6} - \dfrac{1}{3}$ Multiplication is performed first.

$= \dfrac{5}{6} - \dfrac{2}{6}$ Build $\dfrac{1}{3}$ to a denominator of 6.

$= \dfrac{3}{6}$ Subtract.

$= \dfrac{1}{2}$ Simplify.

WARM-UP

B. Simplify: $\dfrac{3}{8} \div \dfrac{5}{16} \cdot \dfrac{1}{3}$

B. Simplify: $\dfrac{1}{6} \div \dfrac{2}{5} \cdot \dfrac{3}{5}$

$\dfrac{1}{6} \div \dfrac{2}{5} \cdot \dfrac{3}{5} = \dfrac{1}{6} \cdot \dfrac{5}{2} \cdot \dfrac{3}{5}$ Division is performed first, as it appears from left to right.

$= \dfrac{15}{60}$ Multiply from left to right.

$= \dfrac{1}{4}$ Simplify.

WARM-UP

C. Simplify: $\left(\dfrac{2}{5}\right)^2 \cdot \dfrac{5}{8} - \dfrac{1}{20}$

C. Simplify: $\left(\dfrac{2}{3}\right)^2 \cdot \dfrac{1}{2} - \dfrac{1}{5}$

$\left(\dfrac{2}{3}\right)^2 \cdot \dfrac{1}{2} - \dfrac{1}{5} = \dfrac{\overset{2}{\cancel{4}}}{9} \cdot \dfrac{1}{\underset{1}{\cancel{2}}} - \dfrac{1}{5}$ Exponentiation is done first, then simplify.

$= \dfrac{2}{9} - \dfrac{1}{5}$ Multiply.

$= \dfrac{10}{45} - \dfrac{9}{45}$ Build to a common denominator.

$= \dfrac{1}{45}$ Subtract.

ANSWERS TO WARM-UPS A–C

A. $\dfrac{2}{7}$

B. $\dfrac{2}{5}$

C. $\dfrac{1}{20}$

D. Jorge and Ramona agree to share a job. Jorge works $\frac{5}{8}$ of the job and Ramona $\frac{3}{8}$ of the job. When Jorge got a second job, he decided to give Ramona $\frac{3}{5}$ of his part of the first job. How much of the total job does Ramona have now?

STRATEGY: Find $\frac{3}{5}$ of Jorge's share of the job, $\frac{5}{8}$, and add it to Ramona's share, $\frac{3}{8}$.

$$\frac{3}{8} + \frac{3}{5} \cdot \frac{5}{8} = \frac{3}{8} + \frac{3}{8} \qquad \text{Multiply first.}$$

$$= \frac{6}{8} \qquad \text{Add.}$$

$$= \frac{3}{4} \qquad \text{Simplify.}$$

Ramona's now has $\frac{3}{4}$ of the job.

WARM-UP

D. Jill, Jean, and Joan have equal shares in a gift shop. Jill sells her share. She sells $\frac{1}{4}$ to Jean, and the rest to Joan. What share of the gift shop does Joan now own?

HOW & WHY

■ **OBJECTIVE 2** Find the average of a group of fractions.

To find the average of a set of fractions, divide the sum of the fractions by the number of fractions. The procedure is the same for all numbers.

> **To find the average of a set of numbers**
>
> 1. Add the numbers.
> 2. Divide the sum by the number of numbers in the set.

EXAMPLES E–G

DIRECTIONS: Find the mean (average).

STRATEGY: Find the sum of the set of fractions, then divide by the number of fractions.

E. Find the mean: $\frac{2}{5}, \frac{3}{10},$ and $\frac{1}{2}$

$$\frac{2}{5} + \frac{3}{10} + \frac{1}{2} = \frac{4}{10} + \frac{3}{10} + \frac{5}{10} \qquad \text{Add the three fractions in the set.}$$

$$= \frac{12}{10}, \text{ or } \frac{6}{5}$$

$$\frac{6}{5} \div 3 = \frac{6}{5} \cdot \frac{1}{3} \qquad \text{Divide the sum by 3, the number of fractions in the set.}$$

$$= \frac{2}{5}$$

The mean is $\frac{2}{5}$.

WARM-UP

E. Find the mean: $\frac{1}{6}, \frac{5}{8},$ and $\frac{3}{4}$

ANSWERS TO WARM-UPS D–E

D. Joan owns $\frac{7}{12}$ of the gift shop.

E. $\frac{37}{72}$

F. A class of 12 students takes a 20-problem test. The results are listed in the table. What is the average score?

Number of Students	Fraction of Problems Correct
1	$\dfrac{20}{20}$
2	$\dfrac{19}{20}$
4	$\dfrac{16}{20}$
5	$\dfrac{14}{20}$

F. A class of 10 students takes a 12-problem quiz. The results are listed in the table. What is the average score?

Number of Students	Fraction of Problems Correct
1	$\dfrac{12}{12}$
2	$\dfrac{11}{12}$
3	$\dfrac{10}{12}$
4	$\dfrac{9}{12}$

STRATEGY: To find the class average, add all the grades and divide by 10. There were two scores of $\dfrac{11}{12}$, three scores of $\dfrac{10}{12}$, and four scores of $\dfrac{9}{12}$ in addition to one perfect score of $\dfrac{12}{12}$.

$$\frac{12}{12} + 2\left(\frac{11}{12}\right) + 3\left(\frac{10}{12}\right) + 4\left(\frac{9}{12}\right) \qquad \text{Find the sum of the 10 scores.}$$

$$\left(\frac{12}{12} + \frac{22}{12} + \frac{30}{12} + \frac{36}{12}\right) = \frac{100}{12}$$

$$\frac{100}{12} \div 10 = \frac{100}{12} \cdot \frac{1}{10} \qquad \text{Divide the sum by the number of students in the class.}$$

$$= \frac{10}{12}$$

CAUTION

Do not simplify the answer, because the test scores are based on 12.

The class average is $\dfrac{10}{12}$ of the problems correct, or 10 problems correct.

G. Find the average: $2\dfrac{3}{5}, 4\dfrac{1}{4}, 4\dfrac{3}{10}$, and $1\dfrac{1}{20}$

G. Find the average: $9\dfrac{2}{3}, 10\dfrac{5}{6}$, and $11\dfrac{3}{4}$

$$9\frac{2}{3} + 10\frac{5}{6} + 11\frac{3}{4} = 32\frac{1}{4}$$

$$32\frac{1}{4} \div 3 = \frac{129}{4} \cdot \frac{1}{3}$$

$$= \frac{43}{4}$$

$$= 10\frac{3}{4}$$

The average is $10\dfrac{3}{4}$.

ANSWERS TO WARM-UPS F–G

F. The class average is $\dfrac{16}{20}$ of the problems correct. **G.** $3\dfrac{1}{20}$

OBJECTIVE 1 Do any combinations of operations with fractions. (See page 251.)

A *Perform the indicated operations.*

1. $\dfrac{4}{11} + \dfrac{9}{11} - \dfrac{6}{11}$

2. $\dfrac{7}{15} - \dfrac{2}{15} + \dfrac{4}{15}$

3. $\dfrac{5}{17} - \left(\dfrac{1}{17} + \dfrac{2}{17} \right)$

4. $\dfrac{5}{17} - \left(\dfrac{2}{17} - \dfrac{1}{17} \right)$

5. $\dfrac{7}{8} - \dfrac{3}{4} \cdot \dfrac{1}{2}$

6. $\dfrac{1}{6} + \dfrac{1}{2} \div \dfrac{3}{2}$

7. $\dfrac{1}{5} + \dfrac{3}{10} \div \dfrac{1}{2}$

8. $\dfrac{1}{4} \div \dfrac{3}{8} + \dfrac{1}{2}$

9. $\dfrac{3}{5} \cdot \dfrac{1}{4} - \dfrac{1}{8}$

10. $\dfrac{5}{7} \cdot \dfrac{3}{4} - \dfrac{5}{14}$

11. $\dfrac{7}{12} \div \left(\dfrac{1}{3} + \dfrac{1}{3} \right)$

12. $\dfrac{1}{3} \div \left(\dfrac{1}{6} + \dfrac{4}{9} \right)$

13. $\dfrac{7}{10} + \left(\dfrac{1}{5} + \dfrac{1}{5} \right)^2$

14. $\dfrac{3}{4} + \left(\dfrac{1}{2} \right)^2$

B

15. $\dfrac{1}{2} \div \dfrac{2}{3} \cdot \dfrac{5}{6}$

16. $\dfrac{1}{2} \div \left(\dfrac{2}{3} \cdot \dfrac{5}{6} \right)$

17. $\dfrac{11}{12} + \left(\dfrac{3}{8} \cdot \dfrac{1}{2} \right)$

18. $\dfrac{2}{3} - \left(\dfrac{5}{6} \cdot \dfrac{1}{9} \right)$

19. $\dfrac{3}{5} - \dfrac{1}{3} \div \dfrac{3}{4} + \dfrac{3}{10}$

20. $\dfrac{7}{8} - \dfrac{1}{6} \div \dfrac{2}{3} + \dfrac{5}{6}$

21. $\dfrac{5}{8} \cdot \dfrac{1}{3} + \dfrac{1}{2} \div \dfrac{1}{3} - \dfrac{7}{8}$

22. $\dfrac{1}{9} \div \dfrac{1}{2} - \dfrac{2}{3} \cdot \dfrac{1}{4} + \dfrac{1}{9}$

23. $\dfrac{11}{14} + \left(\dfrac{2}{7} \right)^2 - \dfrac{17}{49} \cdot \dfrac{5}{2}$

24. $\dfrac{3}{4} \cdot \dfrac{4}{5} - \dfrac{3}{0} + \left(\dfrac{2}{3} \right)^2$

25. $\dfrac{17}{18} - \dfrac{7}{9} + \dfrac{1}{9} \div \left(\dfrac{4}{3} \right)^2$

26. $\dfrac{19}{25} - \left(\dfrac{2}{5} \right)^2 + \dfrac{3}{5} \div \dfrac{2}{3}$

27. $\dfrac{7}{8} - \left(\dfrac{1}{2} \div \dfrac{2}{5} - \dfrac{3}{8} \right)$

28. $\dfrac{2}{3} + \left(\dfrac{3}{4} \cdot \dfrac{4}{9} + \dfrac{1}{2} \right)$

OBJECTIVE 2 Find the average of a group of fractions. (See page 253.)

A *Find the average.*

29. $\dfrac{4}{11}$ and $\dfrac{8}{11}$

30. $\dfrac{3}{10}$ and $\dfrac{7}{10}$

31. $\dfrac{2}{7}, \dfrac{4}{7}$, and $\dfrac{5}{7}$

32. $\dfrac{3}{5}, \dfrac{4}{5}$, and $\dfrac{2}{5}$

33. $\dfrac{4}{15}, \dfrac{8}{15}$, and $\dfrac{4}{15}$

34. $\dfrac{1}{5}, \dfrac{3}{5}$, and $\dfrac{8}{5}$

35. $\dfrac{1}{5}, \dfrac{2}{5}$, and $\dfrac{7}{15}$

36. $\dfrac{1}{2}, \dfrac{3}{4}$, and $\dfrac{11}{12}$

37. $2\dfrac{1}{3}, 3\dfrac{1}{2}$, and $1\dfrac{5}{6}$

38. $2\dfrac{2}{3}, 4\dfrac{1}{3}$, and $5\dfrac{5}{6}$

B

39. $\dfrac{11}{15}, \dfrac{2}{5}, \dfrac{1}{3}$, and $\dfrac{1}{5}$

40. $\dfrac{1}{6}, \dfrac{5}{12}, \dfrac{1}{4}$, and $\dfrac{1}{2}$

41. $\dfrac{4}{9}, \dfrac{5}{18}, \dfrac{5}{6}, \dfrac{2}{3}$, and $\dfrac{1}{2}$

42. $\dfrac{2}{5}, \dfrac{1}{10}, \dfrac{3}{10}, \dfrac{1}{2}$, and $\dfrac{4}{5}$

43. $3\dfrac{1}{6}, 2\dfrac{1}{3}$, and $4\dfrac{1}{6}$

44. $6\dfrac{3}{7}, 4\dfrac{1}{2}$, and $5\dfrac{9}{14}$

45. $5\dfrac{1}{3}, 6\dfrac{2}{5}$, and $9\dfrac{13}{15}$

46. $2\dfrac{1}{2}, 1\dfrac{1}{6}, 1\dfrac{7}{9}$, and $2\dfrac{1}{3}$

C *Perform the indicated operations.*

47. $\dfrac{14}{15} - \left(\dfrac{4}{5} - \dfrac{1}{2} \div \dfrac{2}{3}\right) \cdot \dfrac{4}{5}$

48. $\dfrac{1}{2} - \left(\dfrac{1}{2} - \dfrac{3}{4} \div \dfrac{12}{5}\right) \cdot \dfrac{1}{6}$

49. $\left(\dfrac{3}{5} + \dfrac{7}{10} \cdot \dfrac{2}{3}\right)\left(\dfrac{3}{2}\right)^2$

50. $\dfrac{7}{8} \div \dfrac{3}{4} \cdot \dfrac{3}{14} \div \left(\dfrac{1}{2}\right)^3$

Find the average.

51. $3\dfrac{2}{3}, 4\dfrac{5}{6}$, and $2\dfrac{5}{9}$

52. $6\dfrac{7}{8}, 8\dfrac{3}{4}$, and $8\dfrac{1}{2}$

53. $\dfrac{7}{15}, 4, \dfrac{4}{5}$, and 7

54. $5, \dfrac{8}{9}, 12$, and $\dfrac{5}{6}$

55. Wayne catches six salmon. The salmon measure $23\dfrac{1}{4}$ in., $31\dfrac{5}{8}$ in., $42\dfrac{3}{4}$ in., $28\dfrac{5}{8}$ in., $35\dfrac{3}{4}$ in., and 40 in. in length. What is the average length of the salmon?

56. Karla, a nurse at Kaiser Hospital, weighs five new babies. They weigh $6\dfrac{1}{2}$ lb, $7\dfrac{3}{4}$ lb, $9\dfrac{3}{8}$ lb, $7\dfrac{1}{2}$ lb, and $8\dfrac{7}{8}$ lb. What is the average weight of the babies?

57. The mothers of a swim team are making Rice Krispie treats for a big meet for the team. Each batch calls for 6 cups of cereal, $\frac{1}{4}$ cup of butter, $1\frac{3}{4}$ cups of chocolate chips, and $3\frac{1}{2}$ cups of marshmallows. They are also making Gorp, which uses $2\frac{1}{2}$ cups of cereal, 4 cups of pretzels, 1 cup of marshmallows, 1 cup of raisins, and $2\frac{1}{2}$ cups of chocolate chips. How much of each ingredient do they need if they intend to make 5 batches of Rice Krispie treats and 10 batches of Gorp?

58. Nellie is making the trellis shown below out of $\frac{1}{2}$-in. copper pipe, which comes in 10-ft lengths. How many 10-ft lengths does she need and how much will she have left over?

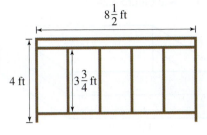

59. The results of the Women's Shot Put for the last five Olympiads are given in the table. What is the length of the average winning throw over the past 20 years? *Hint*: Convert to inches.

Year	Winner	Distance
1992	Svetlana Krivaleva, Unified Team	69' 1$\frac{1}{4}$"
1996	Astrid Kumbernuss, Germany	67' 5$\frac{1}{2}$"
2000	Yanina Korolchik, Belarus	67' 5"
2004	Irina Korzhanenko, Russia	69' 1$\frac{1}{8}$"
2008	Valerie Villi, New Zealand	69' 5$\frac{1}{2}$"

60. Celsuis temperature can be obtained from Fahrenheit temperature by subtracting 32, then multiplying the result by $\frac{5}{9}$. What Celsius temperature corresponds to 212°F (the boiling point of water)?

61. Celsuis temperature can be obtained from Fahrenheit temperature by subtracting 32, then multiplying the result by $\frac{5}{9}$. What Celsius temperature corresponds to $98\frac{3}{5}$°F (normal body temperature)?

62. In a walk for charity, seven people walk $3\frac{1}{8}$ mi, six people walk $2\frac{7}{8}$ mi, nine people walk $3\frac{1}{4}$ mi, and five people walk $6\frac{1}{2}$ mi. What is the total number of miles walked? If the charity raises $2355, what is the average amount raised per mile, rounded to the nearest dollar?

Exercises 63–64 relate to the chapter application. See page 167.

63. Now that Shane has finished his deck, he wants to build planter boxes along one end. Each planter box is 2 ft wide and 4 ft long, and he will build them using 2 × 12s. See Figure 3.24.

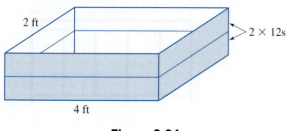

Figure 3.24

How many 2 × 12s will Shane need to construct the sides of one planter box? Assume that the 2 × 12s come in 10-ft lengths.

64. How many cubic inches of potting soil are needed to fill one planter box?

STATE YOUR UNDERSTANDING

65. Write out the order of operations for fractions. How is it different from the order of operations for whole numbers?

66. Must the average of a group of numbers be larger than the smallest number and smaller than the largest number? Why?

CHALLENGE

Perform the indicated operations.

67. $2\frac{5}{8}\left(4\frac{1}{5} - 3\frac{5}{6}\right) \div 2\frac{1}{2}\left(3\frac{1}{7} + 2\frac{1}{5}\right)$

68. $1\frac{2}{5}\left(5\frac{1}{5} - 4\frac{3}{4}\right) \div 4\frac{1}{2}\left(3\frac{1}{7} - 2\frac{1}{3}\right)$

69. The Acme Fish Company pays $1500 per ton for crab. Jerry catches $3\frac{2}{5}$ tons; his brother Joshua catches $1\frac{1}{2}$ times as many as Jerry. Their sister Salicita catches $\frac{7}{8}$ the amount that Joshua does. What is the total amount paid to the three people by Acme Fish Company, to the nearest dollar?

Perform the indicated operations.

70. $3\frac{4}{7} + \frac{9}{14}$

71. $3\frac{4}{7} - \frac{9}{14}$

72. $3\frac{4}{7}\left(\frac{9}{14}\right)$

73. $3\frac{4}{7} \div \frac{9}{14}$

74. $\frac{15}{28} \cdot \frac{21}{45} \cdot \frac{20}{35}$

75. $\frac{9}{15} \cdot \frac{3}{4} \cdot \frac{35}{6}$

76. Find the prime factorization of 650.

77. Find the prime factorization of 975.

78. A coffee table is made of a piece of maple that is $\frac{3}{4}$ in. thick, a piece of chipboard that is $\frac{3}{8}$ in. thick, and a veneer that is $\frac{1}{8}$ in. thick. How thick is the tabletop?

79. Jason walked the following distances during a 5-day walkathon for charity: $4\frac{3}{4}$ mi, $5\frac{1}{2}$ mi, $6\frac{3}{4}$ mi, $3\frac{1}{3}$ mi, and $4\frac{2}{3}$ mi. What was the total distance the Jason walked? If his charity receives $10 per mile walked, how much money did he raise?

KEY CONCEPTS

SECTION 3.1 Proper and Improper Fractions; Mixed Numbers

Definitions and Concepts	Examples
A fraction has the form $\dfrac{\text{numerator}}{\text{denominator}}$.	$\dfrac{4}{81}, \dfrac{20}{3}, \dfrac{25}{25}$
A proper fraction has a smaller numerator than denominator.	$\dfrac{4}{81}$ is a proper fraction.
An improper fraction has a numerator that is not smaller than the denominator.	$\dfrac{20}{3}$ and $\dfrac{25}{25}$ are improper fractions.
A mixed number is the sum of a whole number and a fraction.	$4\dfrac{1}{6}$
To change a mixed number to an improper fraction, • Multiply the whole number by the denominator and add the numerator. • Place the sum over the denominator.	$4\dfrac{1}{6} = \dfrac{4 \cdot 6 + 1}{6} = \dfrac{25}{6}$
To change an improper fraction to a mixed number, • Divide the numerator by the denominator. • The mixed number is the whole number plus the remainder over the divisor.	$\dfrac{38}{7} = 5\dfrac{3}{7}$ because $\begin{array}{r} 5 \\ 7\overline{)38} \\ \underline{35} \\ 3 \end{array}$

SECTION 3.2 Simplifying Fractions

Definitions and Concepts	Examples
A fraction is completely simplified when its numerator and denominator have no common factors.	$\dfrac{6}{7}$ is completely simplified. $\dfrac{10}{12}$ is not completely simplified because both 10 and 12 have a factor of 2.
To simplify a fraction, eliminate all common factors of the numerator and denominator.	$\dfrac{36}{72} = \dfrac{9 \cdot 4}{9 \cdot 8} = \dfrac{4}{8} = \dfrac{4 \cdot 1}{4 \cdot 2} = \dfrac{1}{2}$

SECTION 3.3 Multiplying and Dividing Fractions

Definitions and Concepts	Examples
To multiply fractions, simplify if possible and then multiply numerators and multiply denominators.	$\dfrac{8}{\overset{}{\underset{3}{\cancel{15}}}} \cdot \dfrac{\overset{1}{\cancel{5}}}{7} = \dfrac{8}{21}$
Two fractions are reciprocals if their product is 1.	$\dfrac{3}{4}$ and $\dfrac{4}{3}$ are reciprocals because $\dfrac{3}{4} \cdot \dfrac{4}{3} = 1$.
To divide fractions, multiply the first fraction by the reciprocal of the divisor.	$\dfrac{4}{9} \div \dfrac{5}{3} = \dfrac{4}{\underset{3}{\cancel{9}}} \cdot \dfrac{\overset{1}{\cancel{3}}}{5} = \dfrac{4}{15}$

SECTION 3.4 Multiplying and Dividing Mixed Numbers

Definitions and Concepts	Examples
To multiply or divide mixed numbers, change them to improper fractions first. Then multiply or divide.	$3\dfrac{1}{5} \cdot 2\dfrac{3}{4} = \dfrac{\overset{4}{\cancel{16}}}{5} \cdot \dfrac{11}{\underset{1}{\cancel{4}}} = \dfrac{44}{5} = 8\dfrac{4}{5}$

SECTION 3.5 Building Fractions; Listing in Order; Inequalities

Definitions and Concepts	Examples
Building a fraction is writing an equivalent fraction with a different denominator.	$\dfrac{1}{2} = \dfrac{15}{30}$
To build a fraction, multiply both its numerator and denominator by the same factor.	$\dfrac{3}{5} = \dfrac{3 \cdot 6}{5 \cdot 6} = \dfrac{18}{30}$
To list fractions in order, • Rewrite each fraction with a common denominator. • Order the fractions according to their numerators.	List $\dfrac{3}{5}, \dfrac{7}{10},$ and $\dfrac{5}{8}$ in order from smallest to largest. $\dfrac{3}{5} = \dfrac{24}{40}, \quad \dfrac{7}{10} = \dfrac{28}{40}, \quad \dfrac{5}{8} = \dfrac{25}{40}$ $\dfrac{24}{40} < \dfrac{25}{40} < \dfrac{28}{40},$ so $\dfrac{3}{5} < \dfrac{5}{8} < \dfrac{7}{10}$

SECTION 3.6 Adding Fractions

Definitions and Concepts	Examples
Like fractions have common denominators.	$\dfrac{3}{54}$ and $\dfrac{18}{54}$ are like fractions.
Unlike fractions have different denominators.	$\dfrac{1}{4}$ and $\dfrac{2}{5}$ are unlike fractions.
To add fractions, • Rewrite with common denomionators (if necessary). • Add the numerators and keep the common denominator. • Simplify.	$\dfrac{1}{2} + \dfrac{1}{6} = \dfrac{3}{6} + \dfrac{1}{6} = \dfrac{4}{6} = \dfrac{2}{3}$

SECTION 3.7 Adding Mixed Numbers

Definitions and Concepts	Examples
To add mixed numbers, • Add the whole numbers. • Add the fractions. If this sum is more than 1, change to a mixed number and add again. • Simplify.	$\begin{aligned} 5\dfrac{1}{2} &= 5\dfrac{2}{4} \\ +2\dfrac{3}{4} &= 2\dfrac{3}{4} \\ \hline 7\dfrac{5}{4} &= 7 + 1\dfrac{1}{4} = 8\dfrac{1}{4} \end{aligned}$

SECTION 3.8 Subtracting Fractions

Definitions and Concepts	Examples
To subtract fractions, • Rewrite with common denominators (if necessary). • Subtract the numerators and keep the common denominator. • Simplify.	$\dfrac{71}{72} - \dfrac{31}{90} = \dfrac{355}{360} - \dfrac{124}{360}$ $= \dfrac{231}{360}$ $= \dfrac{77}{120}$

SECTION 3.9 Subtracting Mixed Numbers

Definitions and Concepts	Examples
To subtract mixed numbers, • Subtract the fractions. If the fractions cannot be subtracted, borrow 1 from the whole-number part and add it to the fractional part. Then subtract the fractions. • Subtract the whole numbers. • Simplify.	$\begin{aligned} 17\dfrac{3}{8} &= 17\dfrac{45}{120} = 16 + 1\dfrac{45}{120} = 16\dfrac{165}{120} \\ -12\dfrac{14}{15} &= 12\dfrac{112}{120} \qquad\qquad\qquad = 12\dfrac{112}{120} \\ &\qquad\qquad\qquad\qquad\qquad\quad 4\dfrac{53}{120} \end{aligned}$

SECTION 3.10 Order of Operations; Average

Definitions and Concepts	Examples
The order of operations for fractions is the same as that for whole numbers: • Parentheses • Exponents • Multiplication/Division • Addition/Subtraction	$\left(\dfrac{1}{2} + \dfrac{2}{5}\right) \div \left(\dfrac{2}{3}\right)^2 = \left(\dfrac{5}{10} + \dfrac{4}{10}\right) \div \left(\dfrac{2}{3}\right)^2$ $= \left(\dfrac{9}{10}\right) \div \left(\dfrac{2}{3}\right)^2$ $= \dfrac{9}{10} \div \left(\dfrac{4}{9}\right)$ $= \dfrac{9}{10} \cdot \dfrac{9}{4}$ $= \dfrac{81}{40} = 2\dfrac{1}{40}$
Finding the average of a set of fractions is the same as for whole numbers, • Add the fractions. • Divide by the number of fractions.	Find the average of $\dfrac{1}{4}, \dfrac{1}{3},$ and $\dfrac{1}{6}.$ $\dfrac{1}{4} + \dfrac{1}{3} + \dfrac{1}{6} = \dfrac{3}{12} + \dfrac{4}{12} + \dfrac{2}{12} = \dfrac{9}{12} = \dfrac{3}{4}$ $\dfrac{3}{4} \div 3 = \dfrac{3}{4} \cdot \dfrac{1}{3} = \dfrac{1}{4}$ The average is $\dfrac{1}{4}.$

SECTION 3.1

Write the fraction represented by the figure.

1.

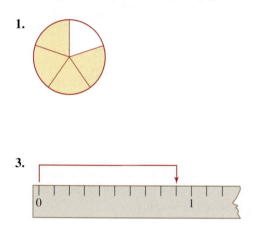

2.

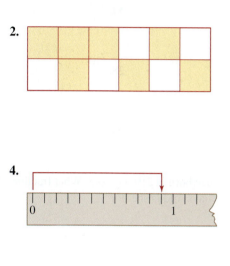

3.

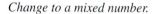

4.

Identify the improper fractions from each list.

5. $\dfrac{8}{3}, \dfrac{11}{12}, \dfrac{9}{9}, \dfrac{22}{19}, \dfrac{3}{20}$

6. $\dfrac{1}{13}, \dfrac{11}{15}, \dfrac{8}{10}, \dfrac{5}{2}, \dfrac{12}{18}$

Change to a mixed number.

7. $\dfrac{82}{11}$

8. $\dfrac{76}{9}$

9. $\dfrac{413}{5}$

10. $\dfrac{344}{7}$

Change to an improper fraction.

11. $6\dfrac{5}{12}$

12. $4\dfrac{3}{7}$

13. $12\dfrac{5}{6}$

14. $9\dfrac{2}{3}$

15. 17

16. 35

17. A food wholesaler packs 24 cans of beans in a case for shipping. Write as a mixed number the number of cases that can be packed if she has 64,435 cans of beans.

18. A golf ball manufacturer packs 15 golf balls in a single package. They then advertise them at the cost for a dozen with three free. Write as a mixed number the number of these packs that can be made from 46,325 golf balls.

SECTION 3.2

Simplify.

19. $\dfrac{24}{32}$

20. $\dfrac{10}{25}$

21. $\dfrac{60}{90}$

22. $\dfrac{18}{24}$

23. $\dfrac{21}{35}$

24. $\dfrac{35}{70}$

25. $\dfrac{102}{6}$

26. $\dfrac{126}{42}$

27. $\dfrac{14}{42}$

28. $\dfrac{30}{45}$

29. $\dfrac{78}{96}$

30. $\dfrac{75}{125}$

31. $\dfrac{26}{130}$

32. $\dfrac{96}{144}$

33. $\dfrac{268}{402}$

34. $\dfrac{630}{1050}$

35. Romoma's newborn is 219 days old. What fraction of a year (365 days) old is she?

36. On a math test, a student answers 42 items correctly and 18 incorrectly. What fraction of the items are answered correctly? Simplify.

SECTION 3.3

Multiply. Simplify completely.

37. $\dfrac{3}{7} \cdot \dfrac{2}{7}$

38. $\dfrac{7}{8} \cdot \dfrac{1}{6}$

39. $\dfrac{3}{5} \cdot \dfrac{6}{11}$

40. $6 \cdot \dfrac{2}{3} \cdot \dfrac{3}{14}$

41. $\dfrac{21}{5} \cdot \dfrac{5}{4} \cdot \dfrac{4}{21}$

42. $\dfrac{28}{35} \cdot \dfrac{3}{8} \cdot \dfrac{5}{9}$

Find the reciprocal.

43. $\dfrac{3}{8}$

44. 5

Divide. Simplify completely.

45. $\dfrac{9}{16} \div \dfrac{7}{8}$

46. $\dfrac{8}{13} \div \dfrac{2}{13}$

47. $\dfrac{15}{18} \div \dfrac{30}{27}$

48. $\dfrac{12}{15} \div \dfrac{15}{8}$

49. $\dfrac{33}{50} \div \dfrac{27}{35}$

50. $\dfrac{32}{45} \div \dfrac{8}{9}$

51. Lois spends half of the family income on rent, utilities, and food. She pays $\dfrac{2}{7}$ of this amount for rent. What fraction of the family income goes for rent?

52. As part of his job at a pet store, Perry feeds each gerbil $\dfrac{1}{8}$ cup of seeds each day. If the seeds come in packages of $\dfrac{5}{4}$ cups, how many gerbils can be fed from one package?

SECTION 3.4

Multiply. Simplify completely and write as a mixed number if possible.

53. $\left(\dfrac{4}{5}\right)\left(2\dfrac{4}{5}\right)$

54. $\left(\dfrac{3}{7}\right)\left(2\dfrac{5}{7}\right)$

55. $\left(4\dfrac{3}{4}\right)\left(3\dfrac{1}{2}\right)$

56. $\left(3\dfrac{1}{3}\right)\left(1\dfrac{4}{5}\right)$

57. $\left(4\dfrac{3}{8}\right)\left(\dfrac{9}{14}\right)$

58. $\left(4\dfrac{4}{9}\right)\left(\dfrac{12}{25}\right)$

59. $\left(3\dfrac{2}{3}\right)\left(\dfrac{15}{22}\right)\left(7\dfrac{1}{2}\right)$

60. $\left(4\dfrac{3}{4}\right)\left(3\dfrac{1}{5}\right)\left(5\dfrac{5}{8}\right)$

Divide. Simplify completely and write as a mixed number if possible.

61. $10 \div 1\dfrac{1}{4}$

62. $4 \div 1\dfrac{1}{4}$

63. $4\dfrac{2}{7} \div 5$

64. $2\dfrac{1}{6} \div 4$

65. $8\dfrac{2}{5} \div 2\dfrac{1}{3}$

66. $3\dfrac{1}{6} \div 4\dfrac{3}{4}$

67. $31\dfrac{1}{3} \div 1\dfrac{1}{9}$

68. $21\dfrac{3}{7} \div 8\dfrac{1}{3}$

69. A corn farmer in Nebraska averages $151\dfrac{3}{5}$ bushels of corn per acre on 205 acres. How many bushels of corn does she harvest?

70. A wildlife survey in a water fowl preserve finds that there are $3\dfrac{1}{3}$ times as many brant geese in the preserve as there are Canada geese. If the survey counts 7740 Canada geese, how many brant geese are there?

SECTION 3.5

Write four fractions equivalent to each of the given fractions by multiplying by $\dfrac{2}{2}, \dfrac{3}{3}, \dfrac{5}{5},$ and $\dfrac{8}{8}$.

71. $\dfrac{2}{3}$

72. $\dfrac{2}{7}$

73. $\dfrac{3}{14}$

74. $\dfrac{4}{11}$

Find the missing numerator.

75. $\dfrac{4}{5} = \dfrac{?}{45}$

76. $\dfrac{6}{7} = \dfrac{?}{56}$

77. $\dfrac{5}{6} = \dfrac{?}{144}$

78. $\dfrac{7}{12} = \dfrac{?}{132}$

List the fractions from smallest to largest.

79. $\dfrac{1}{2}, \dfrac{3}{5}, \dfrac{7}{10}$

80. $\dfrac{1}{4}, \dfrac{5}{12}, \dfrac{3}{8}$

81. $\dfrac{2}{9}, \dfrac{1}{5}, \dfrac{3}{11}$

82. $\dfrac{10}{9}, \dfrac{4}{3}, \dfrac{7}{6}, \dfrac{19}{18}$

83. $\dfrac{3}{7}, \dfrac{5}{14}, \dfrac{11}{28}, \dfrac{3}{8}$

84. $7\dfrac{3}{4}, 7\dfrac{7}{8}, 7\dfrac{5}{6}$

Are the following statements true or false?

85. $\dfrac{3}{14} < \dfrac{5}{14}$

86. $\dfrac{13}{10} > \dfrac{11}{10}$

87. $\dfrac{11}{14} > \dfrac{17}{21}$

88. $\dfrac{12}{17} < \dfrac{14}{19}$

89. Four pickup trucks are advertised in the local car ads. The load capacities listed are $\dfrac{3}{4}$ ton, $\dfrac{5}{8}$ ton, $\dfrac{7}{16}$ ton, and $\dfrac{1}{2}$ ton. Which capacity is the smallest and which is the largest?

90. During 1 week on her diet, Samantha ate five servings of chicken, each containing $\dfrac{3}{16}$ oz of fat. During the same period her brother ate four servings of beef, each containing $\dfrac{6}{25}$ oz of fat. Who ate the greater amount of fat from these entrees?

SECTION 3.6

Add. Simplify completely.

91. $\dfrac{4}{15} + \dfrac{7}{15}$

92. $\dfrac{3}{10} + \dfrac{3}{10}$

93. $\dfrac{2}{9} + \dfrac{2}{9} + \dfrac{2}{9}$

94. $\dfrac{3}{16} + \dfrac{2}{16} + \dfrac{3}{16}$

95. $\dfrac{7}{32} + \dfrac{8}{32} + \dfrac{5}{32}$

96. $\dfrac{7}{30} + \dfrac{7}{30} + \dfrac{6}{30}$

97. $\dfrac{4}{15} + \dfrac{1}{3}$

98. $\dfrac{5}{17} + \dfrac{7}{34}$

99. $\dfrac{3}{35} + \dfrac{8}{21}$

100. $\dfrac{11}{30} + \dfrac{9}{20} + \dfrac{3}{10}$

101. $\dfrac{1}{6} + \dfrac{7}{8} + \dfrac{7}{12}$

102. $\dfrac{7}{15} + \dfrac{11}{30} + \dfrac{5}{6}$

103. An elephant ear bamboo grew $\dfrac{1}{2}$ in. on Tuesday, $\dfrac{3}{8}$ in. on Wednesday, and $\dfrac{1}{4}$ in. on Thursday. How much did the bamboo grow in the 3 days?

104. In order to complete a project, Preston needs $\dfrac{1}{10}$ in. of foam, $\dfrac{3}{10}$ in. of metal, $\dfrac{4}{10}$ in. of wood, and $\dfrac{7}{10}$ in. of plexiglass. What will be the total thickness of this project when these materials are piled up?

SECTION 3.7

Add. Write the results as mixed numbers where possible.

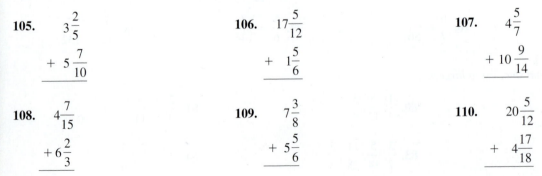

105. $3\dfrac{2}{5}$
$+ \ 5\dfrac{7}{10}$

106. $17\dfrac{5}{12}$
$+ \ 1\dfrac{5}{6}$

107. $4\dfrac{5}{7}$
$+ \ 10\dfrac{9}{14}$

108. $4\dfrac{7}{15}$
$+ 6\dfrac{2}{3}$

109. $7\dfrac{3}{8}$
$+ 5\dfrac{5}{6}$

110. $20\dfrac{5}{12}$
$+ \ 4\dfrac{17}{18}$

111. $14\frac{7}{20}$
$+\ 11\frac{3}{16}$
—————

112. $11\frac{7}{24}$
$+\ 32\frac{7}{18}$
—————

113. $17\frac{1}{5}\ +\ 18\frac{1}{4}\ +\ 19\frac{3}{10}$

114. $18\frac{3}{4}\ +\ 19\ +\ 25\frac{7}{12}$

115. $25\frac{2}{3}\ +\ 16\frac{1}{6}\ +\ 18\frac{3}{4}$

116. $29\frac{7}{8}\ +\ 19\frac{5}{12}\ +\ 32\frac{3}{4}$

117. Russ lost the following pounds during his 5-week diet: $1\frac{3}{4}$ lb, $2\frac{1}{5}$ lb, $\frac{2}{3}$ lb, $1\frac{1}{2}$ lb, and $3\frac{1}{4}$ lb. What was his total weight loss on his diet?

118. On a fishing excursion Roona caught four fish weighing $6\frac{3}{4}$ lb, $1\frac{3}{5}$ lb, $2\frac{2}{3}$ lb, and $5\frac{1}{2}$ lb. What was the total weight of her catch?

SECTION 3.8

Subtract.

119. $\frac{7}{8}-\frac{5}{16}$

120. $\frac{5}{18}-\frac{2}{9}$

121. $\frac{7}{8}-\frac{1}{4}$

122. $\frac{17}{20}-\frac{1}{5}$

123. $\frac{17}{24}-\frac{1}{6}$

124. $\frac{7}{15}-\frac{3}{20}$

125. $\frac{5}{6}-\frac{4}{5}$

126. $\frac{7}{10}-\frac{1}{4}$

127. $\frac{19}{25}-\frac{8}{15}$

128. $\frac{21}{32}-\frac{5}{16}$

129. $\frac{19}{30}-\frac{17}{60}$

130. $\frac{13}{15}-\frac{9}{20}$

131. Wanda finds $\frac{3}{4}$ oz of gold during a day of panning along the Snake River. She gives a $\frac{1}{3}$-oz nugget to Jose, her guide. What fraction of an ounce of gold does she have left?

132. A carpenter planes the thickness of a board from $\frac{13}{16}$ to $\frac{5}{8}$ in. How much is removed?

SECTION 3.9

Subtract. Write the results as mixed numbers where possible.

133. $167\frac{3}{5}$
$-\ 82\frac{3}{10}$
—————

134. $6\frac{5}{6}$
$-3\frac{3}{10}$
—————

135. $26\frac{1}{10}$
$-10\frac{9}{10}$
—————

136. $19\frac{3}{8}$

$-\ 8\frac{5}{8}$

_____

137. $8\frac{2}{3}$

$-5\frac{5}{6}$

_____

138. $76\frac{7}{15}$

$-50\frac{1}{12}$

_____

139. $9\frac{9}{16}$

$-\ 3\frac{5}{12}$

_____

140. $30\frac{7}{16}$

$-22\frac{5}{6}$

_____

141. $9\frac{8}{15} - 4\frac{17}{20}$

142. $33\frac{17}{30} - 25\frac{7}{9}$

143. $11\frac{23}{48} - \frac{23}{24}$

144. $8\frac{11}{20} - \frac{9}{10}$

145. The graph displays the average yearly rainfall for five cities.

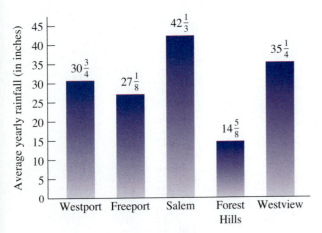

a. How much more rain falls in Westport during a year than in Freeport?

b. In a 10-yr period, how much more rain falls in Salem than in Forest Hills?

146. Using the graph in Exercise 145, if the average rainfall in Westview doubles, how much more rain would it receive than Salem?

SECTION 3.10

Perform the indicated operations.

147. $\frac{1}{4} + \frac{3}{8} \div \frac{1}{2}$

148. $\frac{1}{3} \div \frac{5}{9} + \frac{1}{6}$

149. $\frac{5}{8} \div \frac{3}{4} + \frac{3}{4}$

150. $\frac{1}{3} \div \frac{1}{6} + \frac{4}{9}$

151. $\frac{3}{4} - \left(\frac{1}{2}\right)^2$

152. $\frac{4}{5} - \left(\frac{2}{5}\right)^2$

153. $\left(\frac{9}{8}\right)^2 - \left(\frac{1}{2} \div \frac{4}{5} - \frac{3}{8}\right)$

154. $\left(\frac{1}{2}\right)^2 + \left(\frac{4}{5} \cdot \frac{5}{8} + \frac{2}{3}\right)$

Find the average.

155. $\frac{3}{8}, \frac{1}{4}, \frac{1}{2},$ and $\frac{3}{4}$

156. $\frac{3}{7}, \frac{1}{4}, \frac{2}{7},$ and $\frac{5}{28}$

157. $1\frac{2}{3}$, $1\frac{5}{12}$, $2\frac{1}{2}$, $\frac{3}{4}$, and $\frac{5}{6}$

158. $\frac{2}{3}$, $\frac{5}{12}$, $\frac{1}{2}$, $1\frac{3}{4}$, and $1\frac{5}{6}$

159. A class of 20 students took a 10-problem quiz. Their results were as follows.

Number of Students	Fraction of Problems Correct
3	$\frac{10}{10}$ (all correct)
1	$\frac{9}{10}$
3	$\frac{8}{10}$
5	$\frac{7}{10}$
4	$\frac{6}{10}$
3	$\frac{5}{10}$
1	$\frac{3}{10}$

What is the class average?

160. What is the average of the top six scores in Exercise 159?

TRUE/FALSE CONCEPT REVIEW

Check your understanding of the language of basic mathematics. Tell whether each of the following statements is true (always true) or false (not always true). For each statement you judge to be false, revise it to make a statement that is true.

Answers

1. It is possible to picture an improper fraction using unit regions.

1. _____

2. The fraction $\frac{7}{8}$ written as a mixed number is $1\frac{7}{8}$.

2. _____

3. The whole number 1 can also be written as a proper fraction.

3. _____

4. A fraction is another way of writing a division problem.

4. _____

5. When a fraction is completely simplified, its value is less than 1.

5. _____

6. Every improper fraction can be changed to a mixed number or a whole number.

6. _____

7. Some fractions with large numerators and denominators cannot be further simplified.

7. _____

8. Two mixed numbers can be subtracted without first changing them to improper fractions.

8. _____

9. The reciprocal of an improper fraction is greater than 1.

9. _____

10. The quotient of two nonzero fractions can always be found by multiplication.

10. _____

11. Simplifying fractions is the opposite of building fractions.

11. _____

12. The primary reason for building fractions is so that they will have a common denominator.

12. _____

13. Like fractions have the same numerators.

13. _____

14. Mixed numbers must be changed to improper fractions before adding them.

14. _____

15. It is sometimes necessary to use "borrowing" to subtract mixed numbers as we do when subtracting some whole numbers.

15. _____

16. The order of operations for fractions is the same as the order of operations for whole numbers.

16. _____

17. The average of three nonequivalent fractions is smaller than at least one of the fractions.

17. _____

18. The product of two fractions is sometimes smaller than the two fractions.

18. _____

TEST

Answers

1. Change $\dfrac{61}{3}$ to a mixed number.

1. _____

2. Add: $\dfrac{7}{8} + \dfrac{5}{12}$

2. _____

3. Change $8\dfrac{7}{9}$ to an improper fraction

3. _____

4. List these fractions from the smallest to the largest: $\dfrac{2}{5}, \dfrac{3}{8}, \dfrac{3}{7}$

4. _____

5. Change 11 to an improper fraction.

5. _____

6. Find the missing numerator: $\dfrac{3}{8} = \dfrac{?}{72}$

6. _____

7. Add:

$$5\frac{3}{10}$$
$$+\ 3\frac{5}{6}$$

8. Multiply. Write the result as a mixed number. $\left(3\frac{2}{3}\right)\left(5\frac{1}{9}\right)$

9. Perform the indicated operations: $\frac{1}{2} - \frac{3}{8} \div \frac{3}{4}$

10. Simplify $\frac{68}{102}$ completely.

11. Subtract: $17\frac{4}{5} - 11$

12. Multiply: $\frac{4}{5} \cdot \frac{7}{8} \cdot \frac{15}{21}$

13. Subtract: $\frac{2}{3} - \frac{4}{9}$

14. Divide: $1\frac{2}{9} \div 3\frac{2}{3}$

15. Multiply: $\frac{3}{7} \cdot \frac{4}{5}$

16. Subtract:

$$11\frac{7}{12}$$
$$-\ 4\frac{14}{15}$$

17. Simplify $\frac{220}{352}$ completely.

18. Add: $\frac{1}{35} + \frac{5}{14} + \frac{2}{5}$

19. What is the reciprocal of $3\frac{1}{5}$?

20. What is the reciprocal of $\frac{8}{21}$?

21. Which of these fractions are proper?

$$\frac{7}{8},\ \frac{8}{8},\ \frac{9}{8},\ \frac{7}{9},\ \frac{9}{7},\ \frac{8}{9},\ \frac{9}{9}$$

22. Divide: $\frac{7}{3} \div \frac{8}{9}$

23. Subtract:

$$11\frac{7}{10}$$
$$-\ 3\frac{3}{8}$$

7. _____

8. _____

9. _____

10. _____

11. _____

12. _____

13. _____

14. _____

15. _____

16. _____

17. _____

18. _____

19. _____

20. _____

21. _____

22. _____

23. _____

24. Write the fraction for the shaded part of this figure.

25. Subtract: $11 - 3\dfrac{5}{11}$

26. Add: $\dfrac{4}{15} + \dfrac{8}{15}$

27. Find the average of $1\dfrac{3}{8}, \dfrac{1}{4}, 3\dfrac{1}{2}$, and $2\dfrac{3}{8}$.

28. Multiply: $\left(\dfrac{8}{25}\right)\left(\dfrac{9}{16}\right)$

29. True or false? $\dfrac{5}{7} > \dfrac{11}{16}$

30. Which of the fractions represent the number 1?
$\dfrac{6}{5}, \dfrac{5}{5}, \dfrac{7}{6}, \dfrac{6}{6}, \dfrac{7}{7}, \dfrac{6}{7}, \dfrac{5}{7}$

31. A rail car contains $126\dfrac{1}{2}$ tons of baled hay. A truck that is being used to unload the hay can haul $5\dfrac{3}{4}$ tons in one load. How many truckloads of hay are in the rail car?

32. Jill wants to make up 20 bags of homemade candy for the local bazaar. Each bag will contain $1\dfrac{1}{4}$ lb of candy. How many pounds of candy must she make?

24. _____

25. _____

26. _____

27. _____

28. _____

29. _____

30. _____

31. _____

32. _____

CLASS ACTIVITY 1

Making Pie Charts

In order to make accurate pie charts, it is necessary to use a measuring instrument for angles. One such instrument is called a protractor. It is also necessary to convert the fractions of the components into equivalent fractions with denominators of 360 because there are 360° in a complete circle. Once the circle for the pie chart is drawn, find the center and draw one radius. Starting from that radius, use the protractor to measure the correct angle for each component. Be sure to label each part of the chart.

Make an accurate pie chart for each of the following investment strategies.

1. An aggressive investment strategy allocates $\frac{3}{4}$ of a portfolio to stocks, $\frac{1}{5}$ to bonds, and $\frac{1}{20}$ to money market funds. Make an accurate pie chart for this strategy.

2. A moderate investment strategy allocates $\frac{3}{5}$ of a portfolio to stocks, $\frac{3}{10}$ to bonds, and $\frac{1}{10}$ to money market funds. Make an accurage pie chart for this strategy.

3. A conservative investment strategy allocates $\frac{2}{5}$ of a portfolio to stocks, $\frac{9}{20}$ to bonds, and $\frac{3}{20}$ to money market funds. Make an accurate pie chart for this strategy.

CLASS ACTIVITY 2

Davonna and Latisha are making cookies for a bake sale at their daughters' school. They decide to make chocolate chip cookies and iced sugar cookies.

1. Davonna is going to mix up five times a single recipe of chocolate chip cookies. The recipe calls for $\frac{1}{2}$ c butter, 1 c sugar, 1 egg, $\frac{1}{2}$ tsp vanilla, $1\frac{1}{4}$ c flour, $\frac{1}{4}$ tsp salt, $\frac{1}{2}$ tsp soda, and $\frac{3}{4}$ c chocolate chips. Calculate the total of each ingredient that Davonna needs for all her cookies.

2. Latisha is going to mix up three times a single recipe of sugar cookies. The recipe calls for $2\frac{1}{4}$ c flour, $\frac{3}{4}$ c sugar, $\frac{1}{4}$ tsp baking power, $\frac{1}{2}$ tsp salt, 1 c butter, 1 tsp vanilla, 1 egg, and 1 egg, ingredient that Latisha needs for all her cookies.

3. Latisha also needs to make icing for her cookes. She decides to double the recipe. The recipe calls for 2 egg whites, $\frac{1}{2}$ c sugar, $\frac{1}{8}$ tsp salt, and 1 tsp vanilla. Calculate the total of each ingredient that Latisha needs for her icing.

4. Make a shopping list that includes the total ingredients necessary for Davonna and Latisha's cookies.

5. A bag of chocolate chips contains 2 c of chips. How many bags does Davonna need and how much will be left over?

6. One pound of butter is 2 cups. How many pounds of butter do the two women need for their cookies, and how much will be left over?

One of the major applications of statistics is their value in predicting future occurrences. Before the future can be predicted, statisticians study what has happened in the past and look for patterns. If a pattern can be detected, and it is reasonable to assume that nothing will happen to interrupt the pattern, then it is a relatively easy matter to predict the future simply by continuing the pattern. Insurance companies, for instance, study the occurrences of traffic accidents among various groups of people. Once they have identified a pattern, they use this to predict future accident rates, which in turn are used to set insurance rates. When a group, such as teenaged boys, is identified as having a higher incidence of accidents, their insurance rates are set higher.

Dice

While predicting accident rates is a very complicated endeavor, there are other activities for which the patterns are relatively easy to find. Take, for instance, the act of rolling a die. The die has six sides, marked 1 to 6. Theoretically, each side has an equal chance of ending in the up position after a roll. Fill in the following table by rolling a die 120 times.

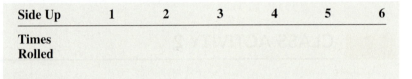

Side Up	1	2	3	4	5	6
Times Rolled						

Theoretically, each side will be rolled the same number of times as the others. Since you rolled the die 120 times and there were six possible outcomes, each side should come up $120 \div 6 = 20$ times. How close to 20 are your outcomes in the table? What do you suppose are reasons for not getting a perfectly distributed table?

Mathematicians are likely to express the relationships in this situation using the concept of *probability*, which is a measure of the likelihood of a particular event occurring. We describe the probability of an event with a fraction. The numerator of the fraction is the number of different ways the desired event can occur and the denominator of the fraction is the total number of possible outcomes. So the probability of rolling a 2 on the die is $\frac{1}{6}$ because there is only one way to roll a 2 but there are 6 possible outcomes when rolling a die. What is the probability of rolling a 5? What is the probability of rolling a 6? Non-mathematicians are more likely to express this relationship using the concept of *odds*. They would say that the odds of rolling a 2 are 1 in 6. This means that for every six times you roll a die, you can expect one of them to result in a 2.

Coin Toss

Suppose you and a friend each flip a coin. What are all the possible joint outcomes? What is the probability of getting two heads? What is the probability of getting two tails? What is the probability of getting one head and one tail? What does it mean if the probability of an event is $\frac{3}{3}$? Is it possible for the probability of an event to be $\frac{5}{4}$? Explain.

Cards

Suppose you pick a card at random out of a deck of playing cards. What is the probability that the card will be the queen of hearts? What is the probability that the card will be a queen? What is the probability that the card will be a heart?

Fill out the following table and try to discover the relationship among these three probabilities.

Probability of a Queen	Probability of a Heart	Probability of the Queen of Hearts

For a card to be the queen of hearts, two conditions must hold true at the same time. The card must be a queen *and* the card must be a heart. Make a guess about the relationship of the probabilities when two conditions must occur simultaneously. Test your guess by considering the probability of drawing a black 7. What are the two conditions that must be true in order for the card to be a black 7? What are their individual probabilities? Was your guess correct?

What two conditions must be true when you draw a red face card? What is the probability of drawing a red face card?

Suppose you pick a card at random out of a deck of playing cards. What is the probability that the card will be a 3 or a 4? What is the probability that the card will be a 3? A 4? Fill in the table to try to discover the relationship between these probabilities.

Probability of a 3	Probability of a 4	Probability of a 3 or a 4

A card is a 3 *or* a 4 if either condition holds. Make a guess about the relationship of the probabilities when either of two conditions must be true. Test your guess by calculating the probability that a card will be a heart or a club. Was your guess correct?

Sometimes a complicated probability is easier to calculate using a backdoor approach. For instance, suppose you needed to calculate the probability that a card drawn is an ace or a 2 or a 3 or a 4 or a 5 or a 6 or a 7 or an 8 or a 9 or a 10 or a jack or a queen. You can certainly add the individual probabilities (what do you get?). However, another way to look at the situation is to ask what is the probability of not getting a king. We reason that if you do not get a king, then you do get one of the desired cards. We calculate this by subtracting the probability of getting a king from 1. This is because 1 must be the sum of all the probabilities that totally define the set (in this case, the sum of the probabilities of getting a king and the probability of getting one of the other cards). Verify that you get the same probability using both methods.

GOOD ADVICE FOR *Studying*

PREPARING FOR TESTS

Differences between Tests and Homework

- You cannot peek at the answer and work backward on a test.
- The skills being tested are all mixed up, instead of just one kind per section.
- Being able to follow directions becomes supremely important.
- There is often a time limit.

Two to Three Days before the Test

- Finish all assigned homework.
- Make a list of vocabulary words.
- Review any concepts that you are still unsure of.

The Day before the Test

- Identify each different kind of problem that will be on the test. Be sure you know what the directions will say and how you will proceed.
- Take a practice test. Your instructor may provide one or you can use the chapter test at the end of the chapter. Check your answers on the test.
- Go back and review your homework and notes for any problems that you are unsure of.
- Get enough sleep.

The Day of the Test

- Eat the meal that falls before your test—include some protein.
- Review the different kinds of problems again and rehearse how to begin each.

One-Half Hour before the Test

- Stop studying.
- Find a quiet place (or take a walk) and tell yourself that you are relaxed and prepared.

© Image Source / Jupiter Images

GOOD ADVICE FOR STUDYING

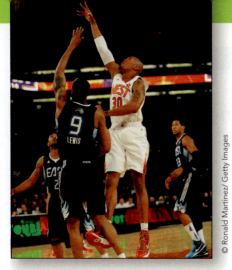
© Ronald Martinez/ Getty Images

Decimals

APPLICATION

Sports hold a universal attraction. People all over the world enjoy a good game. For some sports, it is relatively easy to determine which athlete is the best. In track, downhill skiing, and swimming, for instance, each contestant races against the clock and the fastest time wins. In team sports, it is easy to tell which team wins, but sometimes difficult to determine how the individual athletes compare with one another. In order to make comparisons more objective, we often use sports statistics.

The simplest kind of statistic is to count how many times an athlete performs a particular feat in a single game. In basketball, for instance, it is usual to count the number of points scored, the number of rebounds made, and the number of assists for each player.

Consider the following statistics for the 2009 NBA All-Star Game.

Player	Points Scored	Rebounds	Assists
EAST			
LeBron James	20	5	3
Kevin Garnett	12	4	2
Dwight Howard	13	9	0
Allen Iverson	2	1	3
Dwyane Wade	18	2	5
WEST			
Amar'e Stoudemire	19	6	0
Tim Duncan	6	3	2
Yao Ming	2	3	0
Kobe Bryant	27	4	4
Chris Paul	14	7	14

GROUP DISCUSSION

1. Which player had the best overall statistics? Justify your answer.
2. Which is more important in basketball, rebounds or assists? Explain.
3. Which of the players had the weakest performance? Explain.

4.1 Decimals: Reading, Writing, and Rounding

OBJECTIVES

1. Write word names from place value names and place value names from word names.
2. Round a given decimal.

VOCABULARY

Decimal numbers, more commonly referred to as **decimals,** are another way of writing fractions and mixed numbers. The digits used to write whole numbers and a period called a **decimal point** are used to write place value names for these numbers.

The **number of decimal places** is the number of digits to the right of the decimal point. **Exact decimals** are decimals that show exact values. **Approximate decimals** are rounded values.

HOW & WHY

■ **OBJECTIVE 1** Write word names from place value names and place value names from word names.

Decimals are written by using a standard place value in the same way we write whole numbers. Numbers such as 12.65, 0.45, 0.795, 1306.94, and 19.36956 are examples of decimals.
In general, the place value for decimals is

1. The same as whole numbers for digits to the left of the decimal point, and
2. A fraction whose denominator is 10, 100, 1000, and so on, for digits to the right of the decimal point.

The digits to the right of the decimal point have place values of

$$\frac{1}{10^1} = \frac{1}{10} \qquad\qquad = 0.1$$

$$\frac{1}{10^2} = \frac{1}{10 \cdot 10} \qquad = \frac{1}{100} \qquad = 0.01$$

$$\frac{1}{10^3} = \frac{1}{10 \cdot 10 \cdot 10} \qquad = \frac{1}{1000} \qquad = 0.001$$

$$\frac{1}{10^4} = \frac{1}{10 \cdot 10 \cdot 10 \cdot 10} \qquad = \frac{1}{10{,}000} \qquad = 0.0001$$

and so on, in that order from left to right.
Using the ones place as the central position, the place values of a decimal are shown in Figure 4.1.

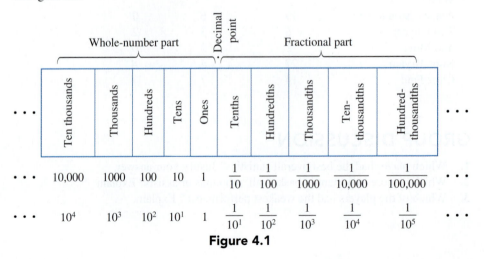

Figure 4.1

Note that the decimal point separates the whole-number part from the fractional part.

$$71.961 = \underbrace{71}_{\text{Whole-number part}} + \underbrace{.961}_{\text{Fraction part}}$$

If the decimal point is not written, as in the case of a whole number, the decimal point is understood to follow the ones place; thus,

$$87 = 87. \qquad 7 = 7. \qquad 725 = 725.$$

We can write an expanded form of the decimal using fractions with denominators that are powers of 10. So

$$71.961 = 70 + 1 + \frac{9}{10} + \frac{6}{100} + \frac{1}{1000}$$

The expanded form can also be written

7 tens + 1 one + 9 tenths + 6 hundredths + 1 thousandth

Table 4.1 shows how to write the word names for 237.58 and 0.723.

TABLE 4.1 Word Names for Decimals

	Number to Left of Decimal Point	Decimal Point	Number to Right of Decimal Point	Place Value of Last Digit
Place value name	237	.	58	$\frac{1}{100}$
Word name of each	Two hundred thirty-seven	and	fifty-eight	hundredths
Word name for decimal	Two hundred thirty-seven and fifty-eight hundredths			
Place value name	0	.	723	$\frac{1}{1000}$
Word name of each	Omit	Omit	Seven hundred twenty-three	thousandths
Word name of decimal	Seven hundred twenty-three thousandths			

For numbers greater than zero and less than one (such as 0.639), it is preferable to write the digit 0 in the ones place.

> **To write the word name for a decimal**
>
> 1. Write the name for the whole number to the left of the decimal point.
> 2. Write the word *and* for the decimal point.
> 3. Write the whole number name for the number to the right of the decimal point.
> 4. Write the place value of the digit farthest to the right.
>
> If the decimal has only zero or no digit to the left of the decimal point, omit steps 1 and 2.

Table 4.2 shows some other numbers and their corresponding word names.

TABLE 4.2 **Word Names for Decimals**

Number	Word Name
76.52	Seventy-six and fifty-two hundredths
0.765	Seven hundred sixty-five thousandths
0.00052	Fifty-two hundred-thousandths
7.005	Seven and five thousandths

> **To write the place value name for a decimal**
>
> 1. Write the whole number. (The number before the word *and*.)
> 2. Write a decimal point for the word *and*.
> 3. Ignoring the place value name, write the name for the number following the word *and*. Insert zeros, if necessary, between the decimal point and the digits following it to ensure that the place on the far right has the correct (given) place value.

So the place value name for six hundred sixteen and eighty-four thousandths is

616	First write the whole number to the left of the word *and*.
616.	Write a decimal point for the word *and*.
616.084	The "whole number" after the word *and* is 84. A zero is inserted to place the 4 in the thousandths place.

EXAMPLES A–D

DIRECTIONS: Write the word name.

STRATEGY: Write the word name for the whole number to the left of the decimal point. Then write the word *and* for the decimal point. Finally, write the word name for the number to the right of the decimal point followed by the place value of the digit farthest to the right.

WARM-UP
A. Write the word name for 0.91.

A. Write the word name for 0.83.

Eighty-three	Write the word name for the number right of the decimal point.
Eighty-three hundredths	Next, write the place value of the digit 3. The word name for 0 in the ones place may be written or omitted. "Zero and eighty-three hundredths" is correct but unnecessary.

WARM-UP
B. Write the word name for 0.091.

B. Write the word name for 0.0027.

Twenty-seven ten-thousandths

ANSWERS TO WARM-UPS A–B
A. ninety-one hundredths
B. ninety-one thousandths

C. Write the word name for 556.43.

Five hundred fifty-six	Write the word name for the whole number left of the decimal point.
Five hundred fifty-six and	Write the word *and* for the decimal point.
Five hundred fifty-six and forty-three	Write the word name for the number right of the decimal point.
Five hundred fifty-six and forty-three hundredths	Write the place value of the digit 3.

WARM-UP

C. Write the word name for 123.053.

D. Janet called an employee to find the measurement of the outside diameter of a new wall clock the company is manufacturing. She asked the employee to check the plans. What is the word name the employee will read to her? The clock is shown.

9.225 in.

The employee will read "Nine and two hundred twenty-five thousandths inches."

WARM-UP

D. The measurement of the outside diameter of another clock is shown in the diagram. What word name will the employee read?

11.375 in.

EXAMPLES E–F

DIRECTIONS: Write the place value name.

STRATEGY: Write the digit symbols for the corresponding words. Replace the word *and* with a decimal point. If necessary, insert zeros.

E. Write the place value name for thirty-eight ten-thousandths.

38	First, write the number for thirty-eight.
.0038	The place value "ten-thousandths" indicates four decimal places, so write two zeros before the numeral thirty-eight and then a decimal point. This puts the numeral 8 in the ten-thousandths place.
0.0038	Since the number is between zero and one, we write a 0 in the ones place.

F. Write the place value name for "seven hundred nine and nine hundred seven ten-thousandths."

709	The whole number part is 709.
709.	Write the decimal point for *and*.
709.0907	The "whole number" after the *and* is 907. A zero is inserted so the numeral 7 is in the ten-thousandths place.

WARM-UP

E. Write the place value name for seventy-six hundred-thousandths.

WARM-UP

F. Write the place value name for two hundred five and five hundred two ten-thousandths.

HOW & WHY

■ **OBJECTIVE 2** Round a given decimal.

Decimals can be either *exact* or *approximate*. For example, decimals that count money are exact. The figure $56.35 shows an exact amount. Most decimals that describe measurements are approximations. For example, 6.1 ft shows a person's height to the nearest tenth of a foot and 1.9 m shows the height to the nearest tenth of a meter, but neither is an exact measure.

ANSWERS TO WARM-UPS C–F

C. one hundred twenty-three and fifty-three thousandths
D. The employee will read "Eleven and three hundred seventy-five thousandths inches."
E. 0.00076 **F.** 205.0502

Decimals are rounded using the same procedure as for whole numbers. Using a ruler (see Figure 4.2), we round 2.563.

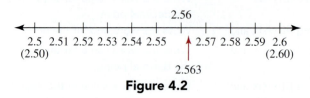

Figure 4.2

To the nearest tenth, 2.563 is rounded to 2.6, because it is closer to 2.6 than to 2.5. Rounded to the nearest hundredth, 2.563 is rounded to 2.56, because it is closer to 2.56 than to 2.57.

To round 21.8573 to the nearest hundredth, without drawing a number line, draw an arrow under the hundredths place to identify the round-off place.

21.8573
   ↑

We must choose between 21.85 and 21.86. Because the digit to the right of the round-off position is 7, the number is more than halfway to 21.86. So we choose the larger number. 21.8573 ≈ 21.86

CAUTION

Do not replace the dropped digits with zeros if the round-off place is to the right of the decimal point. 21.8573 ≈ 21.8600 indicates a round-off position of ten-thousandths.

To round a decimal number to a given place value

1. Draw an arrow under the given place value. (After enough practice, you will be able to round mentally and will not need the arrow.)
2. If the digit to the right of the arrow is 5, 6, 7, 8, or 9, add 1 to the digit above the arrow; that is, round to the larger number.
3. If the digit to the right of the arrow is 0, 1, 2, 3, or 4, keep the digit above the arrow; that is, round to the smaller number.
4. Write whatever zeros are necessary after the arrow so that the number above the arrow has the same place value as the original. See Example H.

This method is sometimes called the "four-five" rule. Although this rounding procedure is the most commonly used, it is not the only way to round. Many government agencies round by *truncation,* that is, by dropping the digits after the decimal point. Thus, $87.32 ≈ $87. It is common for retail stores to round up for any amounts smaller than one cent. Thus, $3.553 ≈ $3.56. There is also a rule for rounding numbers in science, which is sometimes referred to as the "even/odd" rule. You might learn and use a different round-off rule depending on the kind of work you are doing.

EXAMPLES G–J

DIRECTIONS: Round as indicated.

STRATEGY: Draw an arrow under the given place value. Examine the digit to the right of the arrow to determine whether to round up or down.

G. Round 0.7539 to the nearest hundredth.

$0.7539 \approx 0.75$
↑
The digit to the right of the round-off place is 3, so round down.

H. Round 7843.9 to the nearest thousand.

$7843.9 \approx 8000$
↑
Three zeros must be written after the 8 to keep it in the thousands place.

I. Round 537.7 to the nearest unit.

$537.7 \approx 538$
The number to the right of the round-off place is 7, so we round up by adding $537 + 1 = 538$.

J. Round 64.7659 and 9.9897 to the nearest unit, the nearest tenth, the nearest hundredth, and the nearest thousandth.

		Unit		Tenth		Hundredth		Thousandth
64.7659	≈	65	≈	64.8	≈	64.77	≈	64.766
9.9897	≈	10	≈	10.0	≈	9.99	≈	9.990

> **CAUTION**
>
> The zeros following the decimal in 10.0 and 9.990 are necessary to show that the original number was rounded to the nearest tenth and thousandth, respectively.

EXERCISES 4.1

■ **OBJECTIVE 1** Write word names from place value names and place value names from word names. (See page 278.)

A *Write the word name.*

1. 0.26

2. 0.82

3. 0.267

4. 0.943

5. 7.002

6. 4.3007

7. 11.92

8. 32.03

Write the place value name.

9. Forty-two hundredths

10. Sixty-nine hundredths

11. Four hundred nine thousandths

12. Five hundred nineteen thousandths

13. Nine and fifty-nine thousandths

14. Sixteen and six hundredths

15. Three hundred eight ten-thousandths

16. Twelve ten-thousandths

B *Write the word name.*

17. 0.805

18. 8.05

19. 80.05

20. 8.005

21. 61.0203

22. 45.0094

23. 90.003

24. 900.030

Write the place value name.

25. Thirty-five ten-thousandths

26. Thirty-five thousand

27. One thousand eight hundred and twenty-eight thousandths

28. Three hundred and fifteen thousandths

29. Five hundred five and five thousandths

30. Two and two hundred two thousandths

31. Sixty-five and sixty-five thousandths

32. Seven hundred three and three hundred seven thousandths

■ **OBJECTIVE 2** Round a given decimal. (See page 281.)

A *Round to the nearest unit, tenth, and hundredth.*

	Unit	Tenth	Hundredth
33. 35.777			
34. 73.788			
35. 729.638			
36. 922.444			
37. 0.6157			
38. 0.75487			

Round to the nearest cent.

39. $67.4856

40. $27.6372

41. $548.7235

42. $375.7545

B *Round to the nearest ten, hundredth, and thousandth.*

	Ten	Hundredth	Thousandth
43. 35.7834			
44. 61.9639			
45. 86.3278			
46. 212.7364			
47. 0.91486			
48. 0.8049			

Round to the nearest dollar.

49. $72.49 **50.** $38.51 **51.** $7821.51 **52.** $8467.80

C *Exercises 53–56. The graph shows the 2008 property tax rates for various cities in Florida.*

2008 Property Tax Rates per $1000 of Value

53. Write the word name for the tax rate in Gainsville.

54. Write the word name for the tax rate in Cocoa Beach.

55. Round the tax rate for Ft. Lauderdale to the nearest hundredth.

56. Round the tax rate for Miami to the nearest thousandth.

57. Carlos Avilla buys a 5-port USB hub for his computer that has a marked price of $57.79. What word name does he write on the check?

58. Lenora Fong makes a down payment on a used Toyota Corolla. The down payment is $648.89. What word name does she write on the check?

Exercises 59–62. Use the graph of the number line.

59. What is the position of the arrow to the nearest hundredth?

60. What is the position of the arrow to the nearest thousandth?

61. What is the position of the arrow to the nearest tenth?

62. What is the position of the arrow to the nearest unit?

Write the word name.

63. 756.7104

64. 7404.7404

65. The computer at Grant's savings company shows that his account, including the interest he has earned, has a value of $3478.59099921. Round the value of the account to the nearest cent.

66. In doing her homework, Catherine's calculator shows that the answer to a division exercise is 25.69649912. If she is to round the answer to the nearest thousandth, what answer does she report?

Write the place value name.

67. Two hundred thirteen and one thousand, one hundred one ten-thousandths

68. Eleven thousand five and one thousand fifteen hundred-thousandths

69. In 2009, Kuwait had an estimated population density of 130.736 people per square kilometer. Round the population density to the nearest whole person per square kilometer.

70. One text lists the mean (average) distance from Earth to the sun as 92,960,000 miles. To what place does this number appear to have been rounded?

© iStockphoto.com/Günay Mutlu

71. The percent of households in the United States owning pet birds is estimated at 3.90%. To what place does this number appear to have been rounded?

Round to the indicated place value.

	Thousand	Hundredth	Ten-thousandth
72. 3742.80596			
73. 7564.35926			
74. 2190.910053			
75. 78,042.38875			

Exercises 76–79 relate to the chapter application.

76. In January 1906 a Stanley car with a steam engine set a 1-mile speed record by going 127.659 miles per hour. Round this rate to the nearest tenth of a mile per hour.

77. In March 1927 a Sunbeam auto set a 1-mile speed record by going 203.790 mph. What place value was this rate rounded to?

78. In October 1970 a Blue Flame set a 1-mile speed record by going 622.407 mph. Explain why it is incorrect to round the rate to 622.5 mph.

79. On October 15, 1997, Andy Green recorded the first supersonic land 1-mile record in a Thrust SSC with a speed of 763.035 mph. Round this speed to the nearest whole mile per hour and to the nearest tenth of a mile per hour.

STATE YOUR UNDERSTANDING

80. Explain the difference between an exact decimal value and an approximate decimal value. Give an example of each.

81. Explain in words, the meaning of the value of the 4s in the numerals 43.29 and 18.64. Include some comment on how and why the values of the digit 4 are alike and how and why they are different.

82. Consider the decimal represented by abc.defg. Explain how to round this number to the nearest hundredth.

CHALLENGE

83. Round 8.28282828 to the nearest thousandth. Is the rounded value less than or greater than the original value? Write an inequality to illustrate your answer.

84. Round 7.7777777 to the nearest unit. Is the rounded value less than or greater than the original value? Write an inequality to illustrate your answer.

85. Write the place value name for five hundred twenty-two hundred-millionths.

86. Write the word name for 40,715.300519.

87. Make a list of cases in which you think "truncating" is the best way to round. Rounding by truncating means to drop all digit values to the right of the rounding position. For example, $2.77 \approx 2$, $\$19.68 \approx \$19$, $8.66666 \approx 8.6$, and $34,999 \approx 34,000$. Can you find a situation in which such rounding is actually used?

MAINTAIN YOUR SKILLS

Find the missing numerator.

88. $\dfrac{3}{7} = \dfrac{?}{28}$

89. $\dfrac{7}{12} = \dfrac{?}{132}$

90. $\dfrac{42}{125} = \dfrac{?}{10,000}$

91. $\dfrac{19}{40} = \dfrac{?}{1000}$

True or false.

92. $3971 < 3969$

93. $30,951 > 30,899$

List the numbers from smallest to largest.

94. 367, 401, 392, 363, 390

95. 6227, 6223, 6218, 6209, 6215

96. During a recent candy sale, Mia sold 21 boxes of candy each containing 24 candy bars. Chia sold 16 bags of candy, each containing 30 candy bars. Who sold more candy bars?

97. Kobe Bryant hits 17 of 20 field goals attempted. If in the next game he attempts 40 field goals, how many must he hit to have the same shooting percentage?

4.2 Changing Decimals to Fractions; Listing in Order

OBJECTIVES

1. Change a decimal to a fraction.
2. List a set of decimals from smallest to largest.

HOW & WHY

▪ **OBJECTIVE 1** Change a decimal to a fraction.

The word name of a decimal is also the word name of a fraction.
Consider 0.615.

READ: six hundred fifteen thousandths

WRITE: $\dfrac{615}{1000}$

So, $0.615 = \dfrac{615}{1000} = \dfrac{123}{200}$

Some other examples are shown in the Table 4.3.

TABLE 4.3 Decimals to Fractions

Place Value Name	Word Name	Fraction
0.77	Seventy-seven hundredths	$\dfrac{77}{100}$
0.9	Nine tenths	$\dfrac{9}{10}$
0.225	Two hundred twenty-five thousandths	$\dfrac{225}{1000} = \dfrac{9}{40}$
0.48	Forty-eight hundredths	$\dfrac{48}{100} = \dfrac{12}{25}$

> **To change a decimal to a fraction**
>
> 1. Read the decimal word name.
> 2. Write the fraction that has the same value.
> 3. Simplify.

Notice that because of place value, the number of decimal places in a decimal tells us the number of zeros in the denominator of the fraction. This fact can be used as another way to write the fraction or to check that the fraction is correct:

$$4.58 \quad = \quad 4\frac{58}{100} = 4\frac{29}{50} \quad \text{or} \quad 4.58 = \frac{458}{100} = \frac{229}{50} = 4\frac{29}{50}$$

$\underbrace{\text{Two decimal places}} \qquad \underbrace{\text{Two zeros}}$

EXAMPLES A–C

DIRECTIONS: Change the decimal to a fraction or mixed number.

STRATEGY: Say the word name to yourself and write the fraction or mixed number that is equivalent. Simplify.

WARM-UP

A. Change 0.43 to a fraction.

Forty-three hundredths   *Say the word name.*

$\dfrac{43}{100}$ *Write as a fraction.*

A. Change 0.83 to a fraction.

B. Write 0.065 as a fraction.

Sixty-five thousandths   *Say the word name.*

$\dfrac{65}{1000} = \dfrac{13}{200}$ *Write as a fraction and simplify.*

WARM-UP

B. Write 0.875 as a fraction.

C. Write as a mixed number: 23.48

Twenty-three and forty-eight hundredths   *Say the word name.*

$23\dfrac{48}{100} = 23\dfrac{12}{25}$ *Write as a mixed number and simplify.*

WARM-UP

C. Write as a mixed number: 17.95

HOW & WHY

■ **OBJECTIVE 2** List a set of decimals from smallest to largest.

Fractions can be listed in order when they have a common denominator by ordering the numerators. This idea can be extended to decimals when they have the same number of decimal places. For instance, $0.36 = \dfrac{36}{100}$ and $0.47 = \dfrac{47}{100}$ have a common denominator when written in fractional form. Because $\dfrac{36}{100} < \dfrac{47}{100}$, we know that 0.36 is less than 0.47, or $0.36 < 0.47$.

The decimals 0.7 and 0.59 do not have a common denominator. But we can make common denominators by placing a zero after the 7. Thus,

$$0.7 = \frac{7}{10} = \frac{7}{10} \cdot \frac{10}{10} = \frac{70}{100} = 0.70$$

ANSWERS TO WARM-UPS A–C

A. $\dfrac{83}{100}$ **B.** $\dfrac{7}{8}$ **C.** $17\dfrac{19}{20}$

so that

$$0.7 = \frac{70}{100} \quad \text{and} \quad 0.59 = \frac{59}{100}$$

Then, since $\frac{70}{100} > \frac{59}{100}$, we conclude that $0.7 > 0.59$.

We can also use a number line to order decimals. The number line shows that $2.6 < 3.8$ because 2.6 is to the left of 3.8. See Figure 4.3.

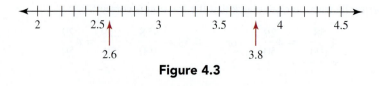

Figure 4.3

Many forms for decimal numbers are equivalent. For example,

$$6.3 = 6.30 = 6.300 = 6.3000 = 6.30000$$
$$0.85 = 0.850 = 0.8500 = 0.85000 = 0.850000$$
$$45.982 = 45.9820 = 45.98200 = 45.982000 = 45.9820000$$

The zeros to the right of the decimal point following the last nonzero digit do not change the value of the decimal. Usually these extra zeros are not written, but they are useful when operating with decimals.

> **To list a set of decimals from smallest to largest**
>
> 1. Make sure that all numbers have the same number of decimal places to the right of the decimal point by writing zeros to the right of the last digit when necessary.
> 2. Write the numbers in order as if they were whole numbers.
> 3. Remove the extra zeros.

EXAMPLES D–F

DIRECTIONS: Is the statement true or false?

STRATEGY: Write each numeral with the same number of decimal places. Compare the values without regard to the decimal point.

WARM-UP
D. True or false:
$0.54 < 0.53$?

D. True or false: $0.84 > 0.81$?

$84 > 81$ is true.   Compare the numbers without regard to the decimal points.

So, $0.84 > 0.81$ is true.

WARM-UP
E. True or false:
$0.487 < 0.4874$?

E. True or false: $0.9331 < 0.933$?

$0.9331 < 0.9330$ Write with the same number of decimal places.
$\quad 9331 < 9330$ is false.   Without regard to the decimal point, 9331 is larger.

So, $0.9331 < 0.933$ is false.

WARM-UP
F. True or false:
$66.7 > 66.683$?

F. True or false: $32.008 > 32.09$?

$32.008 > 32.090$ Write with the same number of decimal places.
$32008 > 32090$ is false.

So, $32.008 > 32.09$ is false.

ANSWERS TO WARM-UPS D–F

D. false   **E.** true   **F.** true

EXAMPLES G–I

DIRECTIONS: List the decimals from smallest to largest.

STRATEGY: Write zeros on the right so that all numbers have the same number of decimal places. Compare the numbers as if they were whole numbers and then remove the extra zeros.

G. List 0.48, 0.472, 0.4734, and 0.484 from smallest to largest.

0.4800   First, write all numbers with the same number of decimal places by
0.4720   inserting zeros on the right.
0.4734
0.4840

0.4720, 0.4734, 0.4800, 0.4840   Second, write the numbers in order as if they were whole numbers.

0.472, 0.4734, 0.48, 0.484   Third, remove the extra zeros.

H. List 8.537, 8.631, 8.5334, and 8.538 from smallest to largest.

8.537 = 8.5370   Step 1
8.631 = 8.6310
8.5334 = 8.5334
8.538 = 8.5380

8.5334, 8.5370, 8.5380, 8.6310   Step 2

8.5334, 8.537, 8.538, 8.631   Step 3

I. Roberto and Raquel both weigh a quarter coin. Roberto measures 5.67 grams and Raquel measures 0.5665 grams. Whose measure is heavier?

5.670, 5.665   Write with the same number of decimal places and compare as whole numbers.

Roberto's weight is heavier since 5670 > 5665.

EXERCISES 4.2

■ **OBJECTIVE 1** Change a decimal to a fraction. (See page 288.)

A *Change each decimal to a fraction and simplify if possible.*

1. 0.83   **2.** 0.37   **3.** 0.65   **4.** 0.6

5. Six hundred fifty-eight thousandths   **6.** Three hundred one thousandths

7. 0.82   **8.** 0.32   **9.** 0.48   **10.** 0.55

B *Change the decimal to a fraction or mixed number and simplify.*

11. 10.41 **12.** 36.39 **13.** 0.125 **14.** 0.575

15. 12.24 **16.** 47.64 **17.** 11.344 **18.** 5.228

19. Seven hundred fifty thousandths **20.** Twenty-five hundred-thousandths

■ **OBJECTIVE 2** List a set of decimals from smallest to largest. (See page 289.)

A *List the set of decimals from smallest to largest.*

21. 0.7, 0.1, 0.4 **22.** 0.07, 0.06, 0.064

23. 0.17, 0.06, 0.24 **24.** 0.46, 0.48, 0.29

25. 3.26, 3.185, 3.179 **26.** 7.18, 7.183, 7.179

Is the statement true or false?

27. $0.38 < 0.3$ **28.** $0.49 < 0.50$ **29.** $10.48 > 10.84$ **30.** $7.78 > 7.87$

B *List the set of decimals from smallest to largest.*

31. 0.0477, 0.047007, 0.047, 0.046, 0.047015 **32.** 1.006, 1.106, 0.1006, 0.10106

33. 0.555, 0.55699, 0.5552, 0.55689 **34.** 7.47, 7.4851, 7.4799, 7.4702

35. 25.005, 25.051, 25.0059, 25.055 **36.** 92.0728, 92.0278, 92.2708, 92.8207

Is the statement true or false?

37. $3.1231 < 3.1213$ **38.** $6.3456 > 6.345$

39. $74.6706 < 74.7046$ **40.** $21.6043 > 21.6403$

C

41. The probability that a flipped coin will come up heads four times in a row is 0.0625. Write this as a reduced fraction.

42. The probability that a flipped coin will come up heads twice and tails once out of three flips is 0.375. Write this as a reduced fraction.

43. The Alpenrose Dairy bids $2.675 per gallon to provide milk to the local school district. Tillamook Dairy puts in a bid of $2.6351, and Circle K Dairy makes a bid of $2.636. Which is the best bid for the school district?

44. Larry loses 3.135 pounds during the week. Karla loses 3.183 pounds and Mitchell loses 3.179 pounds during the same week. Who loses the most weight this week?

Exercises 45–47. The following free-throw records are established in the National Basketball Association: highest percentage made in a season: 0.832, Boston Celtics in 1989–1990; lowest percentage made in a season: 0.635, Philadelphia in 1967–1968; lowest percentage made by both teams in a single game: 0.405, Miami vs. Charlotte in 2005.

45. Write a simplified fraction to show the highest percentage of free throws made in a season.

46. Write a simplified fraction to show the lowest percentage of free throws made in a season.

47. Write a simplified fraction to show the lowest percentage of free throws made in a game by both teams.

Change the decimal to a fraction or mixed number and simplify.

48. 0.1775

49. 0.8375

50. 403.304

51. 25.025

52. Gerry may choose a 0.055 raise in pay or a $\frac{1}{18}$ increase. Which value will yield more money? Compare in fraction form.

53. A chemistry class requires 0.547 ml of acid for each student. Norado has 0.55 ml of acid. Does she need more or less acid?

List the decimals from smallest to largest.

54. 0.00829, 0.0083001, 0.0082, 0.0083, 0.0083015

55. 3.0007, 3.002, 3.00077, 3.00092, 3.00202

56. 36.567, 36.549, 36.509, 36.557, 36.495, 36.7066

57. 82.86, 83.01, 82.85, 82.58, 83.15, 83.55, 82.80, 82.78

58. Lee is pouring a concrete patio in his back yard. Lee needs 2.375 cubic yards of concrete for his patio. Change the amount of concrete to a mixed number and simplify.

59. For a bow for a prom dress, Maria may choose 0.725 yd or $\frac{5}{7}$ yd for the same price. Which should she choose to get the most ribbon? Compare in fraction form.

© dendong/Shutterstock.com

60. One synodic day on Jupiter (midday to midday) is about 9.925933 hours, while one sidereal day (measured by apparent star movements) is about 9.925 hours. Which is longer?

61. In 2008, the population density of Belgium was 341.1622 and the population density of Rwanda was 341.408, both measured in people per square kilometer. Which country had fewer people per square kilometer?

62. Betty Crocker cake mixes, when prepared as directed, have the following decimal fraction of the calories per slice from fat: Apple Cinnamon, 0.36; Butter Pecan, 0.4; Butter Recipe/Chocolate, 0.43; Chocolate Chip, 0.42; Spice, 0.38; and Golden Vanilla, 0.45. If each slice contains 280 calories, which cake has the most calories from fat? Fewest calories from fat?

63. Hash brown potatoes have the following number of fat grams per serving: frozen plain, 7.95 g; frozen with butter sauce, 8.9 g; and homemade with vegetable oil, 10.85 g. Write the fat grams as simplified mixed numbers. Which serving of hash browns has the least amount of fat?

Exercises 64–67 relate to the chapter application.

64. At the end of the 2008–2009 NBA season, the six division leaders won the given decimal fraction of their games: Boston, 0.756; Cleveland, 0.805; Orlando, 0.720; Denver, 0.659; Los Angels, 0.793; San Antonio, 0.659. Rank the teams from best record to worst.

65. For the 2008–2009 NBA season, Shaquille O'Neal of the Phoenix Suns had the top field goal percentage in the NBA. That year, O'Neal made 0.609 of his field goal attempts. Explain this record as a fraction. What fraction of his field goals did he miss?

66. The table displays the batting champions in the National and America Leagues for 2004–2008.

67. Sort the table in Exercise 66 so that the averages are displayed from lowest to highest by league.

National League

Year	Name	Team	Average
2004	Barry Bonds	San Francisco	0.362
2005	Derrek Lee	Chicago	0.335
2006	Freddy Sanchez	Pittsburgh	0.344
2007	Matt Holliday	Colorado	0.340
2008	Chipper Jones	Atlanta	0.364

American League

Year	Name	Team	Average
2004	Ichiro Suzuki	Seattle	0.372
2005	Michael Young	Texas	0.331
2006	Joe Mauer	Minnesota	0.347
2007	Maglio Ordonez	Detriot	0.363
2008	Joe Mauer	Minnesota	0.328

Which player had the highest batting average in the 5-year period?

68. Explain how the number line can be a good visual aid for determining which of two numerals has the larger value.

CHALLENGE

69. Change 0.44, 0.404, and 0.04044 to fractions and reduce.

70. Determine whether each statement is true or false.

a. $7.44 < 7\dfrac{7}{18}$ **b.** $8.6 > 8\dfrac{5}{9}$

c. $3\dfrac{2}{7} < 3.285$ **d.** $9\dfrac{3}{11} > 9.271$

MAINTAIN YOUR SKILLS

Add or subtract.

71. $479 + 3712 + 93 + 7225$

72. $75,881 + 3007 + 45,772 + 306$

73. $34,748 - 27,963$

74. $123,007 - 17,558$

75. $\dfrac{1}{2} + \dfrac{3}{4} + \dfrac{1}{8}$

76. $\dfrac{1}{3} + \dfrac{7}{12} + \dfrac{5}{6}$

77. $\dfrac{25}{64} - \dfrac{3}{8}$

78. $\dfrac{17}{20} - \dfrac{7}{12}$

79. Pedro counted the attendance at the seven-screen MetroPlex Movie Theater for Friday evening. He had the following counts by screen: #1, 456; #2, 389; #3, 1034; #4, 672; #5, 843; #6, 467; #7, 732. How many people attended the theater that Friday night?

80. Joanna has $1078 in her bank account. She writes checks for $54, $103, $152, $25, and $456. What balance does she now have in her account?

4.3 Adding and Subtracting Decimals

HOW & WHY

■ **OBJECTIVE 1** Add decimals.

What is the sum of 27.3 + 42.5? We make use of the expanded form of the decimal to explain addition.

$$
\begin{array}{rl}
27.3 = & 2 \text{ tens} + 7 \text{ ones} + 3 \text{ tenths} \\
+\ 42.5 = & 4 \text{ tens} + 2 \text{ ones} + 5 \text{ tenths} \\
\hline
& 6 \text{ tens} + 9 \text{ ones} + 8 \text{ tenths} = 69.8
\end{array}
$$

We use the same principle for adding decimals that we use for whole numbers—that is, we add like units. The vertical form gives us a natural grouping of the tens, ones, and tenths. By inserting zeros so all the numbers have the same number of decimal places, we write the addition 6.4 + 23.9 + 7.67 as 6.40 + 23.90 + 7.67.

$$
\begin{array}{r}
6.40 \\
23.90 \\
+\ 7.67 \\
\hline
37.97
\end{array}
$$

> **To add decimals**
>
> 1. Write in columns with the decimal points aligned. Insert extra zeros to help align the place values.
> 2. Add the decimals as if they were whole numbers.
> 3. Align the decimal point in the sum with those above.

EXAMPLES A–C

DIRECTIONS: Add.

STRATEGY: Write each numeral with the same number of decimal places, align the decimal points, and add.

A. Add: 8.2 + 56.93 + 38 + 0.08

$$
\begin{array}{r}
8.20 \\
56.93 \\
38.00 \\
+\ 0.08 \\
\hline
103.21
\end{array}
$$

Write each numeral with two decimal places. The extra zeros help line up the place values.

CALCULATOR EXAMPLE:

B. Add: 6.7934 + 0.0884 + 34.7 + 382.7330

STRATEGY: The extra zeros do not need to be inserted. The calculator will automatically align the place values when adding.

The sum is 424.3148.

C. Wanda goes to Target and buys the following: greeting cards, $4.65; Diet Coke, $2.45; lamp, $25.99; camera, $42.64; and dishwashing soap, $3.86. What is the total cost of Wanda's purchase?

STRATEGY: Add the prices of each item.

$$
\begin{array}{r}
\$4.65 \\
2.45 \\
25.99 \\
42.64 \\
+\ 3.86 \\
\hline
\$79.59
\end{array}
$$

Wanda spends $79.59 at Target.

WARM-UP

C. Cheryl goes to Nordstrom Rack and buys the following: shoes, $56.29; hose, $15.95; pants, $29.95; and jacket, $47.83. How much does Cheryl spend at Nordstrom?

HOW & WHY

■ OBJECTIVE 2 Subtract decimals.

What is the difference $8.68 - 4.37$? To find the difference, we write the numbers in column form, aligning the decimal points. Now subtract as if they are whole numbers.

$$
\begin{array}{r}
8.68 \\
-4.37 \\
\hline
4.31
\end{array}
$$
    The decimal point in the difference is aligned with those above.

When necessary, we can regroup, or borrow, as with whole numbers. What is the difference $7.835 - 3.918$?

$$
\begin{array}{r}
7.835 \\
-3.918
\end{array}
$$

We need to borrow 1 from the hundredths column (1 hundredth = 10 thousandths) and we need to borrow 1 from the ones column (1 one = 10 tenths).

$$
\begin{array}{r}
{\scriptstyle 6\ 18\ 2\ 15} \\
7.8\,3\,5 \\
-3.9\,1\,8 \\
\hline
3.9\,1\,7
\end{array}
$$

So the difference is 3.917.

Sometimes it is necessary to write zeros on the right so the numbers have the same number of decimal places. See Example E.

To subtract decimals

1. Write the decimals in columns with the decimal points aligned. Insert extra zeros to align the place values.
2. Subtract the decimals as if they are whole numbers.
3. Align the decimal point in the difference with those above.

ANSWER TO WARM-UP C

C. Cheryl spends $150.02 at Nordstom.

EXAMPLES D–H

DIRECTIONS: Subtract.

STRATEGY: Write each numeral with the same number of decimal places, align the decimal points, and subtract.

D. Subtract: $21.573 - 5.392$

$$\begin{array}{r} 21.573 \\ -\ 5.392 \end{array}$$ Line up the decimal points so the place values are aligned.

$$\begin{array}{r} \overset{4\ \ 17}{21.5\cancel{7}3} \\ -\ 5.392 \end{array}$$ Borrow 1 tenth from the 5 in the tenths place.
(1 tenth = 10 hundredths)

$$\begin{array}{r} \overset{1\ 11\ 4\ 17}{2\cancel{1}.573} \\ -\ 5.392 \\ \hline 16.181 \end{array}$$ Borrow 1 ten from the 2 in the tens place.
(1 ten = 10 ones)

CHECK:
$$\begin{array}{r} 5.392 \\ +16.181 \\ \hline 21.573 \end{array}$$ Check by adding.

The difference is 16.181.

E. Subtract 4.75 from 8.

$$\begin{array}{r} 8.00 \\ -\ 4.75 \end{array}$$ We write 8 as 8.00 so that both numerals will have the same number of decimal places.

$$\begin{array}{r} \overset{7\ 10}{8.\cancel{0}0} \\ -\ 4.75 \end{array}$$ We need to borrow to subtract in the hundredths place. Since there is a 0 in the tenths place, we start by borrowing 1 from the ones place. (1 one = 10 tenths)

$$\begin{array}{r} \overset{9\ 10}{\cancel{10}} \\ \overset{7\ \cancel{10}}{8.\cancel{0}0} \\ -\ 4.75 \\ \hline 3.25 \end{array}$$ Now borrow 1 tenth to add to the hundredths place. (1 tenth = 10 hundredths) Subtract.

CHECK:
$$\begin{array}{r} 3.25 \\ +\ 4.75 \\ \hline 8.00 \end{array}$$

The difference is 3.25.

F. Find the difference of 9.271 and 5.738. Round to the nearest tenth.

$$\begin{array}{r} \overset{8\ 12\ 6\ 11}{9.2\cancel{7}\cancel{1}} \\ -5.738 \\ \hline 3.533 \end{array}$$ The check is left for the student.

The difference is 3.5 to the nearest tenth

CAUTION

Do not round before subtracting. Note the difference if we do:

$9.3 - 5.7 = 3.6$

CALCULATOR EXAMPLE:

G. Subtract: $759.3471 - 569.458$.

STRATEGY: The calculator automatically lines up the decimal points.
The difference is 189.8891.

H. Marta purchases antibiotics for her son. The antibiotics cost $47.59. She gives the clerk three $20 bills. How much change does she get?

STRATEGY: Since three $20 bills are worth $60, subtract the cost of the antibiotics from $60.

$$\begin{array}{r} \$60.00 \\ -\$47.59 \\ \hline \$12.41 \end{array}$$

Marta gets $12.41 in change.

Clerks without a cash register sometimes make change by counting backward, that is, by adding to $47.59 the amount necessary to equal $60.

$47.59 + 1$ penny $= \$47.60$
$47.60 + 4$ dimes $= \$48.00$
$48.00 + 2$ dollars $= \$50.00$
$50.00 + 1$ ten dollar bill $= \$60.00$

So the change is $\$0.01 + \$0.40 + \$2.00 + \$10.00 = \$12.41$.

EXERCISES 4.3

◼ OBJECTIVE 1 Add decimals. (See page 296.)

A *Add.*

1. $0.7 + 0.7$

2. $0.6 + 0.5$

3. $3.7 + 2.2$

4. $7.6 + 2.9$

5. $1.6 + 5.5 + 8.7$

6. $6.7 + 2.3 + 4.6$

7. $34.8 + 5.29$

8. $22.9 + 7.67$

9. To add 7.6, 6.7821, 9.752, and 61, first rewrite each with _____ decimal places.

10. The sum of 6.7, 10.56, 5.993, and 45.72 has _____ decimal places.

B

11.
$$\begin{array}{r} 21.3 \\ + 34.567 \\ \hline \end{array}$$

12.
$$\begin{array}{r} 37.8 \\ + 9.45 \\ \hline \end{array}$$

13.
$$\begin{array}{r} 5.24 \\ 0.66 \\ 19.7 \\ + 6.08 \\ \hline \end{array}$$

14.
$$\begin{array}{r} 37.57 \\ 7.38 \\ 33.9 \\ + 9.75 \\ \hline \end{array}$$

15. $2.337 + 0.672 + 4.056$

16. $9.445 + 5.772 + 0.822$

17. $0.0017 + 1.007 + 7 + 1.071$

18. $1.0304 + 1.4003 + 1.34 + 0.403$

19. $67.062 + 74.007 + 7.16 + 9.256$

20. $58.009 + 6.46 + 7.082 + 63.88$

21. $0.0781 + 0.00932 + 0.07639 + 0.00759$

22. $7.006 + 0.9341 + 0.003952 + 4.0444$

23. $0.067 + 0.456 + 0.0964 + 0.5321 + 0.112$

24. $4.005 + 0.875 + 3.96 + 7.832 + 4.009$

25.
$$\begin{array}{r} 7.8 \\ 35.664 \\ + 76.9236 \\ \hline \end{array}$$

26.
$$\begin{array}{r} 15.07 \\ 189.981 \\ + 6904.4063 \\ \hline \end{array}$$

27.
$$\begin{array}{r} 75.995 \\ 24.9 \\ + 694.447 \\ \hline \end{array}$$

28.
$$\begin{array}{r} 314.143 \\ 712.217 \\ + 333.444 \\ \hline \end{array}$$

29. Find the sum of 23.07, 6.7, 0.468, and 8.03.

30. Find the sum of 1.8772, 3.987, 0.87, and 6.469.

■ **OBJECTIVE 2** Subtract decimals. (See page 297.)

A *Subtract*

31. $0.7 - 0.4$

32. $5.8 - 5.6$

33. $8.6 - 2.5$

34. $0.64 - 0.53$

35.
$$\begin{array}{r} 6.45 \\ -2.35 \\ \hline \end{array}$$

36.
$$\begin{array}{r} 36.29 \\ - 5.17 \\ \hline \end{array}$$

37.
$$\begin{array}{r} 45.42 \\ -27.38 \\ \hline \end{array}$$

38.
$$\begin{array}{r} 55.44 \\ -37.26 \\ \hline \end{array}$$

39. Subtract 11.14 from 32.01.

40. Find the difference of 23.465 and 9.9.

B *Subtract.*

41.
$$\begin{array}{r} 0.723 \\ -0.457 \\ \hline \end{array}$$

42.
$$\begin{array}{r} 7.403 \\ -3.625 \\ \hline \end{array}$$

43.
$$\begin{array}{r} 4.623 \\ -2.379 \\ \hline \end{array}$$

44.
$$\begin{array}{r} 6.843 \\ -2.568 \\ \hline \end{array}$$

45. $0.831 - 0.462$

46. $0.067 - 0.049$

47. $33.456 - 29.457$

48. $7.598 - 4.7732$

49. $327.58 - 245.674$

50. $506.5065 - 341.341$

51.
$$\begin{array}{r} 0.0952 \\ -0.06434 \\ \hline \end{array}$$

52.
$$\begin{array}{r} 0.0066784 \\ -0.005662 \\ \hline \end{array}$$

53. $41.8341 - 34.6152$

54. $7.9342 - 2.78932$

55. $0.075 - 0.0023$

56. $0.00675 - 0.000984$

57. Subtract 56.78 from 61.02.

58. Subtract 6.607 from 11.5.

59. Find the difference of 11.978 and 11.789

60. Find the difference of 74.707 and 52.465

C *Perform the indicated operations.*

61. $0.0643 + 0.8143 + 0.513 - (0.4083 + 0.7114)$

62. $7.619 + 13.048 - (1.699 + 2.539 + 4.87)$

63. $9.056 - (5.55 + 2.62) - 0.0894$

64. $17.084 - (5.229 - 1.661) + 7.564$

65. On a vacation trip, Manuel stopped for gas four times. The first time, he bought 19.2 gallons. At the second station he bought 21.9 gallons, and at the third, he bought 20.4 gallons. At the last stop, he bought 23.7 gallons. How much gas did he buy on the trip?

66. Heather wrote five checks in the amounts of $63.78, $44.56, $394.06, $11.25, and $67.85. She has $595.94 in her checking account. Does she have enough money to cover the five checks?

67. Find the sum of 457.386, 423.9, 606.777, 29.42, and 171.874. Round the sum to the nearest tenth.

68. Find the sum of 641.85, 312.963, 18.4936, 29.0049, and 6.1945. Round the sum to the nearest hundredth.

Exercises 69–71. The table shows the top six gross state products, in trillions of dollars in 2008.

California	1.847	Florida	0.744
Texas	1.22	Illinois	0.63
New York	1.144	Pennsylvania	0.553

69. Find the total of the gross state products of all six states in the table.

70. How much more is the gross state product for California than the one for Pennsylvania?

71. Find the total of the gross state products for the states on the east coast

72. Doris makes a gross salary (before deductions) of $3565 per month. She has the following monthly deductions: federal income tax, $320.85; state income tax, $192.51; Social Security, $196.07; Medicare, $42.78; retirement contribution, $106.95; union dues, $45; and health insurance, $214.35. Find her actual take-home (net) pay.

73. Jack goes shopping with $72 in cash. He pays $7.98 for a T-shirt, $5.29 for a latte, and $27.85 for a sweater. On the way home, he buys gas with the rest of his money. How much did he spend on gas?

74. In 2004, the average interest rate on a 30-year home mortgage was 6.159%. In 2009, the average interest rate was 4.759%. What was the drop in interest rate?

75. What is the total cost of a cart of groceries that contains bread for $3.09, bananas for $1.49, cheese for $2.50, cereal for $4.39, coffee for $7.99, and meat for $9.27?

Exercises 76–77. The table shows the lengths of railway tunnels in various countries.

World's Longest Railway Tunnels

Tunnel	Length (km)	Country
Seikan	53.91	Japan
English Channel Tunnel	49.95	UK–France
Dai-shimizu	22.53	Japan

76. How much longer is the longest tunnel than the second longest tunnel?

77. What is the total length of the Japanese tunnels?

Exercises 78–80. The table shows projections for the number of families without children under 18.

Projected Number of Families without Children under 18

	Year			
	1995	2000	2005	2010
Families without Children under 18 (in millions)	35.8	38.6	42.0	45.7

SOURCE: U.S. Census Bureau

78. What is the projected change in the number of families in the United States without children under 18 between 1995 and 2010?

79. Which 5-year period is projected for the largest change?

80. What could explain the increase indicated in the table?

81. How high from the ground level is the top of the tree shown below? Round to the nearest foot.

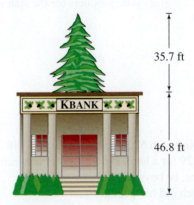

35.7 ft

KBANK

46.8 ft

82. Find the length of the pictured piston skirt (*A*) if the other dimensions are as follows: *B* = 0.3125 in., *C* = 0.250 in., *D* = 0.3125 in., *E* = 0.250 in., *F* = 0.3125 in., *G* = 0.375 in., *H* = 0.3125.

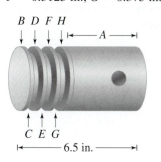

83. What is the center-to-center distance, *A*, between the holes in the diagram?

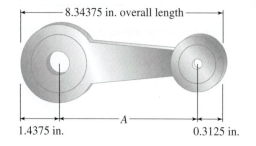

84. Find the total length of the pictured connecting bar.

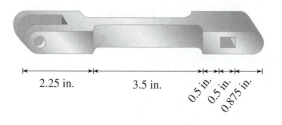

Exercises 85–89 relate to the chapter application.

85. Muthoni runs a race in 12.16 seconds, whereas Sera runs the same race in 11.382 seconds. How much faster is Sera?

86. A skier posts a race time of 1.257 minutes. A second skier posts a time of 1.32 minutes. The third skier completes the race in 1.2378 minutes. Find the difference between the fastest and the slowest times.

87. A college men's 4-×-100-m relay track team has runners with individual times of 9.35 sec, 9.91 sec, 10.04 sec, and 9.65 sec. What is the time for the relay?

88. A high-school girls' swim team has a 200-yd freestyle relay, in which swimmers have times of 21.79 sec, 22.64 sec, 22.38 sec, and 23.13 sec. What is the time for the relay?

89. A high-school women's track coach knows that the rival school's team in the 4-×-100-m relay has a time of 52.78 sec. If the coach knows that her top three sprinters have times of 12.83, 13.22, and 13.56 sec, how fast does the fourth sprinter need to be in order to beat the rival school's relay team?

90. Explain the procedure for adding 2.005, 8.2, 0.00004, and 3.

91. Explain the similarities between subtracting decimals and subtracting fractions.

92. Copy the table and fill it in.

Operation on Decimals	Procedure	Example
Addition		
Subtraction		

CHALLENGE

93. How many 5.83s must be added to have a sum that is greater than 150?

94. Find the missing number in the sequence: 0.4, 0.8, 1.3, _____ , 2.6, 3.4, 4.3, 5.3.

95. Find the missing number in the sequence: 0.2, 0.19, 0.188, _____ , 0.18766, 0.187655.

96. Which number in the following group is 11.1 less than 989.989: 999.999, 989.999, 988.889, 979.889, or 978.889?

97. Write the difference between $6\frac{7}{16}$ and 5.99 in decimal form.

98. Round the sum of 9.8989, 8.9898, 7.987, and 6.866 to the nearest tenth.

99. Write the sum of $2\frac{1}{2}$, $3\frac{1}{8}$, $4\frac{3}{4}$, 2.25, 3.5, and 4.8 in both fraction and decimal form.

Multiply.

100. 62(217)

101. 703(557)

102. 6921
$\times$ 415

103. (83)(27)(19)

104. $\dfrac{1}{5} \cdot \dfrac{9}{10}$

105. $\dfrac{36}{75} \cdot \dfrac{15}{16} \cdot \dfrac{40}{27}$

106. $4\dfrac{2}{5} \cdot 2\dfrac{4}{5}$

107. $\left(4\dfrac{1}{2}\right)\left(5\dfrac{3}{5}\right)$

108. A nursery plants one seedling per square foot of ground. How many seedlings can be planted in a rectangular plot of ground that measures 310 ft by 442 ft?

109. Harry and David puts 24 pears in its Royal Golden Pear Box. How many pears are needed to fill an order for 345 Royal Golden Pear Boxes?

GETTING READY FOR ALGEBRA
HOW & WHY

We solve equations that involve addition and subtraction of decimals in the same way as equations with whole numbers and fractions.

> **To solve an equation using addition or subtraction**
>
> 1. Add the same number to both sides of the equation to isolate the variable, or
> 2. Subtract the same number from both sides of the equation to isolate the variable.

EXAMPLES A–E

DIRECTIONS: Solve.

STRATEGY: Isolate the variable by adding or subtracting the same number to or from both sides.

WARM-UP

A. $11.7 = p + 4.2$

A. $7.8 = x + 5.6$

$$7.8 = x + 5.6$$
$$7.8 - 5.6 = x + 5.6 - 5.6$$

Eliminate by the addition by subtracting 5.6 from both sides of the equation. Since subtraction is the inverse of addition, the variable will be isolated. Simplify.

$$2.2 = x$$

CHECK: $7.8 = 2.2 + 5.6$ Substitute 2.2 for x in the original equation. Simplify.
$7.8 = 7.8$ True.

The solution is $x = 2.2$.

WARM-UP

B. $t - 13.6 = 29.5$

B. $z - 14.9 = 32.7$

$$z - 14.9 = 32.7$$
$$\underline{+\ 14.9 = +14.9}$$
$$z = 47.6$$

Eliminate by the subtraction by adding 14.9 to both sides of the equation. Because addition is the inverse of subtraction, the variable will be isolated.

CHECK: $47.6 - 14.9 = 32.7$ Substitute 47.6 for z in the original equation and simplify.

$32.7 = 32.7$ True.

The solution is $z = 47.6$.

WARM-UP

C. $c + 56.785 = 62$

C. $b + 17.325 = 34.6$

$$b + 17.325 = 34.6$$
$$b + 17.325 - 17.325 = 34.6 - 17.325$$
$$b = 17.275$$

Subtract 17.325 from both sides and simplify.

CHECK: $17.275 + 17.325 = 34.6$ Substitute 17.275 for b in the original equation.
$34.6 = 34.6$ True.

The solution is $b = 17.275$.

WARM-UP

D. $w - 33.17 = 12.455$

D. $y - 6.233 = 8.005$

$$y - 6.233 = 8.005$$
$$\underline{+\ 6.233 = +6.233}$$
$$y = 14.238$$

Add 6.233 to both sides and simplify.

CHECK: $14.238 - 6.233 = 8.005$ Substitute 14.238 for y in the original equation.
$8.005 = 8.005$ True.

The solution is $y = 14.238$.

ANSWERS TO WARM-UPS A–D

A. $p = 7.5$ **B.** $t = 43.1$
C. $c = 5.215$ **D.** $w = 45.625$

E. The price of a graphing calculator decreased by $19.30 over the past year. What was the price a year ago if the calculator now sells for $81.95?

First write the English version of the equation:

(cost last year) − (decrease in cost) = cost this year

Let x represent the cost last year.

$$x - 19.30 = 81.95 \quad \text{Translate to algebra.}$$
$$x - 19.30 + 19.30 = 81.95 + 19.30 \quad \text{Add 19.30 to both sides.}$$
$$x = 101.25 \quad \text{Simplify.}$$

Because $101.25 - 19.30 = 81.95$, the cost of the calculator last year was $101.25.

WARM-UP

E. A farmer practicing "sustainable" farming reduced his soil erosion by 1.58 tons in 1 year. If he lost 3.94 tons of topsoil this year to erosion, how many tons did he lose last year?

ANSWER TO WARM-UP E

E. The former lost 5.52 tons of topsoil to erosion.

EXERCISES

Solve.

1. $16.3 = x + 5.2$

2. $6.904 = x + 3.5$

3. $y - 0.64 = 13.19$

4. $w - 0.08 = 0.713$

5. $t + 0.03 = 0.514$

6. $x + 14.7 = 28.43$

7. $x - 7.3 = 5.21$

8. $y - 9.3 = 0.42$

9. $7.33 = w + 0.13$

10. $14 = x + 7.6$

11. $t - 8.37 = 0.08$

12. $w - 0.03 = 0.451$

13. $5.78 = a + 1.94$

14. $55.9 = w - 11.8$

15. $6.6 = x - 9.57$

16. $7 = 5.9 + x$

17. $a + 82.3 = 100$

18. $b + 45.76 = 93$

19. $s - 2.5 = 4.773$

20. $r - 6.7 = 5.217$

21. $c + 432.8 = 1029.16$

22. $d - 316.72 = 606.5$

23. The price of an energy-efficient hot-water heater decreased by $52.75 over the past 2 years. What was the price 2 years ago if the heater now sells for $374.98?

24. In one state the use of household biodegradable cleaners increased by 2444.67 lb per month because of state laws banning phosphates. How many pounds of these cleaners were used before the new laws if the average use now is 5780.5 lb?

25. The selling price of a personal computer is $1033.95. If the cost is $875.29, what is the markup?

26. The selling price of a new tire is $128.95. If the markup on the tire is $37.84, what is the cost (to the store) of the tire?

27. A shopper needs to buy a bus pass and some groceries. The shopper has $61 with which to make both purchases. If the bus pass costs $24, write and solve an equation that represents the shopper's situation. How much can the shopper spend on groceries?

28. In a math class, the final grade is determined by adding the test scores and the homework scores. If a student has a homework score of 18 and it takes a total of 90 to receive a grade of A, what total test score must the student have to receive a grade of A? Write an equation and solve it to determine the answer.

OBJECTIVE

Multiply decimals.

HOW & WHY

■ **OBJECTIVE**   Multiply decimals.

The "multiplication table" for decimals is the same as for whole numbers. In fact, decimals are multiplied the same way as whole numbers with one exception: the location of the decimal point in the product. To discover the rule for the location of the decimal point, we use what we already know about multiplication of fractions. First, change the decimal form to fractional form to find the product. Next, change the product back to decimal form and observe the number of decimal places in the product. Consider the examples in Table 4.4. We see that the product in decimal form has the same number of decimal places as the total number of places in the decimal factors.

TABLE 4.4 **Multiplication of Decimals**

Decimal Form	Fractional Form	Product of Fractions	Product as a Decimal	Number of Decimal Places in Product
0.3×0.8	$\dfrac{3}{10} \times \dfrac{8}{10}$	$\dfrac{24}{100}$	0.24	Two
11.2×0.07	$\dfrac{112}{10} \times \dfrac{7}{100}$	$\dfrac{784}{1000}$	0.784	Three
0.02×0.13	$\dfrac{2}{100} \times \dfrac{13}{100}$	$\dfrac{26}{10,000}$	0.0026	Four

The shortcut is to multiply the numbers and insert the decimal point. If necessary, insert zeros so that there are enough decimal places. The product of 0.2×0.3 has two decimal places, because tenths multiplied by tenths yields hundredths.

$$0.2 \times 0.3 = 0.06 \quad \text{because} \quad \frac{2}{10} \times \frac{3}{10} = \frac{6}{100}$$

> **To multiply decimals**
>
> 1. Multiply the numbers as if they were whole numbers.
> 2. Locate the decimal point by counting the number of decimal places (to the right of the decimal point) in both factors. The total of these two counts is the number of decimal places the product must have.
> 3. If necessary, zeros are inserted to the *left of the numeral* so there are enough decimal places (see Example D).

When multiplying decimals, it is not necessary to align the decimal points in the decimals being multiplied.

EXAMPLES A–F

DIRECTIONS: Multiply.

STRATEGY: First multiply the numbers, ignoring the decimal points. Place the decimal point in the product by counting the number of decimal places in the two factors. Insert zeros if necessary to produce the number of required places.

A. Multiply: (0.8)(29)

(0.8)(29) = 23.2   Multiply 8 and 29. (8 × 29 = 232) The total number of decimal places in both factors is one (1), so there is one decimal place in the product.

So, (0.8)(29) = 23.2.

B. Find the product of 0.9 and 0.64.

(0.9)(0.64) = 0.576   Multiply 9 and 64. (9 × 64 = 576) There are three decimal places in the product because the total number of places in the factors is three.

So the product of 0.9 and 0.64 is 0.576.

C. Find the product of 9.73 and 6.8.

$$\begin{array}{r} 9.73 \\ \times\ 6.8 \\ \hline 7784 \\ 58380 \\ \hline 66.164 \end{array}$$

Multiply the numbers as if they were whole numbers. There are three decimal places in the product.

So the product of 9.63 and 6.8 is 66.164.

D. Multiply 7.9 times 0.0004.

$$\begin{array}{r} 7.9 \\ \times\ 0.0004 \\ \hline 0.00316 \end{array}$$

Because 7.9 has one decimal place and 0.0004 has four decimal places, the product must have five decimal places. We must insert two zeros to the left so there are enough places in the answer.

So 7.9 times 0.0004 is 0.00316.

CALCULATOR EXAMPLE:

E. Find the product: (9.86)(107.375)

STRATEGY: The calculator will automatically place the decimal point in the correct position.

The product is 1058.7175.

F. If exactly eight strips of metal, each 3.875 inches wide, are to be cut from a piece of sheet metal, what is the smallest (in width) piece of sheet metal that can be used?

STRATEGY: To find the width of the piece of sheet metal, we multiply the width of one of the strips by the number of strips needed.

$$\begin{array}{r} 3.875 \\ \times\ \quad 8 \\ \hline 31.000 \end{array}$$

The extra zeros can be dropped.

The piece must be 31 inches wide.

EXERCISES 4.4

■ **OBJECTIVE**   Multiply decimals. (See page 308.)

A   *Multiply.*

1.   0.5
    $\times\ 9$

2.   0.8
    $\times\ 3$

3.   1.9
   $\times\ 5$

4.   3.4
    $\times\ 7$

5. 8×0.09

6. 0.03×3

7. 0.8×0.8

8. 0.7×0.5

9. 0.06×0.6

10. 0.7×0.004

11. $0.18 \times (0.7)$

12. $1.7 \times (0.07)$

13. The number of decimal places in the product of 3.511 and 6.2 is _____.

14. In the product $0.34 \times ? = 0.408$, the number of decimal places in the missing factor is _____.

B   *Multiply.*

15.   7.72
   $\times\ 0.008$

16.   3.47
   $\times\ 0.0065$

17.   2.44
   $\times\ 4.7$

18.   5.36
   $\times\ 4.9$

19.   6.84
   $\times\ \ 0.42$

20.   4.99
   $\times\ 0.37$

21.   5.92
   $\times\ \ 2.04$

22.   0.084
   $\times\ \ 6.9$

23.   42.7
   $\times\ 0.53$

24.   38.5
   $\times\ 0.21$

25.   0.356
   $\times\ 0.067$

26.   0.567
   $\times\ \ 0.036$

27.   0.0416
   $\times\ \ 4.02$

28.   0.00831
   $\times\ \ \ 6.73$

29.   0.825
   $\times\ 0.0054$

30.   0.575
   $\times\ 0.00378$

31. Find the product of 8.54 and 3.78.

32. Find the product of 6.68 and 4.33.

33. Multiply: $4.4(0.6)(0.48)$

34. Multiply: $9.3(5.7)(0.26)$

Multiply and round as indicated.

35. $32(0.846)$ to the nearest tenth.

36. $680(0.0731)$ to the nearest hundredth.

37. $64.85(34.26)$ to the nearest tenth.

38. $2592.3(44.72)$ to the nearest ten.

39. $16.93(31.47)$ to the nearest hundredth.

40. $0.046(0.9523)$ to the nearest thousandth.

C   *Multiply. Round the product to the nearest hundredth.*

41. $(34.06)(23.75)(0.134)$

42. $(0.056)(67.8)(21.115)$

Exercises 43–47. The table shows the amount of gas purchased by Grant and the price he paid per gallon for five fill-ups.

Gasoline Purchases

Number of Gallons	Price per Gallon
19.7	$2.399
21.4	$2.419
18.6	$2.559
20.9	$2.399
18.4	$2.629

43. What is the total number of gallons of gas that Grant purchased?

44. To the nearest cent, how much did he pay for the second fill-up?

45. To the nearest cent, how much did he pay for the fifth fill-up?

46. To the nearest cent, what is the total amount he paid for the five fill-ups?

47. At which price per gallon did he pay the least for his fill-up?

Multiply.

48. (469.5)(7.12)

49. (78.95)(3.65)

50. (313.17)(8.73)

51. (15.8)(580.04)

Multiply.

52. (7.85)(3.52)(27.89) Round to the nearest hundredth.

53. (4.57)(234.7)(21.042) Round to the nearest thousandth.

54. (7.4)(5.12)(0.88)(13.2) Round to the nearest tenth.

55. (20.4)(0.48)(8.02)(50.4) Round to the nearest hundredth.

56. Joe earns $16.45 per hour. How much does he earn if he works 38.5 hours in 1 week? Round to the nearest cent.

57. Central Grocery has a sale on T-bone steaks at $6.99 per pound. Sonya buys 4.35 pounds of the steak for dinner party. What did she pay for the steak? The store rounds prices to the nearest cent.

Exercises 58–61. The table shows the cost of renting a car from a local agency.

Cost of Car Rentals

Type of Car	Cost per Day	Price per Mile Driven over an Alloted 100 Miles per Day
Compact	$19.95	$0.165
Midsize	$26.95	$0.24
Full-size	$30.00	$0.275

58. What does it cost to rent a compact car for 4 days if it is driven 435 miles?

59. What does it cost to rent a midsize car for 3 days if it is driven 710 miles?

60. What does it cost to rent a full-size car for 5 days if it is driven 1050 miles?

61. Which costs less, renting a full-size car for 3 days and driving it 425 miles or a midsize car for 2 days and driving it 625 miles? How much less does it cost?

62. Tiffany can choose any of the following ways to finance her new car. Which method is the least expensive in the long run?

$750 down and $315.54 per month for 6 years
$550 down and $362.57 per month for 5 years
$475 down and $435.42 per month for 4 years

63. A new freezer–refrigerator is advertised at three different stores as follows:

Store 1: $80 down and $91.95 per month for 18 months
Store 2: $125 down and $67.25 per month for 24 months
Store 3: $350 down and $119.55 per month for 12 month

Which store is selling the freezer–refrigerator for the least total cost?

Exercises 64–66. The table shows calories expended for some physical activities.

Calorie Expenditure for Selected Physical Activities

Activity	Step Aerobics	Running (7 min/mile)	Cycling (10 mph)	Walking (4.5 mph)
Calories per pound of body weight per minute	0.070	0.102	0.050	0.045

64. Vanessa weighs 145 lb and does 75 min of step aerobics per week. How many calories does she burn per week?

65. Steve weighs 187 lb and runs 25 min five times per week at a 7 min/mi pace. How many calories does he burn per week?

66. Sephra weighs 143 lb and likes to walk daily at 4.5 mph. Her friend Dana weighs 128 lb and prefers to cycle daily at 10 mph. If both women exercise the same amount of time, who burns more calories?

67. An order of 43 bars of steel is delivered to a machine shop. Each bar is 17.325 ft long Find the total linear feet of steel in the order.

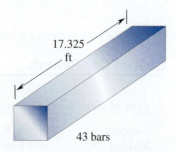

17.325 ft

43 bars

68. From a table in a machinist's handbook, it is determined that hexagon steel bars 1.325 in. across weigh 4.3 lb per running foot. Using this constant, find the weight of a 1.325 in. hexagon steel bar that is 22.56 ft long

69. In 1995, the per capita consumption of beef was 79.4 lb. In 2005, the consumption was 63.7 lb. In 2015, the amount consumed is projected to be 59 lb. per person. Compute the total weight of beef consumed by a family of four using the rates for each of these years. Discuss the reasons for the change in consumption.

70. The fat content in a 3-oz serving of meat and fish is as follows: beef rib, 7.4 g; beef top round, 3.4 g; beef top sirloin, 4.8 g; dark meat chicken without skin, 8.3 g; white meat chicken without skin, 3.8 g; pink salmon, 3.8 g; and Atlantic cod, 0.7 g. Which contains the most grams of fat: 3 servings of beef ribs, 6 servings of beef top round, 4 servings of beef top sirloin, 2 servings of dark meat chicken, 6 servings of white meat chicken, 5 servings of pink salmon, or 25 servings of Atlantic cod?

71. Older models of toilets use 5.5 gallons of water per flush. Models made in the 1970s use 3.5 gallons per flush. The new low-flow models use 1.55 gallons per flush. Assume each person flushes the toilet an average of five times per day. Determine the amount of water used in a town with a population of 41,782 in 1 day for each type of toilet. How much water is saved using the low-flow model as opposed to the pre-1970s model?

72. The Camburns live in Las Vegas, Nevada. Their house is assessed at $344,780. The property tax rate for state, county, and schools is $2.1836 per thousand dollars of assessed evaluation for 2004–2005. Find what they owe in property taxes.

73. Find the property tax on the Sanchez estate that is assessed at $1,461,200. The tax rate in the area is $2.775 per thousand dollars of assessment. Round to the nearest dollar.

Exercises 74–76 relate to the chapter application.

74. In the 2009 World Championships in Athletics, Usain Bolt of Jamaica set a new world record for 100 m, with a time of 9.58 seconds. If he could continue that rate for another 100 m, what would his time be in the 200 m?

75. In the 2009 World Championships in Athletics, Usain Boltset a new world record for 200 m, with a time of 19.19 sec. Assuming he ran the first 100 m in his record time of 9.58 sec, how long did the second 100 m take him?

76. In the 2009 World Championships in Athletics, Shelly-Ann Fraser of Jamaica won the 100 m with a time of 10.73 sec. If she could continue that rate, what would her time for the 400 m be?

© Pete Niesen/Shutterstock.com

77. Explain how to determine the number of decimal places needed in the product of two decimals.

78. Suppose you use a calculator to multiply (0.006)(3.2)(68) and get 13.056. Explain, using placement of the decimal point in a product, how you can tell that at least one of the numbers was entered incorrectly.

CHALLENGE

79. What is the smallest whole number you can multiply 0.74 by to get a product that is greater than 82?

80. What is the largest whole number you can multiply 0.53 by to get a product that is less than 47?

81. Find the missing number in the following sequence: 2.1, 0.42. 0.126, 0.0504, _____ .

82. Find the missing number in the following sequence: 3.1, 0.31, _____ , 0.0000031, 0.0000000031.

83. Fill in the missing number so that $0.3(1.5 + 2.7 - \square) = 0.36$.

MAINTAIN YOUR SKILLS

Multiply or divide as indicated.

84. 337(100)

85. 82(10,000)

86. 235,800 ÷ 100

87. 22,000,000 ÷ 10,000

88. 48(1,000,000)

89. 692×10^3

90. $55,000 \div 10^3$

91. $4,760,000 \div 10^4$

92. 84×10^8

93. $4,210,000,000 \div 10^6$

Multiplying and Dividing by Powers of 10; Scientific Notation

VOCABULARY

Recall that a **power of 10** is the value obtained when 10 is written with an exponent.

Scientific notation is a special way to write numbers as a product using a number between 1 and 10 and a power of 10.

OBJECTIVES

1. Multiply or divide a number by a power of 10.
2. Write a number in scientific notation or change a number in scientific notation to its place value name.

HOW & WHY

■ **OBJECTIVE 1** Multiply or divide a number by a power of 10.

The shortcut used in Section 1.5 for multiplying and dividing by a power of 10 works in a similar way with decimals. Consider the following products:

$$
\begin{array}{r} 0.8 \\ \times 10 \\ \hline 0 \\ 8\,0 \\ \hline 8.0 = 8 \end{array}
\qquad
\begin{array}{r} 0.63 \\ \times\ 10 \\ \hline 0 \\ 6\,30 \\ \hline 6.30 = 6.3 \end{array}
\qquad
\begin{array}{r} 9.36 \\ \times\ \ 10 \\ \hline 0 \\ 93\,60 \\ \hline 93.60 = 93.6 \end{array}
$$

Note in each case that multiplying a decimal by 10 has the effect of moving the decimal point one place to the right.

Because $100 = 10 \cdot 10$, multiplying by 100 is the same as multiplying by 10 two times in succession. So, multiplying by 100 has the effect of moving the decimal point two places to the right. For instance,

$$(0.42)(100) = (0.42)(10 \cdot 10) = (0.42 \cdot 10) \cdot 10 = 4.2 \cdot 10 = 42$$

Because $1000 = 10 \cdot 10 \cdot 10$, the decimal point will move three places to the right when multiplying by 1000. Because $10{,}000 = 10 \cdot 10 \cdot 10 \cdot 10$, the decimal point will move four places to the right when multiplying by 10,000, and so on in the same pattern:

$$(0.05682)(10{,}000) = 568.2$$

Zeros may have to be placed on the right in order to move the correct number of decimal places:

$$(6.3)(1000) = 6.300 = 6300$$

In this problem, two zeros are placed on the right.

Because multiplying a decimal by 10 has the effect of moving the decimal point one place to the right, dividing a number by 10 must move the decimal point one place to the left. Again, we are using the fact that multiplication and division are inverse operations. Division by 100 will move the decimal point two places to the left, and so on. Thus,

$$739.5 \div 100 = 739.5 = 7.395$$

$$0.596 \div 10{,}000 = 0.0000596 \qquad \text{Four zeros are placed on the left so that the decimal point may be moved four places to the left.}$$

> ### To multiply a number by a power of 10
>
> Move the decimal point to the right. The number of places to move is shown by the number of zeros in the power of 10.

EXAMPLES A–G

DIRECTIONS: Multiply or divide as indicated.

STRATEGY: To multiply by a power of 10, move the decimal point to the right, inserting zeros as needed. To divide by a power of 10, move the decimal point to the left, inserting zeros as needed. The exponent of 10 specifies the number of places to move the decimal point.

WARM-UP

A. Multiply: 11.05(10)

A. Multiply: 55.283(10)

$55.283(10) = 552.83$ Multiplying by 10 moves the decimal point one place to the right.

So, $55.283(10) = 552.83$.

WARM-UP

B. Multiply: 0.137(100)

B. Multiply: 0.057(100)

$0.057(100) = 5.7$ Multiplying by 100 moves the decimal point two places to the right.

So, $0.057(100) = 5.7$.

WARM-UP

C. Find the product of 34.78 and 10^3.

C. Find the product of 8.57 and 10^4.

$8.57(10^4) = 85,700$ Multiplying by 10^4 moves the decimal point four places to the right. Two zeros must be inserted on the right to make the move.

So the product of 8.57 and 10^4 is 85,700.

WARM-UP

D. Divide: $662 \div 10$

D. Divide: $9.02 \div 10$

$9.02 \div 10 = 0.902$ Dividing by 10 moves the decimal point one place to the left.

So, $9.02 \div 10 = 0.902$.

WARM-UP

E. Divide: $49.16 \div 100$

E. Divide: $760.1 \div 100$

$760.1 \div 100 = 7.601$ Dividing by 100 moves the decimal point two places to the left.

So, $760.1 \div 100 = 7.601$.

WARM-UP

F. Find the quotient:
$7.339 \div 10^5$

F. Find the quotient: $12.8 \div 10^3$

$12.8 \div 10^3 = 0.0128$ Move the decimal point three places to the left. A zero is inserted on the left so we can make the move.

So, $12.8 \div 10^3 = 0.0128$.

WARM-UP

G. Ten thousand sheets of clear plastic are 5 in. thick. How thick is each sheet? (This is the thickness of some household plastic wrap.)

G. Bi-Mart orders 1000 boxes of chocolates for their Valentine's Day sales. The total cost to Bi-Mart was $19,950. What did Bi-Mart pay per box of chocolates?

$19,950 \div 1000 = 19.95$ To find the cost paid per box, divide the total cost by the number of boxes.

Bi-Mart paid $19.95 per box of chocolates.

ANSWERS TO WARM-UPS A–G

A. 110.5   **B.** 13.7   **C.** 34,780
D. 66.2   **E.** 0.4916   **F.** 0.00007339
G. Each sheet of plastic is 0.0005 in. thick.

HOW & WHY

■ **OBJECTIVE 2** Write a number in scientific notation or change a number in scientific notation to its place value name.

Scientific notation is widely used in science, technology, and industry to write large and small numbers. Every "scientific calculator" has a key for entering numbers in scientific notation. This notation makes it possible for a calculator or computer to deal with much larger or smaller numbers than those that take up 8, 9, or 10 spaces on the display.

For example, see Table 4.5.

Scientific Notation

A number in scientific notation is written as the product of two numbers. The first number is between 1 and 10 (including 1 but not 10) and the second number is a power of 10. The multiplication is indicated using an "×" symbol.

TABLE 4.5 Scientific Notation

Word Form	Place Value (Numeral Form)	Scientific Notation	Calculator or Computer Display		
One million	1,000,000	1×10^6	1. 06	or	1. **E** 6
Five billion	5,000,000,000	5×10^9	5. 09	or	5. **E** 9
One trillion, three billion	1,003,000,000,000	1.003×10^{12}	1.003 12 or 1.003 **E** 12		

Small numbers are shown by writing the power of 10 using a negative exponent. (This is the first time that we have used negative numbers. You probably have run into them before. For instance, when reporting temperatures, a reading of 10 degrees above zero is written $+10$. While a reading of 10 degrees below zero is written -10. (You will learn more about negative numbers in Chapter 8.) For now, remember that multiplying by a negative power of 10 is the same as dividing by a power of 10, which means you will be moving the decimal point to the left. See Table 4.6.

TABLE 4.6 Scientific Notation

Word Name	Place Value Name	Scientific Notation	Calculator or Computer Display
Eight thousandths	0.008	8×10^{-3}	8. -03 or 8. **E** -3
Seven ten-millionths	0.0000007	7×10^{-7}	7. -07 or 7. **E** -7
Fourteen hundred-billionths	0.00000000014	1.4×10^{-10}	1.4 -10 or 1.4 **E** -10

The shortcut for multiplying by a power of 10 is to move the decimal to the right, and the shortcut for dividing by a power of 10 is to move the decimal point to the left.

To write a number in scientific notation

1. Move the decimal point right or left so that only one digit remains to the left of the decimal point. The result will be a number between 1 and 10. If the choice is 1 or 10 itself, use 1.
2. Multiply the decimal found in step 1 by a power of 10. The exponent of 10 to use is one that will make the new product equal to the original number.
   a. If you had to move the decimal to the left, multiply by the same number of 10s as the number of places moved.
   b. If you had to move the decimal to the right, divide (by writing a negative exponent) by the same number of 10s as the number of places moved.

To change from scientific notation to place value name

1. If the exponent of 10 is positive, multiply by as many 10s (move the decimal point to the right as many places) as the exponent shows.
2. If the exponent of 10 is negative, divide by as many 10s (move the decimal point to the left as many places) as the exponent shows.

For numbers larger than 1

Place value name:	15,000	7,300,000	18,500,000,000	
Numbers between 1 and 10:	1.5	7.3	1.85	Move the decimal (which is after the units place) to the left until the number is between 1 and 10 (one digit to the left of the decimal).
Scientific notation:	1.5×10^4	7.3×10^6	1.85×10^{10}	Multiply each by a power of 10 that shows how many places left the decimal moved, or how many places you would have to move to the right to recover the original number.

For numbers smaller than 1

Place value name:	0.000074	0.00000009	0.0000000000267	
Numbers between 1 and 10:	7.4	9.	2.67	Move the decimal to the right until the number is between 1 and 10.
Scientific notation:	7.4×10^{-5}	9×10^{-8}	2.67×10^{-11}	Divide each by a power of 10 that shows how many places right the decimal moved. Show this division by a negative power of 10.

It is important to note that scientific notation is not rounding. The scientific notation has exactly the same value as the original name.

EXAMPLES H–J

DIRECTIONS: Write in scientific notation.

STRATEGY: Move the decimal point so that there is one digit to the left. Multiply or divide this number by the appropriate power of 10 so the value is the same as the original number.

WARM-UP

H. Write in scientific notation: 123,000,000

H. 46,700,000

4.67 is between 1 and 10.

> Move the decimal until the number is between 1 and 10. Moving the decimal left is equivalent to dividing by 10 for each place.

$4.67 \times 10,000,000$ is 46,700,000.

> To recover the original number, we multiply by 10 seven times.

$46,700,000 = 4.67 \times 10^7$

I. 0.00000039

3.9 is between 1 and 10.

Move the decimal until the number is between 1 and 10. Moving the decimal right is equivalent to multiplying by 10 for each place.

$3.9 \div 10,000,000$ is 0.00000039.

To recover the value of the original number, we divide by 10 seven times.

$0.00000039 = 3.9 \times 10^{-7}$

J. One organization estimates there are approximately 46,000,000 people in the world who suffer from autism. Write this number in scientific notation.

4.6 is between 1 and 10.
$4.6 \times 10,000,000 = 46,000,000$
$46,000,000 = 4.6 \times 10^7$

In scientific notation the number of people with autism is 4.6×10^7.

EXAMPLES K–L

DIRECTIONS: Write the place value name.

STRATEGY: If the exponent is positive, move the decimal point to the right as many places as shown in the exponent. If the exponent is negative, move the decimal point to the left as many places as shown by the exponent.

K. Write the place value name for 5.72×10^{-7}.

$5.72 \times 10^{-7} = 0.000000572$

The exponent is negative, so move the decimal point seven places to the left. That is, divide by 10 seven times.

L. Write the place value name for 1.004×10^8.

$1.004 \times 10^8 = 100,400,000$

The exponent is positive, so move the decimal point eight places to the right; that is, multiply by 10 eight times.

EXERCISES 4.5

■ **OBJECTIVE 1** Multiply or divide a number by a power of 10. (See page 315.)

A *Multiply or divide.*

1. $19.3 \div 10$

2. $18.65 \div 10$

3. $92.6(100)$

4. $0.236(100)$

5. $(1.3557)(1000)$

6. $(0.0421)(1000)$

7. $\dfrac{58.18}{100}$

8. $\dfrac{456.71}{1000}$

9. $\dfrac{8325}{100}$

10. $\dfrac{4538.2}{1000}$

11. 0.107×10^4

12. 7.32×10^5

13. To multiply 4.56 by 10^4, move the decimal point four places to the _____.

14. To divide 4.56 by 10^5, move the decimal point five places to the _____.

B *Multiply or divide.*

15. $(78.324)(1000)$

16. $17.66(100)$

17. $99.7 \div 10$

18. $672.86 \div 1000$

19. $57.9(1000)$

20. $0.0364(10)$

21. $\dfrac{9077.5}{10,000}$

22. $\dfrac{6351.42}{100,000}$

23. $28.73(100,000)$

24. $16.33(1,000,000)$

25. $6056.32 \div 100$

26. $33.07 \div 1000$

27. $32.76 \div 100,000$

28. $134.134 \div 1,000,000$

■ **OBJECTIVE 2** Write a number in scientific notation or change a number in scientific notation to its place value name. (See page 317.)

A *Write in scientific notation.*

29. 750,000

30. 19,300

31. 0.000091

32. 0.0000385

33. 4195.3

34. 82710.3

Write in place value form.

35. 12×10^5

36. 3×10^6

37. 4×10^{-3}

38. 1×10^{-7}

39. 9.43×10^5

40. 8.12×10^5

B *Write in scientific notation.*

41. 43,700

42. 81,930,000

43. 0.000000587

44. 0.0000642

45. 0.0000000000684

46. 0.00000000555

47. 64.004

48. 3496.701

Write in place value notation.

49. 7.341×10^{-5}

50. 9.37×10^{-6}

51. 1.77×10^9

52. 7.43×10^8

53. 3.11×10^{-8}

54. 5.6×10^{-9}

55. 1.48×10^{-8}

56. 5.11166×10^6

C

57. Max's Tire Store buys 100 tires that cost $39.68 each. What is the total cost of the tires?

58. If Mae's Tire Store buys 100 tires for a total cost of $5278, what is the cost of each tire?

59. Ms. James buys 100 acres of land at a cost of $3100 per acre. What is the total cost of her land?

60. If 1000 concrete blocks weigh 11,100 lb, how much does each block weigh?

61. The total land area of Earth is approximately 52,000,000 square miles. What is the total area written in scientific notation?

62. A local computer store offers a small computer with 260 MB (2,662,240 bytes) of memory. Write the number of bytes in scientific notation.

63. A nanometer can be used to measure very short lengths. One nanometer is equal to 0.000000001 of a meter. Write this length in scientific notation.

64. The speed of light is approximately 671,000,000 miles per hour. Write this speed in scientific notation.

65. The time it takes light to travel 1 mile is approximately 0.000054 second. Write this time in scientific notation.

66. Earth is approximately 1.5×10^8 kilometers from the sun. Write this distance in place value form.

67. The shortest wavelength of visible light is approximately 4×10^{-5} cm. Write this length in place value form.

68. A sheet of paper is approximately 1.3×10^{-3} in. thick. Write the thickness in place value form.

69. A family in the Northeast used 3.467×10^8 BTUs of energy during 2010. A family in the Midwest used 3.521×10^8 BTUs in the same year. A family in the South used 2.783×10^8 BTUs, and a family in the West used 2.552×10^8 BTUs. Write the total energy usage for the four families in place value form.

70. In 2010, the per capita consumption of fish was 15.1 pounds. In the same year, the per capita consumption of poultry was 82.6 pounds and that of red meat was 118.3 pounds. Write the total amount in each category consumed by 100,000 people in scientific notation.

71. The population of Cargill Cove was approximately 100,000 in 2010. During the year, the community consumed a total of 3,060,000 gallons of milk. What was the per capita consumption of milk in Cargill Cove in 2010?

72. Driving a 2009 Honda Civic Hybrid for 10,000 miles generates 4658 pounds of carbon dioxide. How much carbon dioxide does the 2009 Honda Civic Hybrid generate per mile?

Exercises 73–75 relate to the chapter application. In baseball, a hitter's batting average is calculated by dividing the number of hits by the number of times at bat. Mathematically, this number is always between zero and 1.

© Ken Brown/Shutterstock.com

73. In 1923, Babe Ruth led Major League Baseball with a batting average of 0.393. However, players and fans would say that Ruth has an average of "three hundred ninety-three." Mathematically, what are they doing to the actual number?

74. Explain why the highest possible batting average is 1.0.

75. The major league player with the highest season batting average in the past century was Roger Hornsby of St. Louis. In 1924 he batted 424. Change this to the mathematically calculated number of his batting average.

STATE YOUR UNDERSTANDING

76. Find a pair of numbers whose product is larger than 10 trillion. Explain how scientific notation makes it possible to multiply these factors on a calculator. Why is it not possible without scientific notation?

CHALLENGE

77. A parsec is a unit of measure used to determine distance between stars. One parsec is approximately 206,265 times the average distance of Earth from the sun. If the average distance from Earth to the sun is approximately 93,000,000 miles, find the approximate length of one parsec. Write the length in scientific notation. Round the number in scientific notation to the nearest hundredth.

78. Light will travel approximately 5,866,000,000,000 miles in 1 year. Approximately how far will light travel in 11 years? Write the distance in scientific notation. Round the number in scientific notation to the nearest thousandth.

Simplify.

79. $\dfrac{(3.25 \times 10^{-3})(2.4 \times 10^3)}{(4.8 \times 10^{-4})(2.5 \times 10^{-3})}$

80. $\dfrac{(3.25 \times 10^{-7})(2.4 \times 10^6)}{(4.8 \times 10^4)(2.5 \times 10^{-3})}$

81. Find the product of these four numbers 5.5×10^{-7}, 8.1×10^{12}, 2×10^5, and 1.5×10^{-9} Write the product in both scientific notation and place value notation.

MAINTAIN YOUR SKILLS

Divide.

82. $42\overline{)7938}$

83. $59\overline{)18{,}408}$

84. $216\overline{)66{,}744}$

85. $\dfrac{745}{12}$

86. $\dfrac{5936}{37}$

87. Find the quotient of 630,828 and 243.

88. Find the quotient of 146,457 and 416.

89. Find the quotient of 6,542,851 and 711. Round to the nearest ten.

90. Find the perimeter of a rectangular field that is 312 ft long and 125 ft wide.

91. The area of a rectangle is 1008 in^2. If the length of the rectangle is 4 ft, find the width.

SECTION

Dividing Decimals; Average, Median, and Mode **4.6**

HOW & WHY

OBJECTIVES

1. Divide decimals.
2. Find the average, median, or mode of a set of decimals.

■ **OBJECTIVE 1** Divide decimals.

Division of decimals is the same as division of whole numbers, with one difference. The difference is the location of the decimal point in the quotient.

As with multiplication, we examine the fractional form of division to discover the method of placing the decimal point in the quotient. First, change the decimal form to fractional form to find the quotient. Next, change the quotient to decimal form. Consider the information in Table 4.7.

TABLE 4.7 Division by a Whole Number

Decimal Form	Fractional Form	Division Fractional Form	Division Decimal Form
3)$\overline{0.36}$	$\dfrac{36}{100} \div 3$	$\dfrac{36}{100} \div 3 = \dfrac{\overset{12}{\cancel{36}}}{100} \cdot \dfrac{1}{\underset{1}{\cancel{3}}} = \dfrac{12}{100}$	$\dfrac{0.12}{3)\overline{0.36}}$
8)$\overline{0.72}$	$\dfrac{72}{100} \div 8$	$\dfrac{72}{100} \div 8 = \dfrac{\overset{9}{\cancel{72}}}{100} \cdot \dfrac{1}{\underset{1}{\cancel{8}}} = \dfrac{9}{100}$	$\dfrac{0.09}{8)\overline{0.72}}$
5)$\overline{0.3}$	$\dfrac{3}{10} \div 5$	$\dfrac{3}{10} \div 5 = \dfrac{3}{10} \cdot \dfrac{1}{5} = \dfrac{3}{50} = \dfrac{6}{100}$	$\dfrac{0.06}{5)\overline{0.3}}$

We can see from Table 4.7 that the decimal point for the quotient of a decimal and a whole number is written directly above the decimal point in the dividend. It may be necessary to insert zeros to do the division. See Example B.

When a decimal is divided by 7, the division process may not have a remainder of zero at any step:

```
    0.97
7)6.85
  6 3
  ───
   55
   49
   ──
    6
```

At this step we can write zeros to the right of the digit 5, since 6.85 = 6.850 = 6.8500 = 6.85000 = 6.850000.

```
    0.97857
7)6.85000
  6 3
  ───
   55
   49
   ──
    60
    56
    ──
     40
     35
     ──
      50
      49
      ──
       1
```

It appears that we might go on inserting zeros and continue endlessly. This is indeed what happens. Such decimals are called "nonterminating, repeating decimals." For example, the quotient of this division is sometimes written

0.97857142857142...   or   0.97$\overline{857142}$

The bar written above the sequence of digits 857142 indicates that these digits are repeated endlessly.

In practical applications we stop the division process one place value beyond the accuracy required by the situation and then round. Therefore,

$$
\begin{array}{r}
0.97 \\
7{\overline{\smash{\big)}\,6.85}} \\
\underline{6\ 3} \\
55 \\
\underline{49} \\
6 \quad \text{Stop}
\end{array}
\qquad\qquad
\begin{array}{r}
0.9785 \\
7{\overline{\smash{\big)}\,6.8500}} \\
\underline{6\ 3} \\
55 \\
\underline{49} \\
60 \\
\underline{56} \\
40 \\
\underline{35} \\
5 \quad \text{Stop}
\end{array}
$$

$6.85 \div 7 \approx 1.0$ rounded to the nearest tenth.	$6.85 \div 7 \approx 0.979$ rounded to the nearest thousandth.

Now let's examine division when the divisor is also a decimal. We will use what we already know about division with a whole-number divisor. See Table 4.8.

TABLE 4.8 Division by a Decimal

Decimal Form	Conversion to a Whole Number Divisior	Decimal Form of the Division
$0.3{\overline{\smash{\big)}\,0.36}}$	$\dfrac{0.36}{0.3} \cdot \dfrac{10}{10} = \dfrac{3.6}{3} = 1.2$	$\begin{array}{r}1.2\\0.3{\overline{\smash{\big)}\,0.3\ 6}}\end{array}$
$0.4{\overline{\smash{\big)}\,1.52}}$	$\dfrac{1.52}{0.4} \cdot \dfrac{10}{10} = \dfrac{15.2}{4} = 3.8$	$\begin{array}{r}3.8\\0.4{\overline{\smash{\big)}\,1.5\ 2}}\end{array}$
$0.08{\overline{\smash{\big)}\,0.72}}$	$\dfrac{0.72}{0.08} \cdot \dfrac{100}{100} = \dfrac{72}{8} = 9$	$\begin{array}{r}9.\\0.08{\overline{\smash{\big)}\,0.72}}\end{array}$
$0.25{\overline{\smash{\big)}\,0.3}}$	$\dfrac{0.3}{0.25} \cdot \dfrac{100}{100} = \dfrac{30}{25} = 1.2$	$\begin{array}{r}1.2\\0.25{\overline{\smash{\big)}\,0.30\ 0}}\end{array}$
$0.006{\overline{\smash{\big)}\,4.8}}$	$\dfrac{4.8}{0.006} \cdot \dfrac{1000}{1000} = \dfrac{4800}{6} = 800$	$\begin{array}{r}800.\\0.006{\overline{\smash{\big)}\,4.800}}\end{array}$

We see from the Table 4.8 that we move the decimal point in both the divisor and the dividend the number of places to make the divisor a whole number. Then divide as before.

To divide two numbers

1. If the divisor is not a whole number, move the decimal point in both the divisor and dividend to the right the number of places necessary to make the divisor a whole number.
2. Place the decimal point in the quotient above the decimal point in the dividend.
3. Divide as if both numbers are whole numbers.
4. Round to the given place value. (If no round-off place is given, divide until the remainder is zero or round as appropriate in the problem. For instance, in problems with money, round to the nearest cent.)

EXAMPLES A–G

DIRECTIONS: Divide. Round as indicated.

STRATEGY: If the divisor is not a whole number, move the decimal point in both the divisor and the dividend to the right the number of places necessary to make the divisor a whole number. The decimal point in the quotient is found by writing it directly above the decimal (as moved) in the dividend.

WARM-UP

A. Divide: $14\overline{)94.962}$

A. Divide: $32\overline{)43.0592}$

```
        1.3456
  32)43.0592
     32
     11 0
      9 6
      1 45
      1 28
        179
        160
        192
        192
          0
```

The numerals in the answer are lined up in columns that have the same place value as those in the dividend.

CHECK:
```
      1.3456
  ×      32
     26 912
    403 68
    43.0592
```

So the quotient is 1.3456.

> **CAUTION**
>
> Write the decimal point for the quotient directly above the decimal point in the dividend.

WARM-UP

B. Find the quotient of 42.39 and 18.

B. Find the quotient of 7.41 and 6.

STRATEGY: Recall that the quotient of a and b can be written $a \div b$, $\dfrac{a}{b}$, or $b\overline{)a}$.

```
      1.23
  6)7.41
    6
    1 4
    1 2
      21
      18
       3
```

Here the remainder is not zero, so the division is not complete. We write a zero on the right (7.410) without changing the value of the dividend and continue dividing.

WARM-UP

A. Divide: $14\overline{)94.962}$

WARM-UP

B. Find the quotient of 42.39 and 18.

```
   1.235
6)7.410
   6
   1 4
   1 2
     21
     18
      30
      30
       0
```
Both the quotient (1.235) and the rewritten dividend (7.410) have three decimal places. Check by multiplying 6×1.235:

```
  1.235
×     6
  7.410
```

The quotient is 1.235.

C. Divide 689.4 by 42 and round the quotient to the nearest hundredth.

```
    16.414
42)689.400
   42
   269
   252
    17 4
    16 8
       60
       42
      180
      168
       12
```

It is necessary to place two zeros on the right in order to round to the hundredths place, since the division must be carried out one place past the place value to round.

The quotient is approximately 16.41.

D. Divide $35.058 \div 0.27$ and round the quotient to the nearest tenth.

```
0.27)35.058
```

First move both decimal points two places to the right so the divisor is the whole number 27. The same result is obtained by multiplying both divisor and dividend by 100.

$$\frac{35.058}{0.27} \times \frac{100}{100} = \frac{3505.8}{27}$$

```
     129.84
27)3505.80
   27
   80
   54
   265
   243
   228
   216
   120
   108
    12
```

The number of zeros you place on the right depends on either the directions for rounding or your own choice of the number of places. Here we find the approximate quotient to the nearest tenth.

The quotient is approximately 129.8.

WARM-UP

C. Divide 725.6 by 48 and round the quotient to the nearest hundredth.

WARM-UP

D. Divide $46.72 \div 0.34$ and round the quotient to the nearest tenth.

ANSWERS TO WARM-UPS C–D

C. 15.12
D. 137.4

WARM-UP

E. Divide 0.85697 by 0.083 and round the quotient to the nearest thousandth.

E. Divide 0.57395 by 0.067 and round the quotient to the nearest thousandth.

$$0.067\overline{)0.57395}$$

```
        8.5664
67)573.9500
    536
    37 9
    33 5
     4 45
     4 02
       430
       402
       280
       268
        12
```

Move both decimals three places to the right. It is necessary to insert two zeros on the right in order to round to the the thousandths place.

The quotient is approximately 8.566.

CALCULATOR EXAMPLE:

WARM-UP

F. Find the quotient of 805.5086 and 34.561, and round to the nearest thousandth.

F. Find the quotient of 2105.144 and 68.37 and round to the nearest thousandth.

$2105.144 \div 68.37 \approx 30.79046$

The quotient is 30.790, to the nearest thousandth.

WARM-UP

G. What is the unit price of 4.6 oz of instant coffee if it costs $3.99? Round to the nearest tenth of a cent.

G. What is the cost per ounce of a 1-1b package of spaghetti that costs $1.74? This is called the "unit price" and is used for comparing prices. Many stores are required to show this price for the food they sell.

STRATEGY: To find the unit price (cost per ounce), we divide the cost, 174 cents, by the number of ounces. Because there are 16 ounces per pound, we divide by 16. Round to the nearest tenth of a cent.

```
       10.87
16)174.00
   16
   14
   00
   14 0
   12 8
    120
    112
      8
```

The spaghetti costs approximately $10.9¢ per ounce.

HOW & WHY

■ **OBJECTIVE 2** Find the average, median, or mode of a set of decimals.

The method for finding the average, median, or mode of a set of decimals is the same as that for whole numbers and fractions.

ANSWERS TO WARM-UPS E–G

E. 10.325 **F.** 23.307
G. The unit price of coffee is $0.867, or 86.7¢ per ounce.

To find the average (mean) of a set of numbers

1. Add the numbers.
2. Divide the sum by the number of numbers in the set.

> **To find the median of a set of numbers**
>
> 1. List the numbers in order from smallest to largest.
> 2. If there is an odd number of numbers in the set, the median is the middle number.
> 3. If there is an even number of numbers in the set, the median is the average (mean) of the two middle numbers.

> **To find the mode of a set of numbers**
>
> 1. Find the number or numbers that occur most often.
> 2. If all the numbers occur the same number of times, there is no mode.

EXAMPLES H–J

DIRECTIONS: Find the average, median, or mode.

STRATEGY: Use the same procedures as for whole numbers and fractions.

H. Find the average of 0.75, 0.43, 3.77, and 2.23.

$$0.75 + 0.43 + 3.77 + 2.23 = 7.18 \quad \text{First, add the numbers.}$$
$$7.18 \div 4 = 1.795 \quad \text{Second, divide by 4, the number of numbers.}$$

So the average of 0.75, 0.43, 3.77, and 2.23 is 1.795.

I. Pedro's grocery bills for the past 5 weeks were

Week 1: $155.72
Week 2: $172.25
Week 3: $134.62
Week 4: $210.40
Week 5: $187.91

What are the average and median costs of Pedro's groceries per week for the 5 weeks?

Average:

$$
\begin{array}{r}
155.72 \\
172.25 \\
134.62 \\
210.40 \\
+187.91 \\
\hline
860.90
\end{array}
$$
Add the weekly totals and divide by 5, the number of weeks.

$860.90 \div 5 = 172.18$

Median:

134.62, 155.72, 172.25, 187.91, 210.40   List the numbers from smallest to largest.

172.25   The median is the middle number.

Pedro's average weekly cost for groceries is $172.18, and the median cost is $172.25.

WARM-UP

H. Find the average of 7.3, 0.66, 10.8, 4.11, and 1.32.

WARM-UP

I. Mary's weekly car expenses, including parking, for the past 6 weeks were

Week 1: $37.95
Week 2: $43.15
Week 3: $28.65
Week 4: $59.14
Week 5: $61.72
Week 6: $50.73

What are the average and median weekly expenses for Mary during the 6 weeks?

ANSWERS TO WARM-UPS H–I

H. 4.838.

I. Mary's average weekly car expense is $46.89, and the median expense is $46.94.

J. Over a 7-month period, Carlos was able to save the following amount each month: $117.25, $115.50, $166.20, $105.00, $151.70, $158.80, and $105.00. Find the average, median, and mode for the 7-month period.

J. For a 7-day period, the recorded price for a share of Intel stock on each day was $18.76, $18.64, $18.81, $19.05, $18.77, $18.45, and $18.77. Find the average, median, and mode for the 7-day period.

Average

18.76 + 18.64 + 18.81 + 19.05 + 18.77 + 18.45 + 18.77 = 131.25   Add the numbers.

131.25 ÷ 7 = 18.75   Divide by the number of numbers.

Median

18.45, 18.64, 18.76, 18.77, 18.77, 18.81, 19.05   List the numbers from smallest to largest.

18.77   The middle number is the median.

Mode

18.77   The number that appears most often.

For the 7-day period the average price for a share of Intel is $18.75, the median price is $18.77, and the mode price is $18.77.

ANSWER TO WARM-UP J

J. Carlos' average savings is about $131.35, the median savings is $117.25, and the mode of the savings is $105.00.

EXERCISES 4.6

■ **OBJECTIVE 1** Divide decimals. (See page 323.)

A *Divide.*

1. $8\overline{)6.4}$

2. $8\overline{)4.8}$

3. $4\overline{)19.6}$

4. $5\overline{)35.5}$

5. $0.1\overline{)32.67}$

6. $0.01\overline{)5.05}$

7. $393.9 \div 0.13$

8. $55.22 \div 0.11$

9. $60\overline{)331.8}$

10. $50\overline{)211.5}$

11. $28.35 \div 36$

12. $5.238 \div 36$

13. To divide 27.8 by 0.6, we first multiply both the dividend and the divisor by 10 so we are dividing by a _____.

14. To divide 0.4763 by 0.287, we first multiply both the dividend and the divisor by _____.

B *Divide.*

Divide and round to the nearest tenth.

15. $6\overline{)7.23}$

16. $7\overline{)0.5734}$

17. $1.6\overline{)10.551}$

18. $6.9\overline{)49.381}$

Divide and round to the nearest hundredth.

19. $6\overline{)0.5934}$

20. $8\overline{)0.0693}$

21. $0.7\overline{)5.687}$

22. $0.6\overline{)5.723}$

23. $0.793 \div 0.413$

24. $0.6341 \div 0.0285$

25. $25 \div 0.0552$

26. $76 \div 0.08659$

Divide and round to the nearest thousandth.

27. $2.15\overline{)19.68}$

28. $41.6\overline{)83.126}$

29. $4.16\overline{)0.06849}$

30. $2.74\overline{)0.6602}$

31. $0.13 \div 0.009$

32. $0.39 \div 0.0087$

■ **OBJECTIVE 2** Find the average, median, or mode of a set of decimals. (See page 328.)

A *Find the average.*

33. 8.3, 5.9

34. 3.6, 8.2

35. 5.7, 10.2

36. 21.7, 36.3

37. 10.6, 8.4, 2.9

38. 7.2, 8.8, 0.8

39. 12.1, 12.5, 12.6

40. 7.9, 15.2, 8.7

41. 12.5, 7.1, 16.7, 2.8

42. 14.3, 20.6, 16.7, 11.2

B *Find the average, median, and the mode.*

43. 7.8, 9.08, 3.9, 5.7

44. 4.87, 6.93, 4.1, 9.6

45. 15.8, 23.64, 22.46, 23.64, 18.7

46. 9.4, 6.48, 12.2, 6.48, 8.6

47. 14.3, 15.4, 7.6, 17.4, 21.6

48. 57.8, 36.9, 48.9, 51.9, 63.7

49. 0.675, 0.431, 0.662, 0.904

50. 0.261, 0.773, 0.663, 0.308

51. 0.5066, 0.6055, 0.5506, 0.5066, 0.6505

52. 2.67, 11.326, 17.53, 22.344, 22.344

C

53. The stock of Microsoft Corporation closed at $24.64, $24.64, $24.55, $24.69, and $24.68 during one week in 2009. What was the average closing price of the stock?

54. A consumer watchdog group priced a box of a certain type of cereal at six different grocery stores. They found the following prices: $4.19, $4.42, $4.25, $3.99, $4.05, and $4.59. What are the average and median selling prices of a box of cereal? Round to the nearest cent.

55. Find the quotient of 17.43 and 0.19, and round to the nearest hundredth.

56. Find the quotient of 1.706 and 77, and round to the nearest hundredth.

Exercises 57–62. The table shows some prices from a grocery store.

Item	Quantity	Price	Item	Quantity	Price
Apples	4 lb	$3.79	Potatoes	10 lb	$3.59
Blueberries	3 pt	$7.59	Pork chops	3.5 lb	$6.97
BBQ sauce	18 oz	$1.69	Cod fillet	2.6 lb	$10.38

57. Find the unit price (price per pound) of apples. Round to the nearest tenth of a cent.

58. Find the unit price (price per ounce) of BBQ sauce. Round to the nearest tenth of a cent.

59. Find the unit price of pork chops. Round to the nearest tenth of a cent.

60. Find the unit price of potatoes. Round to the nearest tenth of a cent.

61. Using the unit price, find the cost of a 1.5-lb cod fillet. Round to the nearest cent.

62. Using the unit price, find the cost of 11 pints of blueberries. Round to the nearest cent.

63. Two hundred fifty-six alumni of Miami University donated $245,610 to the university. To the nearest cent, what was the average donation?

64. The Adams family had the following natural gas bills for last year:

January	$176.02	July	$ 10.17
February	69.83	August	14.86
March	43.18	September	18.89
April	38.56	October	23.41
May	12.85	November	63.19
June	29.55	December	161.51

The gas company will allow them to make equal payments this year equal to the monthly average of last year. How much will the payment be? Round to the nearest cent.

65. The average daily temperature by month in Orlando, Florida, measured in degrees Fahrenheit is

January	72	May	88	September	90
February	73	June	91	October	84
March	78	July	92	November	78
April	84	August	92	December	73

To the nearest tenth, what is the average daily temperature over the year? What are the median and mode of the average daily temperatures?

66. Tim Raines, the fullback for the East All-Stars, gained the following yards in six carries: 8.5 yd, 12.8 yd, 3.2 yd, 11 yd, 9.6 yd, and 4 yd. What was the average gain per carry? Round to the nearest tenth of a yard.

67. The price per gallon of the same grade of gasoline at eight different service stations is 2.489, 2.599, 2.409, 2.489, 2.619, 2.599, 2.329, and 2.479. What are the average, median, and mode for the price per gallon at the eight stations? Round to the nearest thousandth.

Exercises 68–72. The table gives the high and low temperatures in cities in the Midwest.

Temperatures in Midwest Cities

City	High (°F)	Low (°F)
Detroit	37	26
Cincinnati	43	31
Chicago	35	30
St. Louis	44	28
Kansas City	33	24
Minneapolis	27	17
Milwaukee	33	28
Rapid City	40	18

68. What was the average high temperature for the cities, to the nearest tenth?

69. What was the average low temperature for the cities, to the nearest tenth?

70. What was the average daily range of temperature for the cities, to the nearest tenth? Did any of the cities have the average daily range?

71. What was the mode for the high temperatures?

72. What was the median of the low temperatures?

73. June drove 766.5 miles on 15.8 gallons of gas in her hybrid car. What is her mileage (miles per gallon)? Round to the nearest mile.

74. A 65-gallon drum of cleaning solvent in an auto repair shop is being used at the rate of 1.94 gallons per day. At this rate, how many full days will the drum last?

75. The Williams Construction Company uses cable that weighs 2.75 pounds per foot. A partly filled spool of the cable is weighed. The cable itself weighs 867 pounds after subtracting the weight of the spool. To the nearest foot, how many feet of cable are on the spool?

76. A plumber connects the sewers of four buildings to the public sewer line. The total bill for the job is $7358.24. What is the average cost for each connection?

77. The pictured 1-ft I beam, weighs 32.7 lb. What is the length of a beam weighing 630.6 lb? Find the length to the nearest tenth of a foot.

78. Allowing 0.125 in. of waste for each cut, how many bushings, which are 1.45 in. in length, can be cut from a 12-in. length of bronze? What is the length of the piece that is left?

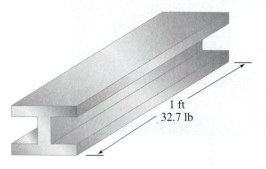

1 ft
32.7 lb

Exercises 79–81. The table shows population and area facts for Scandinavia.

Population and Area in Scandinavia

Country	Area in Square Kilometers	Population, 2009
Denmark	43,095	5,500,510
Finland	336,956	5,250,275
Norway	324,265	4,660,539
Sweden	449,962	9,059,651

79. Which country has the smallest area and which has the smallest population?

80. Population density is the number of people per square kilometer. Calculate the population density for each country, rounded to the nearest hundredth. Add another column to the table with this information.

81. What do you conclude about how crowded the countries are?

Exercises 82–86 relate to the chapter application. In baseball, a pitcher's earned run average (ERA) is calculated by dividing the number of earned runs by the quotient of the number of innings pitched and 9. The lower a pitcher's ERA the better.

82. Suppose a pitcher allowed 34 earned runs in 85 innings of play. Calculate his ERA and round to the nearest hundredth.

83. A pitcher allows 20 earned runs in 110 innings. Calculate his ERA, rounding to the nearest hundredth.

84. A runner's stolen base average is the quotient of the number of bases stolen and the number of attempts. As with the batting averages, this number is usually rounded to the nearest thousandth. Calculate the stolen base average of a runner who stole 18 bases in 29 attempts.

85. A good stolen base average is 0.700 or higher. Express this as a fraction and say in words what the fraction represents.

86. The combined height of the NBA's 348 players at the start of the 2000–2001 season was 2294.67 feet, or about twice the height of the Empire State Building. Find the average height of an NBA player. Round to the nearest tenth.

© Ragne Kabanova/Shutterstock.com

STATE YOUR UNDERSTANDING

87. Describe a procedure for determining the placement of the decimal in a quotient. Include an explanation for the justification of the procedure.

88. Explain how to find the quotient of $4.1448 \div 0.0012$.

89. Copy the table and fill it in.

Division Procedure

Operation on Decimals	Procedure	Example
Division		

CHALLENGE

90. What will be the value of $3000 invested at 6% interest compounded quarterly at the end of 1 year? (Compounded quarterly means that the interest earned for the quarter, the annual interest divided by four, is added to the principal and then earns interest for the next quarter.) How much more is earned by compounding quarterly instead of annually?

91. Perform the indicated operations. Round the result to the nearest hundredth.

$$\frac{8.23}{0.56}$$
$$2.47$$

92. Perform the indicated operations. Round the result to the nearest hundredth.

$$\frac{8.23}{0.56}$$
$$2.47$$

MAINTAIN YOUR SKILLS

Simplify.

93. $\dfrac{95}{114}$

94. $\dfrac{168}{216}$

Write as an improper fraction.

95. $4\dfrac{8}{11}$

96. $18\dfrac{5}{7}$

Write as a mixed number.

97. $\dfrac{215}{12}$

98. $\dfrac{459}{25}$

Find the missing numerator.

99. $\dfrac{17}{25} = \dfrac{?}{100}$

100. $\dfrac{9}{40} = \dfrac{?}{1000}$

Write as a fraction or mixed number and simplify.

101. $24 \div 40$

102. $135 \div 30$

GETTING READY FOR ALGEBRA

HOW & WHY

OBJECTIVE

Solve equations that involve multiplication and division of decimals.

We solve equations that involve multiplication and division of decimals in the same way as equations with whole numbers and fractions.

To solve an equation using multiplication or division

1. Divide both sides of the equation by the same number to isolate the variable, or
2. Multiply both sides of the equation by the same number to isolate the variable.

EXAMPLES A–E

DIRECTIONS: Solve

STRATEGY: Isolate the variable by multiplying or dividing both sides by the same number.

WARM-UP

A. $1.2t = 96$

A. $1.7x = 86.7$

$$\frac{1.7x}{1.7} = \frac{86.7}{1.7}$$

Because x is multiplied by 1.7, we divide both sides by 1.7. The division is usually written in fractional form. Because division is the inverse of multiplication, the variable is isolated.

$x = 51$ Simplify.

CHECK: $1.7(51) = 86.7$ Substitute 51 for x in the original equation and simplify.

$86.7 = 86.7$ True.

The solution is $x = 51$.

WARM-UP

B. $13.5 = \dfrac{r}{9.7}$

B. $9.8 = \dfrac{a}{11.6}$

$$11.6(9.8) = 11.6\left(\frac{a}{11.6}\right)$$

Because a is divided by 11.6, we multiply both sides by 11.6. Because multiplication is the inverse of division, the variable is isolated.

$113.68 = a$

CHECK: $9.8 = \dfrac{113.68}{11.6}$ Substitute 113.68 for a in the original equation and simplify.

$9.8 = 9.8$ True.

The solution is $a = 113.68$.

WARM-UP

C. $0.18a = 1.6632$

C. $34.6y = 186.84$

$$\frac{34.6y}{34.6} = \frac{186.84}{34.6}$$

Divide both sides by 34.6 to eliminate the multiplication and simplify.

$y = 5.4$

CHECK: $34.6(5.4) = 186.84$ Substitute 5.4 for y in the original equation and simplify.

$186.84 = 186.84$ True.

The solution is $y = 5.4$.

ANSWERS TO WARM-UPS A–C

A. $t = 80$ **B.** $130.95 = r$
C. $a = 9.24$

D. $\dfrac{c}{0.508} = 3.1$

WARM-UP

D. $\dfrac{x}{0.481} = 7.2$

$0.508\left(\dfrac{c}{0.508}\right) = 0.508(3.1)$ Multiply both sides by 0.508 and simplify.

$c = 1.5748$

CHECK: $\dfrac{1.5748}{0.508} = 3.1$ Substitute 1.5748 for c in the original equation and simplify.

$3.1 = 3.1$ True.

The solution is $c = 1.5748$.

E. The total number of calories T is given by the formula $T = sC$, where s represents the number of servings and C represents the number of calories per serving. Find the number of calories per serving if 7.5 servings contain 739.5 calories.

First substitute the known values into the formula.

$T = sC$

$739.5 = 7.5C$ Substitute 739.5 for T and 7.5 for s.

$\dfrac{739.5}{7.5} = \dfrac{7.5C}{7.5}$ Divide both sides by 7.5.

$98.6 = C$

Since $7.5(98.6) = 739.5$, the number of calories per serving is 98.6.

WARM-UP

E. Use the formula in Example E to find the number of calories per serving if there is a total of 3253.6 calories in 28 servings.

EXERCISES

Solve.

1. $2.7x = 18.9$

2. $2.3x = 0.782$

3. $0.04y = 12.34$

4. $0.06w = 0.942$

5. $0.9476 = 4.12t$

6. $302.77 = 13.7x$

7. $3.3m = 0.198$

8. $0.008p = 12$

9. $0.016q = 9$

10. $11 = 0.025w$

11. $9 = 0.32h$

12. $2.6x = 35.88$

13. $\dfrac{y}{9.5} = 0.28$

14. $0.07 = \dfrac{b}{0.73}$

15. $0.312 = \dfrac{c}{0.65}$

16. $\dfrac{w}{0.12} = 1.35$

17. $0.0325 = \dfrac{x}{32}$

18. $0.17 = \dfrac{t}{8.23}$

19. $\dfrac{s}{0.07} = 0.345$

20. $\dfrac{y}{16.75} = 2.06$

21. $\dfrac{z}{21.02} = 4.08$

22. $\dfrac{c}{10.7} = 2.055$

23. The total number of calories T is given by the formula $T = sC$, where s represents the number of servings and C represents the number of calories per serving. Find the number of servings if the total number of calories is 3617.9 and there are 157.3 calories per serving.

24. Use the formula in Exercise 23 to find the number of servings if the total number of calories is 10,628.4 and there are 312.6 calories per serving.

25. Ohm's law is given by the formula $E = IR$, where E is the voltage (number of volts), I is the current (number of amperes), and R is the resistance (number of ohms). What is the current in a circuit if the resistance is 22 ohms and the voltage is 209 volts?

26. Use the formula in Exercise 25 to find the current in a circuit if the resistance is 16 ohms and the voltage is 175 volts.

27. Find the length of a rectangle that has a width of 13.6 ft and an area of 250.24 ft^2.

28. Find the width of a rectangular plot of ground that has an area of 3751.44 m^2 and a length of 127.6 m.

29. Each student in a certain instructor's math classes hands in 20 homework assignments. During the term, the instructor has graded a total of 3500 homework assignments. How many students does this instructor have in all her classes? Write and solve an equation to determine the answer.

30. Twenty-four plastic soda bottles were recycled and made into one shirt. At this rate, how many shirts can be made from 910 soda bottles? Write and solve an equation to determine the answer.

Changing Fractions to Decimals

HOW & WHY

■ OBJECTIVE   Change fractions to decimals.

Every decimal can be written as a whole number times the place value of the last digit on the right:

$$0.81 = 81 \times \frac{1}{100} = \frac{81}{100}$$

The fraction has a power of 10 for the denominator. Any fraction that has only prime factors of 2 and/or 5 in the denominator can be written as a decimal by building the denominator to a power of 10.

$$\frac{3}{5} = \frac{3}{5} \cdot \frac{2}{2} = \frac{6}{10} = 0.6$$

$$\frac{11}{20} = \frac{11}{20} \cdot \frac{5}{5} = \frac{55}{100} = 0.55$$

Every fraction can be thought of as a division problem $\left(\frac{3}{5} = 3 \div 5 \right)$. Therefore, a second method for changing fractions to decimals is division. As you discovered in the previous section, many division problems with decimals do not have a zero remainder at any point. If the denominator of a simplified fraction has prime factors other than 2 or 5, the quotient will be a nonterminating decimal. The fraction $\frac{5}{6}$ is an example:

$$\frac{5}{6} = 0.833333333\ldots = 0.8\overline{3}$$

The bar over the 3 indicates that the decimal repeats the number 3 forever. Expressing the decimal form of a fraction using a repeat bar is an exact conversion and is indicated with an equal sign (=). In the exercises for this section, round the division to the indicated decimal place or use the repeat bar as directed.

CAUTION

Be careful to use an equal sign (=) when your conversion is exact and an approximately equal sign(≈)when you have rounded.

To change a fraction to a decimal

Divide the numerator by the denominator.

To change a mixed number to a decimal

Change the fractional part to a decimal and add to the whole-number part.

EXAMPLES A–G

DIRECTIONS: Change the fraction or mixed number to a decimal.

STRATEGY: Divide the numerator by the denominator. Round as indicated. If the number is a mixed number, add the decimal to the whole number.

WARM-UP

A. Change $\dfrac{19}{20}$ to a decimal.

A. Change $\dfrac{37}{40}$ to a decimal.

$$
\begin{array}{r}
0.925 \\
40\overline{)37.000} \\
36\ 0 \\
\hline
1\ 00 \\
80 \\
\hline
200 \\
200 \\
\hline
0
\end{array}
$$

Divide the numerator, 37, by the denominator, 40.

Therefore, $\dfrac{37}{40} = 0.925$.

WARM-UP

B. Change $22\dfrac{8}{50}$ to a decimal.

B. Change $13\dfrac{17}{20}$ to a decimal.

$$
\begin{array}{r}
0.85 \\
20\overline{)17.00} \\
16\ 0 \\
\hline
1\ 00 \\
1\ 00 \\
\hline
0
\end{array}
$$

$13\dfrac{17}{20} = 13.85$ Add the decimal to the whole number.

or

$\dfrac{17}{20} = \dfrac{17}{20} \cdot \dfrac{5}{5} = \dfrac{85}{100} = 0.85$

So $13\dfrac{17}{20} = 13 + 0.85 = 13.85$

A fraction with a denominator that has only 2s or 5s for prime factors ($20 = 2 \cdot 2 \cdot 5$) can be changed to a fraction with a denominator that is a power of 10. This fraction can then be written as a decimal.

WARM-UP

C. Change $\dfrac{21}{32}$ to a decimal.

C. Change $\dfrac{23}{64}$ to a decimal.

$$
\begin{array}{r}
0.359375 \\
64\overline{)23.000000} \\
19\ 2 \\
\hline
3\ 80 \\
3\ 20 \\
\hline
600 \\
576 \\
\hline
240 \\
192 \\
\hline
480 \\
448 \\
\hline
320 \\
320 \\
\hline
0
\end{array}
$$

This fraction can be changed by building to a denominator of 1,000,000, but the factor is not easily recognized, unless we use a calculator.

$\dfrac{23}{64} = \dfrac{23}{64} \cdot \dfrac{15,625}{15,625} = \dfrac{359,375}{1,000,000} = 0.359375$

So $\dfrac{23}{64} = 0.359375$.

ANSWERS TO WARM-UPS A–C

A. 0.95   **B.** 22.16   **C.** 0.65625

CAUTION

Most fractions cannot be changed to terminating decimals, because the denominators contain factors other than 2 and 5. In these cases we round to the indicated place value or use a repeat bar.

D. Change $\dfrac{23}{29}$ to a decimal rounded to the nearest hundredth.

$$\begin{array}{r} 0.793 \\ 29\overline{)23.000} \\ \underline{20\ 3} \\ 2\ 70 \\ \underline{2\ 61} \\ 90 \\ \underline{87} \\ 3 \end{array}$$

Divide 23 by 29. Carry out the division to three decimal places and round to the nearest hundredth.

So $\dfrac{23}{29} \approx 0.79$.

E. Change $\dfrac{13}{33}$ to an exact decimal.

$$\begin{array}{r} 0.3939 \\ 33\overline{)13.0000} \\ \underline{9\ 9} \\ 3\ 10 \\ \underline{2\ 97} \\ 130 \\ \underline{99} \\ 310 \\ \underline{297} \\ 13 \end{array}$$

We see that the division will not have a zero remainder, so we use the repeat bar to show the quotient.

So $\dfrac{13}{33} = 0.\overline{39}$.

CALCULATOR EXAMPLE: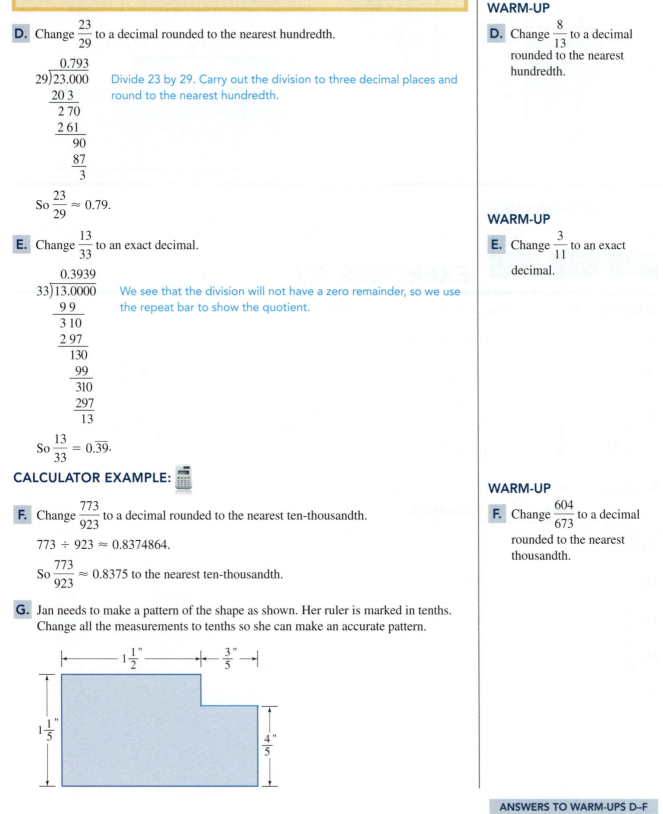

F. Change $\dfrac{773}{923}$ to a decimal rounded to the nearest ten-thousandth.

$773 \div 923 \approx 0.8374864$.

So $\dfrac{773}{923} \approx 0.8375$ to the nearest ten-thousandth.

G. Jan needs to make a pattern of the shape as shown. Her ruler is marked in tenths. Change all the measurements to tenths so she can make an accurate pattern.

$1\frac{1}{2}''$ $\frac{3}{5}''$ $1\frac{1}{5}''$ $\frac{4}{5}''$

WARM-UP

D. Change $\dfrac{8}{13}$ to a decimal rounded to the nearest hundredth.

WARM-UP

E. Change $\dfrac{3}{11}$ to an exact decimal.

WARM-UP

F. Change $\dfrac{604}{673}$ to a decimal rounded to the nearest thousandth.

ANSWERS TO WARM-UPS D–F

D. 0.62   **E.** $0.\overline{27}$ **F.** 0.897

G. Change the measurements on the given pattern to the nearest tenth for use with a ruler marked in tenths.

So that Jan can use her ruler for more accurate measure, each fraction is changed to a decimal.

$$1\frac{1}{2} = 1\frac{5}{10} = 1.5$$

$$\frac{3}{5} = \frac{6}{10} = 0.6$$

$$1\frac{1}{5} = 1\frac{2}{10} = 1.2$$

$$\frac{4}{5} = \frac{8}{10} = 0.8$$

Each fraction and mixed number can be changed by either building each to a denominator of 10 as shown or by dividing the numerator by the denominator. The measurements on the drawing can be labeled:

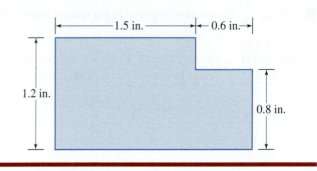

G. The decimal measures are $10\frac{2}{5}$ in. $= 10.4$ in.; $7\frac{15}{16}$ in. ≈ 7.9 in.; and $4\frac{3}{4}$ in. ≈ 4.8 in.

EXERCISES 4.7

■ **OBJECTIVE**   Change fractions to decimals. (See page 339.)

A   *Change the fraction or mixed number to a decimal*

1. $\dfrac{3}{4}$

2. $\dfrac{7}{10}$

3. $\dfrac{3}{20}$

4. $\dfrac{9}{20}$

5. $\dfrac{13}{16}$

6. $\dfrac{23}{32}$

7. $6\dfrac{5}{8}$

8. $6\dfrac{13}{20}$

9. $56\dfrac{73}{125}$

10. $48\dfrac{43}{50}$

B   *Change to a decimal rounded to the indicated place value.*

	Tenth	Hundredth
11. $\dfrac{3}{7}$		
12. $\dfrac{8}{9}$		
13. $\dfrac{5}{12}$		
14. $\dfrac{7}{11}$		
15. $\dfrac{11}{13}$		
16. $\dfrac{11}{14}$		
17. $\dfrac{2}{15}$		
18. $\dfrac{9}{19}$		

	Tenth	Hundredth
19. $7\dfrac{7}{18}$		
20. $46\dfrac{11}{17}$		

Change each of the following fractions to decimals. Use the repeat bar.

21. $\dfrac{9}{11}$ **22.** $\dfrac{7}{22}$ **23.** $\dfrac{1}{12}$ **24.** $\dfrac{5}{18}$

C *Change each of the following fractions to decimals to the nearest indicated place value.*

	Hundredth	Thousandth
25. $\dfrac{9}{79}$		
26. $\dfrac{17}{49}$		
27. $\dfrac{45}{46}$		
28. $\dfrac{83}{99}$		

Change to a decimal. Use the repeat bar.

29. $\dfrac{5}{13}$ **30.** $\dfrac{7}{33}$ **31.** $\dfrac{6}{7}$ **32.** $\dfrac{23}{26}$

33. A piece of blank metal stock is $2\dfrac{3}{8}$ in. in diameter. A micrometer measures in decimal units. If the stock is measured with the micrometer, what will the reading be?

34. A wrist pin is $\dfrac{15}{16}$ in. in diameter. A micrometer measures in decimal units. What is the micrometer reading?

35. Convert the measurements in the figure to decimals.

$\dfrac{3}{8}$ in. $1\dfrac{1}{4}$ in. $\dfrac{1}{2}$ in.

36. Stephen needs $6\dfrac{17}{20}$ in. of chain to secure his garden gate. What is the decimal equivalent?

Change to a decimal. Round as indicated.

	Hundredth	Thousandth	Ten-thousandth
37. $\dfrac{21}{52}$			
38. $27\dfrac{19}{71}$			
39. $16\dfrac{15}{101}$			
40. $\dfrac{888}{2095}$			

41. A remnant of material $1\frac{3}{4}$ yards long costs $7.14. Find the cost per yard of the fabric using fractions. Recalculate the same cost using decimals. Which is easier? Why?

42. An electronics lobbyist works $34\frac{3}{4}$ hours during 1 week. If she is paid $49.40 per hour, compute her gross wages for the week. Did you use decimals or fractions to do the calculation? Why?

43. Michael can run a mile in 6.45 minutes. Convert this decimal to a fraction, and then build the fraction to a denominator of 60 in order to determine his time in minutes and seconds.

44. Ronald is writing a paper for his philosophy class, which must be computer generated. The instructor has specified that all margins should be $1\frac{1}{4}$ inches. The software requires that the margins be specified in decimal form rounded to the nearest tenth of an inch. What number does Ronald specify for the margins?

45. Recall that there are 60 minutes in one hour. So 47 minutes is $\frac{47}{60}$ hr. One day, Anchorage, Alaska, had 19 hours 14 min of daylight. Express the hours of daylight as a decimal, rounded to the nearest hundredth.

Exercises 46–47 relate to the chapter application.

46. In the 2009 World Athletics Championships, Dani Samuels of Australia won the women's discus throw with a toss of 214 ft 8 in. Convert this distance to a mixed number of feet, and then convert it to decimal form.

47. In the 2009 World Athletics Championships, Xue Bai of China won the women's marathon with a time of 2 hr 25 min 15 sec. Convert this time to a mixed number of hours and then convert the time to decimal form.

© Ragne Kabanova/Shutterstock.com

STATE YOUR UNDERSTANDING

48. Write a short paragraph on the uses of decimals and of fractions. Include examples of when fractions are more useful and when decimals are more fitting.

CHALLENGE

49. Which is larger, 0.0012 or $\frac{7}{625}$?

50. Which is larger, 2.5×10^{-4} or $\frac{3}{2000}$?

51. First decide whether the fraction $\dfrac{1.23}{80}$ is more or less than 0.1. Then change the fraction to a decimal. Were you correct in your estimate?

52. First decide whether the fraction $\dfrac{62}{0.125}$ is more or less than 100. Then change the fraction to a decimal. Were you correct in your estimate?

53. Change each of these fractions to decimals rounded to the nearest hundredth: $\dfrac{\frac{5}{16}}{15}, \dfrac{\frac{5}{16}}{15}$.

▮ MAINTAIN YOUR SKILLS

Perform the indicated operations.

54. $7 \cdot 12 \div 4 + 2 - 5$

55. $(9 - 5) \cdot 5 - 14 + 6 \div 2$

56. $6^2 \cdot 4 - 3 \cdot 7 + 15$

57. $(7 - 4)^2 - 9 \div 3 + 7$

58. Estimate the sum of 34, 75, 82, and 91 by rounding to the nearest ten.

59. Estimate the difference of 345 and 271 by rounding to the nearest ten.

60. Estimate the product of 56 and 72 by front rounding both factors.

61. Estimate the product of 265 and 732 by front rounding both factors

62. Mr. Lewis buys 350 books for $60 at an auction. He sells two-fifths of them for $25, 25 books at $1.50 each, 45 books at $1 each, and gives away the rest. How many books does he give away? What is his total profit if his handling cost is $15?

63. John C. Scott Realty sold six houses last week at the following prices: $145,780, $234,700, $195,435, $389,500, $275,000, and $305,677. What was the average sale price of the houses?

Order of Operations; Estimating

SECTION
4.8

HOW & WHY

▮ **OBJECTIVE 1** Do any combination of operations with decimals.

The order of operations for decimals is the same as that for whole numbers and fractions.

OBJECTIVES

1. Do any combination of operations with decimals.
2. Estimate the sum, difference, product, and quotient of decimals.

ORDER OF OPERATIONS

To simplify an expression with more than one operation follow these steps

1. Parentheses—Do the operations within grouping symbols first (parentheses, fraction bar, etc.), in the order given in steps 2, 3, and 4.
2. Exponents—Do the operations indicated by exponents.
3. Multiply and Divide—Do multiplication and division as they appear from left to right.
4. Add and Subtract—Do addition and subtraction as they appear from left to right.

EXAMPLES A–E

DIRECTIONS: Perform the indicated operations.

STRATEGY: Use the same order of operations as for whole numbers and fractions.

WARM-UP

A. Simplify: $0.93 - 0.45(0.62)$

A. Simplify: $0.87 - 0.32(0.35)$

$0.87 - 0.32(0.35) = 0.87 - 0.112$ Multiplication is done first.
$= 0.758$ Subtract.

So $0.87 - 0.32(0.35) = 0.758$.

WARM-UP

B. Simplify: $0.86 \div 0.25(3.05)$

B. Simplify: $5.98 \div 0.23(3.16)$

$5.98 \div 0.23(3.16) = 26(3.16)$ Division is done first, because it occurs first.
$= 82.16$ Multiply.

So $5.98 \div 0.23(3.16) = 82.16$.

WARM-UP

C. Simplify: $(4.5)^2 - (0.7)^3$

C. Simplify: $(5.2)^2 - (1.3)^3$

$(5.2)^2 - (1.3)^3 = 27.04 - 2.197$ Exponents are done first.
$= 24.843$ Subtract.

So $(5.2)^2 - (1.3)^3 = 24.843$.

CALCULATOR EXAMPLE:

WARM-UP

D. Simplify: $102.92 \div 8.3 + (0.67)(34.7) - 21.46$

D. Simplify: $8.736 \div 2.8 + (4.57)(5.9) + 12.67$

STRATEGY: All but the least expensive calculators have algebraic logic. The operations can be entered in the same order as the exercise.

So $8.736 \div 2.8 + (4.57)(5.9) + 12.67 = 42.753$.

WARM-UP

E. Nuyen buys the following tickets for upcoming Pops concerts at the local symphony: 3 tickets at $45.75 each; 2 tickets at $39.50 each; 4 tickets at $42.85 each; 3 tickets at $48.50; and 5 tickets at $40.45. For buying more than 10 tickets, Nuyen gets $3.00 off each ticket purchased. What is Nuyen's total cost for the tickets?

E. Ellen buys the following items at the grocery store: 3 cans of soup at $1.23 each; 2 cans of peas at $0.89 each; 1 carton of orange juice at 2 for $5.00; 3 cans of salmon at $2.79 each; and 1 jar of peanut butter at $3.95. Ellen had a coupon for $2.00 off when you purchase 3 cans of salmon. What did Ellen pay for the groceries?

STRATEGY: Find the sum of the cost of each item and then subtract the coupon savings. To find the cost of each type of food, multiply the unit price by the number of items.

$3(1.23) + 2(0.89) + 1(5.00 \div 2)$ To find the unit price of the orange juice,
$+ 3(2.79) + 1(3.95) - 2.00$ we must divide the price for two by 2.

$3.69 + 1.78 + 2.50 + 8.37$ Multiply and divide.
$+ 3.95 - 2.00$

18.29 Add and subtract.

Ellen spent $18.29 for the groceries.

ANSWERS TO WARM-UPS A–E

A. 0.651
B. 10.492
C. 19.907
D. 14.189
E. Nuyen pays $684.40 for the tickets.

HOW & WHY

■ **OBJECTIVE 2** Estimate the sum, difference, product, and quotient of decimals.

To estimate the sum or difference of decimals, we round the numbers to a specified place value. We then add or subtract these rounded numbers to get the estimate. For example, to estimate the sum of $0.345 + 0.592 + 0.0067$, round each to the nearest tenth.

$$
\begin{aligned}
0.345 &\approx 0.3 \\
0.592 &\approx 0.6 \\
+\ 0.0067 &\approx 0.0 \\
\hline
&0.9
\end{aligned}
$$

So 0.9 is the estimate of the sum. We usually can do the estimation mentally and it serves as a check to see if our actual sum is reasonable. Here the actual sum is 0.9437.

Similarly, we can estimate the difference of two numbers. For instance, Jane found the difference of 0.00934 and 0.00367 to be 0.008973. To check, we estimate the difference by rounding each number to the nearest thousandth,

$$
\begin{aligned}
0.00934 &\approx 0.009 \\
-\ 0.00367 &\approx 0.004 \\
\hline
&0.005
\end{aligned}
$$

So 0.005 is the estimate of the difference. This is not close to Jane's answer, so she needs to subtract again.

$$
\begin{aligned}
0.00934 \\
-0.00367 \\
\hline
0.00567
\end{aligned}
$$

This answer is close to the estimate. Jane may not have aligned the decimal points properly.

EXAMPLES F–I

DIRECTIONS: Estimate the sum or difference.

STRATEGY: Round each number to a specified place value and then add or subtract.

F. Estimate the sum by rounding to the nearest hundredth: $0.012 + 0.067 + 0.065$

$0.01 + 0.07 + 0.07 = 0.15$ Round each number to the nearest hundredth and add.

So the estimated sum is 0.15.

G. Estimate the sum by rounding to the nearest tenth:
$0.0054 + 0.067 + 0.028 + 1.07$

$0.0 + 0.1 + 0.0 + 1.1 = 1.2.$ Round each number to the nearest tenth and add.

The estimated sum is 1.2.

H. Estimate the difference by rounding to the nearest tenth: $0.866 - 0.385$

$0.9 - 0.4 = 0.5$ Round each number to the nearest tenth and subtract.

So the estimated difference is 0.5.

I. Use estimation to see if the following answer is reasonable:

$0.843 - 0.05992 = 0.78308$

I. Use estimation to see if the following answer is reasonable:

$0.0067 - 0.0034 = 0.0023$

$0.007 - 0.003 = 0.004$ Round each number to the nearest thousandth and subtract.

The estimated sum is 0.004, and therefore the answer is not reasonable. So we subtract again.

$0.0067 - 0.0034 = 0.0033$, which is correct.

To estimate the product of decimals, **front round** each number and then multiply. For instance to find the estimated product, $(0.067)(0.0034)$, round to the product, $(0.07)(0.003)$, and then multiply. The estimated product is $(0.07)(0.003) = 0.00021$. If the estimate is close to our calculated product, we will feel comfortable that we have the product correct. In this case our calculated product is 0.0002278.

We estimate a division problem only to verify the correct place value in the quotient. If we front round and then divide the numbers, it could result in an estimate that is as much as 3 units off the correct value. However, the place value will be correct. Find the correct place value of the first nonzero digit in 0.000456 divided by 0.032.

$$0.03 \overline{)0.0005}$$

$$3 \overline{)0.05} \quad \begin{array}{c}.01\end{array}$$ Multiply the divisor and the dividend by 100 so we are dividing by a whole number. Find a partial quotient.

We see that the quotient will have its first nonzero digit in the hundreds place. So given a choice of answers, 0.1425, 0.01425, 0.001425, or 1.425, we choose 0.01425 because the first nonzero digit is in the hundredths place.

EXAMPLES J–M

DIRECTIONS: Estimate the product or quotient.

STRATEGY: Front round each number and then multiply or divide.

J. Estimate the product: $(0.0632)(0.0043)$

$(0.06)(0.004) = 0.00024$ Front round and multiply. Note: there are 5 decimal places in the factors so there must be 5 decimal places in the product.

So the estimated product is 0.00024.

K. Justin calculated $(0.076)(0.02177)$ and got 0.0165452. Estimate the product by front rounding to determine if this is a reasonable answer.

$(0.08)(0.02) = 0.0016$

The estimated product is 0.0016, which is not close to Justin's answer. His answer is not reasonable. The product is 0.00165452.

L. Use estimation to decide if the quotient $0.1677 \div 0.00258$ is (a) 6.5, (b) 0.65, (c) 650, (d) 65, or (e) 0.0065.

$0.2 \div 0.003 = ?$ Front round.

$200 \div 3 \approx 66$ Move the decimal point three places to the right in each number so we are dividing by a whole number and divide.

From the estimated quotient, we see that the first nonzero digit is in the tens place. So the quotient is d, or 65.

M. Jane goes to the store to buy the following items: eggs, $1.29; cereal, $2.89; 3 cans of soup at $.89 each; hamburger, $3.49; 2 cans of fruit at $1.19 each; milk, $2.15; potatoes, $0.79; and bread, $2.79. Jane has $20 to spend, so she will estimate the cost to see if she can afford the items. Can Jane afford all of the items?

Round each price to the nearest dollar and keep a running total:

Item	Actual Cost	Estimated Cost	Running Total
Eggs	$1.29	$1	$1
Cereal	$2.89	$3	$4
Soup	3 × $0.89	3 × $1 = $3	$7
Hamburger	$3.49	$3	$10
Fruit	2 × $1.19	2 × $1 = $2	$12
Milk	$2.15	$2	$14
Potatoes	$0.79	$1	$15
Bread	$2.79	$3	$18

Multiply the rounded cost by 3, the number of cans of soup. Multiply the rounded cost by 2, the number of cans of fruit.

Jane estimates the cost at $18 (the actual cost is $18.45), so she can afford the items.

WARM-UP
M. Pete has $100 on the books at Rock Creek Country Club. He wants to buy the following items: 2 dozen golf balls at $21.95 a dozen; 3 bags of tees at $2.08 each; glove, $5.65; towel, $10.75; cap, $14.78; and 3 pairs of socks at $4.15 each. Round to the nearest dollar to estimate the cost. Can Pete afford all the items?

EXERCISES 4.8

■ **OBJECTIVE 1** Do any combination of operations with decimals. (See page 345.)

A *Perform the indicated operations.*

1. $0.9 - 0.7 + 0.3$

2. $0.8 - 0.2 + 0.4$

3. $0.36 \div 9 - 0.02$

4. $0.56 \div 4 + 0.13$

5. $2.4 - 3(0.7)$

6. $3.6 + 3(0.2)$

7. $6(2.7) + 3(4.4)$

8. $8(1.1) - 0.7(8)$

9. $0.19 + (0.7)^2$

10. $0.52 - (0.4)^2$

B

11. $9.35 - 2.54 + 6.91 - 3.65$

12. $0.89 + 6.98 - 5.67 + 0.09$

13. $9.6 \div 2.4(12.7)$

14. $64.4 \div 9.2(0.55)$

15. $2.28 \div 0.38(0.37)$

16. $(7.5)(3.42) \div 0.15$

17. $(4.6)^2 - 2.6(4.1)$

18. $(6.2)^2 + 2.22 \div 0.37$

19. $(6.7)(1.4)^3 \div 0.7$

20. $(3.1)^3 - (0.8)^2 + 4.5$

■ **OBJECTIVE 2** Estimate the sum, difference, product, and quotient of decimals. (See page 347.)

A *Estimate the sum or difference by rounding to the specified place value.*

21. 0.0749 + 0.0861 + 0.0392, hundredth

22. 0.0056 + 0.00378 + 0.00611, thousandth

23. 0.838 − 0.369, tenth

24. 0.00562 − 0.00347, thousandth

25. 6.299 + 3.0055 + 0.67 + 0.0048, ones

26. 0.67 + 0.345 + 0.0021 + 0.8754, tenth

27. 7.972 − 6.7234, ones

28. 0.0573 − 0.0109, hundredth

Estimate the product by front rounding the factors.

29. 0.00922(0.237)

30. 17.982(3.465)

31. 11.876(4.368)

32. 0.000782(.00194)

Using front rounding to determine the place value of the first nonzero digit in each of the quotients.

33. 2.88 ÷ 0.0462

34. 0.0675 ÷ 0.451

35. 0.0000891 ÷ 3.78

36. 0.000678 ÷ 0.00451

B *Use estimation to see if the following answers are reasonable.*

37. 0.0494 + 0.0663 + 0.07425 = 0.18895

38. 0.00921 − 0.00348 = 0.0573

39. 0.00576(0.0491) = 0.000282816

40. 0.0135 ÷ 0.000027 = 500

C

41. Elmer goes shopping and buys 3 cans of cream-style corn at 89¢ per can, 4 cans of tomato soup at $1.09 per can, 2 bags of corn chips at $2.49 per bag, and 6 candy bars at 59¢ each. How much does Elmer spend?

42. Christie buys school supplies for her children. She buys 6 pads of paper at $1.49 each, 5 pens at $1.19 each, 4 erasers at 59¢ each, and 4 boxes of crayons at $2.49 each. How much does she spend?

43. Using estimation, determine if the answer to 0.0023452 ÷ 0.572 is
(a) 0.041 (b) 4.1 (c) 0.00041 (d) 0.0041 or (e) 0.41

44. Using estimation, determine if the answer to 1.3248 ÷ 0.0032 is
(a) 414 (b) 4.14 (c) 0.0414 (d) 41.4 or (e) 4140

Perform the indicated operations.

45. $(9.9)(4.3) − (5.6)(5.1) + (2.3)^2$

46. $14.7 + 2.49(3.1) − 6.8(1.33) + 34$

47. $32.061 − [(1.1)^3(1.5) + 4.25]$

48. $11.3 − [(2.1)^2 − 3.89]$

49. $3.8(3.46 + 6.89 - 1.27) - 2.25(3.54)$

50. $9.3(10.71 - 5.36 + 0.42) - 5.5(4.18)$

51. Alex multiplies 0.00762 by 0.215 and gets the product 0.0016383. Estimate the product to determine if Alex's answer is reasonable.

52. Catherine divides 0.0064 by 0.0125 and gets the quotient 5.12. Estimate the place value of the largest nonzero place value to see if Catherine's answer is reasonable.

53. Estelle goes to the store to buy a shirt for each of her six grandsons. She finds a style she likes that costs $23.45 each. Estelle has budgeted $120 for the shirts. Using estimation, determine if she has enough money to buy the 6 shirts.

54. Pedro goes to the candy store to buy chocolates for his wife, his mother, and his mother-in-law for Mother's Day. Each 3-lb box of chocolates costs $27.85. Pedro has $80 to buy the chocolates. Estimate the cost to see if Pedro has enough money to buy the boxes of chocolates.

© Angela Jones/Shutterstock.com

55. Estimate the perimeter of the triangle by rounding each measurement to the nearest yard.

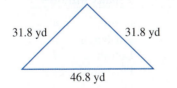

31.8 yd   31.8 yd

46.8 yd

56. Estimate the perimeter of the rectangle by rounding each side to the nearest tenth of an inch.

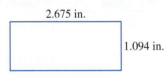

2.675 in.

1.094 in.

57. Rosalie buys her lunch three times a week at the deli near her office. She usually spends around $7.50 for a sandwich, chips, and a drink. In order to save money, she decides to pack the same lunch at home and bring it with her. She estimates that a sandwich will cost her $2.00, a bag of chips, 75¢, and a can of soda, 33¢. Assuming that Rosalie works 48 weeks in a year, what are her savings in bringing her lunch from home for the year?

58. Showers are a major user of hot water. In order to save water and the energy to heat it, many people are installing low-flow showerheads. While a standard showerhead allows a flow of 8 gallons per minute (gpm), low-flow showerheads allow 2.5 gpm and ultralow-flow showerheads allow only 1.6 gpm. Assume Loc takes a 5-minute shower every day. Calculate the amount of water saved in a year by using a low-flow showerhead instead of a standard one. Calculate the amount of water saved in a year by using an ultralow-flow showerhead instead of a low-flow one.

59. Matthew purchased the following items at a big box store in preparation for a fishing trip: fishing pole, $14.88; 4 jars of power bait at $2 each; a fishing rod holder, $12.99; a fishing vest, $16.88; 3 life vests at $12.97 each; 6 packages of snelled hooks at $0.88 each; 3 spools of trilene fishing line at $4.88 each; and 4 fishing lures at $4.88 each. Matthew has a coupon for $14.50 off his purchases. How much did he pay for the items?

60. The wholesale cost of shampoo is $1.11 per bottle, while the wholesale cost of conditioner is $0.89. The Fancy Hair Beauty Salon sells the shampoo for $8.49 a bottle and the conditioner for $8.19 a bottle. What is the net income on the sale of a case, 24 bottles, of each product?

Exercises 61–64 relate to the chapter application.

61. In the 2009 PGA Championship, Y. E. Yang won, and he received $1,350,000. Second place was won by Tiger Woods, who received $810,000. Two players tied for third, and each received $435,000. One player finished fifth, receiving $300,000, and four players tied for sixth place, each receiving $233,000. What was the average earning for the nine players?

62. The table gives the top five salaries in Major League Baseball for 2009.

Player	Team	Salary in millions
Alex Rodriquez	New York Yankees	$27.5
CC Sabathia	New York Yankees	$23.0
Johan Santana	New York Mets	$22.9
Miguel Cabrera	Detroit Tigers	$19.038
Derek Jeter	New York Yankees	$18.9

What was the average salary of these five players, rounded to the nearest dollar? What is the average salary to these five players, rounded to the nearest million dollars?

Exercises 63–64. The table gives a summary of the 2009 Stanley Cup Finals between the Pittsburgh Penguins and the Detroit Red Wings.

Game	Winner (Score)	Shots on Goal		Penalty Minutes	
1	Detroit (3–1)	Detroit	30	Detroit	4
		Pittsburgh	32	Pittsburgh	2
2	Detroit (3–1)	Detroit	26	Detroit	7
		Pittsburgh	32	Pittsburgh	21
3	Pittsburgh (4–2)	Detroit	29	Detroit	6
		Pittsburgh	21	Pittsburgh	4
4	Pittsburgh (4–2)	Detroit	39	Detroit	8
		Pittsburgh	31	Pittsburgh	10
5	Detroit (5–0)	Detroit	29	Detroit	14
		Pittsburgh	22	Pittsburgh	48
6	Pittsburgh (2–1)	Detroit	26	Detroit	4
		Pittsburgh	31	Pittsburgh	4
7	Pittsburgh (2–1)	Detroit	24	Detroit	4
		Pittsburgh	18	Pittsburgh	6

63. How many shots on goal per game did the Red Wings average over the entire series? How many shots on goal per game did the Penguins average?

64. How many minutes of penalty per game did the Red Wings average over the entire series? How many minutes of penalty per game did the Penguins average?

65. Explain the difference between evaluating $0.3(5.1)^2 + 8.3 \div 5$ and $[0.3(5.1)^2 + 8.3] \div 5$. How do the symbols indicate the order of the operations?

CHALLENGE

Insert grouping symbols to make each statement true.

66. $2 \cdot 8.1 \div 5 - 1 = 4.05$

67. $3.62 \div 0.02 + 72.3 \cdot 0.2 = 0.25$

68. $3.62 \div 0.02 + 8.6 \cdot 0.51 = 96.696$

69. $1.4^2 - 0.8^2 = 1.3456$

70. The average of 4.56, 8.23, 16.5, and a missing number is 8.2975. Find the missing number.

71. The body-mass index (BMI) is a technique used by health professionals to assess a person's excess fat and associated risk for heart disease, stroke, hypertension, and diabetes. The BMI is calculated by multiplying a person's weight (in pounds) by 705 and dividing the result by the square of the person's height in inches. The table gives the degree of risk of disease for various BMI values. Calculate your own BMI. Round your calculation to the nearest hundredth. Why are large BMI values associated with more risk for disease? Why are very low values of BMI also associated with greater risk?

BMI	Disease Risk
< 20.00	Moderate to very high
20.00 to 21.99	Low
22.00 to 24.99	Very low
25.00 to 29.99	Low
30.00 to 34.99	Moderate
35.00 to 39.99	High
40 or higher	Very high

SOURCE: *Lifetime Physical Fitness and Wellness* by Hoeger and Hoeger

Change to a decimal.

72. $\dfrac{13}{16}$

73. $\dfrac{27}{32}$

74. $\dfrac{29}{80}$

75. $\dfrac{58}{25}$

Change to a fraction or mixed number and simplify.

76. 0.68

77. 0.408

78. 2.435

79. 6.84

80. The sale price of a upright vacuum cleaner is $69.75. If the sale price was marked down $30.24 from the original price, what was the original price?

81. The price of a Panasonic 17" LCD TV is $588.88. The store is going to put it on sale at a discount of $98.50. What price should the clerk put on the TV for the sale?

GETTING READY FOR ALGEBRA
HOW & WHY

We solve equations that require more than one operation in the same way as equations with whole numbers and fractions.

> **To solve an equation that requires more than one operation**
> 1. Eliminate the addition or subtraction by performing the inverse operation.
> 2. Eliminate the multiplication by dividing both sides by the same number; that is, perform the inverse operation.

EXAMPLES A–C

DIRECTIONS: Solve.

STRATEGY: Isolate the variable by performing the inverse operations.

A. $2.6x + 4.8 = 25.6$

$2.6x + 4.8 - 4.8 = 25.6 - 4.8$ Eliminate the addition by subtracting 4.8 from both sides.

$2.6x = 20.8$

$\dfrac{2.6x}{2.6} = \dfrac{20.8}{2.6}$ Eliminate the multiplication by dividing both sides by 2.6.

$x = 8$

CHECK: $2.6(8) + 4.8 = 25.6$
$20.8 + 4.8 = 25.6$
$25.6 = 25.6$

The solution is $x = 8$.

B. $8.3 = 1.25x + 4.65$

$\begin{array}{r} 8.3 = 1.25x + 4.65 \\ \underline{-4.65 = - 4.65} \\ 3.65 = 1.25x \end{array}$ Subtract 4.65 from both sides.

$\dfrac{3.65}{1.25} = \dfrac{1.25x}{1.25}$ Divide both sides by 1.25.

$2.92 = x$

CHECK: $8.3 = 1.25(2.92) + 4.65$
$8.3 = 3.65 + 4.65$
$8.3 = 8.3$

The solution is $x = 2.92$.

C. Use the formula in Example C to find the Celsius temperature that corresponds to 122.9°F.

C. The formula relating temperature measured in degrees Fahrenheit and degrees Celsius is $F = 1.8C + 32$. Find the Celsius temperature that corresponds to 58.19°F.

First substitute the known values into the formula.

$$F = 1.8C + 32$$
$$58.19 = 1.8C + 32 \qquad \text{Substitute } F = 58.19.$$
$$58.19 - 32 = 1.8C + 32 - 32 \qquad \text{Subtract 32 from both sides.}$$
$$26.19 = 1.8C$$
$$\frac{26.19}{1.8} = \frac{1.8C}{1.8} \qquad \text{Divide both sides by 1.8.}$$
$$14.55 = C$$

Because $1.8(14.55) + 32 = 58.19$, the temperature is 14.55°C.

ANSWER TO WARM-UP C

C. The temperature is 50.5°C.

EXERCISES

Solve.

1. $2.5x - 7.6 = 12.8$

2. $0.25x - 7.3 = 0.95$

3. $1.8x + 6.7 = 12.1$

4. $15w + 0.006 = 49.506$

5. $4.115 = 2.15t + 3.9$

6. $10.175 = 1.25y + 9.3$

7. $0.03x - 18.7 = 3.53$

8. $0.08r - 5.62 = 72.3$

9. $7x + 0.06 = 2.3$

10. $13x + 14.66 = 15.7$

11. $3.65m - 122.2 = 108.115$

12. $22.5t - 657 = 231.75$

13. $5000 = 125y + 2055$

14. $3700 = 48w + 1228$

15. $60p - 253 = 9.5$

16. $17.8 = 0.66y + 7.9$

17. $8.551 = 4.42 + 0.17x$

18. $14 = 0.25w - 8.6$

19. $45 = 1.75h - 1.9$

20. $4000 = 96y + 1772.8$

21. $1375 = 80c + 873$

22. $7632 = 90t - 234$

23. The formula relating temperatures measured in degrees Fahrenheit and degrees Celsius is $F = 1.8C + 32$. Find the Celsius temperature that corresponds to 248°F.

24. Use the formula in Exercise 23 to find the Celsius temperature that corresponds to 45.5°F.

25. The formula for the balance of a loan D is $D + NP = B$, where P represents the monthly payment, N represents the number of payments made, and B represents the amount of money borrowed. Find the number of the monthly payments Gina has made if she borrowed $1764, has a remaining balance of $661.50, and pays $73.50 per month.

26. Use the formula in Exercise 25 to find the number of payments made by Morales if he borrowed $8442, has a balance of $3048.50, and makes a monthly payment of $234.50.

27. Catherine is an auto mechanic. She charges $36 per hour for her labor. The cost of parts needed is in addition to her labor charge. How many hours of labor result from a repair job in which the total bill (including $137.50 for parts) is $749.50? Write and solve an equation to determine the answer.

28. A car rental agency charges $28 per day plus $0.27 per mile to rent one of their cars. Determine how many miles were driven by a customer after a 3-day rental that cost $390.45. Write and solve an equation to determine the answer.

KEY CONCEPTS

SECTION 4.1 Decimals: Reading, Writing, and Rounding

Definitions and Concepts	Examples
Decimal numbers are another way of writing fractions and mixed numbers.	1.3 One and three tenths 2.78 Two and seventy-eight hundredths 5.964 Five and nine hundred sixty-four thousandths
To round a decimal to a given place value, • Mark the given place value. • If the digit on the right is 5 or more, add 1 to the marked place and drop all digits on the right. • If the digit on the right is 4 or less, drop all digits on the right.	Round 4.792 to the nearest tenth $\uparrow$ $4.792 \approx 4.8$ Round 4.792 to the nearest hundredth $\uparrow$ $4.792 \approx 4.79$
• Write zeros on the right if necessary so that the marked digit still has the same place value.	Round 563.79 to the nearest ten $\uparrow$ $563.79 \approx 560$

SECTION 4.2 Changing Decimals to Fractions; Listing in Order

Definitions and Concepts	Examples
To change a decimal to a fraction, • Read the decimal word name. • Write the fraction that has the same name. • Simplify.	0.45 is read "forty-five hundredths" $0.45 = \dfrac{45}{100}$ $\quad\ = \dfrac{9}{20}$
To list decimals in order, • Insert zeros on the right so that all the decimals have the same number of decimal places. • Write the numbers in order as if they were whole numbers. • Remove the extra zeros.	List 1.46, 1.3, and 1.427 in order from smallest to largest. $1.46\ \ = 1.460$ $1.3\ \ \ = 1.300$ $1.427 = 1.427$ $1.300 < 1.427 < 1.460$ So, $1.3 < 1.427 < 1.46$.

SECTION 4.3 Adding and Subtracting Decimals

Definitions and Concepts	Examples
To add or subtract decimals, • Write in columns with the decimal points aligned. Insert zeros on the right if necessary. • Add or subtract. • Align the decimal point in the answer with those above.	$2.67 + 10.9$ $\qquad$ $8.5 - 3.64$ $\begin{array}{r} 2.67 \\ +10.90 \\ \hline 13.57 \end{array}$ $\qquad$ $\begin{array}{r} 8.50 \\ -3.64 \\ \hline 4.86 \end{array}$

SECTION 4.4 Multiplying Decimals

Definitions and Concepts	Examples
To multiply decimals, • Multiply the numbers as if they were whole numbers. • Count the number of decimal places in each factor. The total of the decimal places is the number of decimal places in the product. Insert zeros on the left if necessary.	4.2×0.12 $\qquad$ 0.03×0.007 $\begin{array}{r} 4.2 \\ \times 0.12 \\ \hline 84 \\ 42 \\ \hline 0.504 \end{array}$ $\qquad$ $\begin{array}{r} 0.03 \\ \times 0.007 \\ \hline 0.00021 \end{array}$ (Three decimal places needed) $\qquad$ (Five decimal places needed)

SECTION 4.5 Multiplying and Dividing by Powers of 10; Scientific Notation

Definitions and Concepts	Examples
To multiply by a power of 10, • Move the decimal point to the right the same number of places as there are zeros in the power of 10.	$3.45\,(10,000) = 34,500$ (Move four places right.)
To divide by a power of 10, • Move the decimal point to the left the same number of places as there are zeros in the power of 10.	$3.45 \div 1000 = 0.00345$ (Move three places left.)
Scientific notation is a special way to write numbers as a product of a number between 1 and 10 and a power of 10.	$34,500 = 3.45 \times 10^4$ $0.00345 = 3.45 \times 10^{-3}$

SECTION 4.6 Dividing Decimals; Average, Median, and Mode

Definitions and Concepts	Examples

To divide decimals,
- If the divisor is not a whole number, move the decimal point in both the divisor and the dividend to the right as many places as necessary to make the divisor a whole number.
- Place the decimal point in the quotient above the decimal point in the dividend.
- Divide as if both numbers are whole numbers.
- Round as appropriate.

$$0.04\overline{)5.3} = 004\overline{)530.0}$$

$$\begin{array}{r} 132.5 \\ \hline 4 \\ \hline 13 \\ 12 \\ \hline 10 \\ 8 \\ \hline 20 \\ 20 \\ \hline 0 \end{array}$$

Move two places right.

Finding the average of a set of decimals is the same as for whole numbers:
- Add the numbers.
- Divide by the number of numbers.

Find the average of 5.8, 6.12, and 7.394.

$5.8 + 6.12 + 7.394 = 19.314$

$19.314 \div 3 = 6.438$

The average is 6.438.

Finding the median of a set of decimals is the same as for whole numbers:
- List the numbers in order from smallest to largest.
- If there is an odd number of numbers in the set, the median is the middle number.
- If there is an even number of numbers in the set, the median is the average of the middle two.

Find the median of 5.8, 6.12, 7.394, 9.6, and 7.01.

5.8, 6.12, 7.01, 7.394, 9.6
The median is 7.01.

Finding the mode of a set of decimals is the same as for whole numbers:
- Find the number or numbers that occur most often.
- If all the numbers occur the same number of times, there is no mode.

Find the mode of 5.8, 6.12, 7.03, 6.12, and 8.2.
The mode is 6.12.

SECTION 4.7 Changing Fractions to Decimals

Definitions and Concepts	Examples

To change a fraction to a decimal, divide the numerator by the denominator. Round as appropriate.

Change $\dfrac{5}{8}$ to a decimal.

$$\begin{array}{r} 0.625 \\ \hline 8\overline{)5.000} \\ 4\,8 \\ \hline 20 \\ 16 \\ \hline 40 \\ 40 \end{array}$$

Definitions and Concepts	Examples
The order of operations for decimals is the same as that for whole numbers: • Parentheses • Exponents • Multiplication/Division • Addition/Subtraction	$14.8 - 0.2(8.3 + 4.76)$ $= 14.8 - 0.2 (13.06)$ $= 14.8 - 2.612$ $= 12.188$
To estimate sums or differences, round all numbers to a specified place value.	$0.352 + 0.063 \approx 0.4 + 0.1$ ≈ 0.5
To estimate products, front round each number and multiply.	$(0.352)(0.063) \approx (0.4)(0.06)$ ≈ 0.024

REVIEW EXERCISES

SECTION 4.1

Write the word name.

1. 6.12

2. 0.843

3. 15.058

4. 0.0000027

Write the place value name.

5. Twenty-one and five hundredths

6. Four hundred nine ten-thousandths

7. Four hundred and four hundredths

8. One hundred twenty-five and forty-five thousandths

Exercises 9–11. Round the numbers to the nearest tenth, hundredth, and thousandth.

	Tenth	Hundredth	Thousandth
9. 34.7648			
10. 7.8736			
11. 0.467215			

12. The display on Mary's calculator shows 91.457919 as the result of a division exercise. If she is to round the answer to the nearest thousandth, what answer does she report?

SECTION 4.2

Change the decimal to a fraction or mixed number and simplify.

13. 0.76

14. 7.035

15. 0.00256

16. 0.0545

List the set of decimals from smallest to largest.

17. 0.95, 0.89, 1.01

18. 0.09, 0.093, 0.0899

19. 7.017, 7.022, 0.717, 7.108

20. 34.023, 34.103, 34.0204, 34.0239

Is the statement true or false?

21. 6.1774 < 6.1780

22. 87.0309 > 87.0319

SECTION 4.3

Add.

23.
```
  11.356
   0.67
  13.082
+  9.6
```

24.
```
  12.0678
   7.012
  56.0921
+  0.0045
```

Subtract.

25.
```
  22.0816
−  8.3629
```

26.
```
  54.084
−23.64936
```

27. Find the sum of 3.405, 8.12, 0.0098, 0.3456, 11.3, and 24.9345.

28. Find the difference of 56.7083 and 21.6249.

Find the perimeter of the following figures.

29.

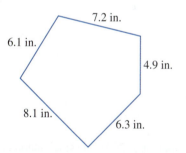

7.2 in.

6.1 in.

4.9 in.

8.1 in.

6.3 in.

30.

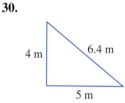

4 m

6.4 m

5 m

31. Hilda makes a gross salary (before deductions) of $6475 per month. She has the following monthly deductions: federal income tax, $1295; state income tax, $582.75; Social Security, $356.12; Medicare, $82.24, retirement contribution, $323.75; union dues, $45; and health insurance, $325.45. Find her actual take-home (net) pay.

32. Mary buys a new television that had a list price of $785.95 for $615.55. How much does she save from the list price?

SECTION 4.4

Multiply.

33.
```
  8.07
× 3.5
```

34.
```
  11.24
×  3.5
```

35.
```
  0.00678
×    3.59
```

36.
```
  12.057
×  8.08
```

37. Multiply: 0.074(2.004). Round to the nearest thousandth.

38. Multiply: (0.0098)(42.7). Round to the nearest hundredth.

39. Multiply: (0.03)(4.12)(0.015). Round to the nearest ten-thousandth.

40. Find the area of the rectangle.

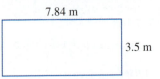

7.84 m

3.5 m

41. Millie selects an upholstery fabric that costs $52.35 per yd. How much will Millie pay for 23.75 yd? Round to the nearest cent.

42. Agnelo can choose any of the following ways to finance his new car. Which method is the least expensive in the long run?

$850 down and $401.64 per month for 5 years
$475 down and $443.10 per month for 54 months
$600 down and $495.30 per month for 4 years

SECTION 4.5

Multiply or divide.

43. $13.765 \div 10^3$

44. 7.023×10^6

45. 0.7321(100,000)

46. $9.503 \div 100$

Write in scientific notation.

47. 0.0078

48. 34.67

49. 0.0000143

50. 65,700.8

Write the place value name.

51. 7×10^7

52. 8.13×10^{-6}

53. 6.41×10^{-2}

54. 3.505×10^3

55. Home Run Sports buys 1000 softball bats for $37,350. What is the average price of a bat?

56. During the bear market of 2001, the stock market at one time was down $50 billion. Write this loss in scientific notation.

SECTION 4.6

Divide.

57. $0.3\overline{)0.0111}$

58. $75\overline{)40.5}$

59. $56.7 \div 0.32$

60. $0.17\overline{)0.01003}$

61. $0.456\overline{)0.38304}$

62. $6.3271 \div 2.015$

Divide and round to the nearest hundredth.

63. $4.7\overline{)332.618}$

64. $0.068\overline{)0.01956}$

65. Two hundred ten employees of Shepard Enterprises donated $13,745.50 to the United Way. To the nearest cent, what was the average donation?

66. Carol drove 375.9 miles on 12.8 gallons of gas. What is her mileage (miles per gallon)? Round to the nearest mile per gallon.

Find the average and median.

67. 4.56, 11.93, 13.4, 1.58, 8.09

68. 61.78, 50.32, 86.3, 95.04

69. 0.5672, 0.6086, 0.3447, 0.5555

70. 14.6, 18.95, 12.9, 23.5, 16.75

71. Tony goes shopping and buys a 3-oz jar of Nescafé Instant Vanilla Roast Coffee for $4.69. What is the unit price of the coffee, rounded to the nearest tenth of a cent?

72. The Metropolis Police Department reported the following number of robberies for the week:

Monday 12 Tuesday 21 Wednesday 5 Thursday 18 Friday 46 Saturday 67 Sunday 17

To the nearest tenth, what is the average number of robberies reported per day?

SECTION 4.7

Change the fraction or mixed number to a decimal.

73. $\dfrac{9}{16}$

74. $\dfrac{7}{20}$

75. $17\dfrac{47}{125}$

Change to a decimal rounded to the indicated place value.

76. $\dfrac{11}{37}$, tenth

77. $\dfrac{57}{93}$, hundredth

78. $\dfrac{54}{61}$, thousandth

Change to a decimal. Use the repeat bar.

79. $\dfrac{9}{13}$

80. $\dfrac{7}{48}$

81. The value of a share of Microsoft is $24\dfrac{9}{32}$. What is the value in decimal form? Round to the nearest hundredth.

82. In a shot put meet where the results were communicated by telephone, the longest put in Georgia was $60\dfrac{11}{16}$ ft. The longest put in Idaho was 60.799 ft. Which state had the winning put?

SECTION 4.8

Perform the indicated operations.

83. $0.65 + 4.29 - 2.71 + 3.04$

84. $13.8 \div 0.12 \times 4.03$

85. $(6.7)^2 - (4.4)(2.93)$

86. $(5.5)(2.4)^3 \div 9.9$

87. $(6.3)(5.08) - (2.6)(0.17) + 2.42$

88. $6.2(3.45 - 2.07 + 0.98) - 3.1(1.45)$

89. Jose did the following addition: $3.67 + 4.874 + 0.0621 + 0.00045 + 1.134 = 9.74055$. Estimate the sum by rounding each addend to the nearest tenth to determine if Jose's answer is reasonable.

90. Sally did the following subtraction: $0.0672 - 0.037612 = 0.0634388$. Estimate the difference by rounding each number to the nearest hundredth to determine if Sally's answer is reasonable.

6. Round to the nearest hundredth: 57.896

6. _____

7. Subtract: $87 - 14.837$

7. _____

8. Change to a mixed number with the fraction part simplified: 18.725

8. _____

9. Write in scientific notation: 0.000000723

9. _____

10. Write as an approximate decimal to the nearest thousandth: $\dfrac{17}{23}$

10. _____

11. Round to the nearest hundred: 72,987.505

11. _____

12. Perform the indicated operations: $2.277 \div 0.33 \times 1.5 + 11.47$

12. _____

13. Subtract: $\begin{array}{r} 305.634 \\ -208.519 \\ \hline \end{array}$

13. _____

14. Change to place value notation: 5.94×10^{-5}

14. _____

15. Write the place value name for nine thousand forty-five and sixty-five thousandths.

15. _____

16. Multiply: 0.000917(100,000)

16. _____

17. Write in scientific notation: 309,720

17. _____

18. Add: $17.98 + 1.467 + 18.92 + 8.37$

18. _____

19. Multiply: 34.4(0.00165)

19. _____

20. Divide: $72\overline{)0.02664}$

20. _____

21. For the first 6 months of 2005, the offering at St. Pius Church was $124,658.95, $110,750.50, $134,897.70, $128,934.55, $141,863.20, and $119,541.10. What was the average monthly offering? Round to the nearest cent.

21. _____

22. Add: $\begin{array}{r} 911.84 \\ 45.507 \\ 6003.62 \\ 7.2 \\ 35.78 \\ +\ 891.361 \\ \hline \end{array}$

22. _____

23. Grant buys 78 assorted flower plants from the local nursery. If the sale price is four plants for $3.48, how much does Grant pay for the flower plants?

23. _____

24. On April 15, 2005, Allen Iverson had the best scoring average per game, with 30.8. How many games had he played in if he scored a total of 2214 points (to the nearest game)?

24. _____

25. In baseball, the slugging percentage is calculated by dividing the number of total bases (a double is worth two bases) by the number of times at bat and then multiplying by 1000. What is the slugging percentage of a player who has 201 bases in 293 times at bat? Round to the nearest whole number.

25. _____

26. Harold and Jerry go on diets. Initially, Harold weighed 267.8 lb and Jerry weighed 209.4 lb. After 1 month of the diet, Harold weighed 254.63 lb and Jerry weighed 196.2 lb. Who lost the most weight and by how much?

26. _____

CLASS ACTIVITY 1

In Olympic diving, seven judges each rate a dive using a whole or half number between 0 and 10. The high and low scores are thrown out and the remaining scores are added together. (If a high or low score occurs more than once, only one is thrown out.) The sum is then multiplied by 0.6 and then by the difficulty factor of the dive to obtain the total points awarded.

1. A driver does a reverse $1\frac{1}{2}$ somersault with $2\frac{1}{2}$ twists, a dive with a difficulty factor of 2.9. She receives scores of 6.0, 6.5, 6.5, 7.0, 6.0, 7.5, and 7.0. What are the total points awarded for the dive?

2. Another diver also does a reverse $1\frac{1}{2}$ somersault with $2\frac{1}{2}$ twists. This diver receives scores of 7.5, 6.5, 7.5, 8.0, 8.0, 7.5, and 8.0. What are the total points awarded for the dive?

3. A cut-through reverse $1\frac{1}{2}$ somersault has a difficulty factor of 2.6. What is the highest number of points possible for this dive?

4. A diver receives 63.96 points for a cut-through reverse $1\frac{1}{2}$ somersault. If four of the five scores that counted toward her dive were 7.5, 8.0, 8.0, and 8.5, what was the fifth?

CLASS ACTIVITY 2

When people drive cars and ride in planes, the vehicle emits carbon dioxide, which is harmful to the atmosphere and contributes to global warming. There are websites that calculate the carbon footprint of various activities and allow the consumer to buy offsets. The offsets are projects that reduce carbon dioxide and "undo" the harm produced by driving or flying.

One such website is TerraPass.com. The website calculates the pounds of carbon dioxide (CO_2) produced for various activities. It then "rounds" up (never down to the next thousand pounds of CO_2. The cost of offsetting each 1000 pounds of CO_2 is $5.95.

1. Leonid, who drives about 15,000 miles per year, is considering buying a new car. He has narrowed his choices to three: a 2000 Ford Explorer 2WD, a 2005 Subaru Legacy wagon AWD, and a 2009 Toyota Prius. According to TerraPass, the vehicles produce the following numbers of pounds of carbon dioxide for every 15,000 miles driven: Ford Explorer, 18,341 lb; Subaru Legacy, 12,759 lb; Toyota Prius, 6244 lb. Calculate the cost of offsetting each car for one year.

2. Holly owns a 1998 Volvo S70. According to TerraPass, driving the Volvo 1000 miles generates 931.6 lb of CO_2. Calulate the cost for Holly to offset driving her Volvo 12,000 miles per year. Holly would like to reduce her carbon footprint by driving less. How much will it save her to reduce her driving to 10,000 miles per year?

3. Jasmine lives in Los Angeles and would like to take her family to visit her brother in Chicago—about 2000 miles away. If she flies, the round trip for each person will generate 1044 lb of CO_2. If she drives her 2005 Honda Civic, the trip one way will generate 1262 lb CO_2. Complete the table.

4. If Jasmine travels alone, what is the least expensive mode of travel (in terms of carbon footprint)? Is this true regardless of the number of people traveling? Explain mathematically

Number of Travelers	Mode of Travel	Total CO₂ Produced	Total Offset Cost
1	Plane		
1	Car		
2	Plane		
2	Car		
3	Plane		
3	Car		
4	Plane		
4	Car		

GROUP PROJECT (2–3 WEEKS)

The NFL keeps many statistics regarding its teams and players. Since quarterbacks play an important part in the overall team effort, much time and attention have been given to keeping statistics on quarterbacks. But all these statistics do not necessarily make it easy to decide which quarterback is the best. Consider the following statistics from the 2004 season.

Highest-Ranked Players in 2004 NFL Season

Player	Passes Attempted	Passes Completed	Yards Gained	Touchdown Passes	Interceptions
Daunte Culpepper, Minnesota	548	379	4717	39	11
Trent Green, Kansas City	556	369	4591	27	17
Peyton Manning, Indiana	497	336	4557	49	10
Jake Plummer, Denver	521	303	4089	27	20
Brett Favre, Green Bay	540	346	4088	30	17

1. Which quarterback deserved to be rated as the top quarterback of the year? Justify your answer.

The NFL has developed a rating system for quarterbacks that combines all of the statistics in the table and gives each quarterback a single numeric "grade" so they can easily be compared. While the exact calculations used by the NFL are complicated, Randolph Taylor of Las Positas College in Livermore, California, has developed the following formula that closely approximates the NFL ratings.

Let A = the number of passes attempted

C = the number of passes completed

Y = the number of yards gained passing

T = the number of touchdowns passed

I = the number of interceptions

$$\text{Rating} = \frac{5}{6}\left(\frac{C}{A}\cdot 100\right) + \frac{25}{6}\left(\frac{Y}{A}\right) + \frac{10}{3}\left(\frac{T}{A}\cdot 100\right) - \frac{25}{6}\left(\frac{I}{A}\cdot 100\right) + \frac{25}{12}$$

2. Use the rating formula to calculate ratings for the quarterbacks in the table. Use your calculator and do not round except at the end, rounding to the nearest hundredth.

4. According to your calculations, who was the best quarterback for the 2004 season?

6. What are the drawbacks to using the rating as the sole measure of a quarterback's performance?

3. Explain why everything in the formula is added except $\frac{25}{6}\left(\frac{I}{A}\cdot 100\right)$.

5. In the 2004 season, Clinton Portis of the Washington Redskins made two attempts at a pass and completed one for 15 yards and a touchdown. He had no interceptions. Calculate his rating and comment on how he compares with the quarterbacks in the table.

7. (Optional) Have your group compile a list of the five all-time best quarterbacks. Find statistics for each of the quarterbacks on your list (use almanacs or the web) and compute their ratings. Comment on your results.

GOOD ADVICE FOR
Studying

TAKING LOW-STRESS TESTS

© Fuse/Jupiter Images

Before the Test

- One-half hour before the test—find a quiet place to physically and mentally relax.
- Arrive at the classroom in time to arrange your tools—pencils, eraser, calculator, scratch paper.
- Remind yourself that you are prepared and will do well—continue to breathe deeply.

When the Test Starts

- Begin with a memory dump—write down formulas, definitions, and any reminders to yourself.
- Read the entire test. Pay attention to directions and point values.
- Begin by doing the problems that you are absolutely sure you can do. This allows your mind to relax and stay focused.
- Next tackle the problems with the highest point values.
- If you are not sure about a problem, mark it and come back to it at the end. Do not allow yourself to spend too much time on any one problem.
- After going through the entire test, go back to any skipped problems. Even if you can't do the problem, write down as many steps as possible. You could get partial credit if you can show your instructor that you can do part of the problem.

In the Last Ten Minutes

- Check that you have answered (or at least attempted) each problem.
- Check that your answers are in the proper format. Applied problems should have sentence answers, including appropriate units.
- Check that you have completely followed the directions.
- Check your math for arithmetic mistakes.

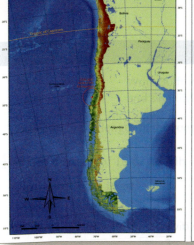

Ratio and Proportion

APPLICATION

From the earliest times, humans have drawn maps to represent the geography of their surroundings. Some maps depict features encountered on a journey, like rivers and mountains. The most useful maps incorporate the concept of scale, or proportion. Simply put, a scaled map accurately preserves relative distances. So if the distance from one city to another is twice the distance from the city to a river in real life, the distance between the cities is twice the distance from the city to a river on the map as well.

The scale of a map depends on how large an area the map covers. In the United States, the scale is often stated as "1 inch represents _____ ." For a street map of a city, the scale could be "1 inch represents 600 yards." The map of an entire state could have a scale of "1 inch represents 45 miles." The map of an entire country could have a scale of "1 inch represents 500 miles." Specific information about the scale is usually given in a corner of the map.

GROUP ACTIVITY

Go to the library and find maps with five different scales. Summarize your findings in the table.

Map Subject	Scale	Width of map (inches)	Width of Map Subject (miles)

67. From a consumer's viewpoint, explain why it is not always an advantage for costs of goods and services to be proportional.

![CHALLENGE]

68. In 1982, approximately 25 California condors were alive. This low population was the result of hunting, habitat loss, and poisoning. The U.S. Fish and Wildlife Service instituted a program that resulted in there being 73 condors alive in 1992. If this increase continues proportionally, predict how many condors will be alive in 2017.

© Kim Worrell/Shutterstock.com

69. The tachometer of a sports car shows the engine speed to be 2800 revolutions per minute. The transmission ratio (engine speed to drive shaft speed) for the car is 2.5 to 1. Find the drive shaft speed.

70. Two families rented a mountain cabin for 19 days at a cost of $1905. The Santini family stayed for 8 days and the Nguyen family stayed for 11 days. How much did it cost each family? Round the rents to the nearest dollar.

MAINTAIN YOUR SKILLS

71. Round 167.8519 to the nearest hundredth and to the nearest hundred.

72. Round 62.3285 to the nearest hundredth and to the nearest thousandth.

73. Compare the decimals 0.01399 and 0.011. Write the result as an inequality.

74. Compare the decimals 0.06 and 0.15. Write the result as an inequality.

75. Lean ground beef is on sale for $1.49 per pound. How much will Mrs. Diado pay for 12 pounds?

76. A barrel of liquid weighs 429.5 lb. If the barrel weighs 22.5 lb and the liquid weighs 7.41 lb per gallon, how many gallons of liquid are in the barrel, to the nearest gallon?

Change each decimal to a simplified fraction.

77. 0.635

78. 0.01125

Change each fraction to a decimal rounded to the nearest thousandth.

79. $\dfrac{345}{561}$

80. $\dfrac{33}{350}$

KEY CONCEPTS

SECTION 5.1 Ratio and Rate

Definitions and Concepts	Examples
A ratio is a comparison of two like measurements by division.	The ratio of the length of a room to its width is $\dfrac{12 \text{ ft}}{9 \text{ ft}} = \dfrac{12}{9} = \dfrac{4}{3}$.
A rate is a comparison of two unlike measurements by division.	The rate of a biker who rides 21 mi in 2 hr is $\dfrac{21 \text{ mi}}{2 \text{ hr}}$.
A unit rate is a rate with a denominator of one unit.	The unit rate of a biker who rides 21 mi in 2 hr is $\dfrac{21 \text{ mi}}{2 \text{ hr}} = \dfrac{10.5 \text{ mi}}{1 \text{ hr}} = 10.5$ mph.

SECTION 5.2 Solving Proportions

Definitions and Concepts	Examples
A proportion is a statement that two ratios are equal.	$\dfrac{6}{12} = \dfrac{1}{2}$ is a proportion.
A proportion is true when the cross products are equal.	$\dfrac{6}{12} = \dfrac{1}{2}$ is true because $6(2) = 12(1)$.
A proportion is false when the cross products are not equal.	$\dfrac{3}{5} = \dfrac{5}{8}$ is false because $3(8) \neq 5(5)$.
To solve a proportion, • Cross multiply. • Do the related division problem to find the missing number.	Solve: $\dfrac{3}{x} = \dfrac{15}{43}$ $3 \cdot 43 = 15x$ $129 = 15x$ $129 \div 15 = x$ $8.6 = x$

SECTION 5.3 Applications of Proportions

Definitions and Concepts	Examples
To solve word problems involving proportions, • Make a table to organize the information. • Write a proportion from the table. • Solve the proportion. • Write the solution, including appropriate units.	If 3 cans of cat food sell for $3.69, how much will 8 cans cost?

	Case I	Case II
Cans	3	8
Cost	$3.69	C

$$\frac{3}{3.69} = \frac{8}{C}$$
$$3C = 29.52$$
$$C = 29.52 \div 3$$
$$C \approx 9.84$$

So 8 cans of cat food will cost $9.84.

REVIEW EXERCISES

SECTION 5.1

Write as a ratio in simplified form.

1. 18 to 90

2. 9 to 54

3. 12 m to 10 m

4. 12 km to 9 km

5. 3 dollar to 80 nickels
   (compare in nickels)

6. 660 ft to 1 mi
   (compare in feet)

7. 16 in. to 2 ft
   (compare in inches)

8. 3 ft to 3 yd
   (compare in feet)

Write a rate and simplify.

9. 9 people to 10 chairs

10. 23 miles to 3 hikes

11. 40 applicants to 15 jobs

12. 10 cars to 6 households

13. 210 books to 45 students

14. 36 buttons to 24 bows

15. 765 people to 27 committees

16. 8780 households to 6 cable companies

Write as a unit rate.

17. 50 mi to 2 hr

18. 60 mi to 4 minutes

19. 90¢ per 10 lb of potatoes

20. $1.17 per 3 lb of broccoli

Write as a unit rate. Round to the nearest tenth.

21. 825 mi per 22 gal

22. 13,266 km per 220 gal

23. $2.10 for 6 croissants

24. $3.75 for 15 oz of cereal

25. One section of the country has 3500 TV sets per 1000 households. Another section has 500 TV sets per 150 households. Are the rates of the TV sets to the number of households the same in both parts of the country?

26. In Pineberg, there are 5000 automobiles per 3750 households. In Firville, there are 6400 automobiles per 4800 households. Are the rates of the number of automobiles to the number of households the same?

SECTION 5.2

True or false?

27. $\dfrac{15}{7} = \dfrac{75}{35}$

28. $\dfrac{2}{3} = \dfrac{26}{39}$

29. $\dfrac{25}{9} = \dfrac{8}{3}$

30. $\dfrac{16}{25} = \dfrac{10}{15}$

31. $\dfrac{31}{35} = \dfrac{6.125}{7}$

32. $\dfrac{9.375}{3} = \dfrac{25}{8}$

Solve.

33. $\dfrac{1}{4} = \dfrac{r}{44}$

34. $\dfrac{1}{3} = \dfrac{s}{18}$

35. $\dfrac{14}{t} = \dfrac{42}{27}$

36. $\dfrac{8}{v} = \dfrac{2}{5}$

37. $\dfrac{f}{9} = \dfrac{3}{45}$

38. $\dfrac{g}{2} = \dfrac{2}{12}$

39. $\dfrac{16}{24} = \dfrac{r}{16}$

40. $\dfrac{s}{10} = \dfrac{15}{16}$

41. $\dfrac{21}{25} = \dfrac{t}{7}$

42. $\dfrac{7}{5} = \dfrac{w}{7}$

Solve. Round to the nearest tenth.

43. $\dfrac{9}{11} = \dfrac{a}{13}$

44. $\dfrac{7}{6} = \dfrac{6}{b}$

45. $\dfrac{16}{7} = \dfrac{c}{12}$

46. $\dfrac{16}{5} = \dfrac{12}{d}$

47. A box of Arm and Hammer laundry detergent that is sufficient for 80 loads of laundry costs $9.99. What is the most that a store brand of detergent can cost if the box is sufficient for 50 loads and is more economical to use than Arm and Hammer? To find the cost, solve the proportion $\dfrac{\$9.99}{80} = \dfrac{c}{50}$, where c represents the cost of the store brand.

48. Available figures show that it takes the use of 18,000,000 gasoline-powered lawn mowers to produce the same amount of air pollution as 3,000,000 new cars. Determine the number of gasoline-powered lawn mowers that will produce the same amount of air pollution as 50,000 new cars. To find the number of lawn mowers, solve the proportion $\dfrac{18,000,000}{3,000,000} = \dfrac{L}{50,000}$, where L represents the number of lawn mowers.

SECTION 5.3

49. For every 2 hr a week that Merle is in class, she plans to spend 5 hr a week doing her homework. If she is in class 15 hr each week, how many hours will she plan to be studying each week?

50. If 16 lb of fertilizer will cover 1500 ft² of lawn, how much fertilizer is needed to cover 2500 ft²?

51. Juan must do 36 hr of work to pay for the tuition for three college credits. If Juan intends to sign up for 15 credit hours in the fall, how many hours will he need to work to pay for his tuition?

52. In Exercise 51, if Juan works 40 hr per week, how many weeks will he need to work to pay for his tuition? (Any part of a week counts as a full week.)

53. Larry sells men's clothing at the University Men's Shop. For $120 in clothing sales, Larry makes $15. How much does he make on a sale of $350 worth of clothing?

54. Dissolving 1.5 lb of salt in 1 gal of water makes a brine solution. At this rate, how many gallons of water are needed to made a brine solution with 12 lb of salt?

Check your understanding of the language of basic mathematics. Tell whether each of the following statements is true (always true) or false (not always true). For each statement you judge to be false, revise it to make a statement that is true.

Answers

1. A fraction can be regarded as a ratio.

1. _____

2. A ratio is a comparison of two numbers or measures usually written as a fraction.

2. _____

3. $\dfrac{18 \text{ miles}}{1 \text{ gallon}} = \dfrac{54 \text{ miles}}{3 \text{ hours}}$

3. _____

4. To solve a proportion, we must know the values of only two of the four numbers.

4. _____

5. If $\dfrac{8}{5} = \dfrac{t}{2}$, then $t = \dfrac{5}{16}$.

5. _____

6. In a proportion, two ratios are equal.

6. _____

7. Three feet and 1 yard are unlike measures.

7. _____

8. Ratios that are rates compare unlike units.

8. _____

9. To determine whether a proportion is true or false, the ratios must have the same units.

9. _____

10. If a fir tree that is 18 ft tall casts a shadow of 17 ft, how tall is a tree that casts a shadow of 25 ft? The following table can be used to solve this problem.

10. _____

	First Tree	**Second Tree**
Height	17	18
Shadow	x	25

_____

TEST

Answers

1. Write a ratio to compare 12 yards to 15 yards.

1. _____

2. On a test, Ken answered 20 of 32 questions correctly. At the same rate, how many would he answer correctly if there were 72 questions on a test?

2. _____

3. Solve the proportion: $\dfrac{4.8}{12} = \dfrac{0.36}{w}$

3. _____

4. Is the following proportion true or false? $\dfrac{16}{35} = \dfrac{24}{51}$

4. _____

5. Solve the proportion: $\dfrac{13}{36} = \dfrac{y}{18}$

5. _____

6. Is the following proportion true or false? $\dfrac{9 \text{ in.}}{2 \text{ ft}} = \dfrac{6 \text{ in.}}{16 \text{ in.}}$

6. _____

7. If Mary is paid $49.14 for 7 hr of work, how much should she be paid for 12 hr of work?

7. _____

8. Write a ratio to compare 8 hr to 3 days (compare in hours).

8. _____

9. There is a canned food sale at the supermarket. A case of 24 cans of peas is priced at $19.68. At the same rate, what is the price of 10 cans?

9. _____

10. If 40 lb of beef contains 7 lb of bones, how many pounds of bones may be expected in 100 lb of beef?

10. _____

11. Solve the proportion: $\dfrac{0.4}{0.5} = \dfrac{0.5}{x}$

11. _____

12. A charter fishing boat has been catching an average of 3 salmon for every 4 people they take fishing. At that rate, how many fish will they catch if over a period of time they take a total of 32 people fishing?

12. _____

13. On a trip home, Jennie used 12.5 gal of gas. The trip odometer on her car registered 295 mi for the trip. She is planning a trip to see a friend who lives 236 mi away. How much gas will Jennie need for the trip?

13. _____

14. Solve the proportion: $\dfrac{a}{8} = \dfrac{4.24}{6.4}$

14. _____

15. Is the following a rate? $\dfrac{130 \text{ mi}}{2 \text{ hr}}$

15. _____

16. If a 20-ft tree casts a 15-ft shadow, how long a shadow is cast by a 14-ft tree?

16. _____

17. What is the population density of a town that is 150 square miles and has 5580 people? Reduce to a 1-square-mile comparison.

17. _____

18. Solve the proportion and round your answer to the nearest hundredth: $\dfrac{4.78}{y} = \dfrac{32.5}{11.2}$

18. _____

19. A landscape firm has a job that it takes a crew of three $4\dfrac{1}{2}$ hr to do. How many of these jobs could the crew of three do in 117 hrs?

19. _____

20. The ratio of males to females in a literature class is 3 to 5. How many females are in a class of 48 students?

20. _____

CLASS ACTIVITY 1

Inflation is the phenomenon in which the prices of goods and services increase, so that the same item costs more this year than last. Economically speaking, this is normal. An official measure of inflation is calculated by the U.S. Department of Labor and it is called the consumer price index (CPI). Over 2 million union wage earners, 47.8 million Social Security beneficiaries, 4.1 million military and federal civil service retirees, and 22.4 million food stamp recipents have their bene-fits tied to the CPI. The table gives CPI values for selected years.

Year	1915	1925	1935	1945	1955	1965	1975	1985	1995	2005	2008
CPI	10.1	17.5	13.7	18.0	26.8	31.5	53.8	107.6	152.4	195.3	215.3

SOURCE: U.S. Department of Labor, Bureau of Labor Statistics

The CPI provides a method of calculating the price of an object or service in one year if the price is known for another year. This is done by using the proportion

$$\frac{\text{price in year } A}{\text{price in year } B} = \frac{\text{CPI in year } A}{\text{CPI in year } B}$$

1. Joe started work for the government in 1945 at an annual salary of $5200. What salary would be equivalent to this in 2005?

2. Buster bought an engagement ring in 1925. His granddaughter had the ring appraised in 1995 at $1400. What was the original price?

3. If a loaf bread costs $2.49 in 2008, what would it have cost in 1915?

4. Tom bought a boat in 1985 for $16,000. A fire destroyed it completely in 2008. Tom's insurance policy specifies that all losses will be replaced. How much does the insurance pay for Tom to replace his boat?

5. Which ten-year period showed a decrease in the CPI? What happened to explain this?

6. Which ten-year period showed the greatest increase? What happened to explain this?

CLASS ACTIVITY 2

1. In geometry, similar triangles are triangles that have exactly the same angle measures. Pick out a pair of similar triangles from the following.

A.

B.

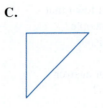

C.

D.

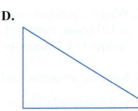

2. Similar triangles have the same shape but are different sizes. They have the property that corresponding sides are proportional. This means that the following proportion is always true:

$$\frac{\text{side 1 in Triangle } A}{\text{side 1 in Triangle } B} = \frac{\text{side 2 in Triangle } A}{\text{side 2 in Triangle } B}$$

Consider the two triangles, A and B.

A.

10 cm

A

7 cm

B.

16 cm

B

x cm

We consider the top side as side 1, and the right side as side 2.

To find the length of side 2 in triangle B, solve the proportion $\dfrac{10}{16} = \dfrac{7}{x}$.

3. Find the unknown sides in triangle D.

A.

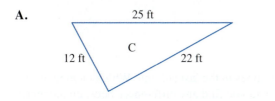

25 ft

C

12 ft

22 ft

B.

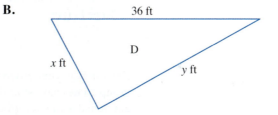

36 ft

D

x ft

y ft

4. It's a sunny day and a math class has been sent outside to determine the height of the flagpole in front of the administration building. They have a 2-m tape measure. They begin by measuring the shadow of the flagpole, which is 475 cm. Then they measure one of the class members and his shadow. He is 215 cm tall and his shadow is 142 cm.

   a. Draw similar triangles for this situation, and label. How do you know your triangles are similar?

   b. How tall is the flagpole?

The human body is the source of many common proportions. Artists have long studied the human figure in order to portray it accurately. Your group will be investigating how each member compares to the standard and how various artists have used the standards.

Most adult bodies can be divided into eight equal portions. The first section is from the top of the head to the chin. Next is from the chin to the bottom of the sternum. The third section is from the sternum to the navel, and the fourth is from the navel to the bottom of the torso. The bottom of the torso to the bottom of the knee is two sections long, and the bottom of the knee to the bottom of the foot is the last two sections. (Actually, these last two sections are a little short. Most people agree that the body is actually closer to 7.5 sections, but because this is hard to judge proportionally, we use eight sections and leave the bottom one short.)

Complete the following table for each group member.

Section	Length (in cm)	Ratio of Section to Head (Actual)	Ratio of Section to Head (Expected)
Head			
Chin to sternum			
Sternum to navel			
Navel to torso bottom			
Torso bottom to knee			
Knee to foot			

Explain how your group arrived at the values in the last column. Which member of the group comes closest to the standards? Did you find any differences between the males and females in your group? Either draw a body using the standard proportions, or get a copy of a figure from a painting and analyze how close the artist came to the standards.

A slightly different method of dividing the upper torso is to start at the bottom of the torso and divide into thirds at the waist and the shoulders. In this method, there is a pronounced difference between males and females. In females, the middle third between the waist and shoulders is actually shorter than the other two. In males, the bottom third from waist to bottom of the torso is shorter than the others. For each member of your group, fill out the following table.

Section	Length (in cm)	Ratio of Section to Entire Upper Torso (Actual)	Ratio of Section to Entire Upper Torso (Expected)
Head to shoulders			
Shoulders to waist			
Waist to bottom of torso			

Explain how your group arrived at the values in the last column (these will depend on gender). Which member of your group comes closest to the standards? Either draw a body using the standard proportions, or get a copy of a figure from a painting and analyze how close the artist came to the standards.

Children have different body proportions than adults, and these proportions change with the age of the child. Measure three children who are the same age. Use their head measurement as one unit, and compute the ratio of head to entire body. How close are the three children's ratios to each other? Before the Renaissance, artists usually depicted children as miniature adults. This means that the proportions fit those in the first table rather than those you just discovered. Find a painting from before the Renaissance that contains a child. Calculate the child's proportions and comment on them. Be sure to reference the painting you use.

GOOD ADVICE FOR *Studying*

EVALUATING YOUR TEST PERFORMANCE

© Getty Images/Photos.com/Jupiter Images

When you get your test back, use it to improve your future performance.

Make Sure You Have the Correct Answer to Every Problem

- If the instructor reviews the test in class, take notes.
- After class, go back and work every problem you missed.
- See your instructor during office hours or go to the tutoring center.

Careless and Procedural Errors

- Make sure to spend the last 10 minutes of the test checking accuracy–use a calculator if permitted.
- Pay particular attention to following the directions completely.

Time Management Errors

- Subtract 10 minutes from the available time. Reserve these for checking at the end.
- Divide the remaining time into blocks for each page or section of the test.
- Assign more time to problems worth more points.

Application Errors—Confusion About Which Procedure to Use

- For each different type of problem covered by the test, make a 3 × 5 card with the directions and a sample problem.
- On the back of the card, write the proper procedure for solving the problem.
- Mix up the order of the cards, and review until you can link the procedure to the problem.

Concept Errors–Failure to Fully Grasp an Underlying Concept

- Make an appointment with your instructor for extra help during office hours.
- Go to the tutoring center. Many have DVDs available for extra help.
- Re-read your text. Do the homework again.
- Use online resources that come with your text.

© iStockphoto.com/Don Bayley

Percent

APPLICATION

The price we pay for everyday items such as food and clothing is theoretically simple. The manufacturer of the item sets the price based on how much it costs to produce and adds a small profit. The manufacturer then sells the item to a retail store, which in turn marks it up and sells it to you, the consumer. But as you know, it is rarely as simple as that. The price you actually pay for an item also depends on the time of year, the availability of raw materials, the amount of competition among manufacturers of comparable items, the economic circumstances of the retailer, the geographic location of the retailer, and many other factors.

GROUP DISCUSSION

Select a common item whose price is affected by the following factors:

1. Time of year
2. Economic circumstances of the retailer
3. Competition of comparable products

Discuss how the factor varies and how the price of the item is affected. For each factor, make a plausible bar graph that shows the change in price as the factor varies. (You may estimate price levels.)

The Meaning of Percent

VOCABULARY

When ratios are used to compare numbers, the denominator is called the

base unit. In comparing 70 to 100 $\left(\text{as the ratio } \dfrac{70}{100}\right)$, 100 is the base unit.

The **percent comparison,** or just the **percent,** is a ratio with a base unit of 100.

The percent $\dfrac{70}{100} = (70)\left(\dfrac{1}{100}\right)$ is usually written 70%. The symbol % is

read "percent," and % $= \dfrac{1}{100} = 0.01$.

HOW & WHY

■ **OBJECTIVE** Write a percent to express a comparison of two numbers.

The word *percent* means "by the hundred." It is from the Roman word *percentum*. In Rome, taxes were collected by the hundred. For example, if you had 100 cattle, the tax collector might take 14 of them to pay your taxes. Hence, 14 per 100, or 14 percent, would be the tax rate.

Look at Figure 6.1 to see an illustration of the concept of "by the hundred." The base unit is 100, and 34 of the 100 parts are shaded. The ratio of shaded parts to total parts is $\dfrac{34}{100} = 34 \times \left(\dfrac{1}{100}\right) = 34\%$. We say that 34% of the unit is shaded.

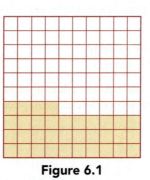

Figure 6.1

Figure 6.1 also illustrates that if the numerator is smaller than the denominator, then not all of the base unit is shaded, and hence the comparison is less than 100%. If the numerator equals the denominator, the entire unit is shaded and the comparison is 100%. If the numerator is larger than the denominator, more than one entire unit is shaded, and the comparison is more than 100%.

Any ratio of two numbers can be converted to a percent, even when the base unit is not 100. Compare 11 to 20. The ratio is $\dfrac{11}{20}$. Now find the equivalent ratio with a denominator of 100.

$$\frac{11}{20} = \frac{55}{100} = 55 \cdot \frac{1}{100} = 55\%.$$

If the equivalent ratio with a denominator of 100 cannot be found easily, solve as a proportion. See Example F.

> ### To find the percent comparison of two numbers
> 1. Write the ratio of the first number to the base number.
> 2. Find the equivalent ratio with denominator 100.
> 3. $\dfrac{\text{numerator}}{100} = \text{numerator} \cdot \dfrac{1}{100} = \text{numerator} \%$

EXAMPLES A–C

DIRECTIONS: Write the percent of each region that is shaded.

STRATEGY: (1) Count the number of parts in each unit. (2) Count the number of parts that are shaded. (3) Write the ratio of these as a fraction and build the fraction to a denominator of 100. (4) Write the percent using the numerator in step 3.

A. What percent of the unit is shaded?

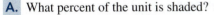

1. 100 parts in the region
2. 59 parts are shaded.
3. $\dfrac{59}{100}$
4. 59%

So 59% of the region is shaded.

B. What percent of the region is shaded?

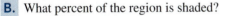

1. 8 parts in the region
2. 8 parts are shaded.
3. $\dfrac{8}{8} = \dfrac{100}{100}$
4. 100%

So 100% of the region is shaded.

WARM-UP

A. What percent of the unit is shaded?

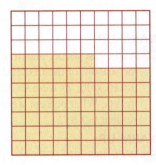

WARM-UP

B. What percent of the region is shaded?

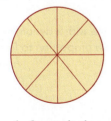

WARM-UP

C. What percent of the region is shaded?

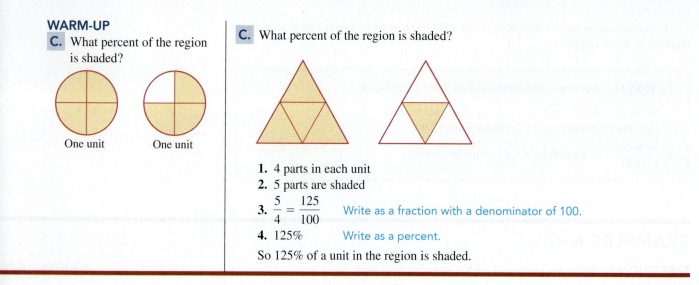

One unit   One unit

C. What percent of the region is shaded?

1. 4 parts in each unit
2. 5 parts are shaded
3. $\dfrac{5}{4} = \dfrac{125}{100}$ Write as a fraction with a denominator of 100.
4. 125%   Write as a percent.

So 125% of a unit in the region is shaded.

EXAMPLES D–H

DIRECTIONS: Write the percent for the comparison.

Write the comparison in fraction form. Build the fraction to hundredths or solve a proportion and write the percent using the numerator.

WARM-UP

D. At the last soccer match of the season, of the first 100 tickets sold, 77 were student tickets. What percent were student tickets?

D. At a football game, 22 children are among the first 100 fans to enter. What percent of the first 100 fans are children?

$$\dfrac{22}{100} = 22 \cdot \dfrac{1}{100} = 22\%$$ The comparison of children to first 100 fans is 22 to 100. Write the fraction and change to a percent.

So 22% of the first 100 fans are children.

WARM-UP

E. Write the ratio of 7 to 5 as a percent.

E. Write the ratio of 8 to 5 as a percent.

$$\dfrac{8}{5} = \dfrac{160}{100}$$ Write the ratio and build to a fraction which has a denominator of 100.

$$= 160 \cdot \dfrac{1}{100}$$

$$= 160\%$$ Change to a percent.

So the ratio of 8 to 5 is 160%.

WARM-UP

F. Write the ratio of 10 to 12 as a percent.

F. Write the ratio of 15 to 21 as a percent.

$$\dfrac{15}{21} = \dfrac{R}{100}$$ Because we cannot build the fraction to one with a denominator of 100 using whole numbers, we write a proportion to find the percent.

$$15(100) = 21R$$ Cross multiply.

$$\dfrac{1500}{21} = R$$

$$71\dfrac{3}{7} = R$$

So, $\dfrac{15}{21} = \dfrac{71\frac{3}{7}}{100}$

$\qquad\quad = 71\frac{3}{7} \cdot \dfrac{1}{100}$

$\qquad\quad = 71\frac{3}{7}\%$

So the ratio of 15 to 21 is $71\frac{3}{7}\%$.

CALCULATOR EXAMPLE:

G. Compare 208 to 1280 as a percent.

$\dfrac{208}{1280} = \dfrac{R}{100}$ Write as a proportion.

$1280R = 208(100)$

$\qquad R = 208(100) \div 1280$ Solve.

$\qquad R = 16.25$ Evaluate using a calculator.

So 208 is 16.25% of 1280.

H. During a campaign to lose weight, the 180 participants lost a total of 4158 lb. If they weighed collectively 37,800 lb before the campaign, what percent of their weight was lost?

$\dfrac{4158}{37,800} = \dfrac{11}{100}$ Write the ratio comparison and simplify.

$\qquad\quad = 11 \cdot \dfrac{1}{100}$

$\qquad\quad = 11\%$

So 11% of the total weight of the 180 dieters was lost during the campaign.

EXERCISES 6.1

■ **OBJECTIVE**   Write a percent to express a comparison of two numbers. (See page 408.)

A   *What percent of each of the following regions is shaded?*

1.

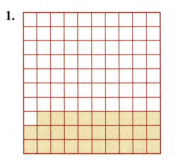

2.

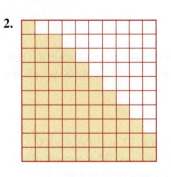

3.

4.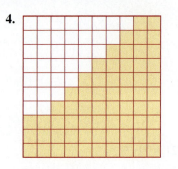

5.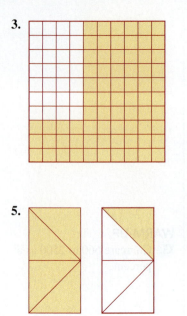

6.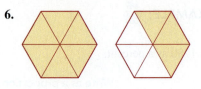

Write an exact percent for these comparisons.

7. 62 of 100

8. 52 per 100

9. 32 to 100

10. 37 to 100

11. 28 per 50

12. 17 per 50

13. 12 of 25

14. 21 to 25

15. 11 per 20

16. 13 per 20

B

17. 13 to 10

18. 450 to 120

19. 313 of 313

20. 92 to 92

21. 30 to 12

22. 44 to 16

23. 85 to 200

24. 65 to 200

25. 15 per 40

26. 83 per 500

27. 70 per 80

28. 98 per 80

29. 180 to 480

30. 29 to 30

31. 68 to 102

32. 8 to 15

C

33. It is estimated that 2% of the U.S. population has red hair. This indicates that _____ out of 100 people are redheads.

34. In a recent election there was a 73% turnout of registered voters. This indicates that _____ out of 100 registered voters turned out to vote.

35. In a recent mail-in election, 82 out of every 100 eligible voters cast their ballots. What percent of the eligible voters exercised their right to vote?

36. Of the people who use mouthwash daily, 63 out of 100 report fewer cavities. Of every 100 people who report, what percent do not report fewer cavities?

Write an exact percent for these comparisons; use fractions when necessary.

37. 129 to 400

38. 204 to 480

39. 175 to 50

40. 213 to 15

41. 115 to 15

42. 64 to 900

43. If a luxury tax is 11 cents per dollar, what percent is this?

44. For every $100 spent on gasoline in Nebraska, the state receives $9.80 tax. What percent of the price of gasoline is the state tax?

45. A bank pays $1.95 interest per year for every $100 in savings. What is the annual interest rate?

46. James has $500 in his savings account. Of that amount, $35 is interest that was paid to him. What percent of the total amount is the interest?

47. Last year, Mr. and Mrs. Johanson were informed that the property tax rate on their home was $1.48 per $100 of the house's assessed value. What percent is the tax rate?

48. Beginning in the early 1970s, women in the armed forces were treated the same as men with respect to training, pay, and rank. As a result, the number of women in the armed forces nearly tripled over the levels of the late 1960s. In the year 2001, about 7.5 out of every 50 officers were women. Express this as a percent.

49. In 1980, the rate of arrests for burglary for juveniles aged 10 to 17 was about 800 arrests per 100,000 juveniles. In 2007, the rate was about 225 per 100,000. What percent of the juveniles population in 1980 and what percent in 2007 were arrested for burglary?

50. In 2007, what percent of the juvenile population was not arrested for burglary? (See Exercise 49.)

51. According to the U. S. Census Bureau, in 2007 one out of every three women aged 25 to 29 had a bachelor's degree or higher. What percent of women 25 to 29 had a bachelor's degree?

52. According to the U. S. Census Bureau, in 2007, 13 out of every 50 men aged 25 to 29 had a bachelor's degree or higher. What percent of men 25 to 29 had a bachelor's degree?

Exercises 53–56 are related to the chapter application.

53. Carol spends $82 on a new outfit. If she has $100, what percent of her money does she spend on the outfit?

54. A graphing calculator originally priced at $100 is on sale for $78. What is the percent of discount? (Discount is the difference between the original price and the sale price.)

55. Mickie bought a TV and makes monthly payments on it. Last year, she paid a total of of $900. Of the total that she paid, $180 was interest. What percent of the total was interest?

56. Pablo buys a suit that was originally priced at $100. He buys it for 35% off the original price. What does he pay for the suit?

STATE YOUR UNDERSTANDING

57. What is a percent? How is it related to fractions and decimals?

58. Explain the difference in meaning of the symbols 25% and 125%. In your explanation, use diagrams to illustrate the meanings. Contrast similarities and differences in the diagrams.

CHALLENGE

59. Write the ratio of 109 to 500 as a fraction and as a percent.

60. Write the ratio of 514 to 800 as a fraction and as a percent.

61. Write the ratio of 776 to 500 as a fraction, as a mixed number, and as a percent.

MAINTAIN YOUR SKILLS

Multiply.

62. 7.83(100)

63. 47.335 × 100

64. 0.00578(1000)

65. 207.8 × 1000

66. 12.45 ÷ 100

67. 0.0672 ÷ 1000

68. $1743 \div 10^4$

69. $0.9003 \div 10^2$

70. Bill goes to the store with $25. He uses his calculator to keep track of the money he is spending. He decides that he could make the following purchases. Is he correct?

Article	Cost
1 loaf of bread	$3.29
2 bottles of V-8 juice	$3.39 each
2 boxes of crackers	$2.69 each
1 package of cheddar cheese	$3.99
2 cartons of orange juice	$2.00 each

71. Ms. Henderson earns $23.85 per hour and works the following hours during 1 month. How much are her monthly earnings?

Week	Hours
1	35
2	30.25
3	25
4	36.75
5	6

Changing Decimals to Percents and Percents to Decimals

HOW & WHY

OBJECTIVES

1. Write a given decimal as a percent.
2. Write a given percent as a decimal.

■ **OBJECTIVE 1** Write a given decimal as a percent.

In multiplication, where one factor is $\dfrac{1}{100}$, the indicated multiplication can be read as a percent; that is, $75\left(\dfrac{1}{100}\right) = 75\%, 0.8\left(\dfrac{1}{100}\right) = 0.8\%$, and $\dfrac{3}{4}\left(\dfrac{1}{100}\right) = \dfrac{3}{4}\%$.

To write a number as a percent, multiply by $100 \cdot \dfrac{1}{100}$, a name for 1. This is shown in Table 6.1.

TABLE 6.1 Change a Decimal to a Percent

Number	Multiply by 1 $100\left(\dfrac{1}{100}\right) = 1$	Multiply by 100	Rename as a Percent
0.74	$0.74(100)\left(\dfrac{1}{100}\right)$	$74 \cdot \left(\dfrac{1}{100}\right)$	74%
0.6	$0.6(100)\left(\dfrac{1}{100}\right)$	$60 \cdot \left(\dfrac{1}{100}\right)$	60%
4	$4(100)\left(\dfrac{1}{100}\right)$	$400 \cdot \left(\dfrac{1}{100}\right)$	400%

In each case the decimal point is moved two places to the right and the percent symbol (%) is inserted.

> ### To change a decimal to a percent
> 1. Move the decimal point two places to the right. (Write zeros on the right if necessary.)
> 2. Write the percent symbol (%) on the right.

EXAMPLES A–F

DIRECTIONS: Change the decimal to a percent.

STRATEGY: Move the decimal point two places to the right and write the percent sign on the right.

WARM-UP **A.** Write 0.73 as a percent.	**A.** Write 0.26 as a percent. $0.26 = 26\%$ Move the decimal point two places to the right and write the percent symbol on the right. So $0.26 = 26\%$.
WARM-UP **B.** Change 0.04 to a percent.	**B.** Change 0.03 to a percent. $0.03 = 003\% = 3\%$ Since the zeros are to the left of 3, we can drop them. So $0.03 = 3\%$.
WARM-UP **C.** Change 0.0023 to a percent.	**C.** Change 0.0011 to a percent. $0.0011 = 000.11\% = 0.11\%$ This is eleven hundredths of one percent. So $0.0011 = 0.11\%$.
WARM-UP **D.** Write 7 as a percent.	**D.** Write 14 as a percent. $14 = 14.00 = 1400\%$ Insert two zeros on the right so we can move two decimal places. Fourteen hundred percent is 14 times 100%. So $14 = 1400\%$.
WARM-UP **E.** Change $0.73\overline{3}$ to a percent.	**E.** Change $0.26\overline{6}$ to a percent. $0.26\overline{6} = 26.\overline{6}\% = 26\frac{2}{3}\%$ The repeating decimal $0.\overline{6} = \frac{2}{3}$. So $0.26\overline{6} = 26.\overline{6}\%$ or $26\frac{2}{3}\%$.
WARM-UP **F.** The tax code lists the tax rate on a zone 3 lot at 0.031. What is the tax rate expressed as a percent?	**F.** The tax rate on a building lot is given as 0.027. What is the tax rate expressed as a percent? $0.027 = 002.7\% = 2.7\%$ So the tax rate expressed as a percent is 2.7%.

ANSWERS TO WARM-UPS A–F

A. 73%   **B.** 4%   **C.** 0.23%

D. 700%   **E.** $73.\overline{3}\%$ or $73\frac{1}{3}\%$

F. The tax rate is 3.1%.

HOW & WHY

■ **OBJECTIVE 2** Write a given percent as a decimal.

The percent symbol indicates multiplication by $\frac{1}{100}$, so

$$55\% = 55 \cdot \frac{1}{100} = \frac{55}{100} = 55 \div 100$$

As we learned in Section 4.5, dividing a number by 100 is done by moving the decimal point two places to the left.

$$55\% = 55 \div 100 = 0.55$$

> **To change a percent to a decimal**
>
> 1. Move the decimal point two places to the left. (Write zeros on the left if necessary.)
> 2. Drop the percent symbol (%).

EXAMPLES G–K

DIRECTIONS: Change the percent to a decimal.

STRATEGY: Move the decimal point two places to the left and drop the percent symbol.

G. Change 28.7% to a decimal.

$28.7\% = 0.287$ Move the decimal point two places left. Drop the percent symbol.

So 28.7% = 0.287.

H. Change 561% to a decimal.

$561\% = 5.61$ A value larger than 100% becomes a mixed number or a whole number.

So 561% = 5.61.

I. Write $77\frac{14}{25}\%$ as a decimal.

$77\frac{14}{25}\% = 77.56\%$ Change the fraction to a decimal.

$\qquad = 0.7756$ Change the percent to a decimal.

So $77\frac{14}{25}\% = 0.7756$

J. Change $33\frac{7}{18}\%$ to a decimal. Round to the nearest thousandth.

$33\frac{7}{18}\% = 33.3\overline{8}\%$ By division, $\frac{7}{18} = 0.3\overline{8}$.

$\qquad = 0.333\overline{8}$ Change to a decimal.

$\qquad \approx 0.334$ Round to the nearest thousandth.

So $33\frac{7}{18}\% \approx 0.334$.

WARM-UP
G. Change 48.3% to a decimal.

WARM-UP
H. Change 833% to a decimal.

WARM-UP
I. Write $19\frac{3}{4}\%$ as a decimal.

WARM-UP
J. Change $21\frac{7}{13}\%$ to a decimal.
Round to the nearest thousandth.

ANSWERS TO WARM-UPS G–J
G. 0.483   **H.** 8.33
I. 0.1975   **J.** 0.215

K. When ordering fresh vegetables, a grocer orders 9.3% more than is needed to allow for spoilage. What decimal is entered into the computer to calculate the amount of extra vegetables to be added to the order?

9.3% = 0.093   Change the percent to a decimal.

So the grocer will enter 0.093 in the computer.

ANSWER TO WARM-UP K

K. The contractor will enter 0.031 in the computer

EXERCISES 6.2

■ OBJECTIVE 1 Write a given decimal as a percent. (See page 415.)

A *Write each decimal as a percent.*

1. 0.47	**2.** 0.83	**3.** 2.32	**4.** 8.64
5. 0.08	**6.** 0.03	**7.** 4.96	**8.** 6.98
9. 19	**10.** 21	**11.** 0.0083	**12.** 0.0017
13. 0.952	**14.** 0.376	**15.** 0.592	**16.** 0.712

B

17. 0.0731	**18.** 0.0716	**19.** 20	**20.** 62
21. 17.81	**22.** 4.311	**23.** 0.00044	**24.** 0.00471
25. 7.1	**26.** 2.39	**27.** 0.8867	**28.** 0.9708
29. $0.811\overline{6}$	**30.** $0.04\overline{3}$	**31.** 0.2409	**32.** 0.61609

■ OBJECTIVE 2 Write a given percent as a decimal. (See page 417.)

A *Write each of the following as a decimal.*

33. 96%	**34.** 47%	**35.** 73%	**36.** 83%
37. 1.35%	**38.** 8.12%	**39.** 908%	**40.** 444%
41. 652.5%	**42.** 560.7%	**43.** 0.0062%	**44.** 0.0048%
45. 0.071%	**46.** 0.672%	**47.** 3940%	**48.** 8643%

B

49. 0.092%

50. 0.0582%

51. 100%

52. 300%

53. 662%

54. 363%

55. $\frac{1}{2}$%

56. $\frac{7}{10}$%

57. $\frac{3}{16}$%

58. $\frac{4}{5}$%

59. $73\frac{3}{4}$%

60. $12\frac{9}{10}$%

61. $\frac{1}{8}$%

62. $\frac{7}{16}$%

63. 413.773%

64. 222.05%

C

65. If the tax rate on a person's income in Colorado was 0.0463, what was the rate expressed as a percent?

66. A 2-year nursing program has a completion rate of 0.734. What is the rate as a percent?

67. A Girl Scout sold 0.36 of her quota of cookies on the first day of the sale. What percent of her cookies did she sell on the first day?

68. The sales tax in Illinois was 0.0625. Express this as a percent.

69. Employees just settled their new contract and got a 3.15% raise over the next 2 years. Express this as a decimal.

70. The CFO of an electronics firm adds 5.35% to the budget as a contingency fund. What decimal part is this?

71. Interest rates are expressed as percents. The Credit Union charged 6.34% interest on new 48-month auto loans. What decimal will they use to compute the interest?

72. What decimal is used to compute the interest on a mortgage that has an interest rate of 5.34%?

Change to a decimal rounded to the nearest thousandth.

73. $\frac{7}{15}$%

74. $55\frac{3}{4}$%

75. $88\frac{7}{12}$%

76. $\frac{75}{7}$%

77. In industrialized countries, 60% of the river pollution is due to agricultural runoff. Change this to a decimal.

78. Recycling aluminum cans consumes 95% less energy than smelting new stocks of metal. Change this to a decimal.

© iStockphoto.com/FotografiaBasica

79. The Westview High School golf team won 0.875 of their matches. Write this as a percent.

80. A WNBA basketball player makes 0.642 of her free throws. Express this as a percent.

81. Over the 2008 season, Kurt Warner of the Arizona Cardinals had a 0.67 completion rate for all passes attempted. Express his completion rate as a percent

82. In 2009, the highest batting average in the American League was 0.371 by Joe Mauer of the Minnesota Twins. The National League's batting leader was Hanley Ramirez of the Florida Marlins with 0.350. Express these as percents.

83. The Bureau of Labor Statistics expects that from 2006 to 2016 there will be about 47,000 new physical therapy jobs. The total number of physical therapy jobs will be 1.272 times the number of jobs in 2006. Express this as a percent.

84. Find today's interest rates for home mortgages for 15- and 30-year fixed-rate loans. Express these as decimals.

85. One mile is about 160.9% of a kilometer. Express this as a decimal.

86. One yard is about 91.4% of a meter. Express this as a decimal.

87. The Moscow subway system has the largest number of riders of any subway system in the world. The New York City subway system has 40.6% of the riders of the Moscow system. Express this as a decimal.

88. The Burj Khalifa in Dubai, is the tallest building in the world. It is about 163% of the height of the Taipei 101 in Taipei, the next tallest building. Express this as a decimal.

89. Pluto is a dwarf planet in our solar system, with a diameter that is about 27.4% of the diameter of Earth. Express this as a decimal.

90. A nautical mile is about 1.15 times the length of a statute (land) mile. Express this as a percent.

91. The amount of Social Security paid by employees is found by multiplying the gross wages by 0.062. The Medicare payment is found by multiplying the gross wages by 0.029. Express the sum of these amounts as a percent.

92. The recommended level for total cholesterol is 200 or less. In 1960, the average cholesterol level for Americans was estimated at 222. By 2007 the average level had dropped 10.4% to 199. Express this percent drop as a decimal.

Exercises 93–95 relate to the chapter application.

93. The sale price of a can of beans is 0.89 of what it was before the sale. Express this as a percent. What "percent off" will the store advertise?

94. Mary spends 0.285 of her monthly income on groceries. What percent of her monthly income is spent on groceries?

95. A box of cereal claims to contain 125% of what it used to contain. Express this as a decimal.

96. Explain how the decimal form and the percent form of a number are related. Give an example of each form.

97. When changing a percent to a decimal, how can you tell when the decimal will be greater than 1?

CHALLENGE

Write as percents.

98. 0.0004507

99. 18,000

Write as percents without using repeating decimals.

100. 0.024 and 0.02$\overline{4}$

101. 0.425 and 0.42$\overline{5}$

102. Change $11\frac{4}{17}\%$ to a decimal rounded to the nearest tenth and the nearest thousandth.

103. Change $56\frac{11}{12}\%$ to a decimal rounded to the nearest tenth and the nearest thousandth.

104. Baseball batting averages are written as decimals. A batter with an average of 238 has hit an average of 238 times out of 1000 times at bat (0.238). Find the batting averages of the top five players in the American and National Leagues. Express these average as percents.

105. It not unusual to read or hear that a person gave 110% for their job, profession, or team. Is it possible to "give" more than 100%? Could it be that they put in 110% more time than was required? Or that they achieved 10% more than any of their coworkers or teammates? What do think people who say this mean?

MAINTAIN YOUR SKILLS

Change to a decimal.

106. $\frac{7}{8}$

107. $\frac{9}{64}$

108. $\frac{19}{16}$

109. $\frac{117}{65}$

Change to a fraction and simplify.

110. 0.715

111. 0.1025

112. Round to the nearest thousandth: 3.87264

113. Round to the nearest ten-thousand: 345,891.62479

114. In 1 week, Greg earns $245. His deductions (income tax, Social Security, and so on) total $38.45. What is his "take home" pay?

115. The cost of gasoline is reduced from $0.695 per liter to $0.629 per liter. How much money is saved on an automobile trip that requires 340 liters?

6.3 Changing Fractions to Percents and Percents to Fractions

HOW & WHY

■ **OBJECTIVE 1** Change a fraction or mixed number to a percent.

We already know how to change fractions to decimals and decimals to percents. We combine the two ideas to change a fraction to a percent.

> **To change a fraction or mixed number to a percent**
>
> 1. Change to a decimal. The decimal is rounded or carried out as directed.
> 2. Change the decimal to a percent.

Unless directed to round, the division is completed or else the quotient is written as a repeating decimal.

EXAMPLES A–G

DIRECTIONS: Change the fraction or mixed number to a percent.

STRATEGY: Change the number to a decimal and then to a percent.

WARM-UP

A. Change $\dfrac{13}{20}$ to a percent.

A. Change $\dfrac{5}{8}$ to a percent.

$\dfrac{5}{8} = 0.625$ Divide 5 by 8 to change the fraction to a decimal.

$\quad = 62.5\%$ Change the decimal to a percent.

So $\dfrac{5}{8} = 62.5\%$.

WARM-UP

B. Write $\dfrac{3}{8}$ as a percent.

B. Write $\dfrac{11}{16}$ as a percent.

$\dfrac{11}{16} = 0.6875$ Change to a decimal.

$\quad = 68.75\%$

So $\dfrac{11}{16} = 68.75\%$.

WARM-UP

C. Change $\dfrac{17}{30}$ to a percent.

C. Change $\dfrac{11}{18}$ to a percent.

$\dfrac{11}{18} = 0.61\overline{1}$ Write $\dfrac{11}{18}$ as a repeating decimal.

$\quad = 61.\overline{1}\%$ Change to a percent.

$\quad = 61\dfrac{1}{9}\%$ The repeating decimal $0.\overline{1} = \dfrac{1}{9}$.

So $\dfrac{11}{18} = 61.\overline{1}\%$, or $61\dfrac{1}{9}\%$.

D. Write $2\dfrac{7}{20}$ as a percent.

$$2\dfrac{7}{20} = 2.35$$
$$= 235\%$$

So $2\dfrac{7}{20} = 235\%$.

E. Change $\dfrac{5}{13}$ to a percent. Round to the nearest tenth of a percent.

CAUTION

One tenth of a percent is a thousandth; that is,

$$\dfrac{1}{10} \text{ of } \dfrac{1}{100} = \dfrac{1}{10} \cdot \dfrac{1}{100} = \dfrac{1}{1000} = 0.001.$$

STRATEGY: To write the percent rounded to the nearest tenth of a percent, we need to change the fraction to a decimal rounded to the nearest thousandth (that is, we round to the third decimal place).

$$\dfrac{5}{13} \approx 0.385 \quad \text{Write as a decimal rounded to the nearest thousandth.}$$
$$\approx 38.5\%$$

So $\dfrac{5}{13} \approx 38.5\%$.

CALCULATOR EXAMPLE: 🖩

F. Write $6\dfrac{159}{295}$ as a percent rounded to the nearest tenth of a percent.

$$\dfrac{159}{295} \approx 0.5389830 \quad \text{First, convert the fraction to a decimal.}$$
$$6\dfrac{159}{295} \approx 6.5389830 \quad \text{Add the whole number.}$$
$$\approx 6.539 \quad \text{Round to the nearest thousandth.}$$
$$\approx 653.9\% \quad \text{Change to percent.}$$

So $6\dfrac{159}{295} \approx 653.9\%$.

G. A motor that needs repair is only turning $\dfrac{9}{16}$ the number of revolutions per minute that is normal. What percent of the normal rate is this?

$$\dfrac{9}{16} = 0.5625$$
$$= 56.25\%$$

So the motor is turning at 56.25% of its normal rate.

HOW & WHY

■ **OBJECTIVE 2** Change percents to fractions or mixed numbers.

The expression 65% is equal to $65 \times \dfrac{1}{100}$. This gives a very efficient method for changing a percent to a fraction. See Example H.

> **To change a percent to a fraction or a mixed number**
>
> 1. Replace the percent symbol (%) with the fraction $\left(\dfrac{1}{100}\right)$.
> 2. If necessary, rewrite the other factor as a fraction.
> 3. Multiply and simplify.

EXAMPLES H–L

DIRECTIONS: Change the percent to a fraction or mixed number.

STRATEGY: Change the percent symbol to the fraction $\dfrac{1}{100}$ and multiply.

WARM-UP

H. Change 45% to a fraction.

H. Change 35% to a fraction.

$$35\% = 35 \cdot \dfrac{1}{100} \qquad \text{Replace the percent symbol (\%) with } \dfrac{1}{100}.$$

$$= \dfrac{35}{100} \qquad \text{Multiply.}$$

$$= \dfrac{7}{20} \qquad \text{Simplify.}$$

> **CAUTION**
> You need to multiply by $\dfrac{1}{100}$, not just write it down.

So $35\% = \dfrac{7}{20}$.

WARM-UP

I. Change 515% to a mixed number.

I. Change 436% to a mixed number.

$$436\% = 436 \cdot \dfrac{1}{100} \qquad \% = \dfrac{1}{100}$$

$$= \dfrac{436}{100} \qquad \text{Multiply.}$$

$$= \dfrac{109}{25} \qquad \text{Simplify.}$$

$$= 4\dfrac{9}{25} \qquad \text{Write as a mixed number.}$$

So $436\% = 4\dfrac{9}{25}$.

J. Change $12\frac{2}{3}\%$ to a fraction.

$$12\frac{2}{3}\% = 12\frac{2}{3} \cdot \frac{1}{100}$$
$$= \frac{38}{3} \cdot \frac{1}{100}$$
$$= \frac{38}{300}$$
$$= \frac{19}{150}$$

So $12\frac{2}{3}\% = \frac{19}{150}$.

K. Change 15.8% to a fraction.

$$15.8\% = 0.158$$
$$= \frac{158}{1000}$$
$$= \frac{79}{500}$$

So $15.8\% = \frac{79}{500}$.

L. A biological study shows that spraying a forest for gypsy moths is 92% successful. What fraction of the moths survive the spraying?

STRATEGY: Subtract the 92% from 100% to find the percent of the moths that survived. Then change the percent that survive to a fraction.

$$100\% - 92\% = 8\%$$
$$8\% = 8 \cdot \frac{1}{100}$$
$$= \frac{8}{100}$$
$$= \frac{2}{25}$$

So $\frac{2}{25}$, or 2 out of 25 gypsy moths, survived the spraying.

WARM-UP

J. Change $4\frac{7}{12}\%$ to a fraction.

WARM-UP

K. Change 14.8% to a fraction.

WARM-UP

L. Greg scores 88% on a math test. What fraction of the questions does he get incorrect?

ANSWERS TO WARM-UPS J–L

J. $\frac{11}{240}$ **K.** $\frac{37}{250}$

L. Greg gets $\frac{3}{25}$ of the questions incorrect.

EXERCISES 6.3

■ **OBJECTIVE 1** Change a fraction or mixed number to a percent. (See page 422.)

A *Change each fraction to a percent.*

1. $\frac{67}{100}$ **2.** $\frac{37}{100}$ **3.** $\frac{37}{50}$ **4.** $\frac{8}{10}$

5. $\frac{17}{20}$ **6.** $\frac{22}{25}$ **7.** $\frac{1}{2}$ **8.** $\frac{3}{5}$

9. $\dfrac{17}{20}$ **10.** $\dfrac{9}{50}$ **11.** $\dfrac{21}{20}$ **12.** $\dfrac{53}{50}$

13. $\dfrac{15}{8}$ **14.** $\dfrac{21}{16}$ **15.** $\dfrac{63}{1000}$ **16.** $\dfrac{247}{1000}$

B   *Change each fraction or mixed number to a percent.*

17. $4\dfrac{3}{5}$ **18.** $6\dfrac{1}{4}$ **19.** $\dfrac{1}{3}$ **20.** $\dfrac{5}{6}$

21. $\dfrac{29}{6}$ **22.** $\dfrac{25}{3}$ **23.** $4\dfrac{11}{12}$ **24.** $8\dfrac{11}{12}$

Change each fraction or mixed number to a percent. Round to the nearest tenth of a percent.

25. $\dfrac{8}{13}$ **26.** $\dfrac{17}{35}$ **27.** $\dfrac{5}{9}$ **28.** $\dfrac{7}{9}$

29. $\dfrac{11}{14}$ **30.** $\dfrac{13}{23}$ **31.** $32\dfrac{11}{29}$ **32.** $56\dfrac{11}{15}$

■ **OBJECTIVE 2**   Change percents to fractions or mixed numbers. (See page 424.)

A   *Change each of the following percents to fractions or mixed numbers.*

33. 12%   **34.** 20%   **35.** 85%   **36.** 5%

37. 130%   **38.** 180%   **39.** 200%   **40.** 100%

41. 84%   **42.** 68%   **43.** 25%   **44.** 80%

45. 45%   **46.** 67%   **47.** 150%   **48.** 225%

B

49. 45.5%   **50.** 24.4%   **51.** 6.8%   **52.** 3.5%

53. 60.5%   **54.** 16.8%   **55.** $\dfrac{1}{4}$%   **56.** $\dfrac{5}{7}$%

57. $2\dfrac{1}{2}$%   **58.** $8\dfrac{3}{4}$%   **59.** $\dfrac{2}{11}$%   **60.** $\dfrac{4}{13}$%

61. $44\frac{1}{4}\%$

62. $28\frac{3}{4}\%$

63. $331\frac{2}{3}\%$

64. $243\frac{1}{3}\%$

C

65. Kobe Bryant made 17 out of 20 free throw attempts in one game. What percent of the free throws did he make?

66. Maureen gets 19 problems correct on a 25-problem test. What percent is correct?

67. In the 2008 presidential election, President Barack Obama received 69,456,897 votes out of the 131,257,328 votes cast for president. What percent of the vote did he receive, to the nearest tenth of a percent?

68. In a supermarket, 2 eggs out of 11 dozen are lost because of cracks. What percent of the eggs must be discarded, to the nearest tenth of a percent?

© Christopher Halloran/Shutterstock.com

69. In the 2008 U. S. presidential election, President Barack Obama received 365 electoral college votes out of a possible 538. What percent, to the nearest tenth of a percent, of the electoral college votes did he receive?

70. Ms. Nyuen was awarded scholarships that will pay for 56% of her college tuition. What fractional part of her tuition will be paid through scholarships?

71. Artis Gilmore is the NBA all-time leader in field goal percentage. Over his career he made 5732 field goals out of 9570 attempts. What percent, to the nearest tenth of a percent, of field goals attempted did he make?

72. The offensive team for the Detroit Lions was on the field 55% of the time during a game with the New York Jets. What fractional part of the game were they on the field?

Change each of the following fractions or mixed numbers to a percent rounded to the nearest hundredth of a percent.

73. $\frac{67}{360}$

74. $\frac{567}{8000}$

75. $1\frac{25}{66}$

76. $27\frac{41}{79}$

Write each of the following as a fraction.

77. $4\frac{4}{9}\%$

78. $7\frac{7}{9}\%$

79. 0.275%

80. 0.975%

81. A vitamin C tablet is listed as fulfilling $\frac{1}{8}$ the recommended daily allowance for vitamin C. Miguel takes 13 of these tablets per day to ward off a cold. What percent of the average recommended allowance is he taking?

82. During the 2008 regular season, the Philadelphia Phillies won 92 games and lost 70. Write a fraction that gives the number of games won compared to the total games played. Convert this to a percent, rounded to the nearest tenth of a percent.

83. Economic factors caused the enrollment at City Community College to be 126% of last year's enrollment. What fraction of last year's enrollment does this represent?

84. A western city had a population in 2010 that was 135% of its population in 2008. Express the percent of the 2008 population as a fraction or mixed number.

85. A census determines that $37\frac{1}{2}\%$ of the residents of a city are age 40 or older and that 45% are age 25 or younger. What fraction of the residents are between the ages of 25 and 40?

86. Jorge invests $26\frac{3}{7}\%$ of his money in money market funds. The rest he puts in common stocks. What fraction of the total investment is in common stocks?

87. The area of the island of St. Croix, one of the Virgin Islands, is 84 mi². The total area of the Virgin Islands is 140 mi². Write a fraction that represents the ratio of the area of St. Croix to the area of the Virgin Islands. Change this fraction to a percent.

88. Burger King's original Double Whopper with cheese contains 69 g of fat. Each gram of fat has 9 calories. If the entire sandwich contains 1060 calories, what percent of the calories come from the fat content? Round to the nearest percent.

89. In 2004, one area of California had 211 smoggy days. What was the percent of smoggy days? In 2009, there were only 167 smoggy days. What was the percent that year? Compare these percents and discuss the possible reasons for this decline. Round the percents to the nearest tenth of a percent.

90. Spraying for mosquitos, in an attempt to eliminate the West Nile Virus, is found to be 85% successful. What fraction of the mosquitos are eliminated?

91. The salmon run in an Oregon stream has dropped to 42% of what it was 10 years ago. What fractional part of the run was lost during the 10 years?

92. The literacy rate in Vietnam is 94%. Convert this to a fraction and explain its meaning.

93. According to recent estimates, the population of Kenya is 0.17% white. Convert this to a fraction and explain its meaning.

Exercises 94–97 relate to the chapter application.

94. Consumer reports indicate that the cost of food is $1\frac{1}{12}$ what it was 1 year ago. Express this as a percent. Round to the nearest tenth of a percent.

95. A department store advertises one-third off the regular price on Monday and an additional one-seventh off the original price on Tuesday. What percent is taken off the original price if the item is purchased on Tuesday? Round to the nearest whole percent.

96. The U S. Department of Agriculture forecasts that the price of eggs will be $\frac{57}{50}$ of the price last year. Express this as a percent.

97. Wendy bought a barbeque grill at WalMart that was on sale for $\frac{2}{3}$ of its original price. What percent off was the grill? Round to the nearest whole percent.

STATE YOUR UNDERSTANDING

98. Explain why not all fractions can be changed to a whole-number percent. What is special about the fractions that can?

99. Name two circumstances that can be described by either a percent or a fraction. Compare the advantages or disadvantages of using percents or fractions.

100. Explain how to change between the fraction and percent forms of a number. Give an example of each.

101. Change $2\frac{4}{13}$% to the nearest tenth of a percent.

102. Change $\frac{6}{13}$% to the nearest hundredth of a percent.

103. Change 0.00025 to a fraction.

104. Change 150.005% to a mixed number.

105. Change 180.04% to a mixed number.

106. Change 0.0005% to a fraction.

107. Keep a record of everything you eat for one day. Use exact amounts as much as possible. With a calorie and fat counter, compute the percent of fat in each item. Then find the percent of fat you consumed that day. The latest recommendations suggest that the fat content not exceed 30% per day. How did you do? Which foods have the highest and which the lowest fat content? Was this a typical day for you?

108. Read the ads for the local department stories in the weekend paper. Record the "% off" in as many ads as you can find. Convert the percents to fractions. In which form is it easier to estimate the savings because of the sale?

MAINTAIN YOUR SKILLS

Change to a fraction or mixed number.

109. 0.84

110. 0.132

111. 4.065

112. 16.48

Change to a decimal.

113. $\frac{33}{40}$

114. $\frac{443}{640}$

Change to a percent.

115. 0.567

116. 5.007

Change to a decimal.

117. 8.13%

118. 112.8%

Fractions, Decimals, Percents: A Review

HOW & WHY

OBJECTIVE

Given a decimal, fraction, or percent, rewrite in an equivalent form.

■ **OBJECTIVE**   Given a decimal, fraction, or percent, rewrite in an equivalent form.

Decimals, fractions, and percents can each be written in the other two forms. We can

write a percent as a decimal and as a fraction, and

write a fraction as a percent and as a decimal, and

write a decimal as a percent and as a fraction.

For example,

$$40\% = 40 \cdot \frac{1}{100} = \frac{40}{100} = \frac{2}{5} \quad \text{and} \quad 40\% = 0.4$$

$$\frac{5}{8} = 5 \div 8 = 0.625 \quad \text{and} \quad \frac{5}{8} = 0.625 = 62.5\%$$

$$0.95 = 95\% \quad \text{and} \quad 0.95 = \frac{95}{100} = \frac{19}{20}$$

Table 6.2 shows some common fractions and their decimal and percent equivalents. Some of the decimals are repeating decimals. Remember that a repeating decimal is shown by the bar over the digits that repeat. These fractions occur often in applications of percents. They should be memorized so that you can recall the patterns when they appear.

TABLE 6.2   **Common Fractions, Decimals, and Percents**

Fraction	Decimal	Percent	Fraction	Decimal	Percent
$\frac{1}{2}$	0.5	50%	$\frac{1}{6}$	$0.1\overline{6}$	$16\frac{2}{3}\%$ or $16.\overline{6}\%$
$\frac{1}{3}$	$0.\overline{3}$	$33\frac{1}{3}\%$ or $33.\overline{3}\%$	$\frac{5}{6}$	$0.8\overline{3}$	$83\frac{1}{3}\%$ or $83.\overline{3}\%$
$\frac{2}{3}$	$0.\overline{6}$	$66\frac{2}{3}\%$ or $66.\overline{6}\%$	$\frac{1}{8}$	0.125	12.5%
$\frac{1}{4}$	0.25	25%	$\frac{3}{8}$	0.375	37.5%
$\frac{3}{4}$	0.75	75%	$\frac{5}{8}$	0.625	62.5%
$\frac{1}{5}$	0.2	20%	$\frac{7}{8}$	0.875	87.5%
$\frac{2}{5}$	0.4	40%			
$\frac{3}{5}$	0.6	60%			
$\frac{4}{5}$	0.8	80%			

EXAMPLES A–B

DIRECTIONS: Fill in the empty spaces with the related percent, decimal, or fraction.

STRATEGY: Use the procedures of the previous sections.

Fraction	Decimal	Percent
$\dfrac{5}{12}$		
		27%
	0.72	
$\dfrac{73}{100}$		
		160%
	1.3	

A.

Fraction	Decimal	Percent
		30%
$\dfrac{5}{16}$		
	0.62	

Fraction	Decimal	Percent
$\dfrac{3}{10}$	0.3	30%
$\dfrac{5}{16}$	0.3125	31.25% or $31\dfrac{1}{4}$%
$\dfrac{31}{50}$	0.62	62%

$$30\% = 0.30 = \frac{30}{100} = \frac{3}{10}$$

$$\frac{5}{16} = 0.3125 = 31.25\% = 31\frac{1}{4}\%$$

$$0.62 = 62\% = \frac{62}{100} = \frac{31}{50}$$

WARM-UP

B. In Example B, write the percent that is recycled as a decimal and as a fraction.

B. The average American uses about 200 lb of plastic a year. Approximately 60% of this is used for packaging and about 5% of it is recycled. Write the percent used for packaging as a decimal and a fraction.

$$60\% = 0.60 = \frac{60}{100} = \frac{3}{5}$$

So 0.6, or $\dfrac{3}{5}$, of the plastic is used for packaging.

ANSWERS TO WARM-UPS A–B

A.

Fraction	Decimal	Percent
$\dfrac{5}{12}$	$0.41\overline{6}$	$41\dfrac{2}{3}$%
$\dfrac{27}{100}$	0.27	27%
$\dfrac{18}{25}$	0.72	72%
$\dfrac{73}{100}$	0.73	73%
$1\dfrac{3}{5}$	1.6	160%
$1\dfrac{3}{10}$	1.3	130%

B. Of the 200 lb of plastic, 0.05, or $\dfrac{1}{20}$, is recycled.

■ **OBJECTIVE** Given a decimal, fraction, or percent, rewrite in an equivalent form. (See page 431.)

Exercises 1–26. Fill in the empty spaces with the related percent, decimal, or fraction.

	Fraction	Decimal	Percent
1.			10%
3.		0.75	
5.	$1\frac{3}{20}$		
7.		0.001	
9.		4.25	
11.			$5\frac{1}{2}\%$
13.			13.5% or $13\frac{1}{2}\%$
15.			$62\frac{1}{2}\%$
17.	$\frac{17}{50}$		
19.		0.08	
21.		0.96	
23.			35%
25.		0.125	

	Fraction	Decimal	Percent
2.			30%
4.	$\frac{9}{10}$		
6.	$\frac{3}{8}$		
8.		1	
10.		0.8	
12.		0.875	
14.		0.6	
16.			50%
18.	$\frac{5}{6}$		
20.	$\frac{2}{3}$		
22.		0.2	
24.			$33.\overline{3}\%$ or $33\frac{1}{3}\%$
26.	$\frac{7}{10}$		

27. Louis goes to the Bon Marché to buy a new swim suit. He finds one on sale for $\frac{1}{4}$ off. What percent is this?

28. Michael and Louise's new baby now weighs $66\frac{2}{3}\%$ more than his birth weight. What fraction is this?

29. During the month of August, Super Value Grocery has a special on sweet corn: Buy 12 ears and get $\frac{1}{6}$ more for 1 cent. During the same time period, Hank's Super Market also runs a special on sweet corn: Buy 12 ears and get 25% more for 1 cent. Which store offers the better deal?

30. Three multivitamins contain the following amounts of the RDA (recommended daily allowance) of calcium: brand A, 15%; brand B, 0.149; and brand C, $\frac{1}{7}$. Which brand contains the most calcium?

31. Nita's supervisor has authorized her to purchase a lot for a small business. Seller A has offered to sell a lot for 96% of the listed price. Nita's supervisor has authorized her to pay no more than 0.965 of the listed price. Seller B has offered Teresa a deal on an equivalent lot that is $\frac{23}{25}$ of the same amount. Who has offered the better deal? Does either or both meet the boss's authorized amount?

32. During the 2008–2009 NBA basketball season, Dallas, at home, made 46.2% of their field goals, New Jersey made $\frac{56}{125}$ of their shots, and Philadelphia shot an average of 0.459. What team had the best field goal statistics?

33. During one softball season, Jane got a hit 35% of the times she was at bat, Stephanie got a hit $\frac{11}{30}$ of the times she was at bat, and Ellie's batting average was 0.361. Which girl had the best batting average?

34. During a promotional sale, Rite-Aid advertises Coppertone sunscreen for $\frac{1}{4}$ off the suggested retail price whereas Walgreen's advertises it for 25% off. Which is the better deal?

35. In Peru, the literacy rate is 89%. In Jamaica, the literacy rate is 851 per 1000 people. Which country has the higher rate?

36. In Bulgaria, the death rate is about 13.3 per 1000 people. In Estonia, the death rate is about 1.31%. Which country has the lower death rate?

Exercises 37–40 relate to the chapter application.

37. George buys a new TV at a 40%-off sale. What fraction is this?

38. A local department store is having its red tag sale. All merchandise will now be 20% off the original price. What decimal is this?

39. Melinda is researching the best place to buy a computer. On the same-priced computer, Family Computers offers $\frac{1}{8}$ off, The Computer Store will give a 12% discount, and Machines Etc. will allow a 0.13 discount. Where does Melinda get the best deal?

40. Randy is trading in his above-ground swimming pool for a larger model. Prices are the same for Model PS+ and Model PT. PS+ holds 11% more water than his old pool, whereas PT holds $\frac{1}{9}$ more water. Which should he use to get the most additional water for his new pool?

© Pokomeda/Shuterstock.com

41. Write a short paragraph with examples that illustrate when to use fractions, when to use decimals, and when to use percents to show comparisons.

CHALLENGE

42. Fill in the table with exact values.

Fraction	Decimal	Percent
$\frac{23}{6}$		
	0.2675	
		$100\frac{5}{8}\%$ or 100.625%

43. Fill in the table with decimal and percent values rounded to the nearest tenth.

Fraction	Decimal	Percent
$\frac{23}{7}$		
	0.8	
		210.5%

Solve the following proportions.

44. $\dfrac{25}{30} = \dfrac{x}{45}$

45. $\dfrac{45}{81} = \dfrac{23}{y}$

46. $\dfrac{x}{72} = \dfrac{40}{9}$

47. $\dfrac{x}{8.3} = \dfrac{117}{260}$

48. $\dfrac{\frac{1}{2}}{100} = \dfrac{A}{21}$

49. $\dfrac{85}{100} = \dfrac{170}{B}$

50. $\dfrac{R}{100} = \dfrac{9}{34}$; round to the nearest tenth.

51. $\dfrac{23}{100} = \dfrac{A}{41.3}$; round to the nearest tenth.

52. Sean drives 536 miles and uses 10.6 gallons of gasoline. At that rate, how many gallons will he need to drive 1150 miles? Round to the nearest tenth of a gallon.

53. The Bacons' house is worth $235,500 and is insured so that the Bacons will be paid four-fifths of the value of any damage. One-third of the value of the house is destroyed by fire. How much insurance money should they collect?

SECTION
6.5 Solving Percent Problems

OBJECTIVES

1. Solve percent problems using the formula.
2. Solve percent problems using a proportion.

VOCABULARY

In the statement "*R* of *B* is *A*,"

R is the **rate** of percent.

B is the **base** unit and follows the word *of*.

A is the **amount** that is compared to *B*.

To solve a percent problem means to do one of the following:

1. Find *A*, given *R* and *B*.

2. Find *B*, given *R* and *A*.

3. Find *R*, given *A* and *B*.

HOW & WHY

■ **OBJECTIVE 1** Solve percent problems using the formula.

We show two methods for solving percent problems. We refer to these as

The percent formula, $R \times B = A$ (see Examples A–E).

The proportion method (see Examples F–H).

In each method we must identify the rate of percent (R), the base (B), and the amount (A). To help determine these, keep in mind that

R, the rate of percent, includes the percent symbol (%).

B, the base, follows the words *of* or *percent of.*

A, the amount, sometimes called the *percentage,* is the amount compared to B and follows the word *is*.

The method you choose to solve percent problems should depend on

1. the method your instructor recommends.
2. your major field of study.
3. how you use percent in your day-to-day activities.

What percent of B is A? The word *of* in this context and in other places in mathematics indicates multiplication. The word *is* describes the relationship "is equal to" or "=." Thus, we write,

R of B is A
 ↓   ↓
$R \times B = A$

When solving percent problems, identify the rate (%) first, the base (of) next, and the amount (is) last. For example, what percent of 30 is 6?

R of B is A The rate R is unknown, the base B (B follows the word *of*) is 30, and the amount A (A follows the word *is*) is 6.

$R \times B = A$
  $R(30) = 6$ Substitute 30 for B and 6 for A.
    $R = 6 \div 30$ Divide.
      $= 0.2$
      $= 20\%$ Change to percent.

So, 6 is 20% of 30.

All percent problems can be solved using $R \times B = A$. However, there are two other forms that can speed up the process.

$$A = R \times B \qquad R = A \div B \qquad B = A \div R$$

The triangle in Figure 6.2 is a useful device to help you select the correct form of the formula to use.

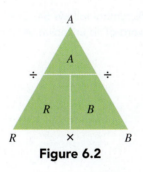

Figure 6.2

When the unknown value is covered, the positions of the uncovered (known) values help us remember which operation to use:

When A is covered, we see $R \times B$, reading from left to right.

When B is covered, we see $A \div R$, reading from top to bottom.

When R is covered, we see $A \div B$, reading from top to bottom.

For example, 34% of what number is 53.04? The rate (R) is 34%, B is unknown (B follows the word *of*), and A (A follows the word *is*) is 53.04. Fill in the triangle and cover B. See Figure 6.3.

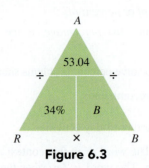

Figure 6.3

Reading from the top, we see that A is divided by R. Therefore,

$B = A \div R$
$\quad = 53.04 \div 0.34$ Substitute 53.04 for A and 0.34 for R.
$\quad = 156$

So 34% of 156 is 53.04.

CAUTION

Remember to use the decimal or fraction form of the rate R when solving percent problems.

EXAMPLES A–E

DIRECTIONS: Solve the percent problems using the formula.

STRATEGY: Identify R, B, and A. Select a formula. Substitute the known values and find the unknown value.

A. 54% of what number is 108?

    54% of B is 108   The % symbol follows 54, so $R = 54\%$. The base B (following the word *of*) is unknown. A follows the word *is*, so $A = 108$.

$B = A \div R$ Cover the unknown, B, in the triangle and read "$A \div R$."
$B = 108 \div 0.54$ Use the decimal form of the rate, 54% = 0.54.
$B = 200$

So 54% of 200 is 108.

B. 75 is what percent of 125?

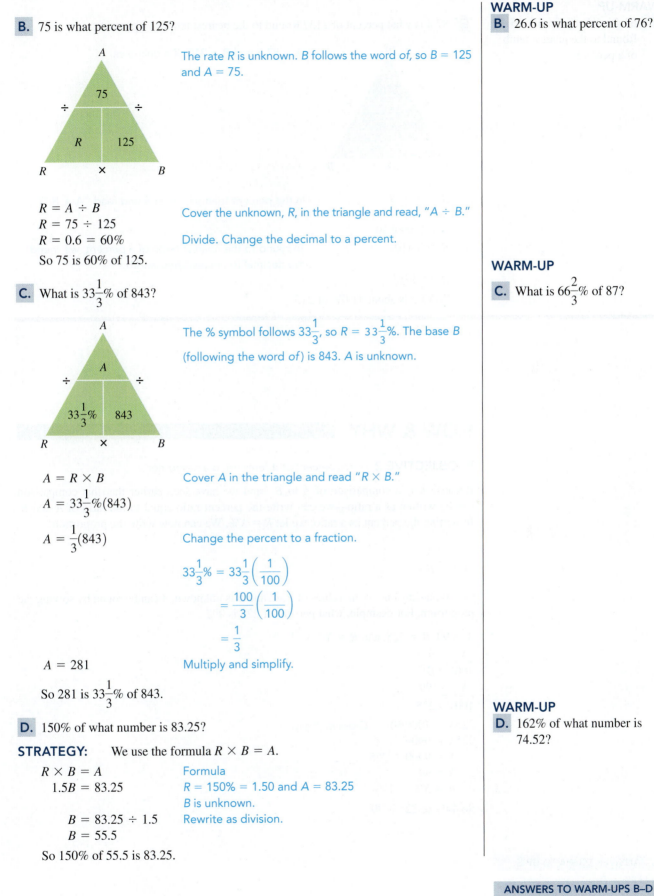

The rate *R* is unknown. *B* follows the word *of*, so *B* = 125 and *A* = 75.

$R = A \div B$
$R = 75 \div 125$
$R = 0.6 = 60\%$

So 75 is 60% of 125.

Cover the unknown, *R*, in the triangle and read, "*A* ÷ *B*."

Divide. Change the decimal to a percent.

C. What is $33\frac{1}{3}\%$ of 843?

The % symbol follows $33\frac{1}{3}$, so $R = 33\frac{1}{3}\%$. The base *B* (following the word *of*) is 843. *A* is unknown.

$A = R \times B$

$A = 33\frac{1}{3}\%(843)$

$A = \frac{1}{3}(843)$

Cover *A* in the triangle and read "*R* × *B*."

Change the percent to a fraction.

$$33\frac{1}{3}\% = 33\frac{1}{3}\left(\frac{1}{100}\right)$$
$$= \frac{100}{3}\left(\frac{1}{100}\right)$$
$$= \frac{1}{3}$$

$A = 281$

Multiply and simplify.

So 281 is $33\frac{1}{3}\%$ of 843.

D. 150% of what number is 83.25?

STRATEGY: We use the formula $R \times B = A$.

$R \times B = A$ Formula
$1.5B = 83.25$ $R = 150\% = 1.50$ and $A = 83.25$
              B is unknown.
$B = 83.25 \div 1.5$ Rewrite as division.
$B = 55.5$

So 150% of 55.5 is 83.25.

E. 68.4 is what percent of 519? Round to the nearest tenth of a percent.

E. 87.3 is what percent of 213? Round to the nearest tenth of a percent.

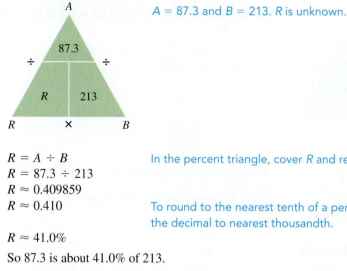

$A = 87.3$ and $B = 213$. R is unknown.

$R = A \div B$
$R = 87.3 \div 213$
$R \approx 0.409859$
$R \approx 0.410$

In the percent triangle, cover R and read "$A \div B$."

To round to the nearest tenth of a percent, we round the decimal to nearest thousandth.

$R \approx 41.0\%$

So 87.3 is about 41.0% of 213.

HOW & WHY

■ **OBJECTIVE 2** Solve percent problems using a proportion.

Because R is a comparison of A to B—and we have seen earlier that this comparison can be written as a ratio—we can write the percent ratio equal to the ratio of A and B. In writing the percent as a ratio, we let $R = X\%$. We can now write the proportion:

$$\frac{X}{100} = \frac{A}{B}.$$

When any one of the values of R, A, and B is unknown, it can be found by solving the proportion. For example, what percent of 225 is 99?

$A = 99$, $B = 225$, and $R = X\% = ?$

$$\frac{X}{100} = \frac{A}{B}$$

$$\frac{X}{100} = \frac{99}{225}$$

$225X = 100(99)$ Cross multiply.
$225X = 9900$
$X = 9900 \div 225$
$X = 44$
$R = X\% = 44\%$

So 44% of 225 is 99.

EXAMPLES F–H

DIRECTIONS: Solve the percent problem using a proportion.

STRATEGY: Write the proportion $\dfrac{X}{100} = \dfrac{A}{B}$, fill in the known values, and solve.

F. 85% of 76 is what number?

$$\frac{X}{100} = \frac{A}{B}$$ Proportion for solving percent problems

$$\frac{85}{100} = \frac{A}{76}$$ $R = X\% = 85\%$, so $X = 85$, $B = 76$, and A is unknown.

$85(76) = 100A$ Cross multiply.

$6460 = 100A$

$6460 \div 100 = A$ Rewrite as division.

$64.6 = A$

So 85% of 76 is 64.6.

G. 132% of _____ is 134.64

$$\frac{X}{100} = \frac{A}{B}$$ Proportion for solving percent problems

$$\frac{132}{100} = \frac{134.64}{B}$$ $R = X\% = 132\%$, so $X = 132$, B is unknown, and $A = 134.64$.

$132(B) = 100(134.64)$ Cross multiply.

$132(B) = 13{,}464$

$B = 13{,}464 \div 132$ Rewrite as division.

$B = 102$

So 132% of 102 is 134.64.

H. 78 is what percent of 327? Round to the nearest tenth of a percent.

$$\frac{X}{100} = \frac{A}{B}$$ Proportion for solving percent problems

$$\frac{X}{100} = \frac{78}{327}$$ $R = X\%$ is unknown, $A = 78$, and $B = 327$.

$327(X) = 100(78)$ Cross multiply.

$327(X) = 7800$ Multiply

$X = 7800 \div 327$ Rewrite as division.

$X \approx 23.85$ Carry out the division to two decimal places.

$X \approx 23.9$ Round to the nearest tenth.

$R \approx 23.9\%$ $R = X\%$.

So 78 is 23.9% of 327, to the nearest tenth of a percent.

EXERCISES 6.5

■ **OBJECTIVE 1** Solve percent problems using the formula. (See page 436.)

■ **OBJECTIVE 2** Solve percent problems using a proportion. (See page 440.)

A *Solve.*

1. 27 is 50% of _____.

2. 18 is 20% of _____.

3. _____ is 60% of 125.

4. _____ is 60% of 80.

5. 12 is _____ % of 4.

6. 8 is _____ % of 2.

7. _____ % of 200 is 150.

8. _____ % of 88 is 22.

9. 80% of _____ is 32.

10. 30% of _____ is 18.

11. 30% of 91 is _____ .

12. _____ is 55% of 72.

13. 96 is _____ % of 120.

14. _____ % of 75 is 15.

15. 39% of _____ is 39.

16. 62 is _____ % of 62.

17. $\frac{1}{3}$% of 600 is _____ .

18. $\frac{2}{9}$% of 2700 is _____ .

19. 130% of 90 is _____ .

20. 140% of 70 is _____ .

B

21. 17.5% of 70 is _____ .

22. 57.5% of 110 is _____ .

23. 3.3 is _____ % of 60.

24. 0.13 is _____ % of 65.

25. 497.8 is 76% of _____ .

26. 162 is 18% of _____ .

27. 45.5% of 80 is _____ .

28. 17.2% of 55 is _____ .

29. 39% of _____ is 105.3.

30. 73% of _____ is 83.22.

31. 124% of _____ is 328.6.

32. 205% of _____ is 750.3.

33. 96 is _____ % of 125.

34. 135 is _____ % of 160.

35. 6.14% of 350 is _____ .

36. 12.85% of 980 is _____ .

37. 2.05 is _____ % of 3.28.

38. 6.09 is _____ % of 17.4.

39. $11\frac{1}{9}$% of 1845 is _____ .

40. $16\frac{2}{3}$% of 3522 is _____ .

C

41. What percent of 85 is 41? Round to the nearest tenth of a percent.

42. What percent of 666 is 247? Round to the nearest tenth of a percent.

43. Eighty-two is 24.8% of what number? Round to the nearest hundredth.

44. Forty-one is 35.2% of what number? Round to the nearest hundredth.

45. Thirty-two and seven tenths percent of 695 is what number?

46. Seventy-three and twelve hundredths percent of 35 is what number?

47. Thirty-seven is what percent of 156? Round to the nearest tenth of a percent.

48. Two hundred thirty-two is what percent of 124? Round to the nearest tenth of a percent.

Solve.

49. $5\frac{1}{8}$% of $8\frac{1}{3}$ is _____. (as a fraction)

50. _____ is $2\frac{5}{16}$% of $6\frac{2}{5}$. (as a decimal)

51. $8\frac{7}{15}$% of 1350 is _____. (as a mixed number)

52. $23\frac{5}{11}$% of _____ is 87.5. (as a decimal rounded to the nearest hundredth)

53. _____% of 34.76 is 45.87. (round to the nearest tenth of a percent)

54. _____.% of $12\frac{4}{15}$ is $7\frac{2}{3}$. (as a mixed number)

STATE YOUR UNDERSTANDING

55. Explain the inaccuracies in this statement: "Starbuck industries charges 70¢ for a part that cost them 30¢ to make. They're making 40% profit."

56. Explain how to use the *RAB* triangle to solve percent problems.

CHALLENGE

57. $\frac{1}{2}$% of $45\frac{1}{3}$ is what fraction?

58. $\frac{3}{7}$% of $21\frac{1}{9}$ is what fraction?

59. $\frac{3}{4}$% of $13\frac{1}{3}$ is what decimal?

60. $\frac{2}{7}$% of $8\frac{3}{4}$ is what decimal?

MAINTAIN YOUR SKILLS

Solve the proportions.

61. $\dfrac{18}{24} = \dfrac{x}{60}$

62. $\dfrac{6.2}{2.5} = \dfrac{93}{y}$

63. $\dfrac{a}{\frac{5}{8}} = \dfrac{1\frac{1}{2}}{3\frac{3}{4}}$

64. $\dfrac{1\frac{1}{2}}{t} = \dfrac{5\frac{5}{8}}{1\frac{2}{3}}$

65. $\dfrac{1.4}{0.21} = \dfrac{w}{3.03}$

66. $\dfrac{2.6}{0.07} = \dfrac{3.51}{t}$

Exercises 67–70. On a certain map, $1\frac{3}{4}$ in. represents 70 mi.

67. How many miles are represented by $3\frac{5}{8}$ in.?

68. How many miles are represented by $7\frac{11}{16}$ in.?

69. How many inches are needed to represent 105 mi?

70. How many inches are needed to represent 22 mi?

6.6 Applications of Percents

OBJECTIVES

1. Solve applications involving percent.
2. Find percent of increase and percent of decrease.
3. Read data from a circle graph or construct a circle graph from data.

VOCABULARY

When a value B is increased by an amount A, the rate of percent R, or $\frac{A}{B}$, is called the **percent of increase**.

When a value B is decreased by an amount A the rate of percent R, or $\frac{A}{B}$, is called the **percent of decrease**.

HOW & WHY

■ **OBJECTIVE 1** Solve applications involving percent.

When a word problem is translated to the simpler form "What percent of what is what?" the unknown value can be found using one of the methods from the previous section. For example,

> A census listed the population of Detroit at 4,043,467. The African American population was 1,011,038. What percent of the population of Detroit was African American?

> We first rewrite the problem in the form "What percent of what is what?"

> What percent of the population is African American?

We know that the total population is 4,043,467 and we know the African American population is 1,011,038. Substituting these values we have

> What percent of 4,043,467 is 1,011,038?

Using the percent formula, we know that R is unknown, B is 4,043,467, and that A is 1,011,038. We substitute these values into the triangle and solve for R.

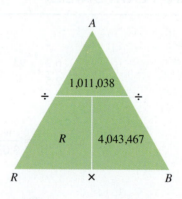

$R = A \div B$
$R = 1{,}011{,}038 \div 4{,}043{,}467$
$R \approx 0.25$ Rounded to the nearest hundredth.
$R \approx 25\%$

So the population of Detroit was about 25% African American.

EXAMPLES A–C

DIRECTIONS: Solve the percent word problem.

STRATEGY: Write the problem in the form "What percent of what is what?" Fill in the known values and find the unknown value.

A. This year the population of Deschutes County is 162% of its population 10 years ago. The population 10 years ago was 142,000. What is the population this year?

STRATEGY: Use the proportion.

162% of 142,000 is the current population.

$$\frac{162}{100} = \frac{A}{142,000}$$

$R = X\% = 162\%$, so $X = 162$, and $B = 142,000$. Substitute these values into the proportion $\frac{X}{100} = \frac{A}{B}$, and solve.

$$142,000(162) = 100A$$
$$23,004,000 = 100A$$
$$23,004,000 \div 100 = A$$
$$230,040 = A$$

Cross multiply.

The population this year is 230,040.

B. In a poll of a group of students, 4 of them say they walk to school, 28 say they ride the bus, 12 ride in car pools, and 8 drive their own cars. What percent of the group rides the bus? Round to the nearest whole pecent.

STRATEGY: Use the proportion.

$4 + 28 + 12 + 8 = 52$ First, find the number in the group.

What % of 52 is 28?   There are 52 in the group, so $B = 52$, and 28 ride in car pools, so $A = 28$.

$$\frac{X}{100} = \frac{28}{52}$$
$$52(X) = 2800$$ Cross multiply.
$$X = 2800 \div 52$$
$$X \approx 54$$ Round to the nearest whole number.
$$R \approx 54\%$$ $R = X\%$.

Approximately 54% of the students ride the bus.

C. In a statistical study of 545 people, 215 said they preferred eating whole wheat bread. What percent of the people surveyed preferred eating whole wheat bread? Round to the nearest whole percent.

STRATEGY: Use the percent formula to solve.

What percent of 545 is 215? $B = 545$ and $A = 215$

$R \times B = A$ Percent formula

$R(545) = 215$ Substitute.

$R = 215 \div 545$

$R \approx 0.39$ Round to the nearest hundredth.

$R \approx 39\%$

So approximately 39% of the people surveyed preferred eating whole wheat bread.

WARM-UP

A. The cost of a certain model of Ford is 135% of what it was 5 years ago. If the cost of the automobile 5 years ago was $20,400, what is the cost today?

WARM-UP

B. A list of grades of students taking biology revealed that 10 students earned an A, 16 earned a B, 21 earned a C, and 8 earned a D. What percent of the students earned a grade of B? Round to the nearest whole percent.

WARM-UP

C. In a similar study of 615 people, 185 said they jog for exercise. What percent of those surveyed jog? Round to the nearest whole percent.

ANSWERS TO WARM-UPS A–C

A. The cost of the Ford today is $27,540.
B. The percent of students earning a B grade is approximately 29%.   **C.** Of the 615 people, 30% jog.

HOW & WHY

■ **OBJECTIVE 2** Find percent of increase and percent of decrease.

To find the percent of increase or decrease, the base *B* is the starting number. The increase or decrease is the amount *A*. For instance, if the population of a city grew from 86,745 to 90,310 in 3 years, the base is 86,745 and the increase is the difference in the populations, $90{,}310 - 86{,}745 = 3565$.

To find the percent of increase in the population, we ask the question "What percent of 86,745 is 3565?" Using the percent equation $R = A \div B$, we have

$R = 3565 \div 86{,}745$ Substitute 3565 for *A* and 86,745 for *B*.
$R \approx 0.0410974$
$R \approx 0.04$ Round to the nearest hundredth.
$R \approx 4\%$ Change to a percent.

So the population increased about 4% in the 3 years.

EXAMPLES D–E

DIRECTIONS: Find the percent of increase or decrease.

STRATEGY: Use one of the two methods to solve for *R*.

WARM-UP
D. At one location the price of a gallon of gasoline went from $1.66 in 2000 to $3.09 in 2010. Find the percent of increase in the price to the nearest tenth of a percent.

D. The cost of a first-class stamp rose from 37¢ in 2002 to 44¢ in May 2009. Find the percent of increase in the price to the nearest tenth of a percent.

STRATEGY: Use the percent formula.

$44¢ - 37¢ = 7¢$ The difference, 7¢, is the amount of increase. The percent of increase is calculated from the amount of increase, 7¢, based on the original amount, 37¢.

What percent of 37¢ is 7¢? $B = 37$ and $A = 7$

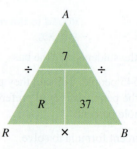

$R = A \div B$
$R = 7 \div 37$
$R \approx 0.189189$
$R \approx 0.189$ Round to the nearest thousandth.
$R \approx 18.9\%$ Change to a percent.

The increase in the price of a stamp was about 18.9%.

ANSWER TO WARM-UP D

D. The price of a gallon of gasoline increased by about 86.1% over the 10-year period.

E. John went on a diet. At the beginning of the diet, he weighed 267 lb. After 6 weeks, he weighed 216 lb. Find the percent of decrease in his weight to the nearest tenth of a percent.

STRATEGY: Use the proportion.

267 lb − 216 lb = 51 lb   The difference, 51 lb, is the amount of decrease from 267 to 216. The percent of decrease is the comparison of the amount, 51 lb, to the original weight, 267 lb.

What percent of 267 is 51? $B = 267$ and $A = 51$

$$\frac{X}{100} = \frac{A}{B}$$

$$\frac{X}{100} = \frac{51}{267}$$ Substitute 51 for A and 267 for B.

$$267X = 51(100)$$
$$X = 5100 \div 267$$
$$X \approx 19.101123$$
$$X \approx 19.1$$ Round to the nearest tenth.
$$R \approx 19.1\%$$ $R = X\%$

So John had a decrease of about 19.1% in his weight.

WARM-UP

E. The price of an LCD televison set decreased from $6734 in 2001 to $3989 in 2005. Find the percent of decrease in the price to the nearest tenth of a percent.

HOW & WHY

■ **OBJECTIVE 3** Read data from a circle graph or construct a circle graph from data.

A circle graph, or pie chart, is used to show how a whole unit is divided into parts. The area of the circle represents the entire unit and each subdivision is represented by a sector. Percents are often used as the unit of measure of the subdivision. Consider the following pie chart (Figure 6.4).

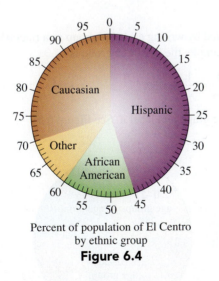

Percent of population of El Centro
by ethnic group
Figure 6.4

From the circle graph we can conclude,

1. the largest ethnic group in El Centro is Hispanic.
2. the Caucasian population is twice the African American population.
3. the African American and Hispanic populations are 60% of the total.

ANSWER TO WARM-UP E

E. The price of an LCD television set decreased about 40.8%.

If the population of El Centro is 125,000, we can also compute the approximate number in each group. For instance, the number of Hispanics is found by

$$R \times B = A \quad R = 45\% = 0.45, B = 125,000$$
$$(0.45)(125,000) = A$$
$$56,250 = A$$

There are approximately 56,250 Hispanics in El Centro.

To construct a circle graph, determine what fractional part or percent each subdivision is, compared to the total. Then draw a circle and divide it accordingly. We can draw a pie chart of the data in Table 6.3.

TABLE 6.3 Population by Age Group

	Age Groups		
	0–21	**22–50**	**Over 50**
Population	14,560	29,120	14,560

Begin by adding two rows and a column to Table 6.3 to create Table 6.4.

TABLE 6.4 Population by Age Group

	Age Group			
	0–21	**22–50**	**Over 50**	**Total**
Population	14,560	29,120	14,560	58,240
Fractional part	$\dfrac{1}{4}$	$\dfrac{1}{2}$	$\dfrac{1}{4}$	1
Percent	25%	50%	25%	100%

The third row is computed by writing each age group as a fraction of the total population and simplifying. For example, the 0–21 age group is

$$\frac{14,560}{58,240} = \frac{1456}{5824} = \frac{364}{1456} = \frac{1}{4}, \text{ or } 25\%$$

Now draw the circle graph and label it. See Figure 6.5.

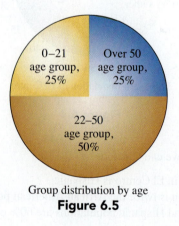

Group distribution by age
Figure 6.5

Sometimes circle graphs are drawn using 1 degree as the unit of measure for the sectors. This is left for a future course.

EXAMPLE F

DIRECTIONS: Answer the questions associated with the graph.

STRATEGY: Examine the graph to determine the size of the related sector.

F. The sources of City Community College's revenue are displayed in the circle graph.

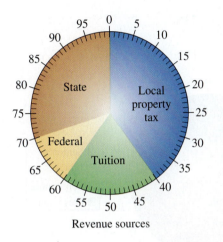

Revenue sources

1. What percent of the revenue is from the federal government?
2. What percent of the revenue is from tuition and property taxes?
3. What percent of the revenue is from federal and state governments?

1. 10%   Read directly from the graph.
2. 60%   Add the percents for tuition and property taxes.
3. 40%   Add the percents for state and federal sources.

So the percent of revenue from the federal government is 10%; the percent from tuition and property taxes is 60%; and the percent from federal and state governments is 40%.

WARM-UP

F. The sales of items at Grocery Mart are displayed in the circle graph.

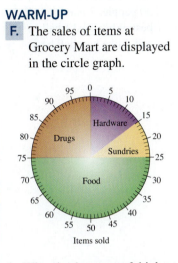

Items sold

1. What is the area of highest sales?
2. What percent of total sales is from sundries and drugs?
3. What percent of total sales is from food and hardware?

ANSWER TO WARM-UP F

F. 1. The highest sales are in food.
2. Sundries and drugs account for 35% of sales.
3. Food and hardware account for 65% of sales.

EXAMPLE G

DIRECTIONS: Construct a circle graph that illustrates the information.

STRATEGY: Use the information to calculate the percents. Divide the circle accordingly and label.

WARM-UP

G. Construct a circle graph to illustrate that in a survey of 20 people, 7 like football best, 9 like basketball best, and 4 like baseball best.

G. Construct a circle graph to illustrate that of the 45 students in a chemistry lecture class, 5 are seniors, 8 are juniors, 24 are sophomores, and 8 are freshmen.

Seniors: $\dfrac{5}{45} \approx 11\%$ Compute the percents to the nearest whole percent.

Juniors: $\dfrac{8}{45} \approx 18\%$

Sophomores: $\dfrac{24}{45} \approx 53\%$

Freshmen: $\dfrac{8}{45} \approx 18\%$

Construct and label the graph.

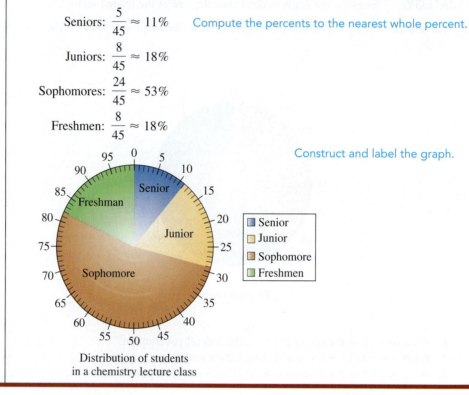

Distribution of students
in a chemistry lecture class

ANSWER TO WARM-UP G

G.

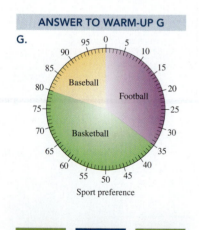

Sport preference

EXERCISES 6.6

■ **OBJECTIVE 1** Solve applications involving percent. (See page 444.)

1. About 16% of Yale's 5240 undergraduate students are from low-income families. To the nearest student, how many undergraduates are from low-income families?

2. Of the 1436 water fowl counted at the Jackson Bottom Wildlife Refuge, 23% were mallard ducks. How many mallard ducks were counted, to the nearest duck?

3. In early 2010, the estimated population of Los Angeles County was 10,395,000. The population was about 44.8% Hispanic or Latino. How many people of Hispanic or Latino heritage were living in Los Angeles County? Round to the nearest 1000.

4. To pass a test to qualify for a job interview, Whitney must score at least 75%. If there are 60 questions on the test, how many must she get correct to score 75%?

5. Alexa answered 27 problems correctly on a 34-problem test. What is her percent score to the nearest whole-number percent?

6. Vera's house is valued at $345,000 and rents for $17,760 per year. What percent of the value of the house is the annual income from rent? Round to the nearest tenth of a percent.

7. Eddie and his family went to a restaurant for dinner. The dinner check was $43.21. He left the waiter a tip of $7. What percent of the check was the tip? Round to the nearest whole-number percent.

8. Adams High School's lacrosse team finished the season with a record of 14 wins and 6 losses. What percent of the games played were won?

9. In preparing a mixture of concrete, Debbie uses 300 pounds of gravel, 100 pounds of cement, and 200 pounds of sand. What percent of the mixture is sand?

10. K-Mart advertises children's T-shirts, regularly $7.99, for $5. What percent off is this? Round to the nearest whole percent.

11. Delplanche Farms has 1180 acres of land in crops. They have 360 acres in soybeans, 410 acres in sweet corn, 225 acres in clover hay, and the rest in wheat. What percent of the acreage is in soybeans? Round to the nearest whole-number percent.

12. A Barnes and Noble store sold 231 fictional books, 135 books on politics, 83 self-help books, and 46 cookbooks in 1 day. What percent of the books sold are books on politics? Round to the nearest whole-number percent.

13. The city Orlando, Florida, had a 2009/2010 proposed budget of $864,030,000. Of this budget, $360,372,000 was the General Fund Budget. What percent of the total budget is the General Fund Budget? Round to the nearest tenth of a percent.

14. Texas has a total land mass of 261,914 square miles. Alaska has a land mass of 570,374 square miles. What percent of the land mass of Alaska is the land mass of Texas? Round to the nearest tenth of a percent.

15. During Mickey Mantle's career in baseball, he was at bat 8102 times and got a hit 2415 times. What percent of the times at bat did he get a hit? Round to the nearest tenth of a percent.

16. It is estimated that Canada spends 10% of its gross domestic product (GDP) on health care annually. In 2008 Canada's GDP was estimated at $1.3 trillion. How much did Canada spend on health care?

17. According to the Organization for Economic Cooperation and Development, 4% of Japan's population have a body mass index over 30—a mark of obesity. If approximately 5 million Japanese people have a body mass index over 30, what is the population of Japan?

18. According to the Organization for Economic Cooperation and Development, 34% of the United States' population have a body mass index over 30—a mark of obesity. If the population of the United States 307 million, approximately how many American have a body mass index over 30? Round to the nearest million people.

19. The manager of a fruit stand lost $16\frac{2}{3}\%$ of his bananas to spoilage and sold the rest. He discarded four boxes of bananas in 2 weeks. How many boxes did he have in stock?

© Gustavo Miguel Fernandes/Shutterstock.com

20. Frank sells magazine subscriptions. He keeps 18% of the cost of each subscription. How many dollars' worth of subscriptions must he sell to earn $510.48?

21. An appliance salesperson keeps 14% of the cost of each refrigerator she sells. How many refrigerators that cost $1145 each must she sell to earn $641.20?

22. The town of Verboort has a population of 17,850, of which 48% is male. Of the men, 32% are 40 years or older. How many men are there in Verboort who are younger than 40?

23. Mary Ann lost 25.6 lb on her diet. She now weighs 146.5 lb. What percent of her original weight did she lose on this diet? Round to the nearest tenth of a percent.

24. During a 6-year period, the cost of maintaining a diesel engine averages $2650 and the cost of maintaining a gasoline engine averages $4600. What is the percent of savings for the diesel compared with the gasoline engine? Round to the nearest whole percent.

25. The label in the figure shows the nutrition facts for one serving of Toasted Oatmeal. Use the information on the label to determine the recommended daily intake of (a) total fat, (b) sodium, (c) potassium, and (d) dietary fiber. Use the percentages for cereal alone. Round to the nearest whole number.

Nutrition Facts

Serving Size 1 cup (49g)
Servings per Container about 9

Amount Per Serving

	Cereal alone	with 1/2 cup Vitamin A&D Fortified Skim Milk
Calories	190	230
Calories from Fat	25	25
	% Daily Values**	
Total Fat 2.5g*	4%	4%
Saturated Fat 0.5g	3%	3%
Polyunsaturated Fat 0.5g		
Monounsaturated Fat 1g		
Cholesterol 0mg	0%	0%
Sodium 220mg	9%	12%
Potassium 180mg	5%	11%
Total Carbohydrate 39g	13%	15%
Dietary Fiber 3g	14%	14%
Soluble Fiber 1g		
Insoluble Fiber 2g		
Sugars 11g		
Other Carbohydrates 24g		
Protein 5g		

26. The table shows the calories per serving of the item along with the number of fat grams per serving.

Calories and Fat Grams per Serving

Item	Calories per Serving	Fat Grams per Serving
Light mayonnaise	50	4.5
Cocktail peanuts	170	14
Wheat crackers	120	4
Cream sandwich cookies	110	2.5

Each fat gram is equivalent to 10 calories. Find the percent of calories that are from fat for each item. Round to the nearest whole percent.

27. In 1900, there were about 10.3 million foreign-born residents of the United States, which represented about 13.6% of the total population. What was the U.S. population in 1900?

28. In 2009, it was estimated that 15.1% of the population of the United States was Hispanic. If the population was 307,212,123, how many of them were Hispanic? Round to the nearest person.

29. Recent figures show that TV viewing is at an all-time high. It estimated that last year the households with TV watched TV for 8.23 hours per day. It is projected that TV viewing will increase this year by 1.5%. To the nearest hundredth, how long will the viewing time be this year?

30. Based on a survey, approximately 23% of TV sets in the United States do not receive cable. If there are about 304,000,000 sets in the United States, how many do not get cable? Round to the nearest million.

■ **OBJECTIVE 2** Find percent of increase and percent of decrease. (See page 446.)

Fill in the table. Calculate the amount to the nearest whole number and the percent to the nearest tenth of a percent.

	Amount	New Amount	Increase or Decrease	Percent of Increase or Decrease
31.	345	415		
32.	1275	1095		
33.	764		Increase of 124	
34.	4050		Decrease of 1255	
35.	2900			Increase of 15%
36.	900			Decrease of 45%

37. The Denver Nuggets's average attendance per home game dropped from 17,364 in 2007–2008 to 17,223 in 2008–2009. What was the percent of decrease? Round to the nearest tenth of a percent.

38. The city of Portland currently charges $10 per year for a permit for sidewalk cafes. The city council has recommended that the permits be increased to $150 per year. What percent increase is this?

39. According to the Office of Immigration Statistics, 1,052,415 persons received legal permanent resident status in 2007 in California. In 2008, the number was 1,107,126. What was the percent of incease? Round to the nearest tenth of a percent.

40. One day the Standard & Poor 500 Index declined 1.24 points to 1090.78. What was the level of the index on the previous day? What percent decrease is this? Round to the nearest tenth of a percent.

41. Jose's Roth IRA grew, due to contributions and interest, from $18,678 on January 1, 2009, to $27,467 on December 31, 2009. What was the percent of increase in the IRA? Round to the nearest tenth of a percent.

42. The population of Virginia was 7,769,089 in 2008, a 9.7% increase over its 2000 population. What was the population of Virginia in 2000? Round to the nearest person.

43. In one western city, 330 homeless families used winter shelter during 2007–2008. In the winter of 2008–2009, 620 homeless families used the shelters. What was the percent of increase? Round to the nearest tenth of a percent.

44. One model of a van gets 17.4 miles per gallon when driven at 50 miles per hour. When the van is driven at 60 miles per hour, the mileage decreases to 13.2 mpg. What is the percent of decrease in mileage at the faster speed? Round to the nearest percent.

45. Wheat production in the United States fell from 2.34 billion bushels in 2003 to 2.2 billion in 2004. What was the percent of decrease in production of wheat? Round to the nearest tenth of a percent.

46. The average salary at Funco Industries went from $34,783 in 2009 to $36,725 in 2010. Find the percent of increase in the average salary. Round to the nearest tenth of a percent.

47. Good driving habits can increase mileage and save on gas. If good driving causes a car's mileage to go from 27.5 mpg to 29.6 mpg, what is the percent of increased mileage? Round to the nearest tenth of a percent.

48. The change in average credit card debt from 2005 to 2008 among low-and middle-income households for people 65 and older was a 26% increase to $10,235. What was the average credit card debt in 2005? Round to the nearest dollar.

49. According to the National Association of Home Builders, in 1978 the median size of new homes was 1650 ft². By 2008, the median size of new homes was 2224 ft². What was the percent increase? Round to the nearest tenth of a percent.

50. In 1900, tuition at Harvard University was $3000 per year. In 1999, Harvard tuition had risen to $22,054. What percent of increase is this? In 2009, the tuition grew to $33,696. What percent of increase is this over the 1999 tuition?

■ **OBJECTIVE 3** Read data from a cricle graph or construct a circle graph from data. (See page 447.)

For Exercises 51–53, the figure shows the ethnic distribution of the population of Texas.

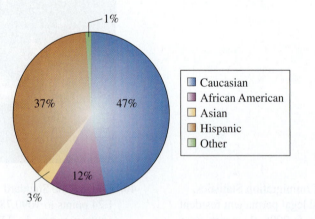

51. Which identifiable ethnic group has the smallest population in Texas?

52. What percent of the population of Texas is nonwhite?

53. What is the second largest ethnic population in Texas?

For Exercises 54–56, the figure shows the number of new car sales by some manufacturers.

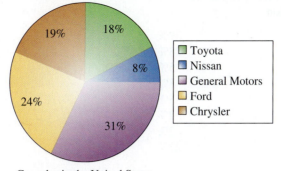

- ☐ Toyota
- ☐ Nissan
- ☐ General Motors
- ☐ Ford
- ☐ Chrysler

Car sales in the United States

54. Which car company sold the greatest number of cars?

55. Which company sold more cars, Nissan or Chrysler?

56. What percent of the cars sold were not GM or Ford models?

57. In a family of three children, there are eight possibilities of boy–girl combinations. One possibility is that they are all girls. Another possibility is that they are all boys. There are three ways for the family to have two girls and a boy. There are also three ways for the family to have two boys and a girl. Use the following circle to make a circle graph to illustrate this information.

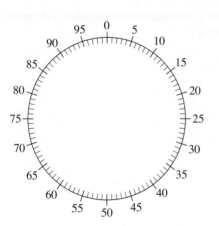

58. In 2000, Orlando, Florida, had the following percentages of households: married couples, 50%; other families, 16%; people living alone, 25%; and other non-family households, 9%. Use the circle to make a circle graph illustrating this information.

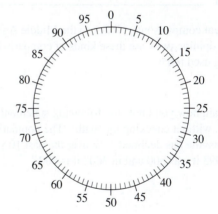

59. The leading causes of death in the United States are as follows. Use the following circle to make a circle graph to illustrate this information.

Coronary heart diseases	39%
Cancer	23%
Lower respiratory infections	5%
Diabetes	3%
Influenza and pneumonia	3%
Alzheimers	2%
Motor vehicle accidents	2%
Other causes	23%

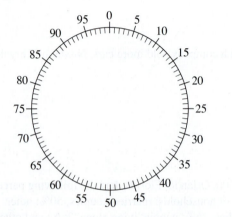

STATE YOUR UNDERSTANDING

60. Percent comparisons go back to the Middle Ages. In your opinion, why are these kinds of comparisons still being used today?

61. When a population doubles, what is the percent of increase?

62. Explain how you know the following statement is false, without checking the math. "The population of Allensville has declined 12% over the past 10 years. In 1993 it was 500 and in 2003 it is 600."

CHALLENGE

63. Carol's baby weighed $7\frac{3}{4}$ lb when he was born. On his first birthday, he weighed $24\frac{3}{8}$ lb. What was the percent of increase during the year? Round to the nearest whole percent.

64. Jose purchases a car for $12,500. He makes a down payment of $1500. His payments are $265 per month for 48 months. What percent of the purchase cost does Jose pay for the car, including the interest? Round to the nearest tenth of a percent.

65. Matilda buys the following items at Safeway: a can of peas, $0.89; Wheaties, $2.46; butter, $1.78; Ivory Soap, $1.15; broom, $4.97; steak, $6.78; chicken, $3.21; milk, $1.89; eggs, $1.78; bread, $1.56; peanut butter, $2.35; stamps, $6.80; potatoes, $1.98; lettuce, $2.07; and orange juice; $2.89. What percent of the cost was in nonfood items? What percent of the cost was in meat products? Round to the nearest tenth of a percent.

66. Theo weighed in at 208.2 lb at the beginning of a weight loss program involving exercise and diet. After three months in the program, his weight had dropped an average of 9.8 lb per month. What was the percent of decrease during this time? Round to the nearest whole percent.

MAINTAIN YOUR SKILLS

67. Sally and Rita are partners. How much does each receive of the income if they are to share $12,600 in a ratio of 6 to 4 (Sally 6, Rita 4)?

68. A 9.6-m board is to be cut in two pieces in a ratio of 6 to 2. What is the length of each board after the cut?

69. Joe buys a new car for $2100 down and $321 per month for 48 months. What is the total amount paid for the car?

70. Pia has a savings account balance of $3892. She deposits $74 per month for a year. What is her new account balance, not including interest earned?

71. An engine with a displacement of 400 cubic inches develops 260 horsepower. How much horsepower is developed by an engine with a displacement of 175 cubic inches?

72. Harry pays federal income taxes of $11,450 on an income of $57,250. In the same tax bracket, what would be the tax on an income of $48,000?

73. Peter attended 18 of the 20 G.E.D. classes held last month. What percent of the classes did he attend?

74. A family spends $120 for food out of a budget of $500. What percent goes for food?

75. There are 20 problems on an algebra test. What is the percent score for 17 problems correct?

76. There are 23 questions on a test for volleyball rules. What is the percent score for 18 correct answers, to the nearest percent?

6.7 Sales Tax, Discounts, and Commissions

OBJECTIVES

1. Solve applications involving sales tax.
2. Solve applications involving discount.
3. Solve applications involving commission.

VOCABULARY

The **original price** of an article is the price the business sets to sell the article. When a store has a sale, the **amount of discount** is the amount subtracted from the original price (regular price). **Sales tax** is the amount charged on the final sale price to finance state and/or city programs. **Net income** is the amount left after expenses have been paid. **Commission** is the money salespeople earn based on the dollar value of the goods sold.

HOW & WHY

■ **OBJECTIVE 1** Solve applications involving sales tax.

Most states and some cities charge a tax on certain items when purchased. Stores collect this tax and send it on to the governmental unit. The amount of the sales tax is added to the purchase price to get the final cost to the buyer.

A sales tax is a percent of the purchase price. For example, a 7.5% sales tax on a purchase price of $100 is found by taking 7.5% of $100.

Sales tax = 7.5% of $100 = $0.075 \times \$100 = \$7.50$

The total cost to the customer would be

Purchase price + Sales tax = Total cost
$100   + $7.50   = $107.50

So the total cost to the customer is $107.50.

EXAMPLES A–B

DIRECTIONS: Solve sales-tax-related applications.

STRATEGY: Use the equations: Sales tax = Sales tax rate × Purchase price and Total cost = Purchase price + Sales tax.

WARM-UP

A. Find the sales tax and the total cost of a dishwasher that has a purchase price of $729 in a city where the sales tax rate is 8.15%.

A. Find the sales tax and the total cost of a Sunraye Slow Cooker that has a purchase price of $89.95 in a city where the sales tax rate is 5.6%.

Sales tax = Sales tax rate × Purchase price
Sales tax = 5.6% × $89.95
        = 0.056($89.95)
        = $5.0372
        ≈ $5.04   Round to the nearest cent.

So the sales tax is $5.04.
To find the total cost to the customer, add the sales tax to the purchase price.

Total cost = $89.95 + $5.04
        = $94.99

So the total cost of the Slow Cooker to the customer is $94.99.

ANSWER TO WARM-UP A

A. The sales tax is $59.41. The final price is $788.41

B. Garcia buys a set of luggage for $335. The cashier charges $353.43 on his credit card. Find the sales tax and the sales tax rate.

To find the sales tax, subtract the purchase price from the total cost.

$353.43 − $335 = $18.43

So the sales tax is $18.43.

To find the sales tax rate we ask the question, "What percent of the purchase price is the sales tax?"

$$R \times B = A$$
$$R(\$335) = \$18.43$$ Substitute 335 for B and 18.43 for A.
$$R = \$18.43 \div \$335$$
$$R \approx 0.0550149$$
$$R \approx 0.055$$
$$R \approx 5.5\%$$ Round to the nearest thousandth.

The sales tax rate is 5.5%. Because all sales tax rates are exact, we can assume the approximation of the rate is due to rounding the tax to the nearest cent.

WARM-UP

B. Juanda buys a 30-in. stainless steel range for $695. The cashier charges $736.70 on her credit card. Find the sales tax and the sales tax rate.

HOW & WHY

■ **OBJECTIVE 2** Solve applications involving discount.

Merchants often discount items to move merchandise. For instance, a store might discount a dress that lists for $125.65 by 30%. This means that the merchant will subtract 30% of the original price to determine the sale price. (Equivalently, the sale price is 70% of the original price.) To find the sale price, we first calculate the amount of the discount using the percent formula:

$R\%$ of list price is the amount of the discount.
30% of $125.65 is the discount.

$$30\% \times \$125.65 = \text{discount}$$
$$0.3(\$125.65) = \text{discount}$$
$$\$37.70 \approx \text{discount}$$ Rounded to the nearest cent.

To find the sale price, subtract the discount from the original price.

$$\$125.65 - \$37.70 = \text{sale price}$$
$$\$87.95 = \text{sale price}$$

So the sale price for the dress is $87.95.

EXAMPLES C–D

DIRECTIONS: Solve discount-related applications.

STRATEGY: Use the equations: Original price − Discount = Sale price and Rate of discount × Original price = Amount of discount.

C. Safeway offers $20 off any purchase of $100 or more. Jan purchases groceries totaling $118.95. What is her total cost and what is the percent of discount?

To find the total cost to Jan, subtract $20 from her purchases.

$118.95 − $20.00 = $98.95

So the groceries cost Jan $98.95.

To find the percent of discount answer the question:

What percent of $118.95 is $20.00? $118.95 is the original cost before the discount.

WARM-UP

C. Safeway offers $20 off any purchase of $100 or more. Gil purchases groceries totaling $149.10. What is his total cost and what is the percent of discount?

ANSWERS TO WARM-UPS B–C

B. The sales tax is $41.70. The sales tax rate is 6%.
C. Gil's groceries cost $129.10. The percent of discount is about 13%.

Now use the formula

$$R \times B = A$$
$$R(\$118.95) = \$20 \qquad \text{Substitute 118.95 for } B \text{ and 20 for } A.$$
$$R = \$20 \div \$118.95$$
$$R \approx 0.1681378$$
$$R \approx 0.17 \qquad \text{Round to the nearest hundredth.}$$
$$R \approx 17\%$$

So Jan received about a 17% discount on the cost of her groceries.

WARM-UP

D. Ralph goes to the sale at the store in Example D and buys a $330 rifle that is discounted 20%. What does Ralph pay for the rifle, including the sales tax, if he buys the rifle during the 6 A.M. to 10 A.M. time period?

D. Floremart Discount advertises a special Saturday morning sale. From 6 A.M. to 10 A.M., all purchases will be discounted an additional 15% off the already discounted prices. At 8:30 A.M., Carol buys a $1029 TV set that is on sale at a 30% discount. Floremart Discount is in a county with a total sales tax of 5.2%. What does Carol pay for the TV set, including sales tax?

First, find the price of the TV after the 30% discount.

The discount is 30% of $1029.

$$\text{Discount} = 0.30 \times \$1029$$
$$= \$308.70$$

$$\text{Sale price} = \text{Original price} - \text{Discount}$$
$$= \$1029 - 308.70$$
$$= \$720.30$$

Now find the additional 15% discount off the sale price.

$$\text{Additional discount} = 15\% \text{ of the sale price.}$$
$$= 0.15 \times \$720.3.$$
$$\approx \$108.05 \qquad \text{Rounded to the nearest cent.}$$

Now, to find the purchase price for Carol, subtract the additional discount from the sale price.

$$\$720.30 - \$108.05 = \$612.25$$

So Carol's purchase price for the TV is $612.25.

Finally, calculate the sales tax and add it on to the purchase price to find the total cost to Carol.

$$\text{Sales tax} = \text{Rate of sale tax} \times \text{Purchase price}$$
$$\text{Sales tax} = 5.2\% \times 612.25$$
$$\approx \$31.84 \qquad \text{Rounded to the nearest cent.}$$

$$\text{Total cost} = \text{Purchase price} + \text{Sales tax}$$
$$= \$612.25 + \$31.84$$
$$= \$644.09$$

Including sales tax, Carol pays $644.09 for the TV set.

HOW & WHY

■ OBJECTIVE 3 Solve applications involving commission.

Commission is the amount of money salespeople earn based on the dollar value of goods sold. Often salespeople earn a base salary plus commission. When this happens, the total salary earned is found by adding the commission to the base salary. For instance, Nuyen earns a base salary of $250 per week plus an 8% commission on her total sales. One week Nuyen has total sales of $2896.75. What are her earnings for the week?

First, find the amount of her commission by multiplying the rate times the total sales.

$$\text{Commission} = \text{Rate} \times \text{Total sales}$$
$$= 8\% \times \$2896.75 \qquad \text{Substitute 8\% for the rate and \$2896.75}$$
$$= 0.08 \times \$2896.75 \qquad \text{for the total sales.}$$
$$= \$231.74$$

So Nuyen earns $231.74 in commission.

ANSWER TO WARM-UP D

D. Ralph pays $236.07 for the rifle, including the sales tax.

Now find her total earnings by adding the commission to her base salary.

Total salary = Base pay + Commission
               = $250 + $231.74   *Substitute $250 for base pay and $231.74*
               = $481.74   *for commission.*

So Nuyen earns $481.74 for the week.

EXAMPLES E–G

DIRECTIONS: Solve commission-related applications.

STRATEGY: Use the equations: Commission = Commission rate × Total sales and Total earnings = Base pay + Commission.

E. Monty works for a medical supply firm on straight commission. He earns 11% commission on all sales. During the month of July, his sales totaled $234,687. How much did Monty earn during July?

Commission = Commission rate × Total sales
               = 11% × $234,687
               = 0.11 × $234,687
               = $25,815.57

Monty earned $25,815.57 during July.

F. Dallas earns $900 per month plus a 4.8% commission on his total sales at Weaver's Used Cars. In November, Dallas had sales totaling $98,650. How much did Dallas earn in November?

First find Dallas's commission earnings.

Commission = Commission rate × Total sales
               = 4.8% × $98,650
               = 0.048 × $98,650
               = $4735.20

Dallas earned $4735.20 in commission in November.

Now find Dallas's total earnings.

Total earnings = Base pay + Commission
               = $900 + $4735.20
               = $5635.20

Dallas's total earnings for November were $5635.20.

G. Mabel is paid a straight commission. Last month, she earned $22,280.40 on total sales of $185,670. What is her rate of commission?

We need to answer the question:
What percent of $185,670 is $22,280.40?

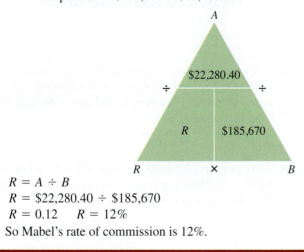

$R = A \div B$
$R = \$22,280.40 \div \$185,670$
$R = 0.12$ $R = 12\%$
So Mabel's rate of commission is 12%.

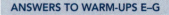

■ **OBJECTIVE 1** Solve applications involving sales tax. (See page 458.)

Fill in the table. Round tax rates to the nearest tenth of a percent.

	Marked Price	Sales Tax Rate	Amount of Tax	Total Cost
1.	$238	5.8%		
2.	$65.70	6.8%		
3.	$90.10	3.8%		
4.	$467	4.7%		
5.	$628		$28.26	
6.	$37.49		$2.32	
7.	$1499			$1552.96
8.	$213.12			$225.05

9. A treadmill costs $895.89 plus a 6.8% sales tax. Find the amount of the sales tax.

10. A snowmobile costs $4786.95 plus a 5.5% sales tax. Find the amount of the sales tax.

© Olga Utlyakova/Shutterstock.com

11. George buys a new suit at the Bon Marché. The suit costs $476.45 plus a 6.3% sales tax. What is the total cost of the suit, including the sales tax?

12. Mildred buys a prom dress for her daughter. The dress costs $214.50 plus a 4.6% sales tax. What is the total cost of the dress, including the sales tax?

13. Wilbur buys a new refrigerator that costs $1075.89. When he pays for the refrigerator, the bill is $1143.67, including the sales tax. Find the sales tax rate to the nearest tenth of a percent.

14. Helen buys an above-ground swimming pool for $2460.61. The total bill for the pool, including sales tax, is $2664.84. Find the sales tax rate to the nearest tenth of a percent.

15. Jim buys a new 42-plasma HDTV priced at $749.99 on sale for $125 off. The city sales tax is 4.8%. What is the final cost to Jim?

16. Terry buys an 8GB MP3 player at $119.95 on sale for $19.99 off. There is a county sales tax of 2.4%. What does the player cost Terry?

17. Malisha buys a Blu-ray DVD player priced at $189.90 less a discount of $22. The state sales tax on the purchase adds $6.05 to the purchase price. What is the sales tax rate?

18. Skylar buys two pair of denim jeans marked at $92.99 less $14 off. In addition, he pays $5.24 state sales tax. What is the sales tax rate?

19. Jennifer buys a new gas barbeque grill priced at $675.95. When she checks out, the total bill is $726.65, including the sales tax. What is the sales tax rate?

20. Hilda buys a new self-propelled lawn mower for $1125. When she checks out, the total bill is $1175.63, including the sales tax. What is the sales tax rate?

■ **OBJECTIVE 2** Solve applications involving discount. (See page 459.)

Fill in the table.

	Original Price	Rate of Discount	Amount of Discount	Sale Price
21.	$75.82	18%		
22.	$25.65	20%		
23.	$320	30%		
24.	$587.50	25%		
25.	$15.95		$0.49	
26.	$249.99		$11.00	
27.	$1798			$1706.30
28.			$320	$1280

29. Joan buys an oil painting with a list price of $189.95 at a 15% discount. What does she pay for the painting?

30. Les Schwab Tires advertises tires at a 20% discount. If a tire has a list price of $93.65, what is the sale price?

31. Larry buys a new Lexus RX 400 SUV for 9% below the sticker price. If the sticker price is $52,890, how much does Larry pay for the Lexus?

32. Helen's Of Course advertises leather jackets for women at a 35% discount. What will Carol pay for a jacket that has a list price of $456.85?

33. Melvin buys a $979.99 Honda DuraPower generator at a discount of 28%. What is the amount of the discount? What is the price Melvin pays for the generator?

34. Tran bought a new rug priced at $785.50 at a discount of 15%. What was the amount of the discount? How much did Tran pay for the rug?

35. Linda and Ron buy a rocker recliner at a 44% discount from the original price plus sales tax. If the list price is $598 and their total bill is $350.95, what are the discount price and the sales tax rate? Find the sales tax rate to the nearest tenth of a percent.

36. Melissa buys a new queen mattress set at a 58% discount from the original price plus sales tax. If the original price is $1299 and Melissa's final bill is $574.50, what is the discount price? Find the sales tax rate to the nearest tenth of a percent.

37. Autogajet offers a new GPS with an original price of $89.99 for $20 off. This weekend, they offer a discount of 10% on their already reduced price. If Heather buys the GPS this weekend, how much will she save off the original price?

38. Exerworld sells an elliptical trainer that originally sold for $1049 at a discount price of $489.99. This weekend, they are advertising an additional 12% off their already reduced price. How much will Louise save off the original price for trainer this weekend?

39. The Hugh TV and Appliance Store regularly sells a TV for $536.95. An advertisement in the paper shows that it is on sale at a discount of 25%. What is the sale price to the nearest cent?

40. A competitor of the store in Exercise 39 has the same TV set on sale. The competitor normally sells the set for $539.95 and has it advertised at a 26% discount. To the nearest cent, what is the sale price of the TV? Which is the better buy and by how much?

41. The Top Company offers a 6% rebate on the purchase of their best model of canopy. If the regular price is $445.60, what is the amount of the rebate to the nearest cent?

42. The Stihl Company is offering a $24 rebate on the purchase of a chain saw that sells for $610.95. What is the percent of the rebate to the nearest whole-number percent?

43. Corduroy overalls that are regularly $42.75 are on sale for 25% off. What is the sale price of the overalls?

44. A pair of New Balance cross trainers, which regularly sells for $107.99, goes on sale for $94.99. What percent off is this? Round to the nearest whole-number percent.

45. Raven buys a leather coat advertised as 45% off at a department store. If the coat was originally marked $249 and the store also has an "additional 24% off all reduced goods" sale, what is the price of the coat? What percent savings does Raven receive compared to the original price?

46. Camera Warehouse offers a high-definition camcorder for $275 off the original price of $889. In addition, they are running an end-of-the-month discount of 8.5% off any purchase. What is the final price of the camcorder and what percent savings does this represent over the original price?

47. A shoe store advertises "Buy one pair, get 50% off a second pair of lesser or equal value." The mother of twin boys buys a pair of basketball shoes priced at $55.99 and a pair of hikers priced at $42.98. How much does she pay for the two pairs of shoes? What percent savings is this to the nearest tenth of a percent?

48. The Klub House advertises on the radio that all merchandise is on sale at 25% off. When you go in to buy a set of golf clubs that originally sold for $1375, you find that the store is giving an additional 10% discount off the original price. What is the price you will pay for the set of clubs?

49. In Exercise 48, if the salesperson says that the 10% discount can only be applied to the sale price, what is the price of the clubs?

50. A store advertises "30% off all clearance items." A boy's knit shirt is on a clearance rack that is marked 20% off. How much is saved on a knit shirt that was originally priced $25.99?

■ OBJECTIVE 3 Solve applications involving commission. (See page 460.)

Fill in the table.

	Sales	Rate of Commission	Commission
51.	$4890	9%	
52.	$11,560	6.5%	
53.	$67,320	8.5%	
54.	$234,810	15%	
55.	$1100		$44
56.	$1,780,450		$62,315.75
57.		9%	$31,212
58.		8.5%	$1426.64

59. Grant earns a 7.5% commission on all of his sales at Computer Universe. If Grant's sales for the week totaled $8340.90, what was his commission?

60. Mayfair Real Estate Co. charges a 5.1% commission on each townhouse it sells. What is the commission on the sale of a $455,999 townhouse? Round to the nearest dollar.

61. Walt's Ticket Agency charges a 7% commission on all ticket sales. What is the commission charged for eight tickets priced at $45.50 each?

62. A medical supplies salesperson earns an 8% commission on all sales. Last month, she had total sales of $345,980. What was her commission?

63. A salesperson at Wheremart earns $500 per week plus a commission of 9% on all sales over $2000. Last week, he had total sales of $4678.50. How much did he earn last week?

64. A salesperson at Goldman's Buick earns a base salary of $1800 per month plus a commission of 2.5% on all sales. If she sold cars totaling $178,740 during June, how much did she earn that month?

65. Carlita earns a 7% commission on all of her sales. How much did she earn last week if her total sales were $6025?

66. Tyler earns a 10.5% commission on all sales. How much did he earn last week if his total sales were $5500?

67. Perry receives a weekly salary of $260 plus a commission of 7.5% on his sales. Last week he earned $705. What were his total sales for the week?

68. Belinda is paid a weekly salary of $205 plus a commission of 9% on her total sales. How much did she earn last week if her total sales were $2250?

69. During one week, Ms. James sold a total of $26,725 worth of hardware to the stores in her territory. She receives a 4% commission on sales of $2000 or less, 5% on that portion of her sales over $2000 and up to $15,000, and 6% on all sales over $15,000. What was her total commission for the week?

70. If Ms. James, in Exercise 69, had sales of $22,455 the next week, how much did she earn in commissions?

71. Does a sales tax represent a percent increase or decrease in the price a consumer pays for an item? Explain.

72. When a salesperson is working on commission, is it to his or her advantage to sell you a modestly priced item or an expensive item? Why?

CHALLENGE

73. Marlene's sales job pays her $1500 per month plus commissions of 9% for sales up to and including $30,000 and 5% for all sales over $30,000. In July Marlene had sales of $18,700 and in August she had sales of $49,500. What was Marlene's salary for each of the two months?

74. Fransica bought a sweater that was originally priced at $136 on Senior Day at the department store. The sweater was on sale at 30% off the original price. On Senior Day, the store offers seniors an additional 25% off the sale price. What did Fransica pay for the sweater? The next day, the store had the same sweater on sale again for 30% off the original price plus a 15% discount on the original price. Maria bought the sweater and used a $5 coupon. What did Maria pay for the sweater? Who got the sweater for the best price and by how much?

MAINTAIN YOUR SKILLS

Add.

75. $\dfrac{11}{24} + \dfrac{13}{36}$

76. $\dfrac{17}{21} + \dfrac{13}{28}$

Subtract.

77. $\dfrac{9}{16} - \dfrac{1}{12}$

78. $13\dfrac{1}{3} - 9\dfrac{5}{12}$

Multiply.

79. $\dfrac{7}{16} \cdot \dfrac{8}{21}$

80. $6\dfrac{5}{6} \cdot 3\dfrac{2}{3}$

Divide.

81. $\dfrac{11}{12} \div \dfrac{33}{52}$

82. $5\dfrac{5}{6} \div 2\dfrac{1}{3}$

83. Find the perimeter of the rectangle.

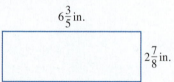

$6\dfrac{3}{5}$ in.

$2\dfrac{7}{8}$ in.

84. Find the area of the rectangle in Exercise 83.

Interest on Loans

OBJECTIVES

1. Calculate simple interest.
2. Calculate compound interest.
3. Solve applications related to credit card payments.

VOCABULARY

Interest is the fee charged for borrowing money. It is usually assessed as a percent of the money borrowed, or the **interest rate**. **Principal** is the amount of money borrowed. When the interest is based on borrowing the money for one year it is called **simple interest**. **Compound interest** occurs when interest is computed on interest already earned. **Interest** is also the money paid for use of your money. Interest is paid on savings and on investments.

HOW & WHY

■ **OBJECTIVE 1** Calculate simple interest.

Simple interest is seldom used these days in the business world. You are more likely to find it when borrowing money from a family member or friend. For instance, Joyce borrows $2000 from her uncle to pay for this year's tuition. She agrees to pay her uncle 4% interest on the money at the end of a year. To find the interest Joyce owes her uncle at the end of the year, we use the equation:

$I = PRT$

that is,

Simple interest = Principal × Interest rate × Time

Here, the principal is $2000, the interest rate is 4%, and the time is 1 year. Substituting, we have

Simple interest = $2000 × 4% × 1
             = $2000 × 0.04 × 1
             = $80

So Joyce owes her uncle $80 in interest at the end of 1 year. She owes a total of $2080.

If the money is borrowed for less than 1 year, the time is expressed as a fraction of a year. See Example B.

EXAMPLES A–B

DIRECTIONS: Find simple interest.

STRATEGY: Use the equation: $I = PRT$.

A. Juan borrows $4500 at 7.5% simple interest to buy a new plasma TV. He agrees to pay back the entire amount at the end of 3 years. How much interest will Juan owe? What is the total amount he will owe at the end of 3 years?

$I = PRT$

Simple interest = Principal × Interest rate × Time
              = $4500 × 7.5% × 3   *Principal is $4500, interest rate is 7.5%,*
              = $4500 × 0.075 × 3   *and the time period is 3 years.*
              = $1012.50

At the end of 3 years, Juan will owe $1012.50 in interest.

To find the total amount Juan will owe, add the interest to the principal.

$$\text{Total amount owed} = \text{Principal} + \text{Interest}$$
$$= \$4500 + \$1012.50$$
$$= \$5512.50$$

So at the end of 3 years, Juan will owe $5512.50.

B. Alex borrows $875 at 4% simple interest to pay for his vacation. He agrees to pay back the entire amount, including interest, at the end of 9 months. How much interest does he pay? What is the total amount he owes after the 9 months?

First find the interest:

$$I = PRT$$
$$\text{Simple interest} = \text{Principal} \times \text{Interest rate} \times \text{Time}$$
$$= \$875 \times 4\% \times 0.75 \qquad \text{Nine months is } \frac{9}{12} \text{ of a year, or } \frac{3}{4} \text{ or } 0.75$$
$$= \$875 \times 0.04 \times 0.75 \qquad \text{of a year.}$$
$$= \$26.25$$

So Alex owes $26.25 in interest.
Now find the total amount owed:

$$\text{Total amount owed} = \text{Principal} + \text{Interest}$$
$$= \$875 + \$26.25$$
$$= \$901.25$$

So Alex owes $26.25 in interest and a total of $901.25.

HOW & WHY

■ **OBJECTIVE 2** Calculate compound interest.

When interest is compounded, the interest earned at the end of one time period is added to the principal and earns interest during the next time period. For instance, if at the beginning of the year you invest $2000 at 5% interest compounded semiannually, your account is first credited with interest in June. During the second half of the year, you will earn interest on both the principal and the interest earned during the first 6 months. To calculate the total interest we first find the simple interest earned after 6 months.

$$I = PRT$$
$$\text{Simple interest} = \text{Principal} \times \text{Interest rate} \times \text{Time}$$
$$= \$2000 \times 5\% \times 6 \text{ mo}$$
$$= \$2000 \times 0.05 \times 0.5 \qquad 6 \text{ mo} = 0.5 \text{ yr}$$
$$= \$50$$

So $50 is earned in interest after 6 months. This amount is added to the principal for the next time period. So for the last 6 months the principal is

$$\text{New principal} = \$2000 + \$50$$
$$= \$2050$$

Now calculate the interest earned during the next 6 months.

$$I = PRT$$
$$\text{Simple interest} = \text{Principal} \times \text{Interest rate} \times \text{Time}$$
$$= \$2050 \times 5\% \times 6 \text{ mo} \qquad \text{The new principal is } \$2050.$$
$$= \$2050 \times 0.05 \times 0.5$$
$$= \$51.25$$

So an additional $51.25 is earned during the last 6 months. To find the balance at the end of the year, add the new interest to the new principal.

Total value at the end of 1 year = $2050 + $51.25
                                 = $2101.25

We can now find the total interest earned by subtracting the original principal from the ending balance.

Total interest earned = Ending balance − Original principal
                      = $2101.25 − $2000
                      = $101.25

So the total interest earned is $101.25.

Simple interest for the year would have been $100 ($I = 2000 \times 0.05 \times 1$), so by compounding the interest semiannually an additional $1.25 was earned. This may not seem like a lot of money, but if the interest was compounded daily and over a number of years, it would amount to a lot of money. For instance, $10,000 invested at 5% simple interest will have a balance of $15,000 at the end of 10 years. However if the $10,000 was invested at 5% compounded daily, it would grow to $16,486.65, which is $1486.65 more than at simple interest.

Computing compound interest can be very tedious, especially as the number of periods increase per year. To ease this burden, accountants have developed compound interest tables that provide a factor to use in calculating the ending balance. Table 6.5 gives the factors for interest rates that are compounded quarterly (4 times per year).

TABLE 6.5 Compound Interest Factors for Quarterly Compounding

Rate	Years					
	1	5	10	15	20	25
2%	1.0202	1.1049	1.2208	1.3489	1.4903	1.6467
3%	1.0303	1.1612	1.3483	1.5657	1.8180	2.1111
4%	1.0406	1.2202	1.4889	1.8167	2.2167	2.7048
5%	1.0509	1.2820	1.6436	2.1072	2.7015	3.4634
6%	1.0614	1.3469	1.8140	2.4432	3.2907	4.4320
7%	1.0719	1.4148	2.0016	2.8318	4.0064	5.6682

To use the table to find the ending balance, choose the row with the appropriate interest rate and the column with the correct number of years. Multiply the original principal by the factor in the table at the intersection of the row and column selected.

Ending balance = Original principal × Compound factor

To calculate the value of $12,000 at 6% interest compounded quarterly for 15 years we write

Ending balance = Original principal × Compound factor
               = $12,000 × 2.4432   Multiply by the compound factor from table.
               = $29,318.40

So $12,000 will grow to $29,318.40 at the end of 15 years.

The total interest earned can be found by subtracting the original principal from the ending balance. The total interest earned is

$29,318.40 − $12,000 = $17,318.40

The account earned $17,318.40 in interest.

Most banks and credit unions compute the interest daily or continuously. For these factors see Appendix D on page 637.

DIRECTIONS: Find compound interest and ending balances.

STRATEGY: Use the equations: Ending balance = Principal × Compound factor

and Interest = Ending balance − Original principal

C. Janis invests $20,000 at 3% interest compounded quarterly for 10 years. Find the value of her investment at the end of the 10 years. Find the interest earned.

C. For his retirement, Usuke invests $18,000 at 7% interest compounded quarterly for 20 years. Find the value of his investment at the end of the 20 years. Find the interest earned.

First find the ending balance:

Ending balance = Principal × Compound factor

  = $18,000 × 4.0064   Multiply by the compound factor found

  = $72,115.20   in Table 6.5.

Now find the interest earned:

Interest = Ending balance − Original principal

  = $72,115.20 − $18,000

  = $54,115.20

So Usuke's retirement investment will be worth $72,115.20, and the interest earned will be $54,115.20.

D. Mitchell invests $1850 at 5% interest compounded daily for 5 years. Find the value of his investment at the end of the 5 years and the interest earned. Use the table in Appendix D.

D. Joan invests $75,000 at 4% interest compounded daily for 10 years. Find the value of her investment at the end of the 10 years and the interest earned. Use the table in Appendix D.

First find the ending balance:

Ending balance = Principal × Compound factor

  = $75,000 × 1.4918   Multiply by the compound factor found in

  = $111,885   Appendix D.

Now find the interest earned:

Interest = Ending balance − Original principal

  = $111,885 − $75,000

  = $36,885

So the value of Joan's investment is $111,885, and the interest earned is $36,885.

HOW & WHY

■ **OBJECTIVE 3** Solve applications related to credit card payments.

Credit card companies and most major department stores charge a fixed interest rate on the unpaid balance in an account. Recently many major credit card companies raised the minimum payment from 2% to 4% of the unpaid balance rounded to the nearest dollar. Of the money in the minimum payment, the credit card company first takes out the interest due and applies the leftover money to reduce the balance. Consider a credit card company that charges 13.49% interest on the unpaid balance. If a card holder has unpaid balance of $2165, we can calculate the minimum payment.

Minimum payment = Unpaid balance × Minimum payment rate

  = $2165 × 4%

  = $2165 × 0.04

  ≈ $87.00   Round to the nearest dollar.

To find the amount of interest in the $87.00 minimum payment, we need to calculate the monthly interest on the credit card balance. To do this we find the simple interest per year on the balance at 13.49% and divide it by 12 to find the monthly interest.

C. Janis's investment is now worth $26,966, and she earned $6966 in interest. **D.** Mitchell's investment is now worth $2375.40, and he earned $525.40 in interest.

$$\text{Monthly interest} = (\text{Unpaid balance} \times \text{Interest rate}) \div 12$$
$$= (\$2165 \times 13.49\%) \div 12$$
$$= (\$2165 \times 0.1349) \div 12$$
$$= \$292.0585 \div 12$$
$$\approx \$24.34 \qquad \text{\color{blue}Round to the nearest cent.}$$

So the interest charged is $24.34. The amount applied to the unpaid balance is found by subtracting the interest charge from the payment:

$$\text{Amount applied to unpaid balance} = \text{Minimum payment} - \text{Interest charge}$$
$$= \$87.00 - \$24.34$$
$$= \$62.66$$

Now find the new unpaid balance:

$$\text{Unpaid balance} = \text{Beginning unpaid balance} - \text{Amount applied to unpaid balance}$$
$$= \$2165 - \$62.66$$
$$= \$2102.34$$

If no further charges are made to the account, next month's payment will be based on the new balance of $2102.34.

If credit card companies receive the payment late, they often add on late fees, which add to the unpaid balance. Some credit card companies will also increase the interest rate on accounts when the payment is received late. For instance, a company that charges a rate of 18.5% may raise the rate to 24.5% for receiving a payment late. This rate then continues for the life of the card.

EXAMPLES E–F

DIRECTIONS: Find the minimum payment, interest paid, and unpaid balance.

STRATEGY: Use the equations:

1. Fixed monthly payments for a set period of time.

   Interest paid = (Monthly payment × Number of months) − Principal

2. Credit and charge card payments.

   Minimum payment = Unpaid balance × Minimum payment rate
   Monthly interest = (Unpaid balance × Interest rate) ÷ 12
   Amount applied to unpaid balance = Minimum payment − Interest charge
        Unpaid balance = Beginning unpaid balance − Amount applied to unpaid balance

E. Millie buys a refrigerator for $1100. She makes a down payment of $100 and agrees to monthly payments of $62.39 for 18 months. How much interest does she pay for the refrigerator by financing the remaining $1000?

$$\text{Interest paid} = (\text{Monthly payment} \times \text{Number of months}) - \text{Principal}$$
$$= (\$62.39 \times 18) - \$1000$$
$$= \$1123.02 - \$1000$$
$$= \$123.02$$

So Millie pays $123.02 in interest.

F. Lucy has an unpaid balance of $1210 on her credit card. The company charges an interest rate of 19.6% and requests a minimum payment of 4% of the unpaid balance. If Lucy makes the minimum payment, calculate the minimum payment, the interest paid, and the new unpaid balance.

F. Ivan has an unpaid balance of $2540 on his credit card. The credit card company charges an interest rate of 15.7% and requires a minimum payment of 4% of the unpaid balance. Calculate the minimum payment. If Ivan makes the minimum payment, calculate the interest paid and the new unpaid balance.

First, find the minimum payment:

Minimum payment = Unpaid balance × Minimum payment rate
= $2540 × 4%
≈ $102.00   Round to the nearest dollar.

So the minimum payment is $102.00.
Now find the amount of interest paid:

Monthly interest = (Unpaid balance × Interest rate) ÷ 12
= ($2540 × 15.7%) ÷ 12
= ($2540 × 0.157) ÷ 12
≈ $33.23   Round to the nearest cent.

So the interest paid is $33.23.
Now find the amount applied to the unpaid balance:

Amount applied to unpaid balance = Minimum payment − Interest charge
= $102.00 − $33.23
= $68.77

Now subtract the amount paid on the unpaid balance to find the new unpaid balance:

Unpaid balance = Beginning unpaid balance − Amount applied to unpaid balance
= $2540 − $68.77
= $2471.23

So Ivan made a minimum payment of $102.00; $33.23 of this payment was interest and the new unpaid balance is $2471.23.

ANSWER TO WARM-UP F

F. Lucy makes a minimum payment of $48; $19.76 is interest and $28.24 is paid on the balance. Her new unpaid balance is $1181.76.

EXERCISES 6.8

OBJECTIVE 1 Calculate simple interest. (See page 467.)

Find the interest and the total amount due on the following simple interest loans.

	Principal	Rate	Time	Interest	Total Amount Due
1.	$10,000	5%	1 year		
2.	$960	7%	1 year		
3.	$5962	8%	1 year		
4.	$42,500	4.5%	1 year		
5.	$32,000	8%	4 years		
6.	$1560	4%	5 years		
7.	$4500	5.5%	9 months		
8.	$850	10%	8 months		

9. Maria invests $2400 for 1 year at 7.5% simple interest. How much interest does she earn?

10. Randolph borrows $6700 at 6.9% simple interest. How much interest does he owe after 1 year?

11. Nancy invests $8500 at 5.5% simple interest for 3 years. How much interest has she earned at the end of the 3 years?

12. Vince invests $13,000 at 6% simple interest for 5 years. How much interest does he earn during this time period?

13. Tyra invests $5765 at 4.5% simple interest for 8 months. How much interest does she earn?

14. Roberto borrows $1700 at 7% simple interest. At the end of 5 months, he pays off the loan and interest. How much does he pay to settle the loan?

15. Janna borrows $2300 at 6.5% simple interest. At the end of 10 months, she pays off the loan and interest. How much does she pay to settle the loan?

16. Oswaldo borrowed $3000 at simple interest, and at the end of 1 year paid off the loan with $3360.
    a. What was the interest paid?
    b. What was the interest rate?

17. Ramon borrows $45,800 at simple interest, and at the end of 1 year he pays off the loan at a cost of $50,838.
    a. What was the interest paid?
    b. What was the interest rate?

■ **OBJECTIVE 2** Calculate compound interest. (See page 468.)

Find the ending balance in the accounts. See Appendix D.

18. The account opens with $6000, earns 5% interest compounded quarterly, and is held for 10 years.

19. The account opens with $21,000, earns 7% interest compounded quarterly, and is held for 5 years.

20. The account opens with $2500, earns 3% interest compounded monthly, and is held for 10 years.

21. The account opens with $60,000, earns 6% interest compounded monthly, and is held for 20 years.

Find the amount of compound interest earned. See Appendix D.

22. The account opens with $7500, earns 4% interest compounded quarterly, and is held for 10 years.

23. The account opens with $9800, earns 5% interest compounded quarterly, and is held for 15 years.

24. The account opens with $36,800, earns 7% interest compounded daily, and is held for 5 years.

25. The account opens with $95,000, earns 6% interest compounded daily, and is held for 15 years.

26. Mary invests $7000 at 6% compounded quarterly. Her sister Catherine invests $7000 at 6% compounded daily. If both sisters hold their accounts for 10 years, how much more interest will Catherine's account earn?

27. Jose invests $15,000 at 4% compounded quarterly. His sister Juanita invests $15,000 at 4% compounded daily. If they both hold their accounts for 15 years, how much more interest will Juanita's account earn?

28. Jim invests $25,000 at 5% simple interest for 10 years. Carol invests $25,000 at 5% compounded daily for 10 years. How much more interest does Carol's account earn?

29. Lucy invests $15,000 at 3% simple interest for 5 years. Carl invests $15,000 at 3% compounded daily for 5 years. How much more interest does Carl's account earn?

30. Felicia's credit card has a balance owed at the end of January of $1346.59. The credit card company charges a rate of 12.98% on the unpaid balance. Felicia makes the minimum payment of $54.
   a. How much of the payment is interest?

   b. How much of the payment goes to pay off the balance?
   c. What is the unpaid balance at the end of February, assuming no additional charges were made?

31. Mark's credit card has a balance owed at the end of May of $3967.10. The credit card company charges a rate of 17.5% on the unpaid balance. Mark makes a payment of $200.
   a. How much of the payment is interest?

   b. How much of the payment goes to pay off the balance?
   c. What is the unpaid balance at the end of June, assuming no additional charges were made?

32. Luis's credit card has a balance owed at the end of June of $2476.10. The credit card company charges a rate of 19.8% on the unpaid balance. Luis makes the minimum payment of $99.
   a. How much of the payment is interest?

   b. How much of the payment goes to pay off the balance?
   c. What is the unpaid balance at the end of July, if Luis uses his card to make $128.54 in additional charges?

33. According to Transunion, one of the three major credit reporting agencies, the average credit card holder in Iowa has an outstanding balance of $4277, the lowest in the nation. Assume that this balance is with a credit card which charges 14.79% interest, and requires a 4% minimum payment.
   a. What is the minimum monthly payment?
   b. How much of the payment is interest?
   c. How much of the payment goes to pay off the balance?
   d. What is the unpaid balance at the end of one month if there are no additional charges?

34. According to Transunion, one of the three major credit reporting companies, the average credit card holder in Alaska has an outstanding balance of $7827, the highest average in the nation. Assume that the balance is with a credit card company which charges 18.24% interest and requires a 4% minimum payment.
   a. What is the minimum monthly payment?

   b. How much of the payment is interest?
   c. How much of the payment goes to pay off the balance?
   d. What is the unpaid balance at the end of one month, assuming no additional charges were made?

35. The average amount of credit card dept in the U.S. is slightly more than $10,000 per household. Assume that Joe Averege has an outstanding balance of $10,000 on a credit card which charges 13% interest and requires a 4% minimum payment. Joe vows to not make any more purchases until his balance drops below $8000.
   a. Complete the table.

	Beginning balance	Minimum payment	Interest paid	Balance reduction	Ending balance
Month 1	$10,000				
Month 2					
Month 3					

   b. Why are the amounts in each column decreasing instead of staying the same from month to month?

   c. Based on the table, do you think Joe will reduce his balance to $8000 by the end of one year? Explain.

36. What is the difference between simple interest and compound interest? Which is more advantageous to you as an investor?

37. Explain how it is possible to have your credit card debt increase while making minimum monthly payments.

CHALLENGE

38. Linn has a balance of $1235.60 on her credit card. The credit card company has an interest rate of 18.6%. The company requires a minimum payment of 4% of the unpaid balance, rounded to the nearest dollar. If Linn makes no additional purchases with her card and makes the minimum payment monthly, what will be her balance at the end of 1 year? How much interest will she have paid?

39. In Exercise 38, if Linn makes a $100 payment each month and makes no additional purchases, how many months will it take to pay off the balance? How much interest will she have paid?

MAINTAIN YOUR SKILLS

Add.

40. $456 + 387 + 1293 + 781$

41. $32.67 + 45.098 + 102.5 + 134.76$

Subtract.

42. $34{,}761 - 29{,}849$

43. $134.56 - 98.235$

Multiply.

44. $(341)(56)$

45. $(56.72)(0.023)$

Divide.

46. $9324 \div 36$

47. $76.4 \div 1.34$ (round to the nearest thousandth)

48. Geri works two jobs each week. Last week, she worked 25 hours at the job paying $7.82 per hour and 23 hours at the job that pays $10.42 per hour. How much did she earn last week?

49. Mary divided her estate equally among her seven nieces and nephews. The executor of the estate received a 5% commission for handling the estate. How much did each niece and nephew receive if the estate was worth $1,456,000?

SECTION 6.1 The Meaning of Percent

Definitions and Concepts	Examples
A percent is a ratio with a base unit (the denominator) of 100. The symbol % means $\frac{1}{100}$ or 0.01. $100\% = \frac{100}{100} = 1$	$75\% = \frac{75}{100}$ If 22 of 100 people are left-handed, what percent is this? $\frac{22}{100} = 22\%$

SECTION 6.2 Changing Decimals to Percents and Percents to Decimals

Definitions and Concepts	Examples
To change a decimal to a percent, • Move the decimal point two places to the right (write zeros on the right if necessary). • Write % on the right. To change a percent to a decimal, • Move the decimal point two places to the left (write zeros on the left if necessary). • Drop the percent symbol (%).	$0.23 = 23\%$ $5.7 = 570\%$ $67\% = 0.67$ $2.8\% = 0.028$

SECTION 6.3 Changing Fractions to Percents and Percents to Fractions

Definitions and Concepts	Examples
To change a fraction or mixed number to a percent, • Change to a decimal. • Change the decimal to a percent. To change a percent to a fraction, • Replace the percent symbol (%) with $\frac{1}{100}$. • If necessary, rewrite the other factor as a fraction. • Multiply and simplify.	$\frac{6}{7} \approx 0.857 \approx 85.7\%$ $3\frac{4}{25} = 3.16 = 316\%$ $45\% = 45 \cdot \frac{1}{100} = \frac{\overset{9}{\cancel{45}}}{\underset{20}{\cancel{100}}} = \frac{9}{20}$ $6.5\% = 6.5 \cdot \frac{1}{100} = \frac{13}{2} \cdot \frac{1}{100} = \frac{13}{200}$

SECTION 6.4 Fractions, Decimals, Percents: A Review

Definitions and Concepts	Examples

Every number has three forms: fraction, decimal, and percent.

Fraction	Decimal	Percent
$\dfrac{7}{10}$	0.7	70%
$5\dfrac{3}{8}$	5.375	537.5%

SECTION 6.5 Solving Percent Problems

Definitions and Concepts	Examples

The percent formula is $R \times B = A$, where R is the rate of percent, B is the base, and A is the amount.

6 is 24% of what number?
$A = 6, R = 0.24, B = ?$

$B = A \div R$
$B = 6 \div 0.24$
$B = 25$

So 6 is 24% of 25.

To solve percent problems using a proportion, set up the proportion.

$\dfrac{X}{100} = \dfrac{A}{B}$, where $R = X\%$

8 is what percent of 200?
$A = 8, B = 200, R = X\% = ?$

$$\frac{8}{200} = \frac{X}{100}$$
$$8 \cdot 100 = 200X$$
$$800 \div 200 = X$$
$$4 = X$$

So 8 is 4% of 200.

SECTION 6.6 Applications of Percents

Definitions and Concepts	Examples

To solve percent applications,
- Restate the problem as a simple percent statement.
- Identify values for A, R, and B.
- Use the percent formula or a proportion.

Of 72 students in Physics 231, 32 are women. What percent of the students are women?

Restate: 32 is what percent of 72?
$32 = A, 72 = B, R = X\% = ?$

$$\frac{32}{72} = \frac{X}{100}$$
$$32 \cdot 100 = 72X$$
$$3200 \div 72 = X$$
$$X \approx 44.4$$

So about 44.4% of the students are women.

When a value B is increased (or decreased) by an amount A, the rate of percent R is called the percent of increase (or decrease).

If a \$230,000 home increases in value to \$245,000 in 1 year, what was the percent of increase?

Increase = \$245,000 − \$230,000 = \$15,000

So \$15,000 is what percent of \$230,000?

$R = A \div B$
$R = \$15,000 \div \$230,000$
$R \approx 0.065$, or 6.5%

Circle graphs convey information about how an entire quantity is composed of its various parts. The size of each sector indicates what percent the part is of the whole.

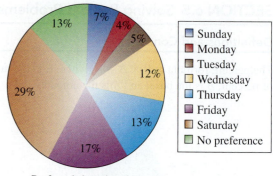

■	Sunday
■	Monday
■	Tuesday
■	Wednesday
■	Thursday
■	Friday
■	Saturday
■	No preference

Preferred shopping day

Most people (29%) prefer to shop on Saturday. Monday and Tuesday are the least favorite days to shop.

SECTION 6.7 Sales Tax, Discounts, and Commissions

Definitions and Concepts	Examples
Sales tax is a percent of the purchase price that is added to the final price.	An electric oven costs \$650. The sales tax is 6%. Find the final purchase price.
Sales tax = Sales tax rate × Purchase price	Find the amount of sales tax, 6% of \$650.

	Sales tax = 6% × \$650
	        = 0.06 × \$650
	        = \$39

Total cost = Purchase price + Sales tax	Add the sales tax to the cost.
	\$650 + \$39 = \$689
	The oven's total cost is \$689.

A discount is a percent of the regular price that is subtracted from the price.

Amount of discount = Rate of discount × Original price

A calculator that sells for $89 is put on sale for 20% off. Find the sale price.

Find the amount of discount, 20% of $89.

Amount of discount = 20% × $89
= $17.8

Sale price = Original price − Discount

$89 − $17.8 = $71.2

The calculator's sale price is $71.20.

A commission is a percent of the value of goods sold that is earned by the salesperson.

Larry earns $400 per month plus a 4% commission. One month, he sold $12,300 worth of appliances. Find his earnings for the month.

Commission = Commission rate × Total sales

Find the amount of the commission, 4% of $12,300.

Commission = 4% × $12,300
= 0.04 × $12,300
= $492

Total earnings = Base pay + Commission

Add the commission to his salary.

$400 + $492 = $892

Larry earns $892 for the month.

SECTION 6.8 Interest on Loans

Definitions and Concepts	Examples
Simple interest is paid once, at the end of the loan.	Scott borrowed $30,000 from his rich aunt. He will pay her 4% simple interest and keep the money for 2 years. How much will he owe his aunt?
Simple interest = Principal × Interest rate × Time	Interest = $30,000 × 4% × 2 = $30,000 × 0.04 × 2 = $2400 Add the interest to the principal. $2400 + $30,000 = $32,400 Scott will owe his aunt $32,400.

Compound interest is paid periodically, so after the first period, interest is paid on interest earned as well as on the principal.

Ending balance = Principal × Compound factor

The compound factor can be found in Appendix D.

Maureen invests $5000 in an account that pays 6% compounded quarterly. How much will be in the account after 3 years?

Multiply the principal by the compound factor.

Ending balance = $5000 × 1.1956
                = $5978

Maureen will have $5978 in her account after 3 years.

Credit card payments are a percent of the balance. The interest owed is paid first, and the remainder is used to reduce the balance.

Minimum payment = Unpaid balance × Minimum payment rate
                  (rounded to the nearest whole dollar)

Monthly interest = (Unpaid balance × Interest rate) ÷ 12

Amount applied to unpaid balance = Minimum payment −
                                    Interest charge

Unpaid balance = Beginning unpaid balance − Amount
                 applied to unpaid balance

Karen has a balance of $876 on her credit card. She must make a 4% minimum payment, and pay 15.99% per year in interest on the balance. Find her new balance.

Find her payment, 4% of $876.

Minimum payment = $876 × 4%
                = $876 × 0.04
                = $35.04

Rounded to the nearest whole dollar the minimum payment is $35.

Find the interest she owes, $\frac{1}{12}$ of 15.99% of $876.

Monthly interest = ($876 × 15.99%) ÷ 12
                 = ($876 × 0.1599) ÷ 12
                 ≈ $11.67

The difference between her total payment and the interest she owes will be used to reduce her balance.

Amount applied to unpaid balance
                = $35.00 − $11.67
                = $23.33

Unpaid balance = $876 − $23.33
                = $852.67

Karen's new balance is $852.67

REVIEW EXERCISES

SECTION 6.1

What percent of each of the following regions is shaded?

1.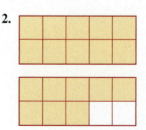

2.

Write an exact percent for these comparisons. Use fractions when necessary.

3. 39 per 50

4. 354 per 120

5. 44 per 77

6. Of teenagers who demonstrated violent behavior, 55 of 100 had used more than one illegal drug during the past year. What percent of the teenagers who demonstrated violent behavior had used illegal drugs?

SECTION 6.2

Write each decimal as a percent.

7. 0.652

8. 0.508

9. 0.00017

10. 73

11. The Phoenix Suns won 0.756 of their 82 league games in 2004–2005. Write this as a percent.

12. The sale price on a new stereo is 0.70 of the original price. Express this as a percent. What "percent off" will the store advertise?

Write each of the following as a decimal.

13. 48%

14. 632%

15. $\frac{1}{16}\%$

16. $81\frac{3}{4}\%$

17. What decimal number is used to compute the interest on a credit card balance that has an interest rate of 18.45%?

18. Thirty-year home mortgages are being offered at 5.72%. What decimal number will be used to compute the interest?

SECTION 6.3

Change each fraction or mixed number to a percent.

19. $\frac{11}{16}$

20. $7\frac{23}{64}$

Change each fraction or mixed number to a percent. Round to the nearest tenth of a percent.

21. $\frac{13}{27}$

22. $9\frac{33}{73}$

23. A Classic League basketball team won 37 of the 46 games they played. What percent of the games did they win, rounded to the nearest hundredth of a percent?

24. Jose got 37 problems correct on a 42-problem exam. What percent of the problems did Jose get correct? Round to the nearest whole percent.

Change each of the following percents to fractions or mixed numbers.

25. 165%

26. 6.4%

27. 32.5%

28. 382%

29. The offensive team for the Chicago Bears was on the field 42% of the time during a recent game with the Green Bay Packers. What fractional part of the game was the Bears defense on the field?

30. Spraying for the gypsy moth is found to be 92.4% effective. What fraction of the gypsy moths are eliminated?

SECTION 6.4

Fill in the table with the related percent, decimal, or fraction.

31–38.

Fraction	Decimal	Percent
$\frac{17}{25}$		
	0.74	
		1.5%
$3\frac{11}{40}$		

SECTION 6.5

Solve.

39. 22% of 455 is _____.

40. 36 is 45% of _____.

41. 17 is _____% of 80.

42. 37 is _____% of 125.

43. 3.4% of 370 is _____.

44. 2385 is 53% of _____.

45. What percent of 677 is 123? Round to the nearest tenth of a percent.

46. Two hundred fifty-four is 154.8% of what number? Round to the nearest hundredth.

SECTION 6.6

47. Last year Melinda had 26.4% of her salary withheld for taxes. If the total amount withheld was $6345.24, what was Melinda's yearly salary?

48. The population of Arlington grew from 3564 to 5721 over the past 5 years. What was the percent of increase in the population? Round to the nearest tenth of a percent.

49. The work force at Omni Plastics grew by 32% over the past 3 years. If the company had 325 employees 3 years ago, how many employees do they have now?

50. Mrs. Hope's third-grade class has the following ethnic distribution: Hispanic, 13; African American, 7; Asian, 5; and Caucasian, 11. What percent of her class is African American? Round to the nearest tenth of a percent.

51. Mr. Jones bought a new Hummer H2 in 2005 for $57,480. At the end of 1 year it had decreased in value to $50,650. What was the percent of decrease? Round to the nearest whole-number percent.

52. To qualify for an interview, Toni had to get a minimum of 70% on a pre-employment test. Toni got 74 out of 110 questions correct. Does Toni qualify for an interview?

Exercises 53–54. The figure shows the grade distribution in an algebra class.

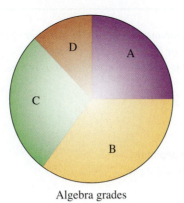

Algebra grades

53. Which grade was received by most students?

54. Were more A and C grades earned than B and D grades?

SECTION 6.7

55. Mary buys a set of golf clubs for $465.75. The store adds on a 6.35% sales tax. What is the total cost of the clubs?

56. Rod buys a new LCD projection TV for $3025. The store charges Rod $3233.58, including sales tax. What is the sales tax rate? Round to the nearest hundredth of a percent.

57. Macys advertises a sale at 25% off on men's suits. The store then offers a coupon that gives an additional 20% off the sale price. What is the cost to the consumer of a suit that was originally priced at $675.90?

58. A salesclerk earns a base salary of $1500 per month plus a commission of 11.5% on all sales over $9500. What is her salary in a month in which her sales totaled $21,300?

59. A pair of New Balance walking shoes, which regularly sells for $110.95, goes on sale for $79.49. What percent off is this? Round to the nearest whole percent.

60. The Bonn offers a ladies' two-piece suit for 30% off the original price of $235. In addition, they offer an early bird special of an additional 15% off the sale price if purchased between 8 A.M. and 10 A.M. After a 4.75% sales tax is added, what is the final cost of the suit if bought during the early bird special?

SECTION 6.8

61. Wanda borrows $5500 from her uncle at 6.5% simple interest for 2 years. How much does Wanda owe her uncle at the end of the 2 years?

62. Larry invests $2000 at 8% compounded monthly. What is the value of his investment after 2 months?

63. Minh has a credit card balance of $1345.60. The credit card company charges 18.6%. If Minh makes a payment of $55 and makes no additional charges, what will be his credit card balance on the next billing?

64. Felicia's credit card has a balance owed at the end of September of $4446.60. The credit card company charges a rate of 19.8% on the unpaid balance. Felicia makes the minimum payment of $178.
  a. How much of the payment is interest?

  b. How much of the payment goes to pay off the balance?
  c. What is the unpaid balance at the end of October, assuming no additional charges were made?

Check your understanding of the language of basic mathematics. Tell whether each of the following statements is true (always true) or false (not always true). For each statement you judge to be false, revise it to make a statement that is true.

Answers

1. *Percent* means per 100.

1. _____

2. The symbol % is read "percent."

2. _____

3. To change a fraction to a percent, move the decimal point in the numerator two places to the left and write the percent symbol.

3. _____

4. To change a decimal to a percent, move the decimal point two places to the left.

4. _____

5. Percent is a ratio.

5. _____

6. In percent, the base unit can be more than 100.

6. _____

7. To change a percent to a decimal, drop the percent symbol and move the decimal point two places to the left.

7. _____

8. A percent can be equal to a whole number.

8. _____

9. To solve a problem written in the form *A* is *R* of *B*, we can use the proportion $\dfrac{B}{A} = \dfrac{X}{100}$, where $R = X\%$.

9. _____

10. To solve the problem "If there is a 5% sales tax on a radio costing $64.49, how much is the tax?" the simpler word form could be "5% of $64.49 is what?"

10. _____

11. $2.5 = 250\%$

11. _____

12. $4\dfrac{3}{4} = 4.75\%$

12. _____

13. $0.009\% = 0.9$

13. _____

14. If 0.4% of *B* is 172, then *B* = 4300.

14. _____

15. If some percent of 64 is 32, then the percent is 50%.

15. _____

16. If $2\dfrac{4}{5}\%$ of 300 is *A*, then $A = 8.4$.

16. _____

17. Two consecutive decreases of 15% is the same as a decrease of 30%.

17. _____

18. If Selma is given a 10% raise on Monday but her salary is cut 10% on Wednesday, her salary is the same as it was Monday before the raise.

18. _____

19. It is possible to increase a city's population by 110%.

19. _____

20. If the price of a stock increases 100% for each of 3 years, the value of $1 of stock is worth $8 at the end of 3 years.

20. _____

21. A 50% growth in population is the same as 150% of the original population.

21. _____

22. $\frac{1}{2}\% = 0.5$

22. _____

23. If interest is compounded quarterly, it means that every 3 months the interest earned is added to the principal and earns interest the next time period.

23. _____

24. If Loretta buys a new toaster for $34.95 and pays $37.22, including sales tax, at the checkout, the sales tax rate is 7.8%.

24. _____

TEST

1. A computer regularly sells for $1495. During a sale, the dealer discounts the price $415.50. What is the percent of discount? Round to the nearest tenth of a percent.

1. _____

2. Write as a percent: 0.03542

2. _____

3. If 56 of every 100 people in a certain town are female, what percent of the population is female?

3. _____

4. What percent of $8\frac{3}{8}$ is $6\frac{1}{4}$? Round to the nearest tenth of a percent.

4. _____

5. Change to a percent: $\frac{27}{32}$

5. _____

6. Two hundred fifty-three percent of what number is 113.85?

6. _____

7. Change to a fraction: $16\frac{8}{13}\%$

7. _____

8. Write as a percent: 0.0078

8. _____

9. What number is 15.6% of 75?

9. _____

10. Change to a percent (to the nearest tenth of a percent): $6\frac{9}{11}$

10. _____

11. Change to a fraction or mixed number: 272%

11. _____

12. Write as a decimal: 7.89%

13. The Adams family spends 17.5% of their monthly income on rent. If their monthly income is $6400, how much do they spend on rent?

14–19. *Complete the following table:*

Fraction	Decimal	Percent
$\dfrac{13}{16}$		
	0.624	
		18.5%

20. 87.6 is _____% of 115.9 (to the nearest tenth of a percent).

21. Write as a decimal: 4.765%

22. The Tire Factory sells a set of tires for $357.85 plus a sales tax of 6.3%. What is the total price charged the customer?

23. Nordstrom offers a sale on Tommy Hilfiger jackets at a discount of 30%. What is the sale price of a jacket that originally sold for $212.95?

24. The population of Nevada grew from 1,998,257 people in 2000 to 2,410,758 people in 2004. What is the percent of increase in the population? Round to the nearest tenth of a percent.

25. The following graph shows the distribution of grades in an American History class. What percent of the class received a B grade? Round to the nearest tenth of a percent.

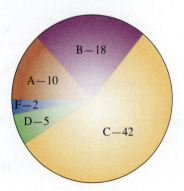

26. Jerry earns $600 per month plus a 6% commission on all his sales. Last month, Jerry sold products totaling $75,850. What were Jerry's earnings last month?

Answers

26. _____

27. Ukiah borrows $6500 from his aunt to attend college. His aunt charges Ukiah 7% simple interest. How much will Ukiah owe his aunt at the end of 1 year?

27. _____

28. If a hamburger contains 780 calories and 37 g of fat, what percent of the calories are from fat? Assume each gram of fat contains 10 calories. Round to the nearest tenth of a percent.

28. _____

29. Greg invests $9000 at 5% interest compounded monthly. What is his investment worth after 10 years?

29. _____

30. Loraine has a balance of $6732.50 on her credit card at the end of April. The credit card company charges 18.5% interest and requires a minimum payment of 4% of the unpaid balance, rounded to the nearest dollar. Loraine makes the minimum payment and charges an additional $234.50 during the next month. What is her credit card balance at the end of May?

30. _____

CLASS ACTIVITY 1

A Tommy Hilfiger sweater at Macys originally cost $86. It went on sale and was marked 25% off but didn't sell. It then was put on a rack for 40% off the marked price. At the President's Day sale, you buy the sweater using a "15% off the lowest price" coupon.

1. What was the price of the sweater when it was 25% off?

2. What was the price of the sweater when it was on the clearance rack?

3. What price did you pay for the sweater? (Round to the nearest penny.)

4. What percent off the original price was the price you paid? (Round to the nearest whole %)

5. Explain why the final price of the sweater is not 25% + 40% + 15% = 80% off the original price.

CLASS ACTIVITY 2

Financial experts uniformly recommend that credit card balances be paid off every month. Unfortunately, only about 33% of consumers do this. Even more problematic, about 25% of consumers say they make only the minimum payment on their credit cards.

Consider Sydney, who uses her credit card to make purchases of $400 per month. She got a new Capital One card that charges 14.45% annual percentage yield (APY) and requires a minimum payment of 4% on the unpaid balance.

1. Fill in the table.

Month	Beginning Balance	Purchases	Total Balance	Payment	Interest Due	Balance Reduction Amount
Jan.	0	400				
Feb.						
March						
April						
May						
June						
July						
August						
Sept.						
Oct.						
Nov.						
Dec.						

2. What is Sydney's balance at the end of the year?

3. Sydney makes a New Year's resolution to pay down her balance She decides to not make any more purchases but continues to make the minimum monthly payment. Fill in the table.

Month	Beginning Balance	Payment	Interest Due	Balance Reduction Amount
Jan.				
Feb.				
March				
April				
May				
June				
July				
August				
Sept.				
Oct.				
Nov.				
Dec.				

4. What was Sydney's balance after the second year?

5. For her next New Year's resolution, Sydney decides that she must make more than minimum payments, so she determines that she will pay $400 each month. Fill the table.

Month	Beginning Balance	Payment	Interest Due	Balance Reduction Amount
Jan.				
Feb.				
March				
April				
May				
June				
July				
August				
Sept.				
Oct.				
Nov.				
Dec.				

6. When does Sydney pay off her balance? How much interest has she paid over the entire period?

7. What have you learned about credit card debt?

Have you ever found yourself short of cash? Everyone does at some time. The business world has many "solutions" for people who need quick money, but each comes with a price in the form of fees and/or interest charged. An informed consumer chooses one option or another after careful consideration of all the costs. Let's look at a hypothetical situation.

The transmission in LaRonda's car needs fixing and she absolutely must have her car to get to and from work. The shop (which does not accept credit cards) gives her a bill for $250, but she only has $43 in her checking account. LaRonda has a good job, but it is 2 weeks until payday. LaRonda can think of three possible solutions to her dilemma.

1. Write the shop a check that she knows will bounce, and straighten everything out as soon as she gets paid.
2. Go to a Payday Loan store and have them hold her check until she gets paid.
3. Get a cash advance on her Visa card, and repay it when she gets paid.

LaRonda begins her research on each of her three options. If she bounces a check, the shop will charge her $30 and so will her bank. In addition, the bank charges her $7 per day for a continuous negative balance beginning on the fourth day of her negative balance. The Payday Loan store will charge her $20 for every $100 she borrows for a 2-week duration. Her Visa card charges her a 3% cash advance fee and then charges her 20.99% APR until she pays it back.

Assuming that LaRonda borrows $250 and pays off her debt in 14 days, complete the following table.

Plan	Amount Borrowed	Fees and/or Interest Paid	Fees as % of Amount Borrowed	Total Payback
Bounce a check				
Payday Loan				
Cash advance				

1. Analyze the pros and cons of each option. Include considerations besides total cost.
2. What seems to be the best course of action for LaRonda and why?
3. Can you think of another option for LaRonda that would save her money? Outline your option and do a cost analysis as in the table.

The APR of a loan is a figure that allows consumers to compare loan fees. The APR is the interest rate paid if the loan is held for 1 year. Complete the table again, assuming that LaRonda borrows $250, holds it for 1 year, and then pays off her debt. Include your fourth option in the table.

Plan	Amount Borrowed	Fees and/or Interest Paid	Fees as % of Amount Borrowed	Total Payback
Bounce a check				
Payday Loan				
Cash advance				
Your solution				

4. Comment on the APRs associated with the various options. Did any of them surprise you?
5. If you need a short-term loan, what is a reasonable interest rate?

GOOD ADVICE FOR *Studying*

EVALUATING YOUR COURSE PERFORMANCE

© Andresr/Shutterstock.com

Keep a Record Sheet of All Assignments

- Use your syllabus to make list of all assignments for your class for the entire term.
- Note the weight of each assignment.
- Note due dates as they become available.
- Record your graded assignments as you receive them.
- Every two weeks or so, calculate your grade.
- Update your list if your instructor changes or adds assignments.
- Keep your record sheet and all graded work together.

Sample Assignment Sheet

Assignment	Weight	Due Date	Score
Homework	25%	ongoing	
Test 1	20%	10/13	
Test 2	20%	11/17	
Project	15%	11/28	
Final exam	20%	12/15	

If You Are Not Satisfied with Your Performance

- Make an appointment with your instructor to discuss your progress. Ask for suggestions for how you can improve your performance.
- Evaluate your effort. Do you need to spend more time doing homework? Do you need to prepare for tests differently? Are you missing assignments or turning in assignments late?
- Consider getting outside help. Study with a friend, use the CD that came with your text, seek out online resources, or go to the tutoring center.
- Do not accept a poor grade as inevitable. Everyone can learn math with sufficient and sustained effort.

© Bruce Burkhardt/CORBIS

Measurement and Geometry

APPLICATION

Paul and Barbara have just purchased a row house in the Georgetown section of Washington, D.C. The backyard is rather small and completely fenced. They decide to take out all the grass and put in a brick patio and formal rose garden. The plans for the patio and garden are shown here.

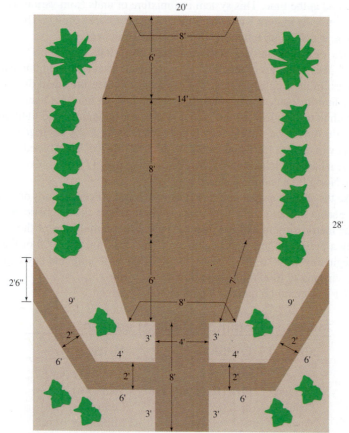

OBJECTIVES

1. Recognize and use appropriate units of length from the English and metric measuring systems.
2. Convert units of length:
   a. Unit fractions within the same system.
   b. Unit fractions between systems.
   c. Moving the decimal point in the metric system.
3. Perform operations on measurements.

VOCABULARY

A **unit of measure** is the name of a fixed quantity that is used as a standard.

A **measurement** is a number together with a unit of measure.

Equivalent measurements are measures of the same amount but using different units.

A **unit fraction** is a fraction whose numerator and denominator are equivalent measurements.

The **English system** is the measurement system commonly used in the United States.

The **metric system** is the measurement system used by most of the world.

HOW & WHY

■ **OBJECTIVE 1** Recognize and use appropriate units of length from the English and metric measuring systems.

One of the main ways of describing an object is to give its measurements. We measure how long an object is, how much it weighs, how much space it occupies, how long it has existed, how hot it is, and so forth. Units of measure are universally defined so that we all mean the same thing when we use a measurement. There are two major systems of measurement in use in the United States. One is the English system, so named because we adopted what was used in England at the time. This system is a mixture of units from various countries and cultures, and it is the system with which most Americans are familiar. The second is the metric system, which is currently used by almost the entire world.

Measures of length answer questions such as "How long?" or "How tall?" or "How deep?" We need units of length to measure small distances, medium distances, and long distances. In the English system, we use inches to measure small distances, feet to measure medium distances, and miles to measure long distances. Other, less common units of length in the English system include yards, rods, fathoms, and light-years.

The metric system was invented by French scientists in 1799. Their goal was to make a system that was easy to learn and would be used worldwide. They based the system for length on the meter and related it to Earth by defining it as 1/10,000,000 of the distance between the North Pole and the equator.

To make the system easy to use, the scientists based all conversions on powers of 10 and gave the same suffix to all units of measure for the same characteristic. So all units of length in the metric system end in "-meter." Furthermore, multiples of the base unit are indicated by a prefix. So any unit beginning with "kilo-" means 1000 of the base unit. Any unit beginning with "centi-" means 1/100 of the base unit, and any unit beginning with "milli-" means 1/1000 of the base unit.

Table 7.1 lists common units of length in both systems.

Meter

A meter is currently defined by international treaty in terms of wavelengths of the orange-red radiation of an isotope of krypton (^{86}Kr) in a vacuum.

TABLE 7.1 **Common Units of Length**

Size	English	Metric
Large	Mile (mi)	Kilometer (km)
Medium	Foot (ft) or yard (yd)	Meter (m)
Small	Inch (in.)	Centimeter (cm)
Tiny		Millimeter (mm)

EXAMPLES A–C

DIRECTIONS: Write both an English unit and a metric unit to measure the following.

STRATEGY: Decide on the size of the object and pick the appropriate units.

A. The distance from Baltimore, Maryland, to Washington, D.C.

This is a long distance, so it is measured in miles or kilometers.

B. The width of a calculator.

This is a small distance, so it is measured in inches or centimeters.

C. The width of a dining room.

This is a medium distance, so it is measured in feet or meters.

WARM-UP
A. The distance from San Francisco to Los Angeles.

WARM-UP
B. The width of a camera.

WARM-UP
C. The length of a bedroom.

HOW & WHY

■ **OBJECTIVE 2a** Convert units of length—unit fractions within the same system.

Using Table 7.2, we can convert measurements in each system.

For a slick converter, check out http://www.onlineconversion.com

TABLE 7.2 Equivalent Length Measurements

English	Metric
12 inches (in.) = 1 foot (ft)	1000 millimeters (mm) = 1 meter (m)
3 feet (ft) = 1 yard (yd)	100 centimeters (cm) = 1 meter (m)
5280 feet (ft) = 1 mile (mi)	1000 meters (m) = 1 kilometer (km)

Because 12 in. = 1 ft, they are equivalent measurements. A fraction using one of these as the numerator and the other as the denominator is equivalent to 1 because the numerator and denominator are equal.

$$\frac{12 \text{ in.}}{1 \text{ ft}} = \frac{1 \text{ ft}}{12 \text{ in.}} = 1$$

To convert a measurement from one unit to another, we multiply by the unit fraction: $\frac{\text{desired unit of measure}}{\text{original unit of measure}}$. Because we are multiplying by 1, the measurement is unchanged but the units are different.

For example, to convert 60 in. to feet, we choose a unit fraction that has feet, the desired units in the numerator, and inches, the original units, in the denominator. In this case, we use $\frac{1 \text{ ft}}{12 \text{ in.}}$ for the conversion.

$$60 \text{ in.} = \frac{60 \text{ in.}}{1} \cdot \frac{1 \text{ ft}}{12 \text{ in.}} \qquad \text{Multiply by the appropriate unit fraction.}$$

$$= \frac{60 \text{ ft}}{12} \qquad \text{Multiply.}$$

$$= 5 \text{ ft} \qquad \text{Simplify.}$$

In some cases, it is necessary to multiply by more than one unit fraction to get the desired results. For example, to convert 7.8 km to centimeters we use the unit fraction $\frac{1000 \text{ m}}{1 \text{ km}}$ to convert the kilometers to meters, and then $\frac{100 \text{ cm}}{1 \text{ m}}$ to convert the meters to centimeters.

ANSWERS TO WARM-UPS A–C

A. miles or kilometers **B.** inches or centimeters **C.** feet or meters

$$7.8 \text{ km} = \frac{7.8 \text{ km}}{1} \cdot \frac{1000 \text{ m}}{1 \text{ km}} \cdot \frac{100 \text{ cm}}{1 \text{ m}} \qquad \text{Multiply by the appropriate unit fractions.}$$
$$= 780{,}000 \text{ cm} \qquad \text{Simplify.}$$

> ### To convert units of length
> 1. Multiply by the unit fraction that has the desired units in the numerator and the original units in the denominator.
> 2. Simplify.

EXAMPLES D–E

DIRECTIONS: Convert units of measure.

STRATEGY: Multiply the given measure by the appropriate unit fraction(s) and simplify.

WARM-UP
D. Convert 56 cm to meters.

D. Convert 32 m to kilometers.

$$32 \text{ m} = \frac{32 \text{ m}}{1} \cdot \frac{1 \text{ km}}{1000 \text{ m}} \qquad \text{Multiply by the appropriate unit fraction.}$$
$$= \frac{32 \text{ km}}{1000} \qquad \text{Multiply.}$$
$$= 0.032 \text{ km} \qquad \text{Simplify.}$$

So 32 m = 0.032 km.

WARM-UP
E. Convert 12 mi to inches.

E. Convert 7.8 mi to inches.

$$7.8 \text{ mi} = \frac{7.8 \text{ mi}}{1} \cdot \frac{5280 \text{ ft}}{1 \text{ mi}} \cdot \frac{12 \text{ in.}}{1 \text{ ft}} \qquad \text{Convert miles to feet and then feet to inches.}$$
$$= 494{,}208 \text{ in.} \qquad \text{Simplify.}$$

So 7.8 mi = 494,208 in.

HOW & WHY

■ OBJECTIVE 2b Convert units of length—unit fractions between systems.

Sometimes we need to convert lengths from one system to another. Most conversions between systems are approximations, as in Table 7.3. The method of using unit fractions for converting is the same as when converting within the same system.

TABLE 7.3 **Length Conversions between English and Metric Systems**

1 inch = 2.54 centimeters	1 centimeter ≈ 0.3937 inch
1 foot ≈ 0.3048 meter	1 meter ≈ 3.2808 feet
1 yard ≈ 0.9144 meter	1 meter ≈ 1.0936 yards
1 mile ≈ 1.6093 kilometers	1 kilometer ≈ 0.6214 mile

Because of the rounding in Table 7.3, we get approximate measurements when moving from one system to another. For everyday measurements, this is not a problem because we usually only measure to the nearest tenth or hundredth. However, in a laboratory or industrial setting, where more precision is necessary, you may need to use

ANSWERS TO WARM-UPS D–E
D. 0.56 m   **E.** 760,320 in.

more accurate conversions than those in Table 7.3. In changing measures from one system to another, we will always use a unit fraction with denominator of 1. For example, when converting inches to centimeters we use $\dfrac{2.54 \text{ cm}}{1 \text{ in.}}$, whereas when converting from centimeters to inches we use $\dfrac{0.3937 \text{ in.}}{1 \text{ cm}}$. There are two reasons for this: (1) It is easier to multiply than to divide decimals when not using a calculator, and (2) It is possible for the results to vary when using different approximating unit fractions.

> **CAUTION**
>
> Because the conversions in Table 7.3 are all rounded to the nearest ten-thousandth, we cannot expect accuracy beyond the ten-thousandths place when using them.

EXAMPLES F–G

DIRECTIONS: Convert the units of measure. Round to the nearest hundredth.

STRATEGY: Multiply the given measure by the appropriate unit fraction(s) and simplify.

F. Convert 5 yd to meters.

$$5 \text{ yd} \approx \frac{5 \text{ yd}}{1} \cdot \frac{0.9144 \text{ m}}{1 \text{ yd}} \qquad \text{Multiply by the appropriate unit fraction.}$$

$$\approx 4.572 \text{ m}$$

So 5 yd ≈ 4.57 m.

G. Convert 1.6 km to feet.

STRATEGY: Convert kilometers to miles and then miles to feet.

$$1.6 \text{ km} \approx \frac{1.6 \text{ km}}{1} \cdot \frac{0.6214 \text{ mi}}{1 \text{ km}} \cdot \frac{5280 \text{ ft}}{1 \text{ mi}} \qquad \begin{array}{l}\text{Convert kilometers to miles, then miles} \\ \text{to feet.} \\ \text{Simplify.}\end{array}$$

$$\approx 5249.5872 \text{ ft}$$

So 1.6 km ≈ 5249.59 ft.

WARM-UP
F. Convert 125 cm to inches.

WARM-UP
G. Convert 6 ft to centimeters.

HOW & WHY

■ **OBJECTIVE 2c** Convert units of length—moving the decimal point in the metric system.

The metric system is based on powers of 10 and it is easy to multiply and divide by powers of 10. Therefore, we can shortcut metric-to-metric conversions by multiplying or dividing by powers of 10.

We have already seen how kilometers, centimeters, and millimeters relate to the base unit, the meter. Other, less frequently used units are also part of the metric system. Consider the following.

Kilometer	Hecto-meter	Deka-meter	Meter	Deci-meter	Centi-meter	Millimeter
km	hm	dam	m	dm	cm	mm
1000 m	100 m	10 m	1 m	0.1 m	0.01 m	0.001 m

ANSWERS TO WARM-UPS F–G
F. 49.21 in. **G.** 182.88 cm

To convert within the metric system, simply determine how many places you must move to get from the given unit to the desired unit. For example, to change kilometers to meters we must move three places to the right.

| km | hm | dam | m | dm | cm | mm |

To change 4 km to meters, we multiply 4 km by 10^3 or 1000. So 4 km = 4000 m. To change centimeters to meters, we must move two places to the left.

Kilometer	Hecto-meter	Dekameter	Meter	Deci-meter	Centi-meter	Millimeter
km	hm	dam	m	dm	cm	mm

To change 85 cm to meters, we must divide by 100, or 10^2. So 85 cm = 0.85 m.

> **To convert units within the metric system**
>
> Move the decimal point the same number of places and in the same direction as you do to go from the given units to the desired units on the chart.

EXAMPLES H–I

DIRECTIONS: Convert units as indicated.

STRATEGY: Use the chart for a shortcut.

H. Change 47 m to millimeters.

Meters to millimeters is three places right on the chart. Therefore, we move the decimal three places to the right.

47 m = 47,000 mm

I. Change 37.12 m to kilometers.

Meters to kilometers is three places to the left on the chart. Therefore we move the decimal place three places to the left.

37.12 m = 0.03712 km

HOW & WHY

■ **OBJECTIVE 3** Perform operations on measurements.

When we write "4 inches," we mean four units that are 1 inch in length. Mathematically we can describe this as 4 inches = $4 \cdot (1 \text{ inch})$. This way of interpreting measurements makes it easy to find multiples of measurements. Consider 3 boards, each 5 feet long. The total length of the boards is

$$3 \cdot (5 \text{ feet}) = 3 \cdot 5 \cdot (1 \text{ foot})$$
$$= 15 \cdot (1 \text{ foot})$$
$$= 15 \text{ feet}$$

Similarly, a bolt of ribbon has 5 yards on it. If the ribbon is cut into 10 equal pieces, how long is each piece?

$$\frac{5 \text{ yd}}{10} = \frac{5 \cdot (1 \text{ yd})}{10}$$
$$= \frac{5}{10} \cdot (1 \text{ yd})$$
$$= \frac{1}{2} \cdot (1 \text{ yd})$$
$$= \frac{1}{2} \text{ yd}$$

So each piece is $\frac{1}{2}$ yd long.

> **To multiply or divide a measurement by a number**
>
> Multiply or divide the two numbers and write the unit of measure.

EXAMPLES J–K

DIRECTIONS: Solve.

STRATEGY: Describe each situation with a statement involving measurements and simplify.

J. What is the total width of four parking spaces if the width of one is 3.6 m?

STRATEGY: To find the total width, multiply the width of one space by 4.

Total width $= 4 \cdot (3.6 \text{ m})$
$= 14.4 \text{ m}$

So the total width is 14.4 m.

K. A fence 24 ft long is to be constructed in four sections. How long is each section?

STRATEGY: To find the length of each section divide the total length by 4.

$$\frac{24 \text{ ft}}{4} = \frac{24 \cdot (1 \text{ ft})}{4}$$
$$= 6 \text{ ft}$$

So each section is 6 ft long.

WARM-UP

J. What is the width of seven boxes if each box is 14.5 inches wide?

WARM-UP

K. An 8-km race is divided into five legs. How long is each leg?

The expression "You can't add apples and oranges" applies to adding and subtracting measurements. Only measurements with the same units of measure may be added or subtracted.

$10 \text{ mm} + 6 \text{ mm} = (10 + 6) \text{ mm}$
$= 16 \text{ mm}$

ANSWERS TO WARM-UPS J–K

J. 101.5 in. **K.** 1.6 km

> **CAUTION**
>
> $3\,m + 4\,cm \neq 7\,m$

> **To add or subtract measurements**
>
> 1. If the units of measure are unlike, convert to like measures.
> 2. Add or subtract the numbers and write the unit of measure.

EXAMPLES L–M

DIRECTIONS: Solve.

STRATEGY: Describe each situation with a statement involving measurements, and simplify.

WARM-UP

L. John Daley hit drives of 320 m, 335 m, 308 m, and 341 m at a recent European event. What was the total length of the four drives?

L. On the European Golf Tour, at one match, Ernie Els sank consecutive putts of 3.45 m, 5.21 m, and 4.3 m. What was the total length of the three putts?

STRATEGY: To find the total length, add the three lengths.

$$\text{Total length} = 3.45\,m + 5.21\,m + 4.3\,m$$
$$= (3.45 + 5.21 + 4.3)\,m$$
$$= 12.96\,m$$

So the total length of the three putts was 12.96 m.

WARM-UP

M. One wall of a den measures 10 ft 4 in. The door is 2 ft 8 in. wide. What is the remaining width?

M. If a carpenter cuts a piece of board that is 2 ft 5 in. from a board that is 8 ft 3 in. long, how much board is left? (Disregard the width of the cut.)

STRATEGY: Subtract the length of the cut piece from the length of the board.

```
   8 ft 3 in.
 − 2 ft 5 in.
remaining board
```

```
   7 ft 1 ft 3 in.      Borrow 1 ft from the 8 ft (1 ft = 12 in.).
 − 2 ft      5 in.
```

```
   7 ft 15 in.          1 ft 3 in. = 15 in.
 − 2 ft  5 in.          Subtract.
   5 ft 10 in.
```

There are 5 ft 10 in. of board remaining.

ANSWERS TO WARM-UPS L–M

L. 1304 m   **M.** 7 ft 8 in.

EXERCISES 7.1

■ **OBJECTIVE 1** Recognize and use appropriate units of length from the English and metric measuring systems. (See page 492.)

A *Give both an English unit and a metric unit to measure the following.*

1. The height of a skyscraper

2. The dimensions of a sheet of typing paper

3. The thickness of a fingernail

4. The length of a table

5. The width of a slat on a mini-blind

6. The distance from home to school

7. The height of a tree

8. The cruising altitude of a 747 jet

B

9. The depth of the ruins of the *Titanic*

10. The thickness of a wire

11. The length of a shoelace

12. The width of a human hair

13. The height of the ceiling in a room

14. The distance from the moon to Earth

15. The length of a swimming pool

16. The height of Mt. Everest

■ **OBJECTIVE 2** Convert units of length. (See page 493–495.)

A *Convert as indicated.*

17. 3 yd to inches

18. 34 cm to millimeters

19. 125 m to centimeters

20. 2 mi to inches

21. 12 yd to feet

22. 418 cm to meters

23. 968.5 m to kilometers

24. 321 ft to yards

25. 6 km to centimeters

26. 3 mi to feet

B *Convert as indicated. Round to the nearest hundredth if necessary.*

27. 4573 yd to miles

28. 3000 ft to meters

29. 27 in. to feet

30. 3.2 cm to kilometers

31. 245 in. to meters

32. 157 mi to kilometers

33. 120 km to miles

34. 46 cm to inches

35. 17.5 m to feet

36. 239 in. to meters

■ **OBJECTIVE 3** Perform operations on measurements. (See page 496.)

A *Do the indicated operations.*

37. 11(13 yd)

38. 19 cm + 45 cm

39. 24.7 ft ÷ 5

40. 78 m − 19 m

41. 25(8 mm)

42. 36 mi ÷ 4

43. 27 in. + 45 in. + 13 in. + 31 in.

44. 81 cm − 58 cm

45. 100(26 mi)

46. 4 km + 6 km + 2 km

B *Do the indicated operations. Round to the nearest hundredth if necessary.*

47. (3 yd 2 ft) + 5 ft.

48. 12 m − 318 cm

49. 256 mi ÷ 13

50. 44 km + 227 m

51. 75(321 mm)

52. 645 cm ÷ 11

53. 14 ft 7 in.
  −8 ft 10 in.

54. 8 yd 2 ft 7 in.
  +3 yd 2 ft 10 in.

55. (107 km)23

56. 12 yd − 133 in.

C

57. (7 ft 3 in.) + (2 ft 9 in.) + (9 ft 7 in.)

58. (7 yd 2 ft 5 in.) − (4 yd 2 ft 11 in.)

59. (2 yd 2 ft 1 in.) · 3

60. 7 yd 1 ft 9 in.
     +8 yd 2 ft 5 in.

61. Maria wants to hang a pine garland around her two doors for the hoildays. If the doors measure 7 ft by 40 in., what is the length of the garland she needs to buy?

62. Enrique, who lives in Seattle, is going on vacation to visit several friends who live in Atlanta, Houston, and Los Angeles. The distance between Seattle and Atlanta is 2182 mi. The distance from Atlanta to Houston is 689 mi. The distance from Houston to Los Angeles is 1374 mi, and the distance from Los Angeles to Seattle is 959 mi. How many miles does Enrique fly on his vacation?

63. Rosa is 137 cm tall. Juan is 142 cm tall. Francesca is 123 cm tall. What is the average height of the trio, expressed in meters?

Exercise 64–65. The table lists the highest bridges in the world.

Bridge	Location	Height (m)
Millau Viaduct	France	343
Royal Gorge Bridge	Colorado, United States	321
Pearl Bridge	Japan	298.3
New River Gorge	West Virginia, United States	276
The Great Belt Fixed Link	Denmark	254

64. How much taller is the Millau Viaduct than The Great Belt Fixed Link?

65. How many feet tall is the Millau Viaduct? Round to the nearest foot.

66. During one round of golf, Rick made birdie putts of 5 ft 6 in., 10 ft 8 in., 15 ft 9 in., and 7 ft 2 in. What was the total length of all the birdie putts?

67. The swimming pool at Tualatin Hills is 50 m long. How many meters of lane dividers should be purchased in order to separate the pool into nine lanes?

68. A decorator is wallpapering. If each length of wallpaper is 7 ft 4 in., seven lengths are needed to cover a wall, and four walls are to be covered, how much total wallpaper is needed for the project?

Exercises 69–72. The table lists the longest rivers in the world.

River	Length (mi)
Nile (Africa)	4160
Amazon (South America)	4000
Chang Jiang (Asia)	3964
Ob-Irtysh (Asia)	3362
Huang (Asia)	2903
Congo (Africa)	2900

69. Which, if any, of these figures appear to be estimates? Why?

70. What is the total length of the five longest rivers in the world?

71. The São Francisco River in South America is the twentieth longest river in the world, with a length of 1988 mi. Write a sentence relating its length to that of the Amazon, using multiplication.

72. Write a sentence relating the lengths of the Nile River and the Congo River using addition or subtraction.

Exercises 73–75, refer to the chapter application. See page 491.

73. What are the dimensions of Paul and Barbara's back yard?

74. How wide is the patio? How long is it?

75. How wide are the walkways?

STATE YOUR UNDERSTANDING

76. Give two examples of equivalent measures.

77. Explain how to add or subtract measures.

78. Explain how to multiply or divide a measure by a number.

MAINTAIN YOUR SKILLS

Do the indicated operations.

79. 34.76(100,000)

80. 4.78 ÷ 100,000

81. $\dfrac{3}{8} + \dfrac{5}{24}$

82. $\dfrac{12}{25} \div \dfrac{28}{15}$

83.
$$\begin{array}{r} 5\frac{1}{3} \\ -2\frac{5}{6} \\ \hline \end{array}$$

84.
$$\begin{array}{r} 12\frac{9}{10} \\ +4\frac{5}{8} \\ \hline \end{array}$$

85. $(2.13)^2$

86. $(4.8)(5.2)$

87. Find the average (mean) of 65, 82, 92, 106, and 77.

88. Find the median of 54, 78, 112, and 162.

Measuring Capacity, Weight, and Temperature

VOCABULARY

Units of **capacity** measure liquid quantities.

Units of **weight** measure heaviness.

HOW & WHY

■ **OBJECTIVE 1**   Convert and perform operations on units of capacity.

Measures of capacity answer the question "How much liquid?" We use teaspoons to measure vanilla for a recipe, buy soda in 2-liter bottles, and measure gasoline in gallons or liters. As is true for length, there are units of capacity in both the English and metric systems. The basic unit of capacity in the metric system is the liter (L), which is slightly more than 1 quart. Table 7.4 shows the most commonly used measures of capacity in each system. Table 7.5 shows some equivalencies between the systems.

TABLE 7.4 Measures of Capacity

English	Metric
3 teaspoons (tsp) = 1 tablespoon (Tbsp) 2 cups (c) = 1 pint (pt) 2 pints (pt) = 1 quart (qt) 4 quarts (qt) = 1 gallon (gal)	1000 milliliters (mL) = 1 liter (L) 1000 liters (L) = 1 kiloliter (kL)

TABLE 7.5 Measures of Capacity

English–Metric	Metric–English
1 teaspoon ≈ 4.9289 milliliters 1 quart ≈ 0.9464 liter 1 gallon ≈ 3.7854 liters	1 milliliter ≈ 0.2029 teaspoon 1 liter ≈ 1.0567 quart 1 liter ≈ 0.2642 gallon

To convert units of capacity, we use the same technique of unit fractions that we used for units of length. See Section 7.1.

To convert units of capacity

1. Multiply by the unit fraction, which has the desired units in the numerator and the original units in the denominator.
2. Simplify.

CAUTION

The accuracy of conversions between systems depends on the accuracy of the conversion factors.

EXAMPLES A–B

DIRECTIONS: Convert units of measure. Round to the nearest thousandth.

STRATEGY: Multiply by the unit fraction with the desired units in the numerator and the original units in the denominator. Simplify.

A. Convert 259 milliliters (mL) to liters (L).

$$259 \text{ mL} = \frac{259 \text{ mL}}{1} \cdot \frac{1 \text{ L}}{1000 \text{ mL}}$$ Use appropriate unit fractions.

$$= \frac{259 \text{ L}}{1000}$$ Simplify.

$$= 0.259 \text{ L}$$

So 259 mL = 0.259 L.

ALTERNATIVE SOLUTION: Because this is a metric-to-metric conversion, we can use a chart as a shortcut.

kL	hL	daL	L	dL	cL	mL

To change from milliliters to liters, we move the decimal point three places left. Therefore, 259 mL = 0.259 L.

B. Convert 21 liters(L) to pints (pt).

$$21 \text{ L} \approx \frac{21 \text{ L}}{1} \cdot \frac{1.0567 \text{ qt}}{1 \text{ L}} \cdot \frac{2 \text{ pt}}{1 \text{ qt}}$$ Use appropriate unit fractions.

$$\approx \frac{21(1.0567)(2)}{1} \text{ pt}$$ Simplify. Round to the nearest thousandth.

$$\approx 44.381 \text{ pt}$$

So 21 L ≈ 44.381 pt.

WARM-UP

A. Convert 33 quarts (qt) to gallons (gal).

WARM-UP

B. Convert 35 pints (pt) to milliliters (mL). Round to the nearest mL.

Operations with units of capacity are performed using the same procedures as those for units of length. See Section 7.1.

To multiply or divide a measurement by a number

Multiply or divide the two numbers and write the unit of measure.

To add or subtract measurements

1. If the units of measure are unlike, convert to like measures.
2. Add or subtract the numbers and write the unit of measure.

ANSWERS TO WARM-UPS A–B

A. 8.25 gal   **B.** 16,562 mL

DIRECTIONS: Solve.

STRATEGY: Describe each situation with a statement involving measurements and simplify.

WARM-UP
C. Jeri used five watering cans of water to water her flower pots. If the can holds 1.5 gal of water, how much water did Jeri use?

C. A chemistry instructor has 450 mL of an acid solution that he needs to divide into equal amounts for four lab groups. How much acid does each group receive?

$$\text{Acid for one group } = \frac{450 \text{ mL}}{4} \qquad \text{Divide.}$$

$$= \frac{450}{4} \text{ mL}$$

$$= 112.5 \text{ mL}$$

Each group receives 112.5 mL of acid solution.

WARM-UP
D. Cynthia made 4 pt 1 c of strawberry jam and Bonnie made 7 c of raspberry jam. How much jam did the two women make?

D. Kiesha has 7 gal 3 qt of biodiesel fuel. Tyronne has 5 gal 2 qt of the same fuel. If they combine their fuel, how much will they have?

$$\begin{aligned} \text{Total biodiesel fuel} &= (7 \text{ gal } 3 \text{ qt}) + (5 \text{ gal } 2 \text{ qt}) \\ &= (7 \text{ gal} + 5 \text{ gal}) + (3 \text{ qt} + 2 \text{ qt}) \\ &= 12 \text{ gal} + 5 \text{ qt} \\ &= 12 \text{ gal} + 1 \text{ gal} + 1 \text{ qt} \\ &= 13 \text{ gal } 1 \text{ qt} \end{aligned}$$

The total amount of biodiesel fuel they have is 13 gal 1 qt.

HOW & WHY

■ **OBJECTIVE 2**   Convert and perform operations on units of weight.

Units of weight answer the question "How heavy is it?" In the English system, we use ounces to measure the weight of baked beans in a can, pounds to measure the weight of a person, and tons to measure the weight of a ship. In the metric system, the basic unit that measures heaviness is the gram, which weighs about as much as a small paper clip. We use grams to measure the weight of baked beans in a can, kilograms to measure the weight of a person, and metric tons to measure the weight of a ship. Very small weights like doses of vitamins are measured in milligrams. Table 7.6 lists the most commonly used measures of weight in each system, and Table 7.7 shows some equivalencies between the systems.

TABLE 7.6   Measures of Weight

English System	Metric System
16 ounces (oz) = 1 pound (lb)	1000 milligrams (mg) = 1 gram (g)
2000 pounds (lb) = 1 ton	1000 grams (g) = 1 kilogram (kg)
	1000 kilograms = 1 metric ton

TABLE 7.7   Measures of Weight between Systems

English-Metric	Metric-English
1 ounce ≈ 28.3495 grams	1 gram ≈ 0.0353 ounces
1 pound ≈ 453.5924 grams	1 kilogram ≈ 2.2046 pounds
1 pound ≈ 0.4536 kilograms	

ANSWERS TO WARM-UPS C–D

C. 7.5 gal   **D.** 8 pt

(*Note:* The weight of an object is different from its mass. An object has the same mass everywhere in space. Weight depends on gravity, so an object has different weights on Earth and on the moon. On Earth, gravity is approximately uniform, so weight and mass are often used interchangeably. Technically, a gram is a unit of mass.)

EXAMPLES E–F

DIRECTIONS: Solve as indicated. Round the nearest hundredth.

STRATEGY: Follow the strategies given for each example.

E. Convert 73 kilograms (kg) to pounds (lb).

STRATEGY: Multiply by the unit fraction with pounds in the numerator and kilograms in the denominator, and simplify.

$$73 \text{ kg} \approx \frac{73 \text{ kg}}{1} \cdot \frac{2.2046 \text{ lb}}{1 \text{ kg}} \qquad \text{Use the appropriate unit fraction.}$$
$$\approx 160.9358 \text{ lb} \qquad \text{Simplify.}$$

So 73 kg ≈ 160.94 lb.

F. Chang buys two packages of steak at Krogers. One package contains 3 lb 5 oz of steak and the other one weighs 4 lb 14 oz. What is the total weight of the steak?

STRATEGY: Add the weights of the two packages together.

$$
\begin{array}{r}
3 \text{ lb } 5 \text{ oz} \\
+4 \text{ lb } 14 \text{ oz} \\
\hline
7 \text{ lb } 19 \text{ oz} = 7 \text{ lb} + 1 \text{ lb } 3 \text{ oz} \\
= 8 \text{ lb } 3 \text{ oz}
\end{array}
$$

Chang bought 8 lb 3 oz of steak.

HOW & WHY

■ **OBJECTIVE 3** Convert units of temperature and time.

Temperature is measured in degrees. There are two major scales for measuring temperature. The English system uses the Fahrenheit scale, which sets the freezing point of water at 32°F and the boiling point of water at 212°F. The Fahrenheit scale was developed by a physicist named Gabriel Daniel Fahrenheit in the early 1700s.

The metric system uses the Celsius scale, which sets the freezing point of water at 0°C and the boiling point of water at 100°C. The Celsius scale is named after Swedish astronomer Anders Celsius, who lived in the early 1700s and invented a thermometer using the Celsius scale.

Converting between scales is often done using special conversion formulas.

> ### To convert from Fahrenheit to Celsius
>
> Use the formula: $C = \dfrac{5}{9} \cdot (F - 32)$ or
>
> subtract 32 from the Fahrenheit temperature and multiply by $\dfrac{5}{9}$.

EXAMPLE G

DIRECTIONS: Convert as indicated.

STRATEGY: Use the conversion formula.

WARM-UP

G. What is 59°F on the Celsius scale?

G. What is 86°F on the Celsius scale?

$$C = \frac{5}{9} \cdot (F - 32)$$

$$= \frac{5}{9} \cdot (86 - 32) \qquad \text{Substitute 86 for } F.$$

$$= \frac{5}{9} \cdot 54 \qquad \text{Simplify.}$$

$$= 30$$

So 86°F = 30°C.

To convert from Celsius to Fahrenheit

Use the formula: $F = \frac{9}{5} \cdot C + 32$ or

multiply the Celsius temperature by $\frac{9}{5}$ and add 32.

EXAMPLE H

DIRECTIONS: Convert as indicated.

STRATEGY: Use the conversion formula.

WARM-UP

H. Convert 30°C to degrees Fahrenheit.

H. Convert 50°C to degrees Fahrenheit.

$$F = \frac{9}{5} \cdot C + 32$$

$$= \frac{9}{5} \cdot 50 + 32 \qquad \text{Substitute 50 for } C.$$

$$= 90 + 32 \qquad \text{Simplify.}$$

$$= 122$$

So 50°C = 122°F

The same units of time are used in both the English and metric systems. We are all familiar with seconds, minutes, hours, days, weeks, and years. The computer age has contributed some new units of time that are very short, such as milliseconds $\left(\frac{1}{1000} \text{ of a second}\right)$ and nanoseconds $\left(\frac{1}{1,000,000,000} \text{ of a second}\right)$. Table 7.8 gives the commonly used time conversions.

TABLE 7.8 Time Conversions

60 seconds (sec) = 1 minute (min)	7 days = 1 week
60 minutes (min) = 1 hour (hr)	365 days ≈ 1 year*
24 hours (hr) = 1 day	

*Technically, a solar year (the number of days it takes Earth to make one complete revolution around the sun) is 365.2422 days. Our calendars account for this by having a leap day once every 4 years.

ANSWERS TO WARM-UPS G–H

G. 15°C **H.** 86°F

EXAMPLES I–J

DIRECTIONS: Solve.

STRATEGY: Follow the strategies given for each example.

I. The four members of an 800-m freestyle relay team had individual times of 1 min 54 sec, 1 min 59 sec, 2 min 8 sec, and 1 min 49 sec. What was the total time for the relay team?

STRATEGY: Add the times.

   1 min 54 sec
   1 min 59 sec   *Add the times.*
   2 min 8 sec
   1 min 49 sec
   5 min 170 sec = 5 min + 2 min 50 sec   *Convert seconds to minutes.*
              = 7 min 50 sec

The relay team's time was 7 min 50 sec.

J. Jerry is 8 years old. How many minutes old is he?

STRATEGY: Use the appropriate unit fractions.

$$8 \text{ yr} = \frac{8 \text{ yr}}{1} \cdot \frac{365 \text{ days}}{1 \text{ yr}} \cdot \frac{24 \text{ hr}}{1 \text{ day}} \cdot \frac{60 \text{ min}}{1 \text{ hr}}$$
$$= 4{,}204{,}800 \text{ min}$$

Note: Jerry has lived through 2 leap years, which are not accounted for in the conversion. Therefore, we must add 2 days to the total.

$$2 \text{ days} = \frac{2 \text{ days}}{1} \cdot \frac{24 \text{ hr}}{1 \text{ day}} \cdot \frac{60 \text{ min}}{1 \text{ hr}}$$
$$= 2880 \text{ min}$$

Jerry has been alive 4,204,800 min + 2880 min = 4,207,680 min.

EXERCISES 7.2

■ **OBJECTIVE 1** Convert and perform operations on units of capacity. (See page 502.)

A *Do the indicated operations. Round decimal answers to the nearest hundredth.*

1. 35 oz + 72 oz

2. (4c) · 13

3. (400 mL) ÷ 16

4. 210 L − 154 L

5. (80 gal) ÷ 20

6. 5 kL + 9 kL + 10 kL

7. Convert 34 c to quarts.

8. Convert 34,600 mL to liters.

9. Convert 21 gal to quarts.

10. Convert 500 mL to liters.

B

11. 42(3 lb)

12. 380 kL − 175 kL

13. 33 oz + 49 oz

14. $(8498 \text{ gal}) \div 7$

15. $63 \text{ gal} - 29 \text{ gal} - 4 \text{ gal}$

16. $29 \text{ mL} + 7 \text{ mL} + 19 \text{ mL} + 18 \text{ mL}$

17. Convert 5.5 gal to cups.

18. Convert 1.3 kL to milliliters.

19. Convert 48.4 L to gallons.

20. Convert 572 mL to cups.

■ **OBJECTIVE 2** Convert and perform operations on units of weight. (See page 504.)

A *Do the indicated operations or conversions. Round decimal answers to the nearest hundredth.*

21. $212 \text{ kg} - 157 \text{ kg}$

22. $3 \cdot (18 \text{ oz})$

23. $37 \text{ lb} + 43 \text{ lb}$

24. $(663 \text{ mg}) \div 17$

25. $(62 \text{ lb})(9)$

26. $3 \text{ oz} + 5 \text{ oz} + 7 \text{ oz}$

27. Convert 8 kg to grams.

28. Convert 23 lb to ounces.

29. Convert 8000 mg to grams.

30. Convert 88 oz to pounds.

B

31. $(1360 \text{ oz}) \div 16$

32. $250 \text{ kg} - 149 \text{ kg}$

33. $214 \text{ lb} + 406 \text{ lb}$

34. $16(24 \text{ mg})$

35. $234 \text{ g} + 147 \text{ g} - 158 \text{ g}$

36. $(3 \text{ lb } 14 \text{ oz}) \cdot 3$

37. Convert 72 g to milligrams.

38. Convert 14 oz to pounds.

39. Convert 13 oz to grams.

40. Convert 140 lb to kilograms.

■ **OBJECTIVE 3** Convert units of temperature and time. (See page 505.)

A *Convert as indicated. Round decimals to the nearest tenth.*

41. 32°F to degrees Celsius

42. 30°C to degrees Fahrenheit

43. 104°F to degrees Celsius

44. 20°C to Fahrenheit

45. 68°F to degrees Celsius

46. 15°C to degrees Fahrenheit

47. 8 min to seconds

48. 4 days to hours

49. 19 weeks to days

50. 14 hr to minutes

B

51. 152°F to Celsius

52. 59°C to degrees Fahrenheit

53. 48°F to degrees Celsius

54. 8°C to degrees Fahrenheit

55. 71°C to Fahrenheit

56. 250°F to degrees Celsius

57. 14°C to degrees Fahrenheit

58. 183°F to Celsius

59. 2.6 yr to hours

60. 22 sec to minutes

61. 250 min to days

62. 2300 hr to weeks

C

63. (12 lb 2 oz) + (2 lb 15 oz) + (11 lb 5 oz)

64. (16 gal 3 qt) + (13 gal 3 qt) + (17 gal 2 qt)

65.
$$\begin{array}{r} 2 \text{ gal } 3 \text{ qt } 1 \text{ pt} \\ + 4 \text{ gal } 2 \text{ qt } 1 \text{ pt} \\ \hline \end{array}$$

66.
$$\begin{array}{r} 34 \text{ days } 14 \text{ hr} \\ - 28 \text{ days } 20 \text{ hr} \\ \hline \end{array}$$

67. The Corner Grocery sold 20 lb 6 oz of hamburger on Wednesday, 13 lb 8 oz on Thursday, and 21 lb 9 oz on Friday. How much hamburger was sold during the 3 days?

68. Normal body temperature is considered 98.6°F. What is normal body temperature on the Celsius scale?

69. Alicia, who is a lab assistant, has 300 mL of acid that is to be divided equally among 24 students. How many milliliters will each student receive?

70. A doctor prescribes allergy medication of two tablets, 20 mg each, to be taken three times per day for a full week. How many milligrams of medication will the patient get in a week?

71. The average high temperature in Ann Arbor, Michigan, for July is 84°F. What is the average temperature in degrees Celsius? Find to the nearest tenth of a degree.

72. A Portuguese stew recipe calls for two 15-oz cans of white beans and a 28-oz can of diced tomatoes. How many pounds of beans and tomatoes are in the stew?

73. Kirsten has $\frac{1}{2}$ gal of orange juice. She wants to use it for a brunch where she has 10 guests. How large is each serving if she uses the entire container?

74. The table gives the daily high temperature for Ocala, Florida, for 1 week in December. Find the average temperature for the week.

Daily High Temperatures

Sunday	Monday	Tuesday	Wednesday	Thursday	Friday	Saturday
62°F	54°F	57°F	64°F	68°F	69°F	67°F

75. If a bag contains 397 g of potato chips, how many grams are contained in 4 bags?

© Tund/Shutterstock.com

76. An elevator has a maximum capacity of 2500 lb. A singing group of 8 men and 8 women get on. The average weight of the women is 125 lb, and the average weight of the men is 190 lb. Can they ride safely together?

77. The weight classes for Olympic wrestling (both freestyle and Greco-Roman) are given in the table. Fill in the equivalent pound measures, rounded to the nearest whole pound.

Olympic Wrestling Weight Classes

Kilogams	54	58	63	69	76	85	97	130
Pounds								

78. In the shipping industry, most weights are measured in long tons, which are defined as 2240 lb. The largest tanker in the world is the *Jahre Viking,* which weighs 662,420 (long) tons fully loaded. How much does the *Jahre Viking* weigh in pounds?

79. How much does the *Jahre Viking* weigh in kilograms? (See Exercise 78.) Round to the nearest million.

80. The recommended daily allowance of protein for sedentary individuals is 0.8 g of protein for each kg of body weight. Barbara weighs 145 lb. How much protein should she eat each day in order to get her recommended daily allowance? Round to the nearest gram.

81. One serving of Life cereal has 3 g of protein, and putting $\frac{1}{2}$ c of 1% milk on it adds 4.5 g of protein. If Steve weighs 227 lb and eats only milk and cereal for a day, how many servings does he need in order to get sufficient protein? (See Exercise 80).

82. Scientists give the average surface temperature of Earth as 59°F. They estimate that the temperature increases about 1°F for every 200 ft drop in depth below the surface. What is the temperature 1 mi below the surface?

83. Can you add 4 g to 5 in.? Explain how, or explain why you cannot do it.

84. If 8 in. + 10 in. = 1 ft 6 in., why isn't it true that 8 oz + 10 oz = 1 lb 6 oz?

CHALLENGE

Use the information in the list of conversions to answer Exercises 85–87. Precious metals and gems are measured in troy weight according to the following:

$$
\begin{aligned}
1\ \text{pennyweight (dwt)} &= 24\ \text{grains} \\
1\ \text{ounce troy (oz t)} &= 20\ \text{pennyweights} \\
1\ \text{pound troy (lb t)} &= 12\ \text{ounce troy}
\end{aligned}
$$

85. How many grains are in 1 oz t? How many grains are in 1 lb t?

86. Suppose you have a silver bracelet that you want a jeweler to melt down and combine with the silver of two old rings to create a medallion that weighs 5 oz t 14 dwt. The bracelet weighs 3 oz t 18 dwt and one ring weighs 1 oz t 4 dwt. What does the second ring weigh?

87. Kayla has an ingot of platinum that weighs 2 lb t 8 oz t 17 dwt. She wants to divide it equally among her six grandchildren. How much will each piece weigh?

MAINTAIN YOUR SKILLS

Do the indicated operations.

88. $92 + 36 + 111$

89. $451 + 88 + 309$

90. $13.67 + 3.82$

91. $34.7 + 6.8 + 0.44$

92. $4\dfrac{3}{8} + 2\dfrac{7}{8}$

93. $23\dfrac{7}{9} + 19\dfrac{5}{6}$

94. $\dfrac{4}{9} + \dfrac{3}{8} + \dfrac{11}{12}$

95. $5.2 + 148 + \dfrac{1}{4}$

96. Convert 4 yd 2 ft 11 in. into inches.

97. Convert 56.84 m to kilometers.

7.3 Perimeter

OBJECTIVES

1. Find the perimeter of a polygon.
2. Find the circumference of a circle

VOCABULARY

A **polygon** is any closed figure whose sides are line segments.

Polygons are named according to the number of sides they have. Table 7.9 lists some common polygons.

Quadrilaterals are polygons with four sides. Table 7.10 lists the characteristics of common quadrilaterals.

The **perimeter** of a polygon is the distance around the outside of the polygon.

The **circumference** of a circle is the distance around the circle.

The **radius** of a circle is the distance from the center to any point on the circle.

The **diameter** of a circle is twice the radius.

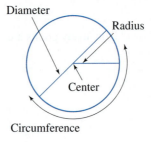

TABLE 7.9 Common Polygons

Number of Sides	Name	Figure	Number of Sides	Name	Figure
3	Triangle ABC		6	Hexagon ABCDEF	
4	Quadrilateral ABCD		8	Octagon ABCDEFGH	
5	Pentagon ABCDE				

TABLE 7.10 **Common Quadrilaterals**

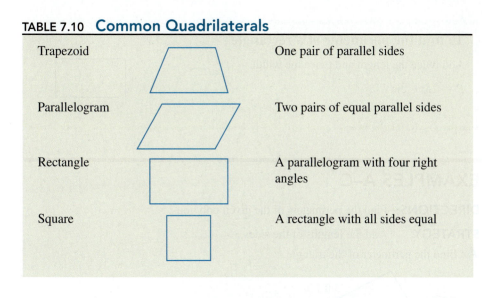

Trapezoid		One pair of parallel sides
Parallelogram		Two pairs of equal parallel sides
Rectangle		A parallelogram with four right angles
Square		A rectangle with all sides equal

HOW & WHY

■ **OBJECTIVE 1** Find the perimeter of a polygon.

The perimeter of a figure can be thought of in terms of the distance traveled by walking around the outside of it or by the length of a fence around the figure. The units of measure used for perimeters are length measures, such as inches, feet, and meters. Perimeter is calculated by adding the length of all the individual sides.

For example, to calculate the perimeter of this figure, we add the lengths of the sides.

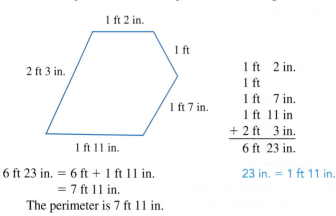

```
  1 ft   2 in.
  1 ft
  1 ft   7 in.
  1 ft  11 in
+ 2 ft   3 in.
  6 ft  23 in.
```

23 in. = 1 ft 11 in.

6 ft 23 in. = 6 ft + 1 ft 11 in.
        = 7 ft 11 in.
The perimeter is 7 ft 11 in.

To find the perimeter of a polygon

Add the lengths of the sides.

To find the perimeter of a square

Multiply the length of one side by 4.

$P = 4s$

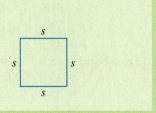

To find the perimeter of a rectangle

Add twice the length and twice the width.

$P = 2\ell + 2w$

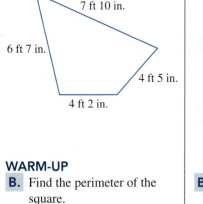

EXAMPLES A–C

DIRECTIONS: Find the perimeters of the given polygons.

STRATEGY: Add the lengths of the sides.

A. Find the perimeter of the triangle.

Triangle with sides: 3 ft 2 in., 5 ft 7 in., 6 ft 10 in.

```
  3 ft    2 in.
  5 ft    7 in.              Add the lengths.
+ 6 ft   10 in.
─────────────────
 14 ft   19 in. = 15 ft 7 in.    19 in. = 1 ft 7 in.
```

The perimeter is 15 ft 7 in.

B. Find the perimeter of the rectangle.

Rectangle: 6 cm by 15 cm

$$P = 2\ell + 2w \qquad \text{Perimeter formula for rectangles}$$
$$= 2\,(15\text{ cm}) + 2(6\text{ cm}) \qquad \text{Substitute.}$$
$$= 30\text{ cm} + 12\text{ cm} \qquad \text{Multiply.}$$
$$= 42\text{ cm} \qquad \text{Add.}$$

The perimeter of the rectangle is 42 cm.

C. A carpenter is replacing the baseboards in a room. The floor of the room is pictured. How many feet of baseboard are needed?

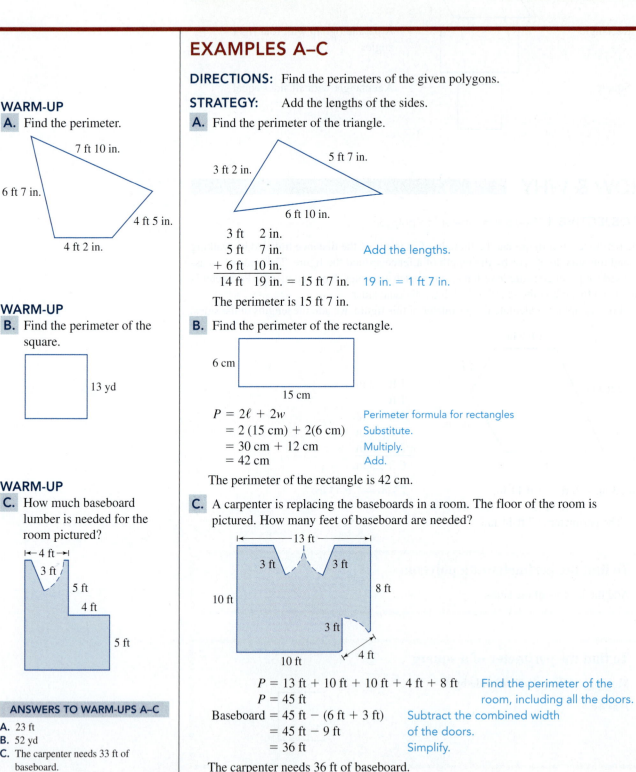

$$P = 13\text{ ft} + 10\text{ ft} + 10\text{ ft} + 4\text{ ft} + 8\text{ ft} \qquad \text{Find the perimeter of the}$$
$$P = 45\text{ ft} \qquad \text{room, including all the doors.}$$
$$\text{Baseboard} = 45\text{ ft} - (6\text{ ft} + 3\text{ ft}) \qquad \text{Subtract the combined width}$$
$$= 45\text{ ft} - 9\text{ ft} \qquad \text{of the doors.}$$
$$= 36\text{ ft} \qquad \text{Simplify.}$$

The carpenter needs 36 ft of baseboard.

WARM-UP

A. Find the perimeter.

Polygon with sides: 7 ft 10 in., 6 ft 7 in., 4 ft 2 in., 4 ft 5 in.

WARM-UP

B. Find the perimeter of the square.

Square: 13 yd

WARM-UP

C. How much baseboard lumber is needed for the room pictured?

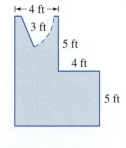

4 ft, 3 ft, 5 ft, 4 ft, 5 ft

ANSWERS TO WARM-UPS A–C

A. 23 ft
B. 52 yd
C. The carpenter needs 33 ft of baseboard.

HOW & WHY

■ **OBJECTIVE 2** Find the circumference of a circle.

Formulas for geometric figures that involve circles contain the number called pi (π). The number π is the quotient of the circumference of (distance around) the circle and its diameter. This number is the same for every circle no matter how large or small. This remarkable fact was discovered over a long period of time, and during that time a large number of approximations have been used. Today, the most commonly used approximations are

$$\pi \approx 3.14 \qquad \text{or} \qquad \pi \approx \frac{22}{7}$$

Calculator note: Scientific and graphing calculators have a $\boxed{\pi}$ key that generates a decimal value for π with 8, 10, or 12 decimal places depending on the calculator. Using the $\boxed{\pi}$ key instead of one of the approximations increases the accuracy of your calculations.

> ## To find the circumference of a circle
>
> If C is the circumference, d is the diameter, and r is the radius of a circle, then the circumference is the product of π and the diameter or the product of π and twice the radius.
>
> $$C = \pi d \qquad \text{or} \qquad C = 2\pi r$$

Because the radius, diameter, and circumference of a circle are all lengths, they are measured in units of length.

EXAMPLES D–E

DIRECTIONS: Find the circumference of the circle.

STRATEGY: Use the formula $C = \pi d$ or $C = 2\pi r$ and substitute.

D. Find the circumference of the circle.

12 in.

$C = 2\pi r$ Formula for circumference.
$C \approx 2(3.14)(12 \text{ in.})$ Substitute. Because we are using an approximation for the value of π, the value of the circumference is also approximation. We show this using $\approx$.
$C \approx 75.36 \text{ in.}$

The circumference is about 75.36 in.

ALTERNATIVE SOLUTION: Use a calculator and the $\boxed{\pi}$ key.

$C = 2\pi r$ — Formula for circumference.

$C = 2\pi (12 \text{ in.})$ — Substitute. Use the $\boxed{\pi}$ key.

$C = 24\pi$ in.

$C \approx 75.39822369$ in. — Change to decimal form.

$C \approx 75.40$ in. — Round to the nearest hundredth.

The circumference is about 75.40 in.

WARM-UP

E. Find the perimeter of the figure.

3 ft

8 ft

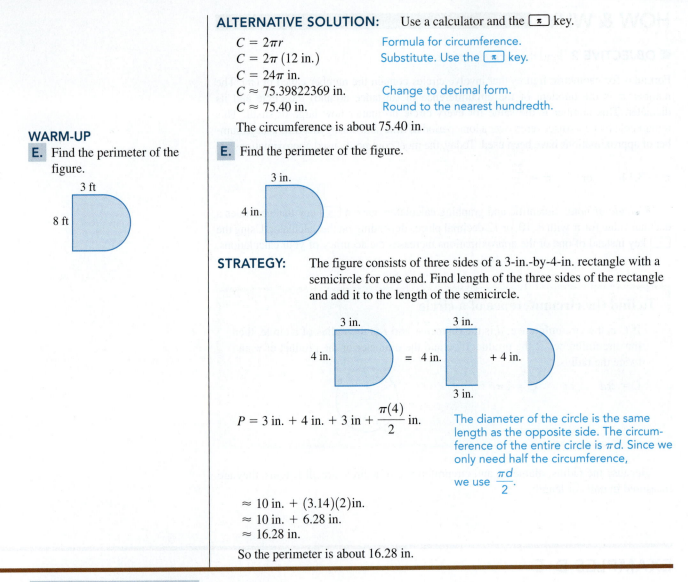

E. Find the perimeter of the figure.

3 in.

4 in.

STRATEGY: The figure consists of three sides of a 3-in.-by-4-in. rectangle with a semicircle for one end. Find length of the three sides of the rectangle and add it to the length of the semicircle.

$$P = 3 \text{ in.} + 4 \text{ in.} + 3 \text{ in} + \frac{\pi(4)}{2} \text{ in.}$$

The diameter of the circle is the same length as the opposite side. The circumference of the entire circle is πd. Since we only need half the circumference, we use $\dfrac{\pi d}{2}$.

$\approx 10 \text{ in.} + (3.14)(2)\text{in.}$

$\approx 10 \text{ in.} + 6.28 \text{ in.}$

$\approx 16.28 \text{ in.}$

So the perimeter is about 16.28 in.

ANSWER TO WARM-UP E

E. 26.56 ft

EXERCISES 7.3

OBJECTIVE 1 Find the perimeter of a polygon. (See page 513.)

A *Find the perimeter of the following polygons.*

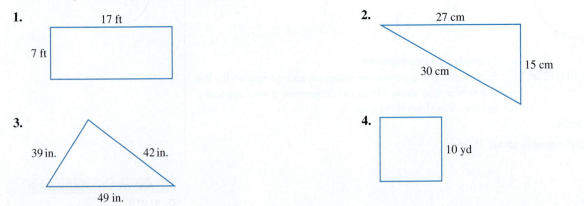

1.
17 ft

7 ft

2.
27 cm

30 cm

15 cm

3.
39 in.   42 in.

49 in.

4.
10 yd

5.

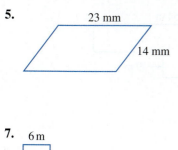
23 mm
14 mm

6.

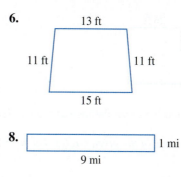
13 ft
11 ft   11 ft
15 ft

7.

6 m
24 m

8.

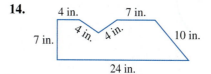
1 mi
9 mi

B

9. Find the perimeter of a triangle with sides of 16 mm, 27 mm, and 40 mm.

10. Find the perimeter of a square with sides of 230 ft.

11. Find the distance around a rectangular play field with a length of 275 ft and a width of 125 feet.

12. Find the distance around a rectangular swimming pool with a width of 20 yd and a length of 25 yd.

Find the perimeter of the following polygons.

13.

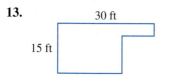
30 ft
15 ft

14.

4 in.   7 in.
7 in. 4 in. 4 in.   10 in.
24 in.

15.

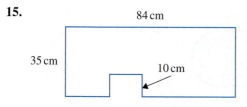
84 cm
35 cm
10 cm

16.

12 mm
9 mm
12 mm   12 mm
9 mm
9 mm

17.

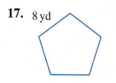
8 yd

18.

6 mi

19.

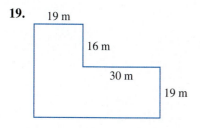
19 m
16 m
30 m
19 m

20.

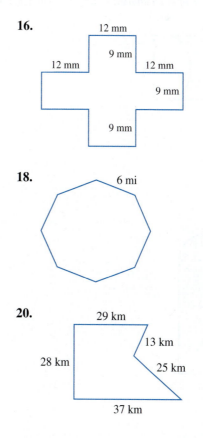
29 km
13 km
28 km
25 km
37 km

21.

22.

OBJECTIVE 2 Find the circumference of a circle. (See page 515.)

A *Find the circumference of the given circles. Let* $\pi \approx 3.14$.

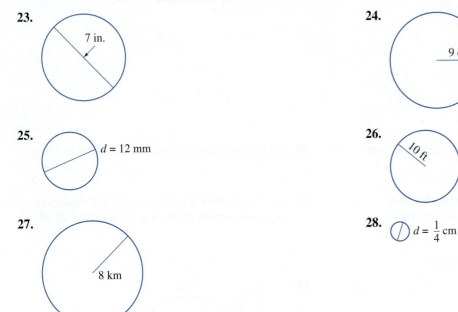

23. 7 in.

24. 9 cm

25. $d = 12$ mm

26. 10 ft

27. 8 km

28. $d = \frac{1}{4}$ cm

B *Find the perimeter of each shaded figure. Use the* $\boxed{\pi}$ *. key, and round to the nearest hundredth.*

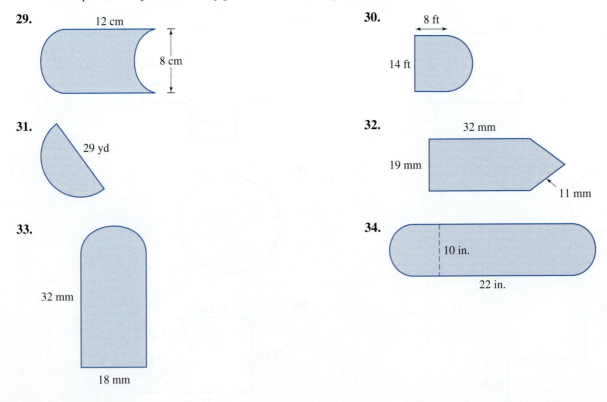

29. 12 cm 8 cm

30. 8 ft 14 ft

31. 29 yd

32. 32 mm 19 mm 11 mm

33. 32 mm 18 mm

34. 10 in. 22 in.

C *Find the perimeter of the shaded regions. Use the π. key, and round decimals to the nearest hundredth when necessary.*

35.

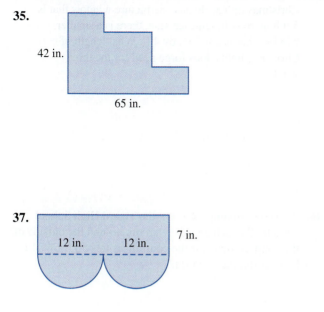

42 in.

65 in.

36.

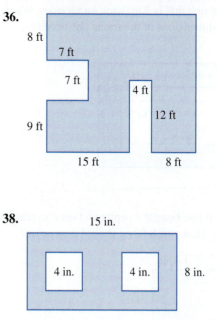

8 ft

7 ft

7 ft

4 ft

9 ft

12 ft

15 ft   8 ft

37.

12 in.   12 in.   7 in.

38.

15 in.

4 in.   4 in.   8 in.

39. How many feet of 6-in.-by-6-in. railroad ties will June need to frame four garden beds that will measure 10 ft by 12 ft including the border?

40. When purchasing molding to make picture frames, it is important to allow enough extra molding to miter the corners. How much extra depends on the width of the molding. Jamie would like to build two frames for photos that are 8 × 10 in. The molding he selected requires an extra $\frac{1}{2}$ in. on each end of each piece of molding for the miter. How much molding does he need for the two frames?

41. A high school track aound the football field is actually two straight lengths that are 125.75 yd each, and two semicircles with a diameter of 60 yd each. How long is the track?

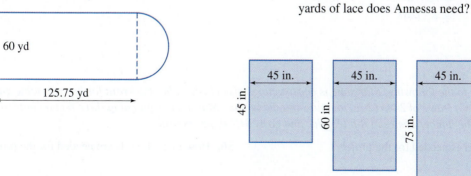

60 yd

125.75 yd

42. Annessa is making a three-tiered bridal veil. Each tier is rectangular, with one of the short ends being gathered into the headpiece. The other three sides of each tier are to be trimmed in antique lace. How many yards of lace does Annessa need?

45 in.

45 in.   45 in.   45 in.

60 in.

75 in.

43. Holli has a watercolor picture that is 14 in. by 20 in. She puts it in a mat that is 3 in. wide on all sides. What are the inside dimensions of the frame she needs to buy?

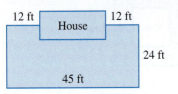

44. Jorge is lining the windows in his living room with Christmas lights. He has one picture window that is 5 ft 8 in. by 4 ft. On each side, there is a smaller window that is 2 ft 6 in. by 4 ft. What length of Christmas lights does Jorge need for the three windows?

45. Jenna and Scott just bought a puppy and need to fence their backyard. How much fence should they order?

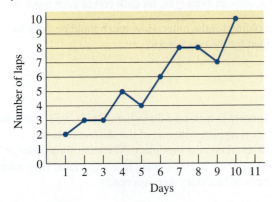

46. As a conditioning exercise, the Tigard High School basketball coach has his team run around the outside of the court 20 times. If the court measures 50 ft by 84 ft, how far did the team run?

47. A high-school football player charted the number of laps he ran around the football field during the first 10 days of practice. If the field measures 120 yd by 53 yd, how far did he run during the 10 days? Convert your answer to the nearest whole mile.

48. A carpenter is putting baseboards in the family room/dining room pictured here. How many feet of baseboard molding are needed?

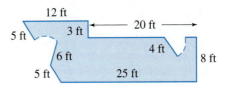

Exercises 49–51 refer to the following. Marylane is constructing raised beds in her backyard for her vegetable garden. Each bed is made from a frame of 2-×-12-in. pressure-treated lumber. Marylane is planning three rectangular frames that measure 4 ft by 8 ft. The lumber comes in 8-ft lengths, and costs $15.49 per board.

49. How much lumber is needed for the project?

50. How many boards are needed for the project?

51. What is the total cost of the lumber for the project?

52. Yolande has lace that is 2 in. wide that she plans to sew around the edge of a circular tablecloth that is 50 in. in diameter. How much lace does she need?

53. Yolande wants to sew a second row of her 2-in.-wide lace around the tablecloth in Exercise 52. How much lace does she need for the second row?

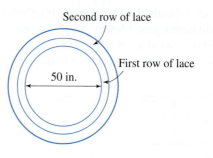

Second row of lace

First row of lace

50 in.

Exercises 54–57 refer to the chapter application. See page 491.

54. What is the shape of the patio?

55. What is the perimeter of the patio?

56. Estimate the perimeter of the flowerbed that wraps around the right side of the patio.

57. Draw a line down the center of the backyard. What do you notice about the two halves? Mathematicians call figures like this symmetrical, because a center line divides the figure into "mirror images." Colonial architecture typically makes use of symmetry.

58. Are rectangles symmetrical around a line drawn lengthwise through the center? Are squares symmetrical?

59. Draw a triangle that is symmetrical, and another that is not symmetrical.

60. Draw a circle that is symmetrical and another that is not symmetrical.

STATE YOUR UNDERSTANDING

61. Explain what perimeter is and how the perimeter of the figure below is determined.

What possible units (both English and metric) would the perimeter be likely to be measured in if the figure is a national park? If the figure is a room in a house? If the figure is a scrap of paper?

62. Is a rectangle a parallelogram? Why or why not?

63. Explain the difference between a square and a rectangle.

64. A farmer wants to build the goat pens pictured below. Each pen will have a gate 2 ft 6 in. wide in the end. What is the total cost of the pens if the fencing is $3.00 per linear foot and each gate is $15.00?

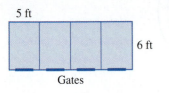

5 ft

6 ft

Gates

MAINTAIN YOUR SKILLS

65. Write the place value name for eight hundred fifty thousand, eight-five.

66. Write the place value name for eight hundred five thousand, eight hundred fifty.

67. Round 32,571,600 to the nearest ten thousand.

68. Find the difference between 733 and 348.

69. Find the product of 673 and 412.

70. Find the quotient of 153,204 and 51.

71. Find the sum of 2 lb 3 oz and 3 lb 14 oz.

72. Find the difference of 3 yd and 5 ft 5 in.

73. Fred scored the following on five algebra tests: 85, 78. 91, 86, and 95. What was his average test score?

74. What is the total cost for a family of one mother and three children to attend a hockey game if adult tickets are $15, student tickets are $10, and the parking fee is $6?

SECTION
7.4 Area

OBJECTIVES

1. Find the area of common polygons and circles.
2. Convert units of area measure.

VOCABULARY

Area is a measure of surface—that is, the amount of space inside a two-dimensional figure. It is measured in square units.

The **base** of a geometric figure is a side parallel to the horizon.

The **altitude** or **height** of a geometric figure is the perpendicular (shortest) distance from the base to the highest point of the figure.

Table 7.11 shows the base and altitude of some common geometric figures.

TABLE 7.11 **Base and Height of Common Geometric Figures**

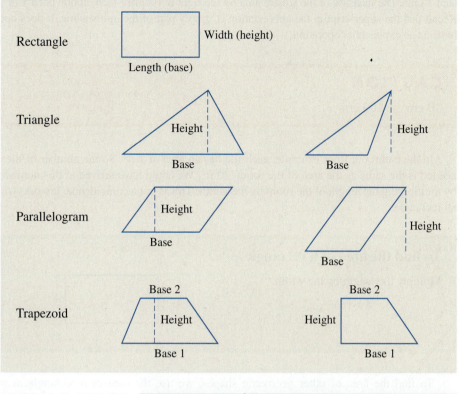

Rectangle — Width (height), Length (base)

Triangle — Height, Base

Parallelogram — Height, Base

Trapezoid — Base 2, Height, Base 1

HOW & WHY

■ **OBJECTIVE 1** Find the area of common polygons and circles.

Suppose you wish to tile a rectangular bathroom floor that measures 5 ft by 6 ft. The tiles are 1-ft-by-1-ft squares. How many tiles do you need?

Using Figure 7.1 as a model of the tiled floor, you can count that 30 tiles are necessary.

Figure 7.1

Area is a measure of the surface—that is, the amount of space inside a two-dimensional figure. It is measured in square units. Square units are literally squares that measure one unit of length on each side. Figure 7.2 shows two examples.

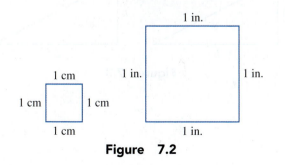

1 cm, 1 cm, 1 cm, 1 cm

1 in., 1 in., 1 in., 1 in.

Figure 7.2

The measure of the square area on the left in Figure 7.2 is 1 square centimeter, abbreviated 1 cm². The measure of the square area on the right is 1 square inch, abbreviated 1 in². Recall that the superscript in the abbreviation is simply part of the unit's name. It does not indicate an exponential operation.

CAUTION

$10 \text{ cm}^2 \neq 100 \text{ cm}$

In the bathroom floor example, each tile has an area of 1 ft². So the number of tiles needed is the same as the area of the room, 30 ft². We could have arrived at this number by multiplying the length of the room by the width. This is not a coincidence. It works for all rectangles.

To find the area of a rectangle

Multiply the length by the width.

$$A = \ell w$$

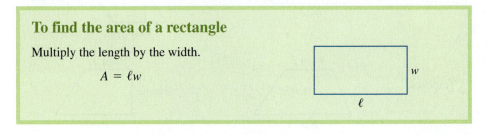

To find the area of other geometric shapes, we use the area of a rectangle as a reference.

Because a square is a special case of a rectangle, with all sides equal, the formula for the area of a square is

$$\begin{aligned} A &= \ell w \\ &= s(s) \\ &= s^2 \end{aligned}$$

To find the area of a square

Square the length of one of the sides.

$$A = s^2$$

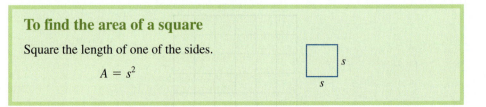

Now let's consider the area of a triangle. We start with a right triangle—that is, a triangle with one 90° angle. See Figure 7.3.

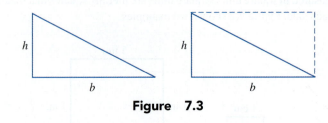

Figure 7.3

The triangle on the left has base b and height h. The figure on the right is a rectangle with length b and width h. According to the formula for rectangles, the area is $A = \ell w$ or $A = bh$. However, the rectangle is made up of two triangles, both of which have a base of b and a height of h. The area of the rectangle (bh) is exactly twice the area of the triangle. So we conclude that the area of the triangle is

$$\frac{1}{2} \cdot bh \quad \text{or} \quad \frac{bh}{2} \quad \text{or} \quad bh \div 2$$

Now let's consider a more general triangle. See Figure 7.4.

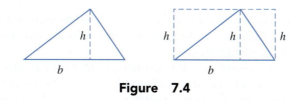

Figure 7.4

Again, the area on the left is a triangle with base b and height h. And the figure on the right is a rectangle with length b and width h. Can you see that the rectangle must be exactly twice the area of the original triangle? So again we conclude that the area of the triangle is

$$\frac{1}{2} \cdot bh \quad \text{or} \quad \frac{bh}{2} \quad \text{or} \quad bh \div 2$$

Recall that height is the perpendicular distance from the base to the highest point of a figure.

To find the area of a triangle

Multiply the base times the height and divide by 2.

$$A = \frac{1}{2} \cdot bh \quad \text{or} \quad A = \frac{bh}{2} \quad \text{or} \quad A = bh \div 2$$

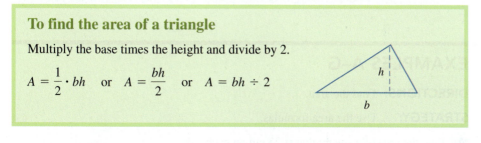

It is possible to use rectangles to find the formulas for the areas of parallelograms and trapezoids. This is left as an exercise. (See Exercise 79.)

To find the area of a parallelogram

Multiply the base times the height.

$$A = bh$$

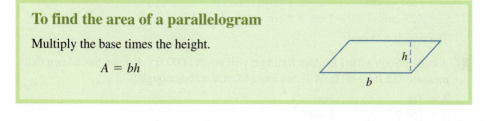

CAUTION

The base of a parallelogram is the same as its longest side, but the height of the parallelogram is *not* the same as the length of its shortest side.

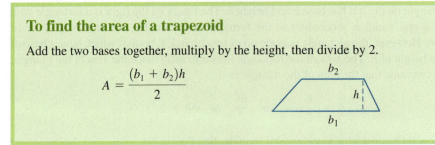

To find the area of a trapezoid

Add the two bases together, multiply by the height, then divide by 2.

$$A = \frac{(b_1 + b_2)h}{2}$$

The lengths, widths, bases, sides, and heights must all be measured using the same units before the area formulas can be applied. If the units are different, convert to a common unit before calculating area.

The formula for the area of a circle is less intuitive than the formulas discussed thus far. Nevertheless, the formula for the area of a circle has been known and used for many years. Even though the sides of a circle are not line segments, we still measure the area in square units.

REMINDER

You can use 3.14 or $\frac{22}{7}$ to approximate π or use the $\boxed{\pi}$ key on your calculator for a more accurate approximation.

To find the area of a circle

Square the radius and multiply by π.

$$A = \pi r^2$$

EXAMPLES A–G

DIRECTIONS: Find the area.

STRATEGY: Use the area formulas.

WARM-UP

A. Find the area of a square that is 18 ft on each side.

A. Find the area of a square that is 25 cm on each side.

$$
\begin{aligned}
A &= s^2 &&\text{Formula.}\\
&= (25 \text{ cm})^2 &&\text{Substitute.}\\
&= (25)^2(1 \text{ cm})^2\\
&= 625 \text{ cm}^2 &&1 \text{ cm} \times 1 \text{ cm} = 1 \text{ cm}^2
\end{aligned}
$$

The area is 625 cm².

WARM-UP

B. A decorator found a 15 yd² remnant of carpet. Will it be enough to carpet a 5-yd-by-4-yd playroom?

B. A bag of Scott's Turf Builder fertilizer will cover 5000 ft². Roman has a lawn that measures 82 ft by 60 ft. Will one bag of fertilizer be enough?

$$
\begin{aligned}
A &= \ell w &&\text{Formula}\\
&= (82 \text{ ft})(60 \text{ ft}) &&\text{Substitute.}\\
&= 4920 \text{ ft}^2 &&1 \text{ ft} \times 1 \text{ ft} = 1 \text{ ft}^2
\end{aligned}
$$

Because 4920 ft² < 5000 ft², 1 bag of fertilizer will be enough.

ANSWERS TO WARM-UPS A–B

A. The area is 324 ft².

B. The remnant will not be enough because the decorator needs 20 yd².

C. Find the area of this triangle.

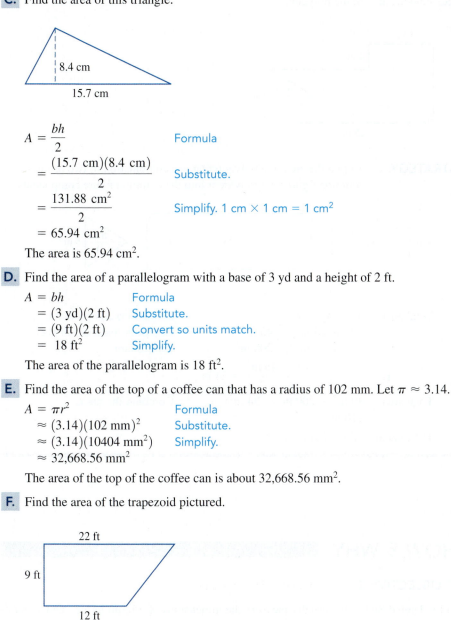

8.4 cm

15.7 cm

$$A = \frac{bh}{2} \qquad \text{Formula}$$
$$= \frac{(15.7 \text{ cm})(8.4 \text{ cm})}{2} \qquad \text{Substitute.}$$
$$= \frac{131.88 \text{ cm}^2}{2} \qquad \text{Simplify. 1 cm} \times 1 \text{ cm} = 1 \text{ cm}^2$$
$$= 65.94 \text{ cm}^2$$

The area is 65.94 cm².

D. Find the area of a parallelogram with a base of 3 yd and a height of 2 ft.

$$A = bh \qquad \text{Formula}$$
$$= (3 \text{ yd})(2 \text{ ft}) \qquad \text{Substitute.}$$
$$= (9 \text{ ft})(2 \text{ ft}) \qquad \text{Convert so units match.}$$
$$= 18 \text{ ft}^2 \qquad \text{Simplify.}$$

The area of the parallelogram is 18 ft².

E. Find the area of the top of a coffee can that has a radius of 102 mm. Let $\pi \approx 3.14$.

$$A = \pi r^2 \qquad \text{Formula}$$
$$\approx (3.14)(102 \text{ mm})^2 \qquad \text{Substitute.}$$
$$\approx (3.14)(10404 \text{ mm}^2) \qquad \text{Simplify.}$$
$$\approx 32{,}668.56 \text{ mm}^2$$

The area of the top of the coffee can is about 32,668.56 mm².

F. Find the area of the trapezoid pictured.

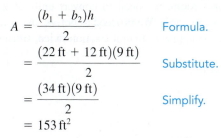

22 ft

9 ft

12 ft

$$A = \frac{(b_1 + b_2)h}{2} \qquad \text{Formula.}$$
$$= \frac{(22 \text{ ft} + 12 \text{ ft})(9 \text{ ft})}{2} \qquad \text{Substitute.}$$
$$= \frac{(34 \text{ ft})(9 \text{ ft})}{2} \qquad \text{Simplify.}$$
$$= 153 \text{ ft}^2$$

The area of the trapezoid is 153 ft².

WARM-UP

C. Find the area of this triangle.

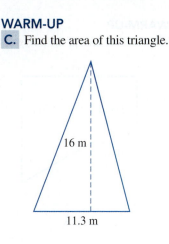

16 m

11.3 m

WARM-UP

D. Find the area of a parallelogram with a base of 4 ft and a height of 18 in.

WARM-UP

E. What is the area of a quarter that has a radius of 13 mm? Let $\pi \approx 3.14$.

WARM-UP

F. Find the area of the trapezoid with bases of 34 m and 42 m and a height of 15 m.

ANSWERS TO WARM-UPS C–F

C. The area is 90.4 m².
D. The area is 864 in², or 6 ft².
E. The area is about 530.66 mm².
F. The area is 570 m².

WARM-UP

G. Find the area of the polygon.

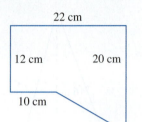

22 cm

12 cm   20 cm

10 cm

G. Find the area of the polygon.

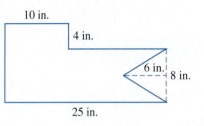

10 in.

4 in.

6 in.

8 in.

25 in.

STRATEGY: To find the area of a polygon that is a combination of two or more common figures, first divide it into the common figure components.

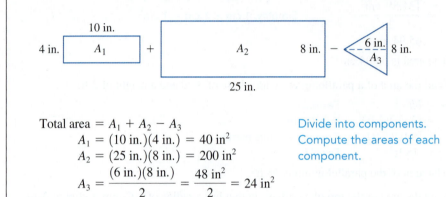

4 in.   10 in. A_1 + A_2 8 in.   −   6 in. A_3 8 in.

25 in.

Total area $= A_1 + A_2 - A_3$ Divide into components.
$A_1 = (10 \text{ in.})(4 \text{ in.}) = 40 \text{ in}^2$ Compute the areas of each
$A_2 = (25 \text{ in.})(8 \text{ in.}) = 200 \text{ in}^2$ component.
$A_3 = \dfrac{(6 \text{ in.})(8 \text{ in.})}{2} = \dfrac{48 \text{ in}^2}{2} = 24 \text{ in}^2$

Total area $= 40 \text{ in}^2 + 200 \text{ in}^2 - 24 \text{ in}^2$ Combine the areas.
           $= 216 \text{ in}^2$

The total area of the figure is 216 in^2.

HOW & WHY

■ **OBJECTIVE 2** Convert units of area measure.

In the United States, we usually measure the dimensions of a room in feet, and so naturally, we measure its area in square feet. However, carpeting is measured in square yards. Therefore, if we want to buy carpeting for a room, we need to convert units of area measure—in this case, from square feet to square yards. We proceed as before, using unit fractions. The only difference is that we use the appropriate unit fraction twice, since we are working with square units.

To change 100 ft² to square yards, we write

$$100 \, \text{ft}^2 = 100(\text{ft})(\text{ft})\left(\frac{1 \, \text{yd}}{3 \, \text{ft}}\right)\left(\frac{1 \, \text{yd}}{3 \, \text{ft}}\right)$$

$$= \frac{100}{9} \, \text{yd}^2$$

$$= 11\frac{1}{9} \, \text{yd}^2$$

EXAMPLES H–I

DIRECTIONS: Convert as indicated.

STRATEGY: Use the appropriate unit fraction as a factor twice.

H. Convert 64 ft² to square meters. Round to the nearest thousandth.

$$64 \text{ ft}^2 \approx 64 \, (\cancel{ft})(\cancel{ft})\left(\frac{0.3048 \text{ m}}{1 \, \cancel{ft}}\right)\left(\frac{0.3048 \text{ m}}{1 \, \cancel{ft}}\right) \qquad \color{blue}{0.3048 \text{ m} \approx 1 \text{ ft}}$$

$$\approx 5.946 \text{ m}^2$$

So 64 ft² ≈ 5.946 m².

I. Convert 378 cm² to square meters.

$$378 \text{ cm}^2 = 378(\cancel{cm})(\cancel{cm})\left(\frac{1 \text{ m}}{100 \, \cancel{cm}}\right)\left(\frac{1 \text{ m}}{100 \, \cancel{cm}}\right) \qquad \color{blue}{100 \text{ cm} = 1 \text{ m}}$$

$$= \frac{378}{10,000} \text{ m}^2 \qquad \color{blue}{\text{Simplify.}}$$

$$= 0.0378 \text{ m}^2$$

So 378 cm² = 0.0378 m².

EXERCISES 7.4

■ **OBJECTIVE 1**   Find the area of common polygons and circles. (See page 523.)

A *Find the area of the following figures. Use* $\pi \approx \dfrac{22}{7}$.

1.
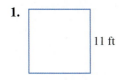
11 ft

2.
13 in.

4 in.

3.
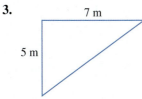
7 m

5 m

4
23 mm

4 mm

5.
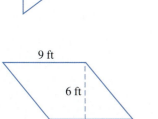
9 ft

6 ft

6.
48 cm

2 cm

7.

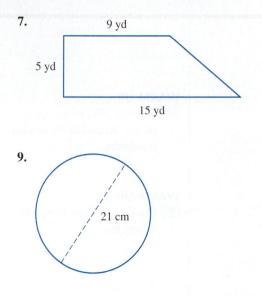

9 yd

5 yd

15 yd

8.

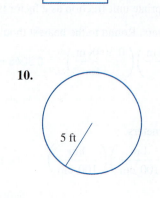

12 mi

9.

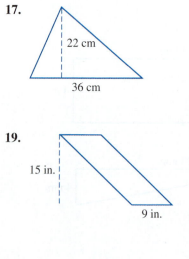

21 cm

10.

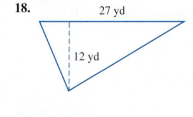

5 ft

B *Find the areas of the figures. Use π ≈ 3.14.*

11. Find the area of a rectangle that has a length of 17 m and a width of 15 m.

12. Find the area of a square with sides of 136 cm.

13. Find the area of a triangle with a base of 20 ft and a height of 17 ft.

14. Find the area of a circle with a radius of 6 m.

15. Find the area of a parallelogram with a base of 54 in. and a height of 30 in.

16. Find the area of a circle with a diameter of 62 cm.

17.

22 cm

36 cm

18.

27 yd

12 yd

19.

15 in.

9 in.

20.

53 ft

78 ft

21.

32 in.

17 in.

48 in.

22.

7 km

13 km

13 km

23.

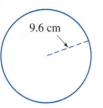

9.6 cm

24.

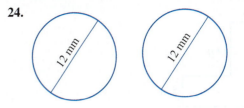

12 mm   12 mm

■ **OBJECTIVE 2** Convert units of area measure. (See page 528.)

A *Convert units as indicated. If necessary, round to the nearest hundredth.*

25. Convert 6,780 in^2 to square feet.

26. Convert 4 cm^2 to square millimeters.

27. Convert 12 yd^2 to square inches.

28. Convert 13 m^2 to square centimeters.

29. Convert 1 mi^2 to square feet.

30. Convert 8 km^2 to square meters.

31. Convert 4056 in^2 to square yards.

32. Convert 2,000,000 cm^2 to square meters.

33. Convert 345 ft^2 to square yards.

34. Convert 8700 mm^2 to square centimeters.

B

35. Convert 44 yd^2 to square feet.

36. Convert 54 km^2 to square centimeters.

37. Convert 15,000,000 ft^2 to square miles.

38. Convert 456,789,123 cm^2 to square meters.

39. Convert 1600 cm^2 to square inches.

40. Convert 100 in^2 to square centimeters.

41. Convert 12 m^2 to square feet.

42. Convert 80 km^2 to square miles.

43. Convert 15 ft^2 to square centimeters.

44. Convert 100 mi^2 to square kilometers.
Let 1 mi$^2 \approx 2.5900$ km^2.

C *Find the areas of the figures.*

45.

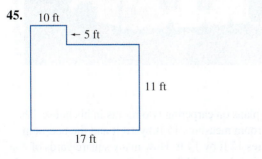

10 ft
← 5 ft
11 ft
17 ft

46.

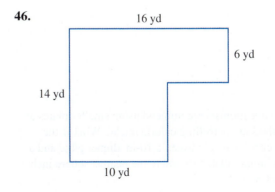

16 yd
6 yd
14 yd
10 yd

47.

15 m

14 m

5 m

22 m

48.

13 ft

10 ft

3 ft

49.

14 yd

6 yd   5 yd

12 yd

40 yd

50.

15 m   15 m

15 m   15 m

50 m

140 m

Find the areas of the shaded regions. Use the ⊡ *key, and round to the nearest hundredth.*

51.

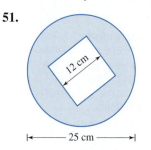

12 cm

25 cm

52.

36 in.

14 in.

5 in.

53. The side of Zakaria's house that has no windows measures 35 ft by 24 ft. If 1 gallon of stain covers 250 ft², how many gallons does he need to stain this side of his house? Round to the nearest tenth of a gallon.

54. The south side of Zakaria's house measures 75 ft by 24 ft and has two windows, each 4 ft by 6 ft. How many gallons of stain does he need for the south side of his house? See Exercise 53. Round to the nearest gallon.

© Rafal Olechowski/Shutterstock.com

55. Some nutritionists recommend using smaller plates as a method of controlling calorie intake. What is the difference in area between a 10-in. dinner plate and a 12-in. dinner plate? Round to the nearest square inch and let $\pi \approx 3.14$.

56. Mario plans on carpeting two rooms in his house. The living room measures 15 ft by 18 ft and the bedroom measures 12 ft by 14 ft. How many square yards of carpet does he need? Round to the nearest square yard.

57. If 1.25 oz of weed killer treats 1 m² of lawn, how many ounces of weed killer will Debbie need to treat a rectangular lawn that measures 15 m by 30 m?

58. To the nearest acre, how many acres are contained in a rectangular plot of ground if the length is 1850 ft and the width is 682 ft? (43,560 ft² = 1 acre.) Round to the nearest whole number.

59. How much glass is needed to replace a set of two sliding glass doors in which each pane measures 3 ft by 6 ft?

60. A rectangular counter in a bathroom measures 18 in. deep by $4\frac{1}{2}$ ft. How many square feet of tile are needed to cover the counter?

61. How many square feet of sheathing are needed for the gable end of a house that has a rise of 9 ft and a span of 36 ft? (See the drawing.)

62. Frederica wants to construct a toolshed that is 6 ft by 9 ft around the base and 8 ft high. How many gallons of paint will be needed to cover the outside of all four walls of the shed? (Assume that 1 gal covers 250 ft².)

63. Maureen is making a quilt for a crib that is 48 in. by 60 in. She is piecing the top using ribbons of fabric that are 2 in. by 60 in. She is using $\frac{1}{4}$-in. seam allowances ($\frac{1}{4}$ in. on each side of the 2-in.-by-60-in. ribbon to sew the ribbons together). How many ribbons of fabric does she need to make the 48-in. width of the quilt? What is the total area of fabric needed to make the quilt?

64. Maureen, see Exercise 63, is going to make half of the ribbons for her baby quilt pink and the other half blue. If the fabric she is using is 36 in. wide, how much of each color should she buy for the quilt? How much fabric will be left over?

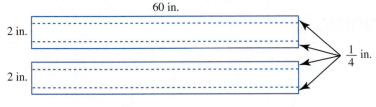

65. April and Larry are building an in-ground circular spa. The spa is 4 ft deep and 8 ft in diameter. They intend to tile the sides and bottom of the spa. How many square feet of tile do they need? (*Hint:* The sides of a cylinder form a rectangle whose length is the circumference of the circle.) Round to the nearest whole number.

66. A window manufacturer is reviewing the plans of a duplex to determine the amount of glass needed to fill the order. The number and size of the windowpanes and sliding glass doors are listed in the table.

	Dimensions	Number Needed
Windows needed	3 ft by 3 ft	4
	3 ft by 4 ft	7
	3 ft by 5 ft	2
	4 ft by 4 ft	2
	5 ft by 6 ft	1
Sliding doors	7 ft by 3 ft	2

How much glass does he need to fill the order?

67. One 2-lb bag of wildflower seed will cover 70 ft². How many bags of seed are needed to cover the region shown here? Do not seed the shaded area.

```
        10 ft          30 ft
      ├──────┼──────────────────┤
      ┌──────┬──────────────────┐
  10 ft│     │   15 ft      ╲
      │  16 ft│  ┌────────┐   8 ft ╲
      │       │  │        │      ╲
      └───────┴──│        │───────┘
              └──┴────────┴──┘
                   20 ft
```

68. How many square yards of carpet are needed to carpet a flight of stairs if each stair is 10 in. by 40 in., the risers are 8 in., and there are 12 stairs and 13 risers in the flight? Round to the nearest square yard.

69. How many 8-ft-by-4-ft sheets of paneling are needed to panel two walls of a den that measure 14 ft by 8 ft and 10 ft by 8 ft?

70. Joy is making a king-sized quilt that measures 72 in. by 85 in. She needs to buy fabric for the back of the quilt. How many yards should she purchase if the fabric is 45 in. wide? Round to the nearest square yard.

Exercises 71–72 refer to the chapter application. See page 491.

71. The contractor who was hired to build the brick patio will begin by pouring a concrete slab. Then he will put the bricks on the slab. The estimate of both the number of bricks and the amount of mortar needed is based on the area of the patio and walkways. Subdivide the patio and the walkways into geometric figures; then calculate the total area to be covered in bricks.

72. The number of bricks and amount of mortar needed also depend on the thickness of the mortar between the bricks. The plans specify a joint thickness of $\frac{1}{4}$ in.

According to industry standards, this will require seven bricks per square foot. Find the total number of bricks required for the patio and walkways.

STATE YOUR UNDERSTANDING

73. What kinds of units measure area? Give examples from both systems.

74. Explain how to calculate the area of the figure below. Do not include the shaded area.

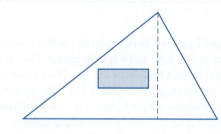

75. Describe how you could approximate the area of a geometric figure using 1-in. squares.

76. Joe is going to tile his kitchen floor. Along the outside he will put black squares that are 6 in. on each side. The next (inside) row will be white squares that are 6 in. on each side. The remaining interior space will be a checkerboard pattern of alternating black and white tiles that are 1 ft on each side. How many tiles of each color will he need for the kitchen floor, which measures 12 ft by 10 ft?

77. A rectangular plot of ground measuring 120 ft by 200 ft is to have a cement walkway 5 ft wide placed around the inside of the perimeter. How much of the area of the plot will be used by the walkway and how much of the area will remain for the lawn?

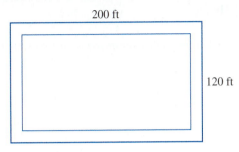

200 ft

120 ft

78. April and Larry decide that their spa should have a built-in ledge to sit on that is 18 in. wide. The ledge will be 2 ft off the bottom. How many square feet of tile do they need for the entire inside of the spa? See Exercise 65. Round to the nearest whole square foot.

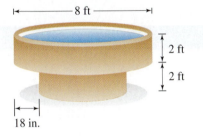

8 ft

2 ft

2 ft

18 in.

79. Multiply 6.708 by 100,000.

80. What percent of 2345 is 653? Round to the nearest whole percent.

81. Simplify: $2(46 - 28) + 50 \div 5$

82. A family of six attends a weaving exhibition. Parking is $4, adult admission is $6, senior admission is $4, and child admission is $3. How much does it cost for two parents, one grandmother, and three children to attend the exhibition?

83. Simplify: $3^2 + 2^3$

84. Solve the proportion: $\dfrac{x}{45} = \dfrac{13}{16}$. Round to the nearest hundredth.

85. Find the average, median, and mode: 16, 17, 18, 19, 16, and 214

86. Find the perimeter of a rectangle that is 4 in. wide and 2 ft long.

87. The table lists calories burned per hour for various activities.

Activity	Calories Burned/Hr
Sitting	50
Slow walking	125
Stacking firewood	350
Hiking	500

Make a bar graph that summarizes this information.

88. Convert 4.5 ft to centimeters. Round to the nearest hundredth.

OBJECTIVES

1. Find the volume of common geometric shapes.
2. Convert units of volume.

VOCABULARY

A **cube** is a three-dimensional geometric solid that has six sides (called **faces**), each of which is a square.

Volume is the name given to the amount of space that is contained inside a three-dimensional object.

HOW & WHY

◼ OBJECTIVE 1 Find the volume of common geometric shapes.

Suppose you have a shoebox that measures 12 in. long by 4 in. wide by 5 in. high that you want to use to store toy blocks that are 1 in. by 1 in. by 1 in. How many blocks will fit in the box? See Figure 7.5.

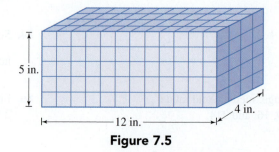

Figure 7.5

In each layer there are 12(4) = 48 blocks and there are 5 layers. Therefore, the box holds 12(4)(5) = 240 blocks.

Volume is a measure of the amount of space contained in a three-dimensional object. Often, volume is measured in cubic units. These units are literally cubes that measure one unit on each side. For example, Figure 7.6 shows a cubic inch (1 in^3) and a cubic centimeter (1 cm^3).

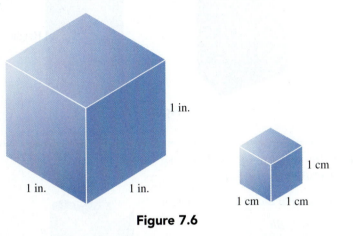

Figure 7.6

The shoebox has a volume of 240 in^3 because exactly 240 blocks, which have a volume of 1 in^3 each, can fit in the box and totally fill it up.

In general, volume can be thought of as the number of cubes that fill up a space. If the space is a rectangular solid, like the shoebox, it is a relatively easy matter to determine the volume by making a layer of cubes that covers the bottom and then deciding how many layers are necessary to fill the box.

Observe that the number of cubes necessary for the bottom layer is the same as the area of the base of the box, ℓw. The number of layers needed is the same as the height of the box, h. So we have the following volume formulas.

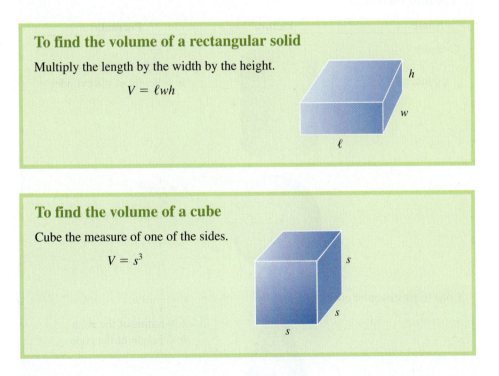

To find the volume of a rectangular solid

Multiply the length by the width by the height.

$$V = \ell w h$$

To find the volume of a cube

Cube the measure of one of the sides.

$$V = s^3$$

The length, width, and height must all be measured using the same units before the volume formulas may be applied. If the units are different, convert to a common unit before calculating the volume.

The principle used for finding the volume of a box can be extended to any solid with sides that are perpendicular to the base. The area of the base gives the number of cubes necessary to make the bottom layer, and the height gives the number of layers necessary to fill the solid. See Figure 7.7.

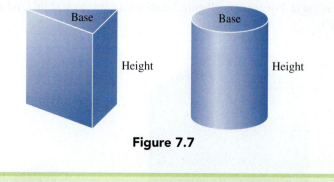

Figure 7.7

To find the volume of a solid with sides perpendicular to the base

Multiply the area of the base by the height.

$$V = Bh$$

where B is the area of the base.

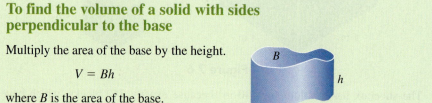

Because cylinders, spheres, and cones contain circles, their volume formulas also contain the number π. Even though these shapes are curved instead of straight, their volumes are measured in cubic units.

Volume formulas

Name	Picture	Formula
Cylinder (right circular cylinder)		$V = \pi r^2 h$ r = radius of the base h = height of the cylinder
Sphere		$V = \dfrac{4}{3}\pi r^3$ r = radius of the sphere
Cone (right cirucular cone)		$V = \dfrac{1}{3}\pi r^2 h$ r = radius of the cone h = height of the cone

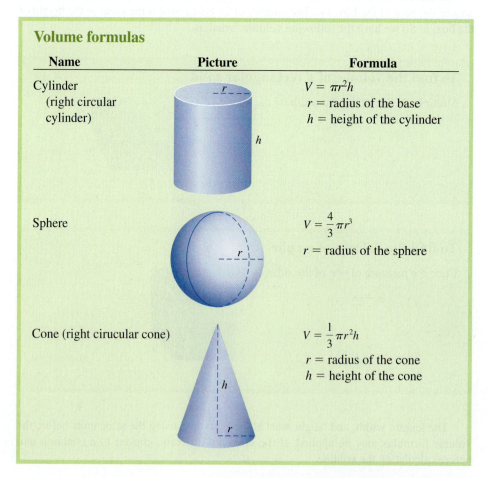

EXAMPLES A–D

DIRECTIONS: Find the volume.

STRATEGY: Use the volume formulas.

A. Find the volume of a cube that is 7 ft on each side.

$V = s^3$ Formula
$= (7\text{ ft})^3$
$= 343\text{ ft}^3$

The volume of the cube is 343 ft^3.

B. How much concrete is needed to pour a step that is 4 ft long, 3 ft wide, and 6 in. deep?

$V = \ell wh$ Formula
$= (4\text{ ft})(3\text{ ft})(6\text{ in.})$ Substitute.
$= (4\text{ ft})(3\text{ ft})(0.5\text{ ft})$ Convert inches to feet.
$= 6\text{ ft}^3$ Simplify.

The step requires 6 ft^3 of concrete.

ALTERNATIVE SOLUTION:

$V = \ell wh$ Formula
$= (4\text{ ft})(3\text{ ft})(6\text{ in.})$ Substitute.
$= (48\text{ in.})(36\text{ in.})(6\text{ in.})$ Convert feet to inches.
$= 10{,}368\text{ in}^3$

The step requires $10{,}368\text{ in}^3$ of concrete.

C. What is the volume of a tank (cylinder) that has a radius of 4 ft and a height of 10 ft? Let $\pi \approx 3.14$.

4 ft

10 ft

$V = \pi r^2 h$

$V = \pi r^2 h$ Formula
$\approx (3.14)(4\text{ ft})^2(10\text{ ft})$ Substitute.
$\approx 502.4\text{ ft}^3$

The volume of the tank is about 502.4 ft^3.

ALTERNATIVE SOLUTION: Use the $\boxed{\pi}$ key and round to the nearest hundredth.

$V = \pi r^2 h$ Formula
$= (\pi)(4\text{ ft})^2(10\text{ ft})$ Substitute.
$\approx 502.7\text{ ft}^3$ Simplify using the $\boxed{\pi}$ key on your calculator.

The volume of the can is about 502.7 ft^3.

D. Find the volume of a right circular cone that has a base diameter of 3 ft and a height of 7 ft. Use the $\boxed{\pi}$, key, and round to the nearest hundredth.

D. Find the volume of the right circular cone. Round to the nearest hundredth.

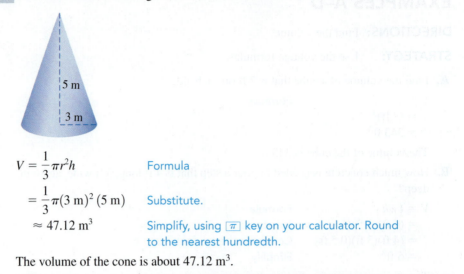

$$V = \frac{1}{3}\pi r^2 h \qquad \text{Formula}$$

$$= \frac{1}{3}\pi (3\text{ m})^2 (5\text{ m}) \qquad \text{Substitute.}$$

$$\approx 47.12\text{ m}^3 \qquad \text{Simplify, using } \boxed{\pi} \text{ key on your calculator. Round to the nearest hundredth.}$$

The volume of the cone is about 47.12 m^3.

HOW & WHY

■ **OBJECTIVE 2** Convert units of volume.

As Example B suggests, it is possible to measure the same volume in both cubic feet and cubic inches. We calculated that the amount of concrete needed for the step is 6 ft^3, or $10,368\text{ in}^3$.

To verify that $6\text{ ft}^3 = 10,368\text{ in}^3$, and to convert units of volume in general, we use unit fractions. Because we are converting cubic units, we use the unit fraction three times.

$$6\text{ ft}^3 = 6\,(\text{ft})\,(\text{ft})\,(\text{ft})$$

$$= 6\left(\text{ft} \cdot \frac{12\text{ in.}}{1\text{ ft}}\right)\left(\text{ft} \cdot \frac{12\text{ in.}}{1\text{ ft}}\right)\left(\text{ft} \cdot \frac{12\text{ in.}}{1\text{ ft}}\right)$$

$$= 6 \cdot 12 \cdot 12 \cdot 12\,(\text{in.})\,(\text{in.})\,(\text{in.})$$

$$= 10,368\text{ in}^3$$

Remember this special relationship:

$1\text{ cm}^3 = 1\text{ mL}$

When measuring the capacity of a solid to hold liquid, we sometimes use special units. Recall that in the English system, liquid capacity is measured in ounces, quarts, and gallons. In the metric system, milliliters, liters, and kiloliters are used. One cubic centimeter measures the same volume as one milliliter. That is, $1\text{ cm}^3 = 1\text{ mL}$. So a container with a volume of 50 cm^3 holds 50 mL of liquid.

EXAMPLES E–H

DIRECTIONS: Convert units of volume.

STRATEGY: Use the appropriate unit fraction three times.

E. Convert 10 ft^3 to cubic inches.

E. Convert 25 yd^3 to cubic feet.

$$25\text{ yd}^3 = 25\text{ yd}^3 \cdot \left(\frac{3\text{ ft}}{1\text{ yd}}\right)\left(\frac{3\text{ ft}}{1\text{ yd}}\right)\left(\frac{3\text{ ft}}{1\text{ yd}}\right) \qquad \text{Use the unit fraction } \frac{3\text{ ft}}{1\text{ yd}}.$$

$$= 25 \cdot 3 \cdot 3 \cdot 3\text{ ft}^3$$

$$= 675\text{ ft}^3$$

So $25\text{ yd}^3 = 675\text{ ft}^3$.

ANSWERS TO WARM-UPS D-E

D. The volume is about 16.49 ft^3.

E. $17,280\text{ in}^3$

F. How many cubic centimeters in 3400 mm³?

$$3400 \text{ mm}^3 = 3400 \text{ mm}^3 \cdot \left(\frac{1 \text{ cm}}{10 \text{ mm}}\right)\left(\frac{1 \text{ cm}}{10 \text{ mm}}\right)\left(\frac{1 \text{ cm}}{10 \text{ mm}}\right)$$

$$= \frac{3400}{10 \cdot 10 \cdot 10} \text{ cm}^3$$

$$= 3.4 \text{ cm}^3$$

So 3400 mm³ = 3.4 cm³.

G. How many milliliters of water does this container hold?

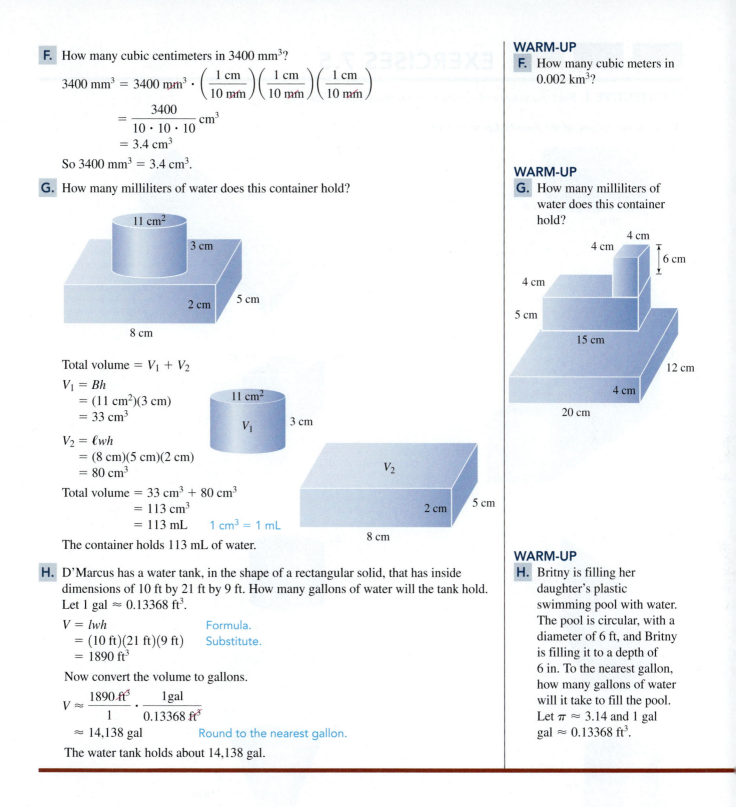

Total volume = $V_1 + V_2$

$V_1 = Bh$
  $= (11 \text{ cm}^2)(3 \text{ cm})$
  $= 33 \text{ cm}^3$

$V_2 = \ell wh$
  $= (8 \text{ cm})(5 \text{ cm})(2 \text{ cm})$
  $= 80 \text{ cm}^3$

Total volume = 33 cm³ + 80 cm³
  $= 113 \text{ cm}^3$
  $= 113 \text{ mL}$ 1 cm³ = 1 mL

The container holds 113 mL of water.

H. D'Marcus has a water tank, in the shape of a rectangular solid, that has inside dimensions of 10 ft by 21 ft by 9 ft. How many gallons of water will the tank hold. Let 1 gal ≈ 0.13368 ft³.

$V = lwh$ Formula.
  $= (10 \text{ ft})(21 \text{ ft})(9 \text{ ft})$ Substitute.
  $= 1890 \text{ ft}^3$

Now convert the volume to gallons.

$$V \approx \frac{1890 \text{ ft}^3}{1} \cdot \frac{1 \text{ gal}}{0.13368 \text{ ft}^3}$$

$\approx 14{,}138 \text{ gal}$ Round to the nearest gallon.

The water tank holds about 14,138 gal.

OBJECTIVE 1 Find the volume of common geometric shapes. (See page 536.)

A *Find the volume of the figures. Let π ≈ 3.14.*

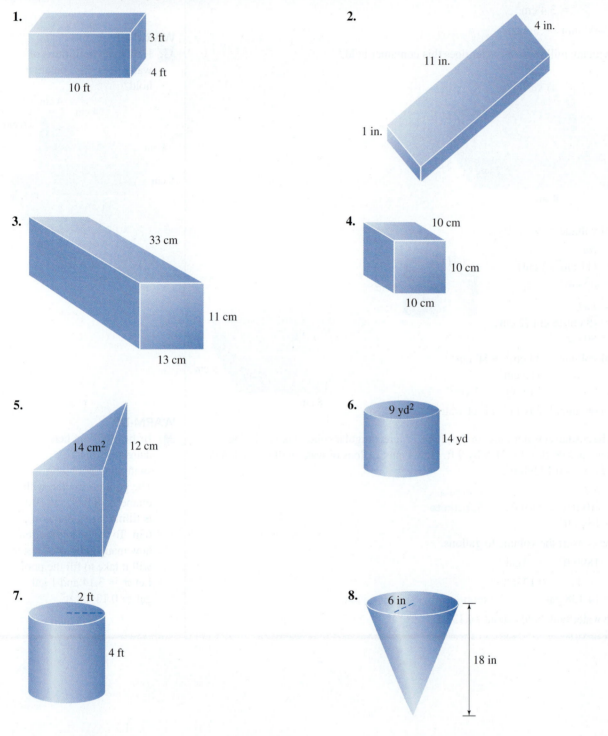

1.
3 ft
4 ft
10 ft

2.
4 in.
11 in.
1 in.

3.
33 cm
11 cm
13 cm

4.
10 cm
10 cm
10 cm

5.
14 cm²
12 cm

6.
9 yd²
14 yd

7.
2 ft
4 ft

8.
6 in
18 in

B

9. How many mL of water will fill up a box that measures 50 cm long, 30 cm wide, and 18 cm high?

10. Find the volume of a cube that measures 22 in. on each side.

11. Find the volume of a cylindrical garbage can that is 2 ft tall and has a base with a diameter of 18 in. Let $\pi \approx 3.14$.

12. Find the volume of two identical fuzzy dice tied to the mirror of a 1957 Chevy if one edge measures 12 cm.

13. Find the volume of the stone soccer ball outside of Dick's Sport Shop. The soccer ball has a diameter of 2 ft. Let $\pi \approx 3.14$ and round to the nearest hundredth.

14. Find the volume of an ice cream cone that is $4\frac{1}{2}$ in. tall and has a top diameter of 4 in. Use $\frac{22}{7}$ for π.

Find the volume of the figure. Use the $\boxed{\pi}$ key and if necessary round to the nearest hundredth.

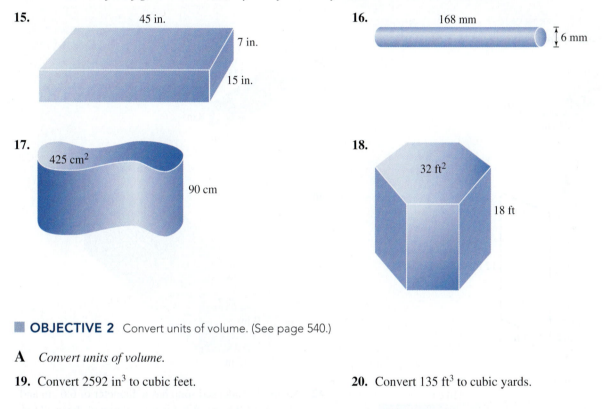

15. 45 in., 7 in., 15 in.

16. 168 mm, 6 mm

17. 425 cm², 90 cm

18. 32 ft², 18 ft

■ **OBJECTIVE 2** Convert units of volume. (See page 540.)

A *Convert units of volume.*

19. Convert 2592 in³ to cubic feet.

20. Convert 135 ft³ to cubic yards.

21. Convert 1 m³ to cubic centimeters.

22. Convert 1 km³ to cubic meters.

23. Convert 1 yd³ to cubic inches.

24. Convert 5 m³ to cubic centimeters.

25. Convert 13,824 in³ to cubic feet.

26. Convert 270 ft³ to cubic yards.

27. Convert 10,000 mm³ to cubic centimeters.

28. Convert 7,000,000,000 m³ to cubic kilometers.

B *Convert. Round decimal values to the nearest hundredth.*

29. Convert 13,560 in³ to cubic feet.

30. Convert 1 mi³ to cubic feet.

31. Convert 350,000 cm³ to cubic meters.

32. Convert 987,654,321 m³ to cubic kilometers.

33. Convert 8.93 km³ to cubic miles.

34. Convert 4 ft³ to cubic centimeters.

35. Find the volume of a granary silo that has a inside base area of 2826 ft² and a height of 14 yards. Find the volume in cubic yards.

36. How many cubic inches of concrete are needed to pour a sidewalk that is 3 ft wide, 4 in. deep, and 54 ft long? Concrete is commonly measured in cubic yards, so convert your answer to cubic yards. (*Hint:* Convert to cubic feet first, and then convert to cubic yards.)

Find the volume of the figures. Let π ≈ 3.14

37.

46 cm

10 cm

24 cm

38.

32 in.

8 in.

10 in.

39.

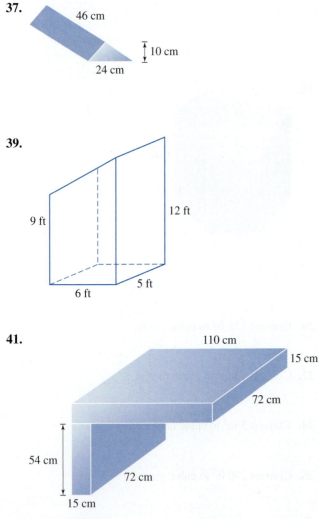

9 ft

12 ft

6 ft

5 ft

40.

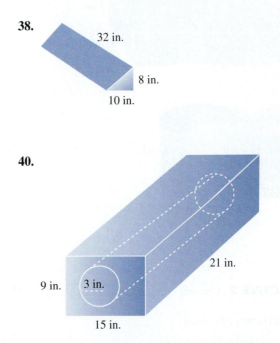

21 in.

9 in.   3 in.

15 in.

41.

110 cm

15 cm

72 cm

54 cm

72 cm

15 cm

42. A can of condensed soup has a diameter of 6.6 cm and a height of 9.8 cm. What is the volume of the can? Let π ≈ 3.14.

SOUP

MUSHROOM

43. A box of pasta shells measures $4\frac{3}{4}$ in. wide, $7\frac{1}{4}$ in. tall, and $2\frac{3}{4}$ in. deep. What is the volume of the pasta box?

44. A farming corporation is building four new grain silos. The inside dimensions of the silos are given in the table.

	Area of Base	Height
Silo A	1800 ft^2	60 ft
Silo B	1200 ft^2	75 ft
Silo C	900 ft^2	80 ft
Silo D	600 ft^2	100 ft

Find the total volume in the four silos.

Exercises 45–49 refer to the chapter application. See page 491.

45. According to industry standards, a joint thickness of $\frac{1}{4}$ in. means the bricklayer will need 9 ft^3 of mortar per 1000 bricks. Find the total amount of mortar needed for the patio and walkways. Round to the nearest cubic foot. (See Exercise 72 in Section 7.4.)

46. The cement subcontractor orders materials based on the total volume of the slab. The industry standard for patios and walkways is 4 to 5 in. of thickness. Because the slab will be topped with bricks, the contractor decides on a thickness of $4\frac{1}{2}$ in. Find the volume of the slab in cubic inches. Convert this to cubic feet, rounding up to the next whole cubic foot. Convert this to cubic yards. Round up to the next whole cubic yard if necessary. (See Exercise 71 in Section 7.4.)

47. Explain why in Exercise 46 it is necessary to round up to the next whole unit, rather than using the rounding rule stated in Section 1.1.

48. The cement contractor in Exercise 46 must first build a wood form that completely outlines the slab. How many linear feet of wood are needed to build the form?

49. The landscaper recommends that Barbara and Paul buy topsoil before planting the garden. They must buy enough to be able to spread the topsoil to a depth of 8 in. How many cubic inches of topsoil do they need? Because soil is usually sold in cubic yards, convert to cubic yards. Round this figure up to the next cubic yard.

50. One gallon is 231 in^3. A cylindrical trash can holds 32 gal. If it is 4 ft tall, what is the area of the base of the trash can?

51. One gallon is 231 in^3. A cylindrical trash can holds 36 gal and is 24 in. in diameter. How tall is the trash can?

52. Norma is buying mushroom compost to mulch her garden. The garden is pictured here. How many cubic yards of compost does she need to mulch the entire garden 4 in. deep? (She cannot buy fractional parts of a cubic yard.)

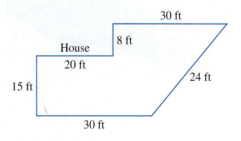

53. A standard 75-gal aquarium measures 48 in. long by 15 in. wide by 24 in. high. Use the fact that 1 gallon is 231 in^3 to calculate the capacity of the aquarium to the nearest tenth of a gallon.

54. A standard 75-gal aquarium measures 48 in. by 15 in. by 24 in. on the outside. The actual amount of water capacity depends on the thickness of the glass (or acrylic) used. If an aquarium has acrylic walls that are $\frac{7}{18}$ in. thick and the aquarium is filled leaving 2 in. of space at the top, what is the capacity of the aquarium? (Recall 1 gallon = 231 in^3.)

55. There are various rules of thumb to calculate the amount of water necessary for the optimum health of fish. One such rule is that each inch of fish requires 3 gal of water. The average freshwater angelfish is 6 in. long. How many angelfish can live in an aquarium filled with 62 gal of water?

STATE YOUR UNDERSTANDING

56. Explain what is meant by "volume." Name three occasions in the past week when the volume of an object was relevant.

57. Explain how to find the volume of the following figure.

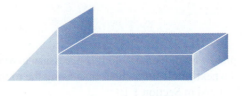

58. Explain why the formula for the volume of a box is a special case of the formula $V = Bh$.

CHALLENGE

59. The Bakers are constructing an in-ground pool in their backyard. The pool will be 15 ft wide and 30 ft long. It will be 3 ft deep for 10 ft at one end. It will then drop to a depth of 10 ft at the other end. How many cubic feet of water are needed to fill the pool?

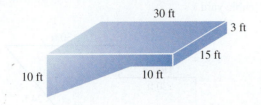

30 ft

3 ft

15 ft

10 ft   10 ft

60. Doug is putting a 5-m-by-20-m swimming pool in his backyard. The pool will be 2 m deep. What is the size of the hole for the pool? The hole will be lined with concrete that is 15 cm thick. How much concrete is needed to line the pool?

61. 9^3

62 $14^2 + 15^2$

63. $2^3 + 4^3 + 5^3$

64. $19^2 - 14^2$

65. $(6^2 + 7^2)^2$

66. $5254 \div 37$

67. $(10^2)(2^2)(3^2)$

68. $7(42 \div 7 + 15 - 21)$

69. Find the area of a rectangle that is 3 in. wide and 2 ft long.

70. A warehouse store sells 5-lb bags of Good & Plenty®. Bob buys a bag and stores the candy in 20-oz jars. How many jars does he need?

Square Roots and the Pythagorean Theorem

VOCABULARY

A **perfect square** is a whole number that is the square of another whole number. For example, $16 = 4^2$, so 16 is a perfect square.

A **square root** of a positive number is a number that is squared to give the original number. The symbol $\sqrt{}$, called a radical sign, is used to name a square root of a number.

$\sqrt{16} = 4$ because $4^2 = 4 \cdot 4 = 16$

A **right triangle** is a triangle with one right (90°) angle.

The **hypotenuse** of a right triangle is the side that is across from the right angle.

The **legs** of a right triangle are the sides adjacent to the right angle.

OBJECTIVES

1. Find the square root of a number.
2. Apply the Pythagorean Theorem.

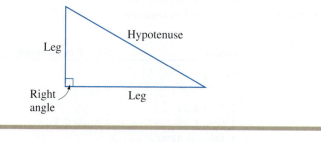

HOW & WHY

■ **OBJECTIVE 1**   Find the square root of a number.

The relationship between squares and square roots is similar to the relationship between multiples and factors. In Chapter 2, we saw that if one number is a multiple of a second number, then the second number is a factor of the first number. It is also true that if one number is the square of a second number, then the second number is a square root of the first. For example, the number 81 is the square of 9, so 9 is a square root of 81. We write this: $9^2 = 81$ so $\sqrt{81} = 9$.

> **To find the square root of a perfect square**
>
> Use guess and check. Square whole numbers until you find the correct value.

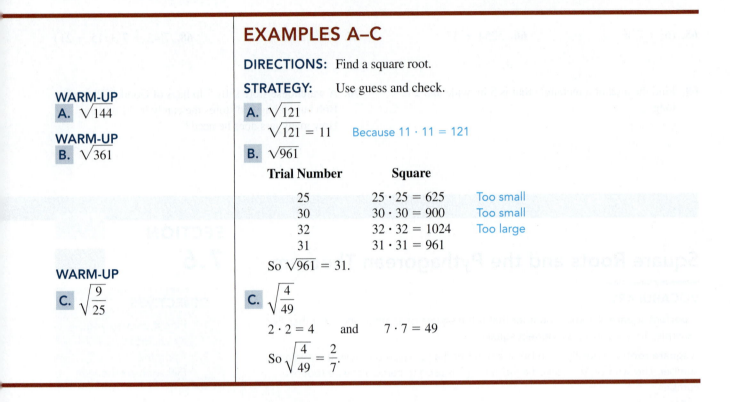

WARM-UP
A. $\sqrt{144}$

WARM-UP
B. $\sqrt{361}$

WARM-UP
C. $\sqrt{\dfrac{9}{25}}$

EXAMPLES A–C

DIRECTIONS: Find a square root.

STRATEGY: Use guess and check.

A. $\sqrt{121}$

$\sqrt{121} = 11$ Because $11 \cdot 11 = 121$

B. $\sqrt{961}$

Trial Number	Square	
25	$25 \cdot 25 = 625$	Too small
30	$30 \cdot 30 = 900$	Too small
32	$32 \cdot 32 = 1024$	Too large
31	$31 \cdot 31 = 961$	

So $\sqrt{961} = 31$.

C. $\sqrt{\dfrac{4}{49}}$

$2 \cdot 2 = 4$ and $7 \cdot 7 = 49$

So $\sqrt{\dfrac{4}{49}} = \dfrac{2}{7}$.

When we need to find the square root of a number that is not a perfect square, the method of guess and check is possible but usually too time-consuming to be practical. To further complicate matters, square roots of nonperfect squares are nonterminating decimals. In these cases, we find an approximation of the desired square root and round to a convenient decimal place. For instance, consider $\sqrt{5}$.

Guess	So $\sqrt{5}$ is between ...
$2^2 = 4$	
$3^2 = 9$	2 and 3
$2.2^2 = 4.84$	
$2.3^2 = 5.29$	2.2 and 2.3
$2.23^2 = 4.9729$	
$2.24^2 = 5.0176$	2.23 and 2.24
$2.236^2 = 4.999696$	
$2.237^2 = 5.004169$	2.236 and 2.237

At this point, we discontinue the process and conclude that $\sqrt{5} \approx 2.24$ to the nearest hundredth.

Because all but the most basic calculators have a square root key, $\boxed{\sqrt{}}$, we usually find the square roots of nonperfect squares by using a calculator. The calculator displays $\sqrt{5} \approx 2.2360679775$, which is consistent with our previous calculation.

Calculator note: Some calculators require that you press the square root key, $\boxed{\sqrt{}}$, before entering the number, and others require that you enter the number first.

ANSWERS TO WARM-UPS A-C

A. 12   **B.** 19   **C.** $\dfrac{3}{5}$

EXAMPLES D–E

DIRECTIONS: Find a square root. Round decimals to the nearest hundredth if necessary.

STRATEGY: Use the square root key, $\boxed{\sqrt{}}$, on a calculator, then round.

D. $\sqrt{106}$

The calculator displays 10.29563014.

So $\sqrt{106} \approx 10.30$, to the nearest hundredth.

E. $\sqrt{345.2}$

The calculator displays 18.57955866.

So $\sqrt{345.2} \approx 18.58$, to the nearest hundredth.

WARM-UP

D. $\sqrt{66}$

WARM-UP

E. $\sqrt{743.25}$

HOW & WHY

◼ **OBJECTIVE 2** Apply the Pythagorean Theorem.

A right triangle is a triangle with one right (90°) angle. Figure 7.8 shows the right angle, legs, and hypotenuse of a right triangle.

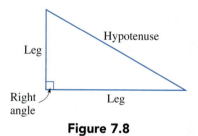

Figure 7.8

Thousands of years ago, a Greek named Pythagoras discovered that the lengths of the sides of a right triangle have a special relationship. This relationship is known as the Pythagorean Theorem.

The Pythagorean Theorem enables us to calculate the length of the third side of a right triangle when we know the lengths of the other two sides. Because the Pythagorean Theorem involves the squares of the sides, when we want the length of a side, we need to use square roots. The following formulas are used to calculate an unknown side of a right triangle.

Pythagorean Theorem

In a right triangle, the sum of the squares of the legs is equal to the square of the hypotenuse

$$(\text{leg})^2 + (\text{leg})^2 = (\text{hypotenuse})^2$$

or

$$a^2 + b^2 = c^2$$

To calculate an unknown side of a right triangle

Use the appropriate formula.

$$\text{Hypotenuse} = \sqrt{(\text{leg}_1)^2 + (\text{leg}_2)^2} \qquad \text{or} \qquad c = \sqrt{a^2 + b^2}$$

$$\text{Leg} = \sqrt{(\text{hypotenuse})^2 - (\text{known leg})^2} \qquad \text{or} \qquad a = \sqrt{c^2 - b^2}$$

ANSWERS TO WARM-UPS D-E

D. 8.12   **E.** 27.26

EXAMPLES F–H

DIRECTIONS: Find the unknown side of the given right triangle.

STRATEGY: Use the formulas. Round decimal values to the nearest hundredth.

WARM-UP

F. Find the approximate length of the hypotenuse.

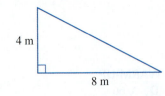

4 m

8 m

F. Find the approximate length of the hypotenuse.

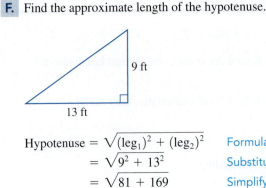

9 ft

13 ft

$$\begin{aligned}
\text{Hypotenuse} &= \sqrt{(\text{leg}_1)^2 + (\text{leg}_2)^2} && \text{Formula}\\
&= \sqrt{9^2 + 13^2} && \text{Substitute 9 and 13 for the lengths.}\\
&= \sqrt{81 + 169} && \text{Simplify.}\\
&= \sqrt{250}\\
&\approx 15.81
\end{aligned}$$

So the length of the hypotenuse is about 15.81 ft.

ALTERNATIVE SOLUTION: Use the formula $c = \sqrt{a^2 + b^2}$.

$$\begin{aligned}
c &= \sqrt{a^2 + b^2}\\
&= \sqrt{9^2 + 13^2} && \text{Substitute 9 for } a \text{ and 13 for } b.\\
&= \sqrt{81 + 169} && \text{Simplify.}\\
&= \sqrt{250}\\
&\approx 15.81
\end{aligned}$$

The hypotenuse is about 15.81 ft.

WARM-UP

G. Find the length of the unknown leg.

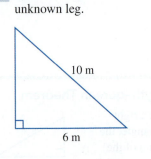

10 m

6 m

G. Find the length of the unknown leg.

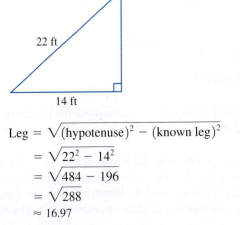

22 ft

14 ft

$$\begin{aligned}
\text{Leg} &= \sqrt{(\text{hypotenuse})^2 - (\text{known leg})^2} && \text{Formula}\\
&= \sqrt{22^2 - 14^2} && \text{Substitute 22 for the length}\\
&= \sqrt{484 - 196} && \text{of the hypotenuse and 14}\\
&= \sqrt{288} && \text{for the length of the}\\
&\approx 16.97 && \text{leg.}
\end{aligned}$$

So the length of the unknown leg is about 16.97 ft.

H. Ted is staking a small tree to keep it growing upright. He plans to attach the rope 3 ft above the ground on the tree and 6 ft from the base of the tree. He needs one extra foot of rope on each end for tying. How much rope does he need?

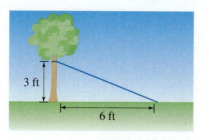

3 ft

6 ft

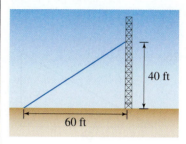

40 ft

60 ft

STRATEGY: The tree trunk, the ground, and the rope form a right triangle, with the rope being the hypotenuse of the triangle. We use the formula for finding the hypotenuse.

$$\text{Hypotenuse} = \sqrt{(\text{leg}_1)^2 + (\text{leg}_2)^2} \quad \text{Formula}$$

$$= \sqrt{3^2 + 6^2} \qquad \text{Substitute 3 and 6 for the lengths of } \text{leg}_1 \text{ and } \text{leg}_2.$$

$$= \sqrt{9 + 36} \qquad \text{Simplify.}$$

$$= \sqrt{45}$$

$$\approx 6.71$$

The rope support is about 6.71 ft long, so Ted needs 8.71 ft, including the extra 2 ft for the ties.

EXERCISES 7.6

■ OBJECTIVE 1 Find the square root of a number. (See page 547.)

A *Simplify.*

1. $\sqrt{49}$

2. $\sqrt{25}$

3. $\sqrt{36}$

4. $\sqrt{64}$

5. $\sqrt{100}$

6. $\sqrt{169}$

7. $\sqrt{900}$

8. $\sqrt{225}$

9. $\sqrt{\dfrac{49}{100}}$

10. $\sqrt{\dfrac{64}{121}}$

11. $\sqrt{0.81}$

12. $\sqrt{0.64}$

B *Simplify. Use the square root key, ☑, on your calculator. Round to the nearest hundredth.*

13. $\sqrt{175}$

14. $\sqrt{888}$

15. $\sqrt{821}$

16. $\sqrt{910}$

17. $\sqrt{13.46}$

18. $\sqrt{50.41}$

19. $\sqrt{\dfrac{14}{23}}$

20. $\sqrt{\dfrac{5}{22}}$

21. $\sqrt{11,500}$

22. $\sqrt{12,000}$

■ **OBJECTIVE 2** Apply the Pythagorean theorem. (See page 549.)

A *Find the unknown side of each right triangle. Round decimals to the nearest hundredth.*

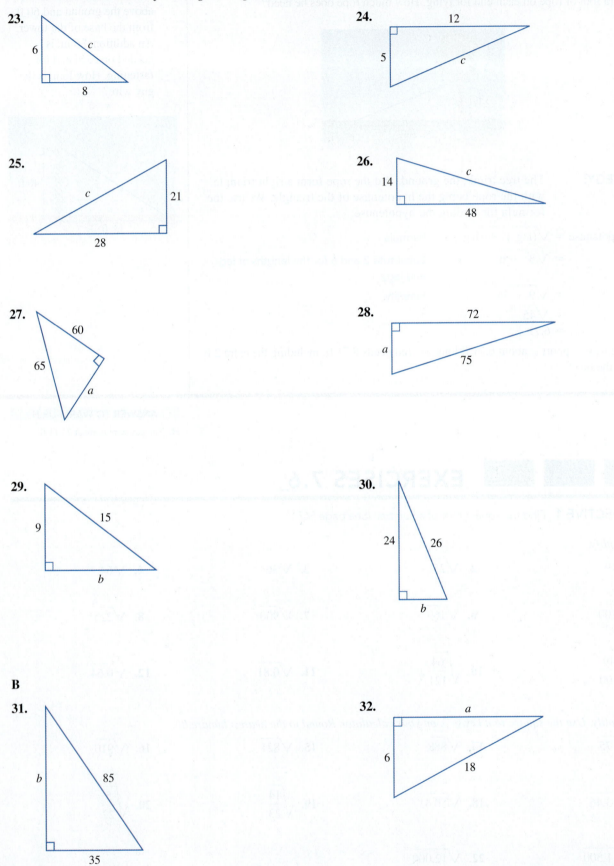

23.

6
c
8

24.

12
5
c

25.

c
21
28

26.

14
c
48

27.

60
65
a

28.

72
a
75

29.

9
15
b

30.

24
26
b

B

31.

b
85
35

32.

a
6
18

33.

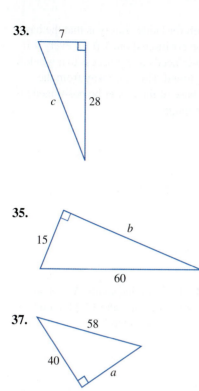

34.

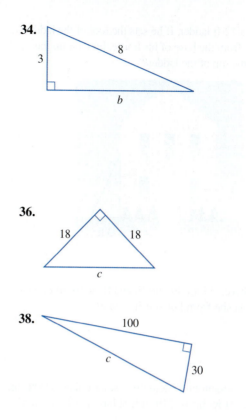

35.

36.

37.

38.

C

39. Find the hypotenuse of a right triangle that has legs of 21 cm and 72 cm.

40. Find the length of one leg of a right triangle if the other leg is 24 in. and the hypotenuse is 30 in.

41. What is the length of the side of a square whose area is 144 ft² ? The formula for the length of a side of a square is $s = \sqrt{A}$, where A is the area.

42. What is the length of the diagonal of a square whose area is 144 ft² ? See Exercise 41.

43. The advertised size of a TV is the measure of the diagonal of the screen. A rear-projection TV has an actual screen size of 42 in. wide by 24 in. tall. What is the advertised size of the TV (rounded to the nearest inch)?

44. A TV advertised as having a 30-in. screen has an actual height of 15 in. How wide is the screen (rounded to the nearest inch)?

45. Paul has a 12-ft ladder. If he sets the foot of the ladder 3 ft away from the base of his house, how far up the house is the top of the ladder?

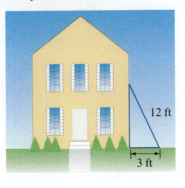

12 ft

3 ft

46. A basic rule of thumb for ladder safety is that the base of the ladder must be positioned out 1 ft for every 4 ft of height. A firefighter needs to get access to a window that is 30 ft off the ground. How far away from the building should the base of the ladder be positioned? Is a 32-ft ladder long enough?

47. Lynzie drives 35 mi due north and then 16 mi due east. How far is she from her starting point?

48. Charles rides his bike 12.7 km due south. Alex starts from the same point and rides his bike 14.2 km due west. How far apart are the two men?

49. A baseball diamond is actually a square that is 90 ft on each side. How far is it from first base across to third base?

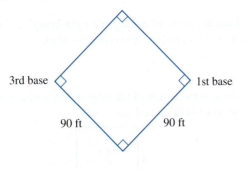

3rd base

1st base

90 ft   90 ft

50. How tall is the tree?

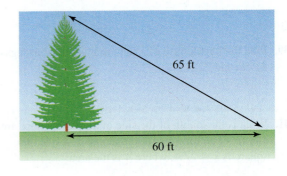

65 ft

60 ft

51. Scott wants to install wire supports for the net on his pickleball court. The net is 36 in. high, and he plans to anchor the supports in the ground 4 ft from the base of the net. Find how much wire he needs for two supports, figuring an additional foot of wire for each support to fasten the ends.

36 in.

4 ft

52. The 7th hole on Jim's favorite golf course has a pond between the tee and the green. If Jim can hit the ball 170 yards, should he hit over the pond or go around it?

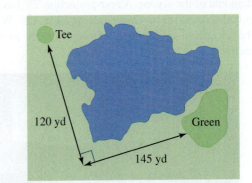

Tee

120 yd   Green

145 yd

53. What is a square root? Explain how to find $\sqrt{289}$ without using a calculator.

54. A carpenter is building a deck on the back of a house. He wants the joists of the deck to be perpendicular to the house, so he attaches a joist to the house and measures 4 ft out on the joist and 3 ft over from the joint on the house. He then wiggles the joist until his two marks are 5 ft apart. At this point, the carpenter knows that the joist is perpendicular (makes a right angle) with the house. Explain how he knows.

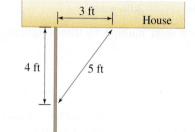

CHALLENGE

55. Sam has a glass rod that is 1 in. in diameter and 35 in. long that he needs to ship across the country. He is having trouble finding a box. His choices are a box that measures 2 ft by 2 ft by 6 in., or a box that measures 18 in. by 32 in. by 5 in. Will Sam's rod fit in either of the boxes?

MAINTAIN YOUR SKILLS

56. What is 35% of 62,000?

57. $48.3 + 562 + 0.931$

58. $\left(4\frac{1}{5}\right)\left(3\frac{6}{7}\right)$

59. Find the prime factorization of 80.

60. $23 - 14\frac{3}{8}$

61. $9.03 \div 0.005$

62. At a certain time of day, a 15-ft tree casts a shadow of 8 ft. How tall is another tree if it casts a shadow of 5 ft?

63. Convert 72.5 m² to square feet. Round to the nearest hundredth.

64. Find the average and median of 3, 18, 29, 31, and 32.

65. Find the volume of a can with a diameter of 8 in. and a height of 10 in. Use the $\boxed{\pi}$ key, and round to the nearest hundredth.

KEY CONCEPTS

SECTION 7.1 Measuring Length

Definitions and Concepts	Examples

Definitions and Concepts

English	**Metric**
inch	centimeter
foot	meter
mile	kilometer

To convert units, multiply by the appropriate unit fraction.

To convert within the metric system, use the chart and move the decimal point the same number of places.

Only like measures can be added or subtracted.

Examples

Convert 60 in. to feet.

$$60 \text{ in.} = \frac{60 \text{ in.}}{1} \cdot \frac{1 \text{ ft}}{12 \text{ in.}}$$
$$= 5 \text{ ft}$$

Convert 25 cm to meters.

km	hm	dam	m	dm	cm	mm

Move the decimal two places to the left.

25 cm = 0.25 m

14 cm + 50 cm = 64 cm
8 ft − 3 ft = 5 ft

SECTION 7.2 Measuring Capacity, Weight, and Temperature

Definitions and Concepts

	English	**Metric**
Capacity	quart	liter
Weight	ounce	gram
	pound	kilogram
Temperature	degrees Fahrenheit	degrees Celsius

To convert units, multiply by the appropriate unit fraction.

To convert temperatures, use the formulas.

$$C = \frac{5}{9} \cdot (F - 32)$$

$$F = \frac{9}{5} \cdot C + 32$$

Examples

Convert 5 qt to liters.

$$5 \text{ qt} \approx \frac{5 \text{ qt}}{1} \cdot \frac{0.9464 \text{ L}}{1 \text{ qt}}$$
$$\approx 4.732 \text{ L}$$

Convert 41°F to degrees Celsius.

$$C = \frac{5}{9} \cdot (41 - 32)$$
$$= \frac{5}{9} \cdot 9$$
$$= 5$$

So 41°F is 5°C.

SECTION 7.3 Perimeter

Definitions and Concepts	Examples

Perimeter is measured in linear units.

To find the perimeter of a polygon, add the measures of all of its sides.

Perimeter of a square:

$$P = 4s$$

Perimeter of a rectangle:

$$P = 2\ell + 2w$$

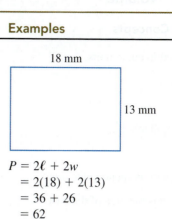

$$
\begin{aligned}
P &= 2\ell + 2w \\
&= 2(18) + 2(13) \\
&= 36 + 26 \\
&= 62
\end{aligned}
$$

The perimeter is 62 mm.

The circumference of a circle is the distance around the circle.

The radius of a circle is the distance from the center to any point on the circle.

The diameter of a circle is twice the radius.

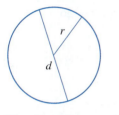

The circumference of a circle:

$$C = 2\pi r \quad \text{or} \quad C = \pi d$$

Find the circumference of

$$
\begin{aligned}
C &= 2\pi r \qquad \pi \approx 3.14 \\
&\approx 2(3.14)(4) \\
&\approx 25.12
\end{aligned}
$$

So the circumference is about 25.12 ft.

SECTION 7.4 Area

Definitions and Concepts	Examples

Area is measured in square units.

Area of a square:

$$A = s^2$$

Area of a rectangle:

$$A = \ell w$$

Area of a triangle:

$$A = \frac{bh}{2}$$

Area of a circle:

$$A = \pi r^2$$

$$
\begin{aligned}
A &= s^2 \\
&= 6^2 \\
&= 36
\end{aligned}
$$

The area is 36 ft².

$$
\begin{aligned}
A &= \frac{bh}{2} \\
&= \frac{20 \cdot 8}{2} \\
&= 80
\end{aligned}
$$

The area is 80 cm².

SECTION 7.5 Volume

Definitions and Concepts	Examples

Definitions and Concepts

Volume is measured in cubic units.

Volume of a cube:

$V = s^3$

Volume of a rectangular solid:

$V = \ell wh$

Volume of a solid with perpendicular sides:

$V = Bh$, where B is the area of the base

The volume of a cylinder:

$V = \pi r^2 h$

Examples

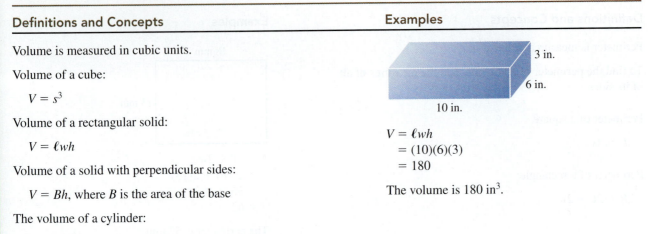

$V = \ell wh$
$\quad = (10)(6)(3)$
$\quad = 180$

The volume is 180 in³.

SECTION 7.6 Square Roots and the Pythagorean Theorem

Definitions and Concepts

A square root of a number is the number that is squared to give the original number.

To find the square root of a number, use the ☑ key on a calculator and round as necessary.

Pythagorean theorem: In a right triangle,

$a^2 + b^2 = c^2$

To find a missing side of a right triangle, use the formulas.

$c = \sqrt{a^2 + b^2}$ or $a = \sqrt{c^2 - b^2}$

Examples

$\sqrt{49} = 7$ because $7 \cdot 7 = 49$

$\sqrt{436} \approx 20.88$

Find c.

$c = \sqrt{3^2 + 4^2}$
$\quad = \sqrt{9 + 16}$
$\quad = \sqrt{25}$
$\quad = 5$

REVIEW EXERCISES

SECTION 7.1

Do the indicated operations.

1. 23 · 2 mm

2. 330 m ÷ 15

3. 5 yd − 12 ft

4. 5 ft + 29 ft + 8 ft

5. (6 ft 2 in.) − (4 ft 10 in.)

Convert as indicated. Round decimal values to the nearest hundredth.

6. 245 cm to meters

7. 4 yd to inches

8. 5 km to miles

9. 15 in. to millimeters

10. Ted is building two plant supports out of copper tubing. Each support requires three vertical tubes that are 8 ft long and six horizontal tubes that are 2 ft long. How much tubing does he need for the two supports?

SECTION 7.2

11. 6 qt · 6

12. (322 hr) ÷ 14

13. 34 gal + 52 gal

14. 34 L − 8 L

15. (6 hr 42 min) + (3 hr 38 min)

Convert as indicated. Round decimal values to the nearest hundredth.

16. 6 g to milligrams

17. 14 lb to ounces

18. 4 L to quarts

19. 95°F to degrees Celsius

20. Gail estimates that she has 45 min to work in her yard for each of the next 5 days. She is building a retaining wall that should take her 4 hr. How much additional time will she need to finish the wall?

SECTION 7.3

Find the perimeter of the following polygons.

21.

22.

23.

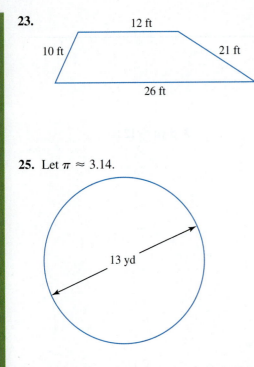

12 ft
10 ft
21 ft
26 ft

24.

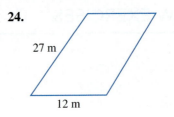

27 m
12 m

25. Let $\pi \approx 3.14$.

13 yd

26.

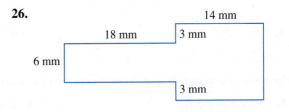

14 mm
18 mm   3 mm
6 mm
3 mm

27. Find the perimeter of a plot of ground in the shape of a parallelogram that has sides of 12 m and 15 m.

28. Find the distance around a circular hot tub that measures 8 ft in diameter. Let $\pi \approx 3.14$.

29. The City of Portland is required by federal law to put a fence around one of its reservoirs. If the reservoir is rectangular with dimensions of 230 ft by 275 ft, how many feet of fencing is required?

30. Larry is buying new baseboards for his den, which measures 9 ft by 12 ft. How much does he need if the door is 4 ft wide?

SECTION 7.4

Find the area of the following. Let $\pi \approx 3.14$.

31.

83 mm
25 mm

32.

25 ft
7 ft
24 ft

33.

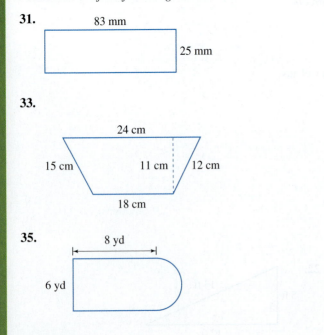

24 cm
15 cm   11 cm   12 cm
18 cm

34.

16 in.
12 in.   8 in.   16 in.
30 in.

35.

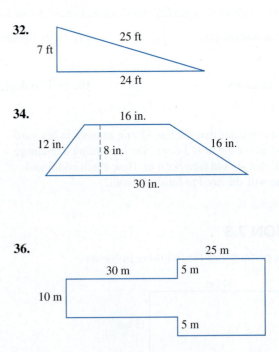

8 yd
6 yd

36.

25 m
30 m   5 m
10 m
5 m

37. What is the area of a square that measures 600 mm on a side?

38. What is the area of a rectangular driveway that is 35 ft by 22 ft?

39. Felicity is painting the walls of her bathroom, which measures 6 ft by 8 ft. The bathroom has an 8-ft ceiling and one 3-ft-by-6-ft door. She has 1 gal of paint that covers 350 ft². Does she have enough paint for two coats?

40. How many square meters are there in the sides and bottom of a rectangular swimming pool that measures 50 m by 30 m and is 3 m deep?

SECTION 7.5

Find the volume of the following figures.

41.

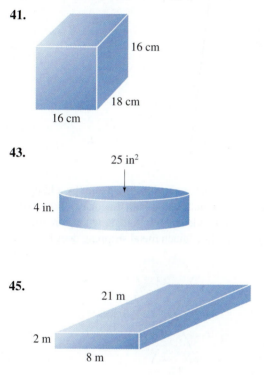

16 cm

18 cm

16 cm

42.

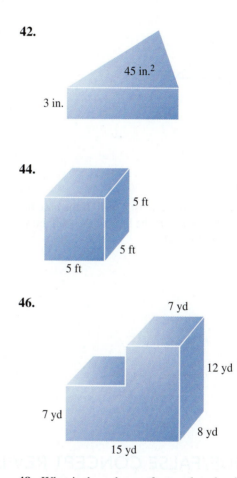

45 in.²

3 in.

43.

25 in²

4 in.

44.

5 ft

5 ft

5 ft

45.

21 m

2 m

8 m

46.

7 yd

12 yd

7 yd

8 yd

15 yd

47. What is the volume of a cylindrical storage bin that has an inside diameter of 45 ft and an inside height of 30 ft? Let $\pi \approx 3.14$. Round to the nearest cubic foot.

48. What is the volume of an enclosed trailer that has inside measurments of 10 ft long, 8 ft wide, and 5 ft high?

49. How many cubic inches are there in a cubic yard?

50. Will a cubic yard of bark dust be sufficient to cover a rose garden that is 20 ft by 6 ft to a depth of 3 in.?

SECTION 7.6

Find the square roots. If necessary, round to the nearest hundredth.

51. $\sqrt{529}$

52. $\sqrt{60}$

53. $\sqrt{\dfrac{81}{49}}$

54. $\sqrt{62.37}$

Find the missing side. Round decimal values to the nearest hundredth.

55.

56.

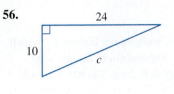

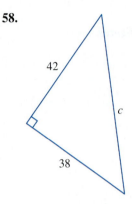

57.

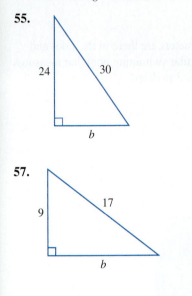

58.

59. A 18-ft ladder is leaning against a wall. If the base of the ladder is 3 ft away from the base of the wall, how high is the top of the ladder?

60. Mark wants to brace a metal storage unit by adding metal strips on the back on both diagonals. If the storage unit is 3 ft wide and 6 ft high, how long are the diagonals and how much metal stripping does he need?

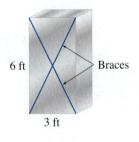

6 ft   Braces

3 ft

  ## TRUE/FALSE CONCEPT REVIEW

Check your understanding of the language of algebra and geometry. Tell whether each of the following statements is true (always true) or false (not always true). For each statement you judge to be false, revise it to make a true statement.

Answers

1. Metric measurements are the most commonly used in the world.

1. _____

2. Equivalent measures have the same number value, as in 7 ft and 7 yd.

2. _____

3. A liter is a measure of capacity or volume.

3. _____

4. The perimeter of a square can be found in square inches or square meters.

4. _____

5. Volume is the measure of the inside of a solid, such as a box or can.

5. _____

6. The volume of a square is $V = s^2$. $V = s^3$

7. The area of a circle is $A = 3.14\, r^2$.

8. It is possible to find equivalent measures without remeasuring the original object.

9. The metric system uses the base 10 place value system.

10. In a right triangle, the longest side is the hypotenuse.

11. $\sqrt{3} = 9$

12. Weight can be measured in pounds, grams, or kilograms.

13. A trapezoid has four sides.

14. One milliliter is equivalent to 1 square centimeter.

15. Measurements can be added or subtracted only if they are all metric measures or all English measures.

16. The distance around a polygon is called the perimeter.

17. Volume is always measured in gallons or liters.

18. The Pythagorean Theorem states that the sum of the lengths of the legs of a right triangle is equal to the length of the hypotenuse.

19. $1\text{ yd}^2 = 3\text{ ft}^2$

20. The prefix kilo- means 1000.

Answers

6. _____
7. _____
8. _____
9. _____
10. _____
11. _____
12. _____
13. _____
14. _____
15. _____
16. _____
17. _____
18. _____
19. _____
20. _____

TEST

Answers

1. 8 m + 329 mm = ? mm

2. Find the perimeter of a rectangle that measures 24 in. by 35 in.

3. Find the volume of a box that is 6 in. high, 20 in. wide, and 12 in. deep.

4. How many square feet of tile are needed to cover the floor of the room pictured below?

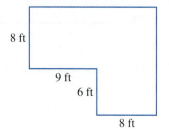

1. _____
2. _____
3. _____
4. _____

5. Find $\sqrt{310}$ to the nearest thousandth.

5. _____

6. Find the perimeter of the figure.

6. _____

8 cm

4 cm

6 cm

3 cm

7. Find the area of the triangle.

7. _____

9 in.   11 in.

5 in.

16 in.

8. Subtract: 6 yd 2 ft
−3 yd 2 ft 5 in.

8. _____

9. Name two units of measure in the English system for weight. Name two units of measure in the metric system for weight.

9. _____

10. Change 6 gal to pints.

10. _____

11. How much molding is needed to trim a picture window 10 ft wide by 8 ft tall and two side windows each measuring 4 ft wide and 8 ft tall? Assume that the corners will be mitered and that the miters require 2 in. extra on each end of each piece of molding.

11. _____

12. Find the volume of the figure.

12. _____

12 in.

5 in.

4 in.

14 in.

13. Find the area of the parallelogram.

13. _____

12 cm

5 cm   4 cm

14. Find the volume of a hot water tank that has a circular base of 4 ft^2 and a height of 6 ft.

14. _____

15. Miss Katlie, a kindergarten teacher, has 192 oz of grape juice to divide equally among her 24 students. How much juice will she give each student?

15. _____

16. The cost of heavy-duty steel landscape edging is $1.20 per linear foot. How much will it cost to line the three garden plots pictured, including the edging between the plots?

9 ft   9 ft   9 ft

25 ft

17. Find the missing side.

8 cm   10 cm

b

18. Find the area of the figure. Round to the nearest tenth. Let $\pi \approx 3.14$.

7 km

11 km

19. The Green Bay Packers condition the team by having them run around the edge of the field four times each hour. A football field measures 100 yd long by 60 yd wide. How far does each player run in a 4-hr practice?

20. Both area and volume describe interior space. Explain how they are different.

21. John's Meat Market has a big sale on steak. They sell 340 lb on Monday, 495 lb on Tuesday, 432 lb on Wednesday, 510 lb on Thursday, and 670 lb on Friday. What is the average number of pounds of steak sold each day?

22. Li is buying lace, with which she plans to edge a circular tablecloth. The tablecloth is 90 in. in diameter. How much lace does she need? If the lace is available in whole-yard lengths only, how much must she buy? Let $\pi \approx 3.14$.

23. Khallil has a 40-lb bag of dog food that is approximately 52 in. long, 16 in. wide, and 5 in. deep. He wants to transfer it to a plastic storage box that is 24 in. long, 18 in. wide, and 12 in. high. Will all of the dog food fit into the box?

24. Macy walks due north for 3.5 mi and then due east for 1.2 mi. How far is she from her starting point?

25. Neela has a gift box that is 12 in. by 14 in. by 4 in. What minimum amount of wrapping paper does she need to wrap the gift?

CLASS ACTIVITY 1

There is another system of measuring precious metals that is left over from the way Roman currency was measured. The troy system of weights derives its name from the city of Troyes in France. As recently as the mid-twentieth century, a pound sterling, the basic currency in Great Britain, was one troy pound of silver.

Here are the basic units of weight in the troy system.

> 1 pennyweight (dwt) = 24 grains
> 1 troy ounce (oz t) = 20 pennyweights
> 1 pound troy (lb t) = 12 ounces troy

1. How many pennyweights are in 1 pound troy? How many grains are in 1 pound troy?

2. Kayla has ingot of platinum that weighs 2 lb t 8 oz t. She wants to divide it equally among her four grandchildren. How much platinum will each grandchild receive?

3. In late 2009, the spot price (price per oz t) of platinum was $1445.00. What is the value of each of Kayla's grandchildren's inheritance?

4. Marcy is taking some old gold jewelry to be melted down and combined into a bracelet. She has a gold chain that weighs 15 dwt, a ring that weighs 1 oz t 4 dwt, and a second ring that weighs 1 oz t 7 dwt. What is the total weight of the gold Marcy has for her bracelet?

5. In late 2009, the spot price (prize per oz t) of gold was $1151.20. What was the value of the gold for Marcy's bracelet?

CLASS ACTIVITY 2

For this activity, you will need a rectangular piece of paper and a ruler marked in both inches and centimeters.

1. Using the ruler, measure your piece of paper as accurately as you can, to the nearest $\frac{1}{8}$ of an inch. Calculate the area of your paper in square inches.

2. Calculate the area of your paper in square centimeters by converting your square inch measurement.

3. Now measure your paper again, this time in centimeters. Give your measurements to the nearset tenth of a centimeter. Calculate the area in square centimeters.

4. How do your answers to part 2 and part 3 compare? Should they be equal? If they are not equal, explain why not.

5. If you needed a piece of a paper with the same area, but with a length of 25.4 cm, what would the width of the new sheet be?

6. Challenge: If you needed a circle that has the same area as your original sheet of paper, what is the radius of the circle? You may choose to work in either square inches or square centimeters.

GROUP PROJECT (2–3 WEEKS)

You are working for a kitchen design firm that has been hired to design a kitchen for the 10-ft by-12-ft room pictured here.

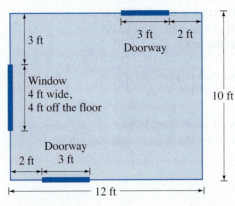

3 ft

3 ft   2 ft
Doorway

Window
4 ft wide,
4 ft off the floor

10 ft

Doorway
2 ft   3 ft

12 ft

The following table lists appliances and dimensions. Some of the appliances are required and others are optional. All of the dimensions are in inches.

Appliance	High	Wide	Deep	Required
Refrigerator	68	30 or 33	30	Yes
Range/oven	30	30	26	Yes
Sink	12	36	22	Yes
Dishwasher	30	24	24	Yes
Trash compactor	30	15	24	No
Built-in microwave	24	24	24	No

The base cabinets are all 30 in. high and 24 in. deep. The widths can be any multiple of 3 from 12 in. to 36 in. Corner units are 36 in. along the wall in each direction. The base cabinets (and the range, dishwasher, and compactor) will all be installed on 4-in. bases that are 20 in. deep.

The wall (upper) cabinets are all 30 in. high and 12 in. deep. Here, too, the widths can be any multiple of 3 from 12 in. to 36 in. Corner units are 24 in. along the wall in each direction.

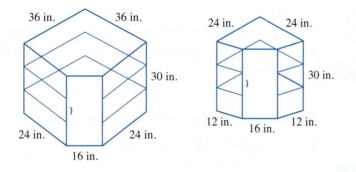

1. The first step is to place the cabinets and appliances. Your client has specified that there must be at least 80 ft³ of cabinet space. Make a scale drawing of the kitchen and indicate the placement of the cabinets and appliances. Show calculations to justify that your plan satisfies that 80-ft³ requirement.

2. Countertops measure 25 in. deep with a 4-in. backsplash. The countertops can either be tile or Formica. If the counters are Formica, there will be a 2-in. facing of Formica. If the counters are tile, the facing will be wood that matches the cabinets. (See the figure.) Calculate the amount of tile and wood needed for the counters and the amount of Formica needed for the counters.

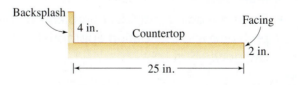

3. The bases under the base cabinets will be covered with a rubber kickplate that is 4 in. high and comes in 8-ft lengths. Calculate the total length of kickplate material needed and the number of lengths of kickplate material necessary to complete the kitchen.

4. Take your plan to a store that sells kitchen appliances, cabinets, and counters. Your goal is to get the best-quality materials for the least amount of money. Prepare at least two cost estimates for the client. Do not include labor in your estimates, but do include the cost of the appliances. Prepare a rationale for each estimate, explaining the choices you made. Which plan will you recommend to the client, and why?

GOOD ADVICE FOR *Studying*

PUTTING IT ALL TOGETHER—PREPARING FOR THE FINAL EXAM

© 2008/Jupiter images

One Week before the Exam

- Double-check that you have turned in all assignments.
- Make an outline of topics covered over the entire course. Use the Key Concepts to help you.
- Organize your homework by topic.
- Work through the review sections for each chapter.
- Identify which topics you need to review and which ones you still remember.

Three to Four Days before the Exam

- Review your notes and homework from the topics you have identified as needing more work.
- Copy all your test/quiz questions onto separate pages, and take all your tests again.
- Check your answers against what you have written on the actual tests.
- Identify any additional topics which need review.

One to Two Days before the Exam

- Make a list of problems and/or skills that you expect to see on the exam.
- Shift from doing individual problems to identifying categories of problems and associating appropriate solving strategies with each category.

Night before and Morning of the Exam

- Get plenty of sleep.
- Eat a good breakfast.

During the Exam

- Practice your relaxation techniques. Tell yourself that you have done the necessary work, and you are prepared.
- Remember to use the test-taking strategies.

Algebra Preview: Signed Numbers

APPLICATION

The National Geographic Society was founded in 1888 "for the increase and diffusion of geographic knowledge." The Society has supported more than 5000 explorations and research projects with the intent of "adding to knowledge of earth, sea, and sky." Among the many facts catalogued about our planet, the Society keeps records of the elevations of various geographic features. Table 8.1 lists the highest and lowest points on each continent.

When measuring, one has to know where to begin, or the zero point. Notice that when measuring elevations, the zero point is chosen to be sea level (Figure 8.1). All elevations compare the high or low point to sea level. Mathematically we represent quantities under the zero point as negative numbers.

TABLE 8.1 **Continental Elevations**

Continent	Highest Point	Feet above Sea Level	Lowest Point	Feet below Sea Level
Africa	Kilimanjaro, Tanzania	19,340	Lake Assal	512
Antarctica	Vinson Massif	16,864	Bentley, Sub-glacial Trench	8327
Asia	Mount Everest, Nepal-Tibet	29,028	Dead Sea, Israel-Jordan	1312
Australia	Mount Kosciusko, New South Wales	7310	Lake Eyre, South Australia	52
Europe	Mount El'brus, Russia	18,510	Caspian Sea, Russia-Azerbaijan	92
North America	Mount McKinley, Alaska	20,320	Death Valley, California	282
South America	Mount Aconcagua, Argentina	22,834	Valdes Peninsula, Argentina	131

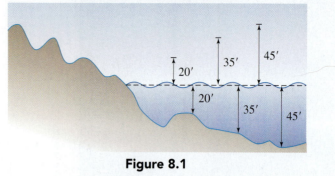

Figure 8.1

What location has the highest continental altitude on Earth? What location has the lowest continental altitude? Is there a location on Earth with a lower altitude? Explain.

8.1 Opposites and Absolute Value

OBJECTIVES

1. Find the opposite of a signed number.
2. Find the absolute value of a signed number.

VOCABULARY

Positive numbers are the numbers of arithmetic and are greater than zero.

Negative numbers are numbers less than zero. Zero is neither positive nor negative. Positive numbers, zero, and negative numbers are called **signed numbers**.

The **opposite** or **additive inverse** of a signed number is the number on the number line that is the same distance from zero but on the opposite side of it. Zero is its own opposite. The opposite of 5 is written –5. This can be read "the opposite of 5" or "negative 5," since they both name the same number.

The **absolute value** of a signed number is the number of units between the number and zero. The expression |6| is read "the absolute value of 6."

HOW & WHY

■ **OBJECTIVE 1** Find the opposite of a signed number.

Expressions such as

6 – 8   9 – 24   11 – 12   and   4 – 134

do not have answers in the numbers of arithmetic. The answer to each is a signed number. Signed numbers (which include both numbers to the right of zero and to the left of zero) are used to represent quantities with opposite characteristics. For instance,

right and left
up and down
above zero and below zero
gain and loss

A few signed numbers are shown on the number line in Figure 8.2.

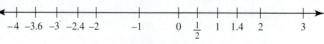

Figure 8.2

The negative numbers are to the left of zero. The negative numbers have a dash, or negative sign, in front of them. The numbers to the right of zero are called positive (and may be written with a plus sign). Zero is neither positive nor negative. Here are some signed numbers.

7	Seven, or positive seven
-3	Negative three
-0.12	Negative twelve hundredths
0	Zero is neither positive nor negative.
$+\dfrac{1}{2}$	One half, or positive one half

Positive and negative numbers are used many ways in the physical world—here are some examples:

Positive	Negative
Temperatures above zero	Temperatures below zero
(83°)	(−11°)
Feet above sea level	Feet below sea level
(6000 ft)	(−150 ft)
Profit	Loss
($94)	(−$51)
Right	Left
(15)	(−12)

In any situation in which quantities can be measured in opposite directions, positive and negative numbers can be used to show direction.

The dash in front of a number is read in two different ways:

-23	The opposite of 23
-23	Negative 23

The opposite of a number is the number on the number line that is the same distance from zero but on the opposite side. To find the opposite of a number, we refer to a number line. See Figure 8.3.

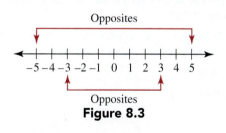

Figure 8.3

$-(5) = -5$	The opposite of positive 5 is negative 5.
$-(-3) = 3$	The opposite of negative 3 is positive 3.
$-0 = 0$	The opposite of 0 is 0.

To find the opposite of a positive number

1. Locate the number on the number line.
2. Count the number of units it is from zero.
3. Count this many units to the left of zero. Where you stop is the opposite of the positive number.

The opposite of a positive number is negative and the opposite of a negative number is positive. The opposite of 0 is 0.

EXAMPLES A–E

DIRECTIONS: Find the opposite.

STRATEGY:　The opposite of a number line the same number of units from zero but on the opposite side of zero.

A. Find the opposite of 14.

STRATEGY:　With no sign in front, the number is positive; 14 is read "fourteen" or "positive fourteen."

$-(14) = ?$　Because 14 is 14 units to the right of zero, the opposite of 14 is 14 units to the left of zero.

$-(14) = -14$

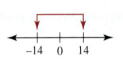

The opposite of 14 is -14.

B. Find the opposite of -28.

STRATEGY:　The number -28 can be thought of in two ways: as a negative number that is 28 units to the left of zero, or as a number that is 28 units on the opposite side of zero from 28; -28 is read "negative 28" or "the opposite of 28."

$-(-28) = ?$　The opposite of negative 28 is written $-(-28)$ and is found 28 units
$-(-28) = 28$　on the opposite side of zero from -28.

The opposite of -28 is 28.

C. Find the opposite of $-(-11)$.

> **CAUTION**
>
> The dash in front of the parentheses is *not* read "negative." Instead, it is read "the opposite of." The dash directly in front of 11 is read "negative" or "the opposite of."

STRATEGY: First find $-(-11)$. Then find the opposite of that value.

$-(-11)$ is "the opposite of negative 11"

$-(-11) = ?$
$-(-11) = 11$

So the opposite of $(-11) = 11$.

$-(-11)$ can also be read "the opposite of the opposite of 11."

The opposite of -11 is 11 units to the right of zero, which is 11.

Continuing, we find the opposite of $-(-11)$, which is written $-[-(-11)]$.

$$-[-(-11)] = -[11]$$
$$= -11$$

The opposite of $-(-11) = -11$.

CALCULATOR EXAMPLE:

D. Find the opposite of 19, -2.3, and 4.7.

STRATEGY: The $\boxed{+/-}$ or the $\boxed{(-)}$ key on the calculator will give the opposite of the number.

Calculator note: Some calculators require that you enter the number and then press the $\boxed{+/-}$ key. Others require that you press the $\boxed{(-)}$ key and then enter the number.

The opposites are -19, 2.3, and -4.7.

E. If 10% of Americans purchased products with no plastic packaging just 10% of the time, approximately 144,000 lb of plastic would be eliminated (taken out of or decreased) from our landfills.

1. Write this decrease as a signed number.
2. Write the opposite of eliminating (decreasing) 144,000 lb of plastic from our landfills as a signed number.

1. Decreases are often represented by negative numbers. Therefore, a decrease of 144,000 lb is $-144,000$ lb.
2. The opposite of a decrease is an increase, so $-(-144,000 \text{ lb})$ is 144,000 lb.

WARM-UP
D. Find the opposite of 13, -7.8, and 7.1.

WARM-UP
E. The energy used to produce a pound of rubber is 15,700 BTU. Recycled rubber requires 4100 BTU less.

1. Express this decrease as a signed number.
2. Express the opposite of a decrease of 4100 BTU as a signed number.

HOW & WHY

■ **OBJECTIVE 2** Find the absolute value of a signed number.

The absolute value of a signed number is the number of units between the number and zero on the number line. Absolute value is defined as the number of units only; direction is not involved. Therefore, the absolute value is never negative. See Figure 8.4.

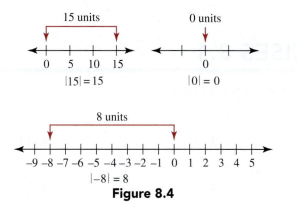

$|15| = 15$

$|0| = 0$

$|-8| = 8$

Figure 8.4

ANSWERS TO WARM-UPS D–E

D. $-13, 7.8, -7.1$ **E.** As a signed number, the decrease is -4100 BTU. As a signed number, the opposite of a decrease of 4100 BTU is 4100 BTU.

From Figure 8.4, we can see that the absolute value of a number and the absolute value of the number's opposite are equal. For example,

$$|45| = |-45| \qquad \text{Because both equal 45.}$$

The absolute value of a positive number is the number itself. The absolute value of a negative number is its opposite. The absolute value of zero is zero.

> ### To find the absolute value of a signed number
>
> The value is
>
> 1. zero if the number is zero.
> 2. the number itself if the number is positive.
> 3. the opposite of the number if the number is negative.

EXAMPLES F–I

DIRECTIONS: Find the absolute value of the number

STRATEGY: If the number is positive or 0, write the number. If the number is negative, write its opposite.

WARM-UP
F. Find the absolute value of 29.
WARM-UP
G. Find the absolute value of −90.

WARM-UP
H. Find the absolute value of −1.

WARM-UP
I. Find the absolute value of $-\dfrac{3}{7}$.

F. Find the absolute value of 17.

$$|17| = 17 \qquad \text{The absolute value of a positive number is the number itself.}$$

G. Find the absolute value of −41.

$$|-41| = -(-41) \qquad \text{The absolute value of a negative number is its opposite.}$$
$$= 41$$

H. Find the absolute value of 0.

$$|0| = 0 \qquad \text{The absolute value of zero is zero.}$$

I. Find the absolute value of $-\dfrac{5}{9}$.

$$\left| -\frac{5}{9} \right| = -\left(-\frac{5}{9} \right) \qquad \text{The absolute value of a negative number is its opposite.}$$
$$= \frac{5}{9}$$

ANSWERS TO WARM-UPS F–I

F. 29 **G.** 90 **H.** 1 **I.** $\dfrac{3}{7}$

EXERCISES 8.1

■ **OBJECTIVE 1** Find the opposite of a signed number. (See page 570.)

A *Find the opposite of the signed number.*

1. −4 **2.** −8 **3.** 5 **4.** 13

5. −2.3 **6.** −3.3 **7.** $\dfrac{3}{5}$ **8.** $\dfrac{3}{4}$

9. $-\dfrac{6}{7}$

10. $-\dfrac{3}{8}$

11. The opposite of _____ is 61.

12. The opposite of -23 is _____.

B *Find the opposite of the signed number.*

13. -42

14. -57

15. -3.78

16. -6.27

17. $\dfrac{17}{5}$

18. $\dfrac{23}{7}$

19. 0.55

20. 0.732

21. 113.8

22. 243.7

23. -0.0123

24. -0.78

■ **OBJECTIVE 2** Find the absolute value of a signed number. (See page 573.)

A *Find the absolute value of the signed number.*

25. $|-4|$

26. $|-11|$

27. $|24|$

28. $|61|$

29. $|-3.17|$

30. $|-4.6|$

31. $\left|\dfrac{5}{11}\right|$

32. $\left|\dfrac{7}{13}\right|$

33. $\left|-\dfrac{3}{11}\right|$

34. $\left|-\dfrac{7}{12}\right|$

35. The absolute value of _____ is 32.

36. The absolute value of _____ is 17.

B *Find the absolute value of a signed number.*

37. $|0.0065|$

38. $|0.0021|$

39. $|-355|$

40. $|-922|$

41. $\left|-\dfrac{11}{3}\right|$

42. $\left|-\dfrac{17}{6}\right|$

43. $\left|-\dfrac{33}{7}\right|$

44. $\left|-\dfrac{18}{5}\right|$

45. $|0|$

46. $|-100|$

47. $|-0.341|$

48. $|-4.96|$

C *Find the value of the following.*

49. Opposite of $\left|-\dfrac{5}{17}\right|$

50. Opposite of $\left|\dfrac{9}{8}\right|$

51. Opposite of $|78|$

52. Opposite of $|-91|$

53. At the New York Stock Exchange, positive and negative numbers are used to record changes in stock prices on the board. What is the opposite of a gain of 1.54?

54. On the NASDAQ, a stock is shown to have taken a loss of three-eighths of a point (-0.375). What is the opposite of this loss?

55. On a thermometer, temperatures above zero are listed as positive and those below zero as negative. What is the opposite of a reading of 14°C?

56. On a thermometer such as the one in Exercise 55, what is the opposite of a reading of 46°C?

57. The modern calendar counts the years after the birth of Christ as positive numbers (A.D. 2001 or +2001). Years before Christ are listed using negative numbers (2011 B.C. or -2011). What is the opposite of 1915 B.C., or -1915?

58. When balancing a checkbook, deposits are considered positive numbers and written checks are considered negative numners. Maya deposits her paycheck of $468.93 into her checking account and then writes checks of $123.76 to the electric company and $98.15 to her cell phone carrier. Represent these transactions as signed numbers.

59. A cyclist travels up a mountain 1415 ft, then turns around and travels down the mountain 1143 ft. Represent each trip as a signed number.

60. An energy audit indicates that the Gates family could reduce their average electric bill by $17.65 per month by doing some minor repairs, insulating their attic and crawl space, and caulking around the windows and other cracks in the siding.
   a. Express this savings as a signed number.

   b. Express the opposite of the savings as a signed number.

61. If 80 miles north is represented by $+80$, how would you represent 80 miles south?

Exercises 62–64 refer to the chapter application. See page 569.

62. Rewrite the continental altitudes in the application using signed numbers.

63. The U.S. Department of Defense has extensive maps of the ocean floors because they are vital information for the country's submarine fleet. The table lists the deepest part and the average depth of the world's major oceans.

Ocean	Deepest Part (ft)	Average Depth (ft)
Pacific	Mariana Trench, 35,840	12,925
Atlantic	Puerto Rico Trench, 28,232	11,730
Indian	Java Trench, 23,376	12,598
Arctic	Eurasia Basin, 17,881	3,407
Mediterranean	Ionian Basin, 16,896	4,926

Rewrite the table using signed numbers.

64. Is the highest point on Earth farther away from sea level than the deepest point in the ocean? Explain. What mathematical concept allows you to answer this question? See Exercise 63.

Simplify

65. $-(-24)$

66. $-(81)$

67. $-(-(-14))$

68. $-(-(33))$

69. The Buffalo Bills are playing a football game against the Seattle Seahawks. On the first play, the Seahawks lose 8 yd. Represent this as a signed number. What is the opposite of a loss of 8 yd? Represent this as a signed number.

70. The Dallas Cowboys and the New York Giants are having an exhibition game in London, England. The Cowboy offensive team runs a gain of 6 yd, a loss of 8 yd, a gain of 21 yd, and a loss of 15 yd. Represent these yardages as signed numbers.

71. The Golden family is on a vacation in the southwestern United States. Consider north and east as positive directions and south and west as negative directions. On 1 day they drive north 137 mi then east 212 mi. The next day they drive west 98 mi then 38 mi south. Represent each of these distances as signed numbers.

STATE YOUR UNDERSTANDING

72. Is zero the only number that is its own opposite? Justify your answer.

73. Is there a set of numbers for which the absolute value of each number is the number itself? If yes, identify that set and tell why this is true.

74. Explain −4. Draw it on a number line. On the number line, use the concepts of opposites and absolute value as they relate to −4. Give an instance in the world when −4 is useful.

CHALLENGE

Simplify.

75. $|16 − 10| − |14 − 9| + 6$

76. $8 − |12 − 8| − |10 − 8| + 2$

77. If n is a negative number, what kind a number is $−n$?

78. For what numbers is $|−n| = n$ always true?

MAINTAIN YOUR SKILLS

Add.

79. $7 + 11 + 32 + 9$

80. $5 + 12 + 21 + 14$

Subtract.

81. $17 − 12$

82. $45 − 23$

Add.

83. 6 m 250 cm
  $+$ 7 m 460 cm
  $= ?$ m

84. 5 L 78 mL
  $+$ 2 L 95 mL
  $= ?$ mL

Subtract.

85. 3 yd 2 ft 8 in.
  $-$ 1 yd 2 ft 11 in.
  $= ?$ in.

86. 7 gal 1 qt 1 pt
  $-$ 4 gal 3 qt 1 pt
  $= ?$ pt

87. A 12-m coil of wire is to be divided into eight equal parts. How many meters are in each part?

88. Some drug doses are measured in grains. For example, a common aspirin tablet contains 5 grains. If there are 15.43 g in a grain, how many grams do two aspirin tablets contain?

Adding Signed Numbers

HOW & WHY

■ **OBJECTIVE**   Add signed numbers.

Positive and negative numbers are used to show opposite quantities:

$+482$ lb may show 482 lb loaded
-577 lb may show 577 lb unloaded
$+27$ dollars may show 27 dollars earned
-19 dollars may show 19 dollars spent

We can use this idea to find the sum of signed numbers. We already know how to add two positive numbers, so we concentrate on adding positive and negative numbers.

Think of 27 dollars earned (positive) and 19 dollars spent (negative). The result is 8 dollars left in your pocket (positive). So

$$27 + (-19) = 8$$

To get this sum, we subtract the absolute value of -19 (19) from the absolute value of 27 (27). The sum is positive.

Think of 23 dollars spent (negative) and 15 dollars earned (positive). The result is that you still owe 8 dollars (negative). So

$$-23 + 15 = -8$$

To get this sum, subtract the absolute value of 15 (15) from the absolute value of -23 (23). The sum is negative.

Think of 5 dollars spent (negative) and another 2 dollars spent (negative). The result is 7 dollars spent (negative). So

$$-5 + (-2) = -7$$

To get this sum we add 5 to 2 (the absolute value of each). The sum is negative.

The results of the examples lead us to the procedure for adding signed numbers.

To add signed numbers

1. If the signs are alike, add their absolute values and use the common sign for the sum.
2. If the signs are not alike, subtract the smaller absolute value from the larger absolute value. The sum will have the sign of the number with the larger absolute value.

As a result of this definition, the sum of a number and its opposite is zero. Thus, $5 + (-5) = 0, 8 + (-8) = 0, -12 + 12 = 0$, and so on.

DIRECTIONS: Add.

STRATEGY: If the signs are alike, add the absolute values and use the common sign for the sum. If the signs are unlike, subtract their absolute values and use the sign of the number with larger absolute value.

A. Add: $45 + (-23)$

STRATEGY: Since the signs are unlike, subtract their absolute values.

$$|45| - |-23| = 45 - 23$$
$$= 22$$

Because the positive number has the larger absolute value, the sum is positive.

So $45 + (-23) = 22$.

B. Find the sum: $-72 + 31$

$$|-72| - |31| = 72 - 31$$
$$= 41$$

$$-72 + 31 = -41 \qquad \text{The sum is negative, because } -72 \text{ has the larger absolute value.}$$

C. Add: $37 + (-53)$

$$|-53| - |37| = 53 - 37$$
$$= 16$$

$$37 + (-53) = -16 \qquad \text{The sum is negative, because } -53 \text{ has the larger absolute value.}$$

D. Add: $-0.12 + (-2.54)$

$$|-0.12| + |-2.54| = 0.12 + 2.54$$
$$= 2.66$$

$$-0.12 + (-2.54) = -2.66 \qquad \text{The numbers are negative, therefore the sum is negative.}$$

E. Find the sum: $\dfrac{5}{8} + \left(-\dfrac{3}{4}\right)$

$$\left|-\dfrac{3}{4}\right| - \left|\dfrac{5}{8}\right| = \dfrac{6}{8} - \dfrac{5}{8}$$
$$= \dfrac{1}{8}$$

$$\dfrac{5}{8} + \left(-\dfrac{3}{4}\right) = -\dfrac{1}{8} \qquad \text{Because the negative number has the larger absolute value, the sum is negative.}$$

F. Find the sum of $-16, 41, -33,$ and -14.

STRATEGY: Where there are more than two numbers to add, it may be easier to add the numbers with the same sign first.

$$\begin{array}{r} -16 \\ -33 \\ -14 \\ \hline -63 \end{array} \qquad \text{Add the negative numbers.}$$

Now add their sum to 41.

$$|-63| - |41| = 63 - 41 \qquad \text{Because the signs are different, find the}$$
$$= 22 \qquad\qquad\qquad \text{difference in their absolute values.}$$

$$-16 + 41 + (-33) + (-14) = -22 \qquad \text{The sum is negative, because } -63 \text{ has the larger absolute value.}$$

G. Add: $-0.4 + 0.7 + (-5.4) + 1.3$

$$\begin{array}{ll} -0.4 & 0.7 \\ \underline{-5.4} & \underline{1.3} \\ -5.8 & 2.0 \end{array}$$

Now add the sums.

$-5.8 + 2.0$ The signs are different so subtract their absolute values.

$$|-5.8| - |2.0| = 5.8 - 2.0$$
$$= 3.8$$

$-0.4 + 0.7 + (-5.4) + 1.3 = -3.8$ Because -5.8 has the larger absolute value, the sum is negative.

H. Add: $-\dfrac{5}{6} + \dfrac{4}{9} + \left(-\dfrac{5}{18}\right) + \left(-\dfrac{7}{18}\right)$

STRATEGY: First add the negative numbers.

$$-\frac{5}{6} + \left(-\frac{5}{18}\right) + \left(-\frac{7}{18}\right) = -\frac{15}{18} + \left(-\frac{5}{18}\right) + \left(-\frac{7}{18}\right)$$
$$= -\frac{27}{18}$$

Add the sum of the negative numbers and the positive number. To do this, find the difference of their absolute values.

$$\left|-\frac{27}{18}\right| - \left|\frac{4}{9}\right| = \frac{27}{18} - \frac{8}{18}$$
$$= \frac{19}{18}$$

$$-\frac{5}{6} + \frac{4}{9} + \left(-\frac{5}{18}\right) + \left(-\frac{7}{18}\right) = -\frac{19}{18}$$ The sum is negative becauset the negative number has the larger absolute value.

CALCULATOR EXAMPLE:

I. Find the sum: $-89 + 77 + (-104) + (-93)$

Use the ⊞ or ⊟ key on the calculator to indicate negative numbers.

The sum is -209.

J. John owns stock that is traded on the NASDAQ. On Monday the stock gains $2.04, on Tuesday it loses $3.15, on Wednesday it loses $1.96, on Thursday it gains $2.21, and on Friday it gains $1.55. What is the net price change in the price of the stock for the week?

STRATEGY: To find the net change in the price of the stock, write the daily changes as signed numbers and find the sum of these numbers.

Monday	gains $2.04	2.04
Tuesday	loses $3.15	-3.15
Wednesday	loses $1.96	-1.96
Thursday	gains $2.21	2.21
Friday	gains $1.55	1.55

$$2.04 + (-3.15) + (-1.96) + 2.21 + 1.55 = 5.8 + (-5.11)$$
$$= 0.69$$

Add the positive numbers and add the negative numbers.

The stock gains $0.69 during the week.

WARM-UP

G. Add: $0.42 + (-5.21) + (-0.94) + 3.65$

WARM-UP

H. $-\dfrac{4}{5} + \dfrac{1}{2} + \left(-\dfrac{3}{4}\right) + \left(-\dfrac{2}{5}\right)$

WARM-UP

I. Find the sum: $-88 + 97 + (-117) + (-65)$

WARM-UP

J. A second stock that John owns has the following changes for a week: Monday, gains $1.34; Tuesday, gains $0.84; Wednesday, loses $4.24; Thursday, loses $2.21; Friday, gains $0.36. What is the net change in the price of the stock for the week?

ANSWERS TO WARM-UPS G–J

G. -2.08 **H.** $-\dfrac{29}{20}$ **I.** -173

J. The stock lost $3.91 ($-3.91$) for the week.

EXERCISES 8.2

OBJECTIVE   Add signed numbers. (See page 579.)

A   *Add.*

1. $-7 + 9$

2. $-7 + 3$

3. $8 + (-6)$

4. $10 + (-6)$

5. $-9 + (-6)$

6. $-7 + (-8)$

7. $-11 + (-3)$

8. $-15 + (-9)$

9. $-11 + 11$

10. $16 + (-16)$

11. $0 + (-12)$

12. $-32 + 0$

13. $-23 + (-23)$

14. $-13 + (-13)$

15. $3 + (-17)$

16. $12 + (-15)$

17. $-19 + (-7)$

18. $-11 + (-22)$

19. $24 + (-17)$

20. $-18 + 31$

B

21. $-5 + (-7) + 6$

22. $-5 + (-8) + (-10)$

23. $-65 + (-43)$

24. $-56 + (-44)$

25. $-98 + 98$

26. $108 + (-108)$

27. $-45 + 72$

28. $53 + (-38)$

29. $-36 + 43 + (-17)$

30. $-37 + 21 + (-9)$

31. $62 + (-56) + (-13)$

32. $-34 + (-18) + 29$

33. $-4.6 + (-3.7)$

34. $-9.7 + (-5.2)$

35. $10.6 + (-7.8)$

36. $-13.6 + 8.4$

37. $-\dfrac{5}{6} + \dfrac{1}{4}$

38. $\dfrac{3}{8} + \left(-\dfrac{1}{2}\right)$

39. $-\dfrac{7}{10} + \left(-\dfrac{7}{15}\right)$

40. $-\dfrac{5}{12} + \left(-\dfrac{2}{9}\right)$

C *Simplify.*

41. $135 + (-256)$

42. $233 + (-332)$

43. $-81 + (-32) + (-76)$

44. $-75 + (-82) + (-71)$

45. $-31 + 28 + (-63) + 36$

46. $-44 + 37 + (-59) + 45$

47. $49 + (-67) + 27 + 72$

48. $81 + (-72) + 33 + 49$

49. $356 + (-762) + (-892) + 541$

50. $-923 + 672 + (-823) + (-247)$

51. Find the sum of 542, -481, and -175.

52. Find the sum of 293, -122, and -211.

Exercises 53–56. The table gives the temperatures recorded by Rover *for a 5-day period at one location on the surface of Mars.*

	Day 1	Day 2	Day 3	Day 4	Day 5
5:00 A.M.	−92°C	−88°C	−115°C	−103°C	−74°C
9:00 A.M.	−57°C	−49°C	−86°C	−93°C	−64°C
1:00 P.M.	−52°C	−33°C	−46°C	−48°C	−10°C
6:00 P.M.	−45°C	−90°C	−102°C	−36°C	−42°C
11:00 P.M.	−107°C	−105°C	−105°C	−98°C	−90°C

53. What is the sum of the temperatures recorded at 9:00 A.M.?

54. What is the sum of the temperatures recorded on day 1?

55. What is the sum of the temperatures recorded at 11:00 P.M.?

56. What is the sum of the temperatures recorded on day 4?

57. An airplane is being reloaded; 963 lb of baggage and mail are removed (-963 lb) and 855 lb of baggage and mail are loaded ($+855$ lb). What net change in weight should the cargo master report?

58. At another stop, the plane in Exercise 57 unloads 2341 lb of baggage and mail and takes on 2567 lb. What net change should the cargo master report?

59. The change in altitude of a plane in flight is measured every 10 min. The figures between 3:00 P.M. and 4:00 P.M. are as follows:

3:00	30,000 ft initially	(+30,000)
3:10	increase of 220 ft	(+220 ft)
3:20	decrease of 200 ft	(−200 ft)
3:30	increase of 55 ft	(+55 ft)
3:40	decrease of 110 ft	(−110 ft)
3:50	decrease of 55 ft	(−55 ft)
4:00	decrease of 40 ft	(−40 ft)

© iStock photo.com/kickers

What is the altitude of the plane at 4 P.M.? (*Hint:* Find the sum of the initial altitude and the six measured changes between 3 and 4 P.M.)

60. What is the final altitude of the airplane in Exercise 59 if it is initially flying at 23,000 ft with the following changes in altitude?

3:00	23,000 ft initially	(+23,000)
3:10	increase of 315 ft	(+315 ft)
3:20	decrease of 825 ft	(−825 ft)
3:30	increase of 75 ft	(+75 ft)
3:40	decrease of 250 ft	(−250 ft)
3:50	decrease of 85 ft	(−85 ft)
4:00	decrease of 70 ft	(−70 ft)

61. The Pacific Northwest Book Depository handles most textbooks for the local schools. On September 1, the inventory is 34,945 volumes. During the month, the company makes the following transactions (positive numbers represent volumes received, negative numbers represent shipments): 3456, −2024, −3854, 612, and −2765. What is the inventory at the end of the month?

62. The Pacific Northwest Book Depository has 18,615 volumes on November 1. During the month, the depository has the following transactions: −4675, 912, −6764, 1612, and −950. What is the inventory at the end of the month?

63. The New England Patriots made the following consecutive plays during a recent football game: an 8-yd loss, a 9-yd gain, and a 7-yd gain. In order to make a first down, the team must have a net gain of at least 10 yards. Did the Patriots get a first down?

64. The Seattle Seahawks have these consecutive plays one Sunday: 12-yd loss, 19-yd gain, and 4-yd gain. Do they get a first down?

65. Nordstrom stock has the following changes in one week: up 0.62, down 0.44, down 1.12, up 0.49, down 0.56.
a. What is the net change for the week?

b. If the stock starts at $34.02 at the beginning of the week, what is the closing price?

66. On a January morning in a small town in upstate New York, the lowest temperature is recorded as 17 degrees below zero. During the following week the daily lowest temperature readings are up 3 degrees, up 7 degrees, down 5 degrees, up 1 degree, no change, and down 3 degrees. What is the low temperature reading for the last day?

67. A new company has the following weekly balances after the first month of business: a loss of $2376, a gain of $5230, a loss of $1055, and a gain of $3278. What is their net gain or loss for the month?

68. Marie decided to play the state lottery for one month. This meant that she played every Wednesday and Saturday for a total of nine times. This is her record: lost $10, won $45, lost $15, lost $12, won $65, lost $12, lost $16, won $10, and lost $6. What is the net result of her playing?

STATE YOUR UNDERSTANDING

69. When adding a positive and a negative number, explain how to determine whether the sum is positive or negative.

CHALLENGE

70. What is the sum of 46 and (−52) increased by 23?

71. What is the sum of (−73) and (−32) added to (−19)?

72. What number added to 47 equals 28?

73. What number added to (−75) equals (−43)?

74. What number added to (−123) equals (−163)?

MAINTAIN YOUR SKILLS

Subtract.

75. 145 − 67

76. 356 − 89

77. $\begin{array}{r} 178 \\ -\ 95 \end{array}$

78. $\begin{array}{r} 897 \\ -\ 698 \end{array}$

79. Find the perimeter of a rectangle that is 42 m long and 29 m wide.

80. Find the circumference of a circle with a radius of 17 cm. The formula is $C = \pi d$, where d is the diameter of the circle. Let $\pi \approx 3.14$.

81. Find the perimeter of a square that is 31 in. on each side.

82. Find the perimeter of a triangle with sides of 65 cm, 78 cm, and 52 cm.

83. Bananas are put on sale for 3 lb for $2.16. What is the cost of 1 lb?

84. The cost of pouring a 3-ft-wide cement sidewalk is estimated to be $21 per square foot. If a walk is to be placed around a rectangular plot of ground that is 42 ft along the width and 50 ft along the length, what is the cost of pouring the walk?

8.3 Subtracting Signed Numbers

HOW & WHY

■ **OBJECTIVE**   Subtract signed numbers.

The expression $11 - 7 = ?$ can be restated using addition: $7 + ? = 11$. We know that $7 + 4 = 11$ and so $11 - 7 = 4$. We use the fact that every subtraction fact can be restated as addition to discover how to subtract signed numbers. Consider $-4 - 8 = ?$ Restating, we have $8 + ? = -4$. We know that $8 + (-12) = -4$ so we conclude that $-4 - 8 = -12$. The expression $-5 - (-3) = ?$ can be restated as $-3 + ? = -5$. Because $-3 + (-2) = -5$, we conclude that $-5 - (-3) = -2$.

To discover the rule for subtraction of signed numbers, compare:

$11 - 7$	and	$11 + (-7)$	Both equal 4.
$-4 - 8$	and	$-4 + (-8)$	Both equal -12.
$-5 - (-3)$	and	$-5 + 3$	Both equal -2.

Every subtraction problem can be worked by asking what number added to the subtrahend will yield the minuend. However, when we look at the second column we see an addition problem that gives the same answer as the original subtraction problem. In each case the opposite (additive inverse) of the number being subtracted is added. Let's look at three more examples.

Answer Obtained by Adding to the Subtrahend	Answer Obtained by Adding the Opposite	
$19 - 7 = 12$	$19 + (-7) = 12$	-7 is the opposite of 7, the number to be subtracted.
$-32 - 45 = -77$	$-32 + (-45) = -77$	-45 is the opposite of 45, the number to be subtracted.
$-21 - (-12) = -9$	$-21 + 12 = -9$	12 is the opposite of -12, the number to be subtracted.

This leads us to the rule for subtracting signed numbers.

> **To subtract signed numbers**
>
> 1. Rewrite as an addition problem by adding the opposite of the number to be subtracted.
> 2. Find the sum.

EXAMPLES A–J

DIRECTIONS: Subtract.

STRATEGY: Add the opposite of the number to be subtracted.

A. Subtract: $42 - 28$

STRATEGY: Rewrite as an addition problem by adding -28, which is the opposite of 28.

$$42 - 28 = 42 + (-28) \qquad \text{Rewrite as addition.}$$
$$= 14 \qquad \text{Add. Because the signs are different, subtract their absolute values and use the sign of the number with the larger absolute value, which is 42.}$$

Because both numbers are positive we can also do the subtraction in the usual manner: $42 - 28 = 14$.

B. Subtract: $-38 - 61$

$$-38 - 61 = -38 + (-61) \qquad \text{Rewrite as addition.}$$
$$= -99 \qquad \text{Add. Because both numbers are negative, add their absolute values and keep the common sign.}$$

C. Find the difference of -54 and -12.

$$-54 - (-12) = -54 + 12 \qquad \text{Rewrite as addition.}$$
$$= -42 \qquad \text{Add.}$$

D. Subtract: $63 - 91$

$$63 - 91 = 63 + (-91) \qquad \text{Rewrite as addition.}$$
$$= -28$$

E. Find the difference: $\dfrac{5}{8} - \left(-\dfrac{2}{3}\right)$

$$\frac{5}{8} - \left(-\frac{2}{3}\right) = \frac{5}{8} + \frac{2}{3} \qquad \text{Rewrite as addition.}$$
$$= \frac{15}{24} + \frac{16}{24} \qquad \text{Write each fraction with a common denominator and add.}$$
$$= \frac{31}{24}$$

F. Subtract: $23 - (-33) - 19$

STRATEGY: Change both subtractions to add the opposite.

$$23 - (-33) - 19 = 23 + 33 + (-19) \qquad \text{Rewrite as addition.}$$
$$= 56 + (-19) \qquad \text{Add 23 and 33.}$$
$$= 37 \qquad \text{Add.}$$

G. Subtract: $-0.34 - (-1.04) - (-3.02) - 0.81$

STRATEGY: Change all subtractions to add the opposite and add.

$$-0.34 - (-1.04) - (-3.02) - 0.81 = -0.34 + 1.04 + 3.02 + (-0.81)$$
$$= 2.91$$

WARM-UP

H. Subtract:

$$-\frac{3}{5} - \left(-\frac{3}{8}\right) - \left(-\frac{7}{10}\right) - \frac{7}{20}$$

H. Subtract: $-\dfrac{3}{4} - \dfrac{7}{8} - \left(-\dfrac{1}{2}\right) - \left(-\dfrac{1}{8}\right)$

$$-\frac{3}{4} - \frac{7}{8} - \left(-\frac{1}{2}\right) - \left(-\frac{1}{8}\right) = -\frac{3}{4} + \left(-\frac{7}{8}\right) + \frac{1}{2} + \frac{1}{8}$$

$$= -\frac{6}{8} + \left(-\frac{7}{8}\right) + \frac{4}{8} + \frac{1}{8}$$

$$= -\frac{8}{8} = -1$$

CALCULATOR EXAMPLE: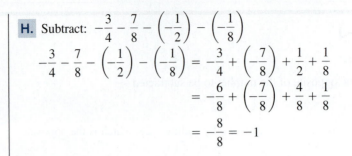

WARM-UP

I. Subtract:

$-483.34 - (-312.43)$

I. Subtract: $-784.63 - (-532.78)$

STRATEGY: The calculator does not require you to change the subtraction to add the opposite.

The difference is -251.85.

WARM-UP

J. One night last winter, the temperature dropped from 24°F to −11°F. What was the difference between the high and the low temperatures?

J. The highest point in North America is Mount McKinley, a peak in central Alaska, which is approximately 20,320 ft above sea level. The lowest point in North America is Death Valley, a deep basin in southeastern California, which is approximately 282 ft below sea level. What is the difference in height between Mount McKinley and Death Valley? (Above sea level is positive and below sea level is negative.)

STRATEGY: To find the difference in height, write each height as a signed number and subtract the lower height from the higher height.

Mount McKinley	20,320 above	20,320
Death Valley	282 below	−282

$20,320 - (-282) = 20,320 + 282$ Change subtraction to add the

$= 20,602$ opposite and add.

The difference in height is approximately 20,602 ft.

ANSWERS TO WARM-UPS H–J

H. $\dfrac{1}{8}$ **I.** −170.91   **J.** The difference in temperatures was 35°F.

EXERCISES 8.3

■ **OBJECTIVE 1** Subtract signed numbers. (See page 586.)

A *Subtract.*

1. $7 - 3$

2. $9 - 5$

3. $-7 - 5$

4. $-8 - 7$

5. $-8 - (-3)$

6. $-7 - (-2)$

7. $11 - 8$

8. $13 - 9$

9. $-12 - 8$

10. $-13 - 6$

11. $-23 - (-14)$

12. $-17 - (-8)$

13. $19 - (-13)$

14. $26 - (-9)$

15. $-23 - 11$

16. $-24 - 12$

17. $-15 - 13$

18. $-33 - 20$

19. $-13 - (-14)$

20. $-34 - (-35)$

21. $-16 - (-16)$

22. $-23 - (-23)$

23. $-9 - 9$

24. $-31 - 31$

25. $-40 - 40$

26. $-14 - 14$

B

27. $72 - (-46)$

28. $54 - (-61)$

29. $-57 - 62$

30. $-82 - 91$

31. $-48 - (-59)$

32. $-66 - (-81)$

33. $-91 - 91$

34. $-43 - 43$

35. $102 - (-102)$

36. $78 - (-78)$

37. $-69 - (-69)$

38. $-83 - (-83)$

39. $134 - (-10)$

40. $164 - (-20)$

41. $132 - (-41)$

42. $173 - (45)$

43. $-6.74 - 3.24$

44. $-13.34 - 9.81$

45. $-4.65 - (-3.21)$

46. $-7.54 - (-8.12)$

47. $-23.43 - 32.71$

48. $-18.63 - (-13.74)$

49. Find the difference between 43 and -73.

50. Find the difference between -88 and -97.

51. Subtract 338 from -349.

52. Subtract 145 from -251.

53. *Rover* records high and low temperatures of $-24°C$ and $-109°C$ for 1 day on the surface of Mars. What is the change in temperature for that day?

54. The surface temperature of one of Jupiter's satellites is measured for 1 week. The highest temperature recorded is $-83°C$ and the lowest is $-145°C$. What is the difference in the extreme temperatures for the week?

55. At the beginning of the month, Joe's bank account had a balance of $782.45. At the end of the month, the account was overdrawn by $13.87 (−$13.87). If there were no deposits during the month, what was the total amount of checks Joe wrote? (*Hint:* Subtract the ending balance from the original balance.)

56. At the beginning of the month, Jack's bank account had a balance of $512.91. At the end of the month, the balance was −$67.11. If there were no deposits, find the amount of checks Jack wrote. (Refer to Exercise 55.)

Exercises 57–59 refer to the chapter application. See page 569.

57. The range of a set of numbers is defined as the difference between the largest and the smallest numbers in the set. Calculate the range of altitude for each continent. Which continent has the smallest range, and what does this mean in physical terms?

58. What is the difference between the lowest point in the Mediterranean and the lowest point in the Atlantic? See Exercise 63 in Section 8.1.

59. Some people consider Mauna Kea, Hawaii, to be the tallest mountain in the world. It rises 33,476 ft from the ocean floor, but is only 13,796 ft above sea level. What is the depth of the ocean floor at this location?

Exercises 60–63 refer to the table below, which shows temperature recordings by a Martian probe for a 5-day period at one location on the surface of Mars.

Temperatures on the Surface of Mars

	Day 1	Day 2	Day 3	Day 4	Day 5
5:00 A.M.	−92°C	−88°C	−115°C	−103°C	−74°C
9:00 A.M.	−57°C	−49°C	−86°C	−93°C	−64°C
1:00 P.M.	−52°C	−33°C	−46°C	−48°C	−10°C
6:00 P.M.	−45°C	−90°C	−102°C	−36°C	−42°C
11:00 P.M.	−107°C	−105°C	−105°C	−98°C	−90°C

60. What is the difference between the high and low temperatures recorded on day 3?

61. What is the difference between the temperatures recorded at 11:00 P.M. on day 3 and day 5?

62. What is the difference between the temperatures recorded at 5:00 A.M. on day 2 and 6:00 P.M. on day 4?

63. What is the difference between the highest and lowest temperatures recorded during the 5 days?

64. Al's bank account had a balance of $318. He writes a check for $412.75. What is his account balance now?

65. Thomas started with $210.34 in his account. He writes a check for $216.75. What is his account balance now?

66. Carol started school owing her mother $18; by school's end she borrowed $123 more from her mother. How does her account with her mother stand now?

67. At the beginning of the month, Janna's bank account had a balance of $467.82. At the end of the month, the account was overdrawn by $9.32. If there were no deposits during the month, what was the total amount of checks Janna wrote?

68. What is the difference in altitude between the highest point and the lowest point in California?

Highest point: Mount Whitney is 14,494 ft above sea level ($+14,494$).

Lowest point: Death Valley is 282 ft below sea level (-282).

69. According to TreasuryDirect.gov, the U.S. national debt was $11.91 trillion in 2009 and $5.67 trillion in 2000. What was the difference in debt over the 9-year period?

70. The New York Jets started on their 40-yd line. After three plays, they were on their 12-yd line. Did they gain or lose yards? Represent their gain or loss as a signed number.

71. In the 2009 PGA championship, Y.E. Yang won with a final score of -8, and Lucas Glover finished fifth with a final score of -2. What was the difference in the scores between first and fifth place?

STATE YOUR UNDERSTANDING

72. Explain the difference between adding and subtracting signed numbers.

73. How would you explain to a 10-year-old how to subtract -8 from 12?

74. Explain why the order in which two numbers are subtracted is important, but the order in which they are added is not.

75. Explain the difference between the problems $5 + (-8)$ and $5 - 8$.

CHALLENGE

76. $-13 - (-54) - |-21|$

77. $-9.46 - [-(-3.22)]$

78. $-76 - (-37) - |-(-55)|$

79. $57 - |-67 - (-51)| - 82$

80. $|17.5 - 13.5| - |21.7 - (-19.6)|$

Multiply.

81. 16(7)

82. 42(13)

83.   156
     × 37

84.   4782
     × 362

85. Find the area of a circle that has a radius of 14 in. (Let $\pi \approx 3.14$.)

86. Find the area of a square that is 16.7 m on each side.

87. Find the area of a rectangle that is 13.6 cm long and 9.4 cm wide.

88. Find the area of a triangle that has a base of 14.4 m and a height of 7.8 m.

89. How many square yards of wall-to-wall carpeting are needed to carpet a rectangular floor that measures 22.5 ft by 30.5 ft?

90. How many square tiles, which are 9 in. on a side, are needed to cover a floor that is 21 ft by 24 ft?

SECTION 8.4 Multiplying Signed Numbers

OBJECTIVES

1. Multiply a positive number and a negative number.
2. Multiply two negative numbers.

HOW & WHY

OBJECTIVE 1 Multiply a positive number and a negative number.

Consider the following multiplications:

$3(4) = 12$
$3(3) = 9$
$3(2) = 6$
$3(1) = 3$
$3(0) = 0$
$3(-1) = ?$
$3(-2) = ?$

Each product is 3 smaller than the one before it. Continuing this pattern,

$3(-1) = -3$ Because $0 - 3 = -3$

and

$3(-2) = -6$ Because $-3 - 3 = -6$

The pattern indicates that the product of a positive and a negative number is negative; that is, the opposite of the product of their absolute values.

> **To find the product of a positive and a negative number**
>
> 1. Find the product of the absolute values.
> 2. Make this product negative.

This is sometimes stated, "The product of two unlike signs is negative."

The commutative property of multiplication dictates that no matter the order in which the positive and negative numbers appear, their product is always negative. Thus,

$3(-4) = -12$ and $-4(3) = -12$

EXAMPLES A–F

DIRECTIONS: Multiply.

STRATEGY: Multiply the absolute values and write the opposite of that product.

A. Find the product: $-7(6)$

$-7(6) = -42$ The product of a positive and a negative number is negative.

B. Find the product: $18(-9)$

$18(-9) = -162$ The product of two factors with unlike signs is negative.

C. Find the product: $5(-3.3)$

$5(-3.3) = -16.5$ The product of two factors with unlike signs is negative.

D. Multiply: $\left(-\dfrac{4}{5}\right)\left(\dfrac{3}{4}\right)$

$\left(-\dfrac{\overset{1}{\cancel{4}}}{5}\right)\left(\dfrac{3}{\underset{1}{\cancel{4}}}\right) = -\dfrac{3}{5}$ Simplify. Find the product of the numerators and the denominators. The product is negative.

E. Multiply: $8(-7)(4)$

$8(-7)(4) = -56(4)$ Multiply the first two factors.

$\qquad\qquad = -224$ Multiply again.

F. In order to attract business, the Family grocery store ran a "loss leader" sale last week. The store sold eggs at a loss of $0.23 ($-0.23) per dozen. If 238 dozen eggs were sold last weekend, what was the total loss from the sale of eggs? Express this loss as a signed number.

$238(-0.23) = -54.74$ To find the total loss, multiply the loss per dozen by the number of dozens sold.

Therefore, the loss, written as a signed number, is $-\$54.74$.

HOW & WHY

■ **OBJECTIVE 2** Multiply two negative numbers.

We use the product of a positive and a negative number to develop a pattern for multiplying two negative numbers.

$$-3(4) = -12$$
$$-3(3) = -9$$
$$-3(2) = -6$$
$$-3(1) = -3$$
$$-3(0) = 0$$
$$-3(-1) = ?$$
$$-3(-2) = ?$$

Each product is three larger than the one before it. Continuing this pattern,

$-3(-1) = 3$ Because $0 + 3 = 3$.

$-3(-2) = 6$ Because $3 + 3 = 6$.

In each case the product is positive.

> **To multiply two negative numbers**
>
> 1. Find the product of the absolute values.
> 2. Make this product positive.

The product of two like signs is positive. When multiplying more than two signed numbers, if there is an even number of negative factors, the product is positive.

EXAMPLES G–M

DIRECTIONS: Multiply.

STRATEGY: Multiply the absolute values; make the product positive.

WARM-UP
G. Multiply: $-12(-8)$

G. Multiply: $-11(-7)$

$-11(-7) = 77$ The product of two negative numbers is positive.

WARM-UP
H. Multiply: $-3.6(-2.7)$

H. Multiply: $-5.2(-0.32)$

$-5.2(-0.32) = 1.664$ The product of two numbers with like signs is positive.

WARM-UP
I. Find the product:
$-111(-6)$

I. Find the product: $-98(-7)$

$-98(-7) = 686$

WARM-UP
J. Find the product of $-\dfrac{7}{12}$
and $-\dfrac{8}{25}$.

J. Find the product of $-\dfrac{2}{7}$ and $-\dfrac{3}{8}$.

$$\left(-\frac{\overset{1}{\cancel{2}}}{7}\right)\left(-\frac{3}{\underset{4}{\cancel{8}}}\right) = \frac{3}{28}$$ Negative times negative is positive.

So the product is $\dfrac{3}{28}$.

WARM-UP
K. Multiply: $-13(-3)(-4)$

K. Multiply: $-16(-3)(-5)$

$-16(-3)(-5) = 48(-5)$ Multiply the first two factors.

$= -240$ Multiply again.

WARM-UP
L. Find the product of -2, 11, 4, -2, and -3.

L. Find the product of -3, 12, 3, -1, and -5.

STRATEGY: There is an odd number of negative factors, therefore the product is negative.

$(-3)(12)(3)(-1)(-5) = -540$

The product is -540.

CALCULATOR EXAMPLE:

WARM-UP
M. Multiply:
$-63(-4.9)(-13.5)$

M. Multiply: $-82(-9.6)(-12.9)$

The product is $-10,154.88$.

ANSWERS TO WARM-UPS G–M

G. 96   **H.** 9.72   **I.** 666   **J.** $\dfrac{14}{75}$

K. -156 **L.** -528 **M.** -4167.45

■ **OBJECTIVE 1** Multiply a positive number and a negative number. (See page 592.)

A *Multiply.*

1. $-2(4)$

2. $4(-5)$

3. $-5(2)$

4. $(-6)(8)$

5. $10(-8)$

6. $-13(5)$

7. $-11(7)$

8. $12(-3)$

9. The product of -5 and _____ is -55.

10. The product of 8 and _____ is -72.

B

11. $-9(34)$

12. $11(-23)$

13. $-17(15)$

14. $23(-18)$

15. $2.5(-3.6)$

16. $-3.4(2.7)$

17. $-0.35(1000)$

18. $2.57(-10,000)$

19. $-\dfrac{2}{3} \cdot \dfrac{3}{8}$

20. $\left(\dfrac{3}{8}\right)\left(-\dfrac{4}{5}\right)$

■ **OBJECTIVE 2** Multiply two negative numbers. (See page 593.)

A *Multiply.*

21. $(-1)(-3)$

22. $(-2)(-4)$

23. $-7(-4)$

24. $-6(-5)$

25. $-11(-9)$

26. $(-4)(-3)$

27. $-12(-5)$

28. $-6(-16)$

29. The product of -6 and _____ is 66.

30. The product of -13 and _____ is 39.

B

31. $-14(-15)$

32. $-23(-17)$

33. $-1.2(-4.5)$

34. $-0.9(-0.72)$

35. $(-5.5)(-4.4)$

36. $(-6.3)(-2.3)$

37. $\left(-\dfrac{3}{14}\right)\left(-\dfrac{7}{9}\right)$

38. $\left(-\dfrac{8}{15}\right)\left(-\dfrac{5}{24}\right)$

39. $(-0.35)(-4.7)$

40. $(-7.2)(-2.1)$

C

41. $(-56)(45)$

42. $(16)(-32)$

43. $(15)(31)$

44. $(-23)(71)$

45. $(-1.4)(-5.1)$

46. $(-2.4)(6.1)$

47. $\left(-\dfrac{9}{16}\right)\left(\dfrac{8}{15}\right)$

48. $\left(-\dfrac{8}{21}\right)\left(-\dfrac{7}{16}\right)$

49. $(-4.01)(3.5)$

50. $(-6.7)(-0.45)$

51. $(-3.19)(-1.7)(0.1)$

52. $-1.3(4.6)(-0.2)$

53. $-2(4)(-1)(0)(-5)$

54. $(-4)(-7)(3)(-8)(0)$

55. $-0.07(0.3)(-10)(100)$

56. $(0.3)(-0.05)(-10)(-10)$

57. $-2(-5)(-6)(-4)(-1)$

58. $9(-2)(-3)(-5)(-4)$

59. $\left(-\dfrac{2}{3}\right)\left(-\dfrac{3}{4}\right)\left(-\dfrac{4}{5}\right)\left(-\dfrac{5}{6}\right)$

60. $\left(-\dfrac{5}{12}\right)\left(\dfrac{7}{8}\right)\left(\dfrac{3}{14}\right)\left(-\dfrac{8}{15}\right)$

61. The formula for converting a temperature measurement from Fahrenheit to Celsius is $C = \dfrac{5}{9}(F - 32)$. What Celsius measure is equal to 5°F?

62. Use the formula in Exercise 61 to find the Celsius measure that is equal to 20°F.

63. While on a diet for 8 consecutive weeks, Ms. Riles averages a weight loss of 2.6 lb each week. If each loss is represented by -2.6 lb, what is her total loss for the 8 weeks, expressed as a signed number?

64. Mr. Riles goes on a diet for 8 consecutive weeks. He averages a loss of 3.2 lb per week. If each loss is represented by -3.2 lb, what is his total loss expressed as a signed number?

65. The Dow Jones Industrial Average sustains 12 straight days of a 1.74 decline. What is the total decline during the 12-day period, expressed as a signed number?

66. The Dow Jones Industrial Average sustains 7 straight days of a 2.33 decline. What is the total decline during the 7-day period, expressed as a signed number?

Simplify.

67. $(15 - 8)(5 - 12)$

68. $(17 - 20)(5 - 9)$

69. $(15 - 21)(13 - 6)$

70. $(25 - 36)(5 - 9)$

71. $(-12 + 30)(-4 - 10)$

72. $(11 - 18)(-13 + 5)$

73. Safeway Inc. offers as a loss leader 10 lb of sugar at a loss of 17¢ per bag $(-17¢)$. If 386 bags are sold during the sale, what is the total loss, expressed as a signed number?

74. Albertsons offers a loss leader of coffee at a loss of 23¢ per 3-lb tin $(-23¢)$. If they sell 412 tins of coffee, find the total loss expressed as a signed number.

75. Winn Dixie's loss leader is a soft drink on which the store loses 6¢ per six-pack. They sell 523 of these six-packs. What is Winn Dixie's total loss expressed as a signed number?

76. Kroger's loss leader is soap powder on which the store loses 28¢ per carton. They sell 264 cartons. What is Kroger's total loss expressed as a signed number?

77. The lowest recorded temprature on Antarctica was $-128.6°F$ in 1983 at Vostok Station. Use the formula $C = \dfrac{5}{9}(F - 32)$ to convert this temprature to degrees Celsius. Round to the nearest degree.

78. A scientist is studying movement of a certain spider within its web. Any movement up is considered to be positive, whereas any movement down is negative. Determine the net movement of a spider that goes up 2 cm five times and down 3 cm twice.

79. A certain junk bond trader purchased 670 shares of stock at 9.34. When she sold her shares, the stock sold for 6.45. What did she pay for the stock? How much money did she receive when she sold this stock? How much did she lose or gain? Represent the loss or gain as a signed number.

80. A company bought 450 items at $1.23 each. They tried to sell them for $2.35 and sold only 42 of them. They lowered the price to $2.10 and sold 105 more. The price was lowered a second time to $1.35 and 85 were sold. Finally they advertised a close-out price of $0.95 and sold the remaining items. Determine the net profit or loss for each price. Did they make a profit or lose money on this item overall?

Exercises 81–82 refer to the chapter application. See page 569.

81. Which continent has a low point that is approximately 10 times the low point of South America?

82. Which continent has a high point that is approximately twice the absolute value of the lowest point?

STATE YOUR UNDERSTANDING

83. Explain why the product of an even number of negative numbers is positive.

84. Explain the procedure for multiplying two signed numbers.

CHALLENGE

Simplify.

85. $|-(-5)|(-9 - [-(-5)])$

86. $|-(-8)|(-8 - [-(-9)])$

87. Find the product of -8 and the opposite of 7.

88. Find the product of the opposite of 12 and the absolute value of -9.

Divide.

89. $66 \div 11$

90. $816 \div 12$

91. $\dfrac{1005}{15}$

92. $54\overline{)14{,}472}$

93. $34\overline{)4876}$, round to the nearest hundredth

94. What is the equivalent piecework wage (dollars per piece) if the hourly wage is $15.86 and the average number of articles completed in 1 hr is 6.1?

95. If carpeting costs $34.75 per square yard installed, what is the cost of wall-to-wall carpeting needed to cover the floor in a rectangular room that is 24 ft wide by 27 ft long?

96. How far does the tip of the hour hand of a clock travel in 6 hr if the length of the hand is 3 in.? Let $\pi \approx 3.14$.

97. How many square feet of sheet metal are needed to make a box without a top that has measurements of 5 ft 6 in. by 4 ft 6 in. by 9 in.?

98. A mini-storage complex has one unit that is 40 ft by 80 ft and rents for $1800 per year. What is the cost of a square foot of storage for a year?

SECTION 8.5 Dividing Signed Numbers

OBJECTIVES

1. Divide a positive number and a negative number.
2. Divide two negative numbers.

HOW & WHY

OBJECTIVE 1 Divide a positive number and a negative number.

To divide two signed numbers, we find the number that when multiplied times the divisor equals the dividend. The expression $-9 \div 3 = ?$ asks $3(?) = -9$; we know $3(-3) = -9$, so $-9 \div 3 = -3$. The expression $24 \div (-6) = ?$ asks $-6(?) = 24$; we know $-6(-4) = 24$, so $24 \div (-6) = -4$.

When dividing unlike signs, we see that the quotient is negative. We use these examples to state how to divide a negative number and a positive number.

> **To divide a positive and a negative number**
>
> 1. Find the quotient of the absolute values.
> 2. Make the quotient negative.

EXAMPLES A–D

DIRECTIONS: Divide

STRATEGY: Divide the absolute values and make the quotient negative.

A. Divide: $32 \div (-8)$

$32 \div (-8) = -4$ The quotient of two numbers with unlike signs is negative.

B. Divide: $(-4.8) \div 3.2$

$(-4.8) \div 3.2 = -1.5$ When dividing unlike signs, the quotient is negative.

C. Divide: $\left(\dfrac{4}{35}\right) \div \left(-\dfrac{4}{5}\right)$

$$\left(\dfrac{4}{35}\right) \div \left(-\dfrac{4}{5}\right) = \left(\dfrac{\overset{1}{\cancel{4}}}{\underset{7}{\cancel{35}}}\right) \cdot \left(-\dfrac{\overset{1}{\cancel{5}}}{\underset{1}{\cancel{4}}}\right)$$ Multiply by the reciprocal.

$$= -\dfrac{1}{7}$$ The product is negative.

D. Over a period of 18 weeks, Mr. Rich loses a total of $4230 ($-4230) in his stock market account. What is his average loss per week, expressed as a signed number?

STRATEGY: To find the average loss per week, divide the total loss by the number of weeks.

$-4230 \div 18 = -235$

Mr. Rich has an average loss of $235 ($-235) per week.

HOW & WHY

■ OBJECTIVE 2 Divide two negative numbers.

To determine how to divide two negative numbers, we again use the relationship to multiplication.

The expression $-21 \div (-7) = ?$ asks $(-7)(?) = -21$; we know that $-7(3) = -21$, so $-21 \div (-7) = 3$. The expression $-30 \div (-6) = ?$ asks $(-6)(?) = -30$; we know that $-6(5) = -30$, so $-30 \div (-6) = 5$. We see that in each case, when dividing two negative numbers, the quotient is positive. These examples lead us to the following rule.

> **To divide two negative numbers**
>
> 1. Find the quotient of the absolute values.
> 2. Make the quotient positive.

EXAMPLES E–G

DIRECTIONS: Divide.

STRATEGY: Find the quotient of the absolute values.

WARM-UP

E. Find the quotient:
$-48 \div (-6)$

E. Find the quotient: $-44 \div (-11)$

$-44 \div (-11) = 4$ The quotient of two negative numbers is positive.

CALCULATOR EXAMPLE:

WARM-UP

F. Find the quotient:
$-22.75 \div (-2.6)$

F. Find the quotient: $-7.74 \div (-3.6)$

$-7.74 \div (-3.6) = 2.15$

WARM-UP

G. Divide:
$$\left(-\frac{9}{30}\right) \text{ by } \left(-\frac{18}{25}\right)$$

G. Divide: $\left(-\dfrac{11}{9}\right)$ by $\left(-\dfrac{22}{27}\right)$

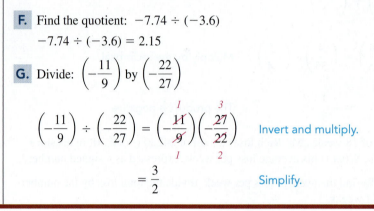

$$\left(-\frac{11}{9}\right) \div \left(-\frac{22}{27}\right) = \left(-\frac{\overset{1}{\cancel{11}}}{\cancel{9}}\right)\left(-\frac{\overset{3}{\cancel{27}}}{\underset{2}{\cancel{22}}}\right) \quad \text{Invert and multiply.}$$

$$= \frac{3}{2} \qquad \text{Simplify.}$$

ANSWERS TO WARM-UPS E–G

E. 8   **F.** 8.75   **G.** $\dfrac{5}{12}$

EXERCISES 8.5

■ **OBJECTIVE 1** Divide a positive number and a negative number. (See page 598.)

A *Divide.*

1. $-10 \div 5$

2. $10 \div (-2)$

3. $-16 \div 4$

4. $15 \div (-3)$

5. $18 \div (-6)$

6. $-18 \div 3$

7. $24 \div (-3)$

8. $-33 \div 11$

9. The quotient of -48 and ———— is -6.

10. The quotient of 70 and ———— is -14.

B

11. $72 \div (-12)$

12. $84 \div (-12)$

13. $6.06 \div (-3)$

14. $3.05 \div (-5)$

15. $-210 \div 6$

16. $-315 \div 9$

17. $\left(-\dfrac{6}{7}\right) \div \dfrac{2}{7}$

18. $\left(-\dfrac{4}{3}\right) \div \dfrac{8}{3}$

19. $0.75 \div (-0.625)$

20. $0.125 \div (-0.625)$

■ **OBJECTIVE 2** Divide two negative numbers. (See page 599.)

A *Divide.*

21. $-10 \div (-5)$ **22.** $-10 \div (-2)$ **23.** $-12 \div (-4)$ **24.** $-14 \div (-2)$

25. $-28 \div (-4)$ **26.** $-32 \div (-4)$ **27.** $-54 \div (-9)$ **28.** $-63 \div (-7)$

29. The quotient of -105 and _____ is 21.   **30.** The quotient of -75 and _____ is 15.

B

31. $-98 \div (-14)$ **32.** $-88 \div (-11)$ **33.** $-96 \div (-12)$

34. $-210 \div (-10)$ **35.** $-12.12 \div (-3)$ **36.** $-18.16 \div (-4)$

37. $\left(-\dfrac{3}{8}\right) \div \left(-\dfrac{3}{4}\right)$ **38.** $\left(-\dfrac{1}{2}\right) \div \left(-\dfrac{5}{8}\right)$ **39.** $-0.65 \div (-0.13)$

40. $-0.056 \div (-0.4)$

C

41. $-540 \div 12$ **42.** $-1071 \div 17$ **43.** $-3364 \div (-29)$

44. $-4872 \div (-48)$ **45.** $3.735 \div (-0.83)$ **46.** $-2.352 \div (-0.42)$

47. $0 \div (-35)$ **48.** $-85 \div 0$ **49.** $-0.26 \div 100$

50. $-0.56 \div (-100)$ **51.** $\dfrac{-16,272}{36}$ **52.** $\dfrac{-34,083}{-63}$

53. Find the quotient of -384 and -24.   **54.** Find the quotient of -357 and 21.

55. The membership of the Burlap Baggers Investment Club takes a loss of \$753.90 $(-\$753.90)$ on the sale of stock. If there are six co-equal members in the club, what is each member's share of the loss, expressed as a signed number?

56. The temperature in Sitka, Alaska, drops from $10°$ above zero $(+10°)$ to $22°$ below zero $(-22°)$ in an 8-hour period. What was the average drop in temperature per hour, expressed as a signed number?

57. Mr. Harkness loses a total of 115 lb in 25 weeks. Express the average weekly loss as a signed number.

58. Ms. Harkness loses a total of 65 lb in 25 weeks. Express the average weekly loss as a signed number.

59. A certain stock loses 45.84 points in 12 days. Express the average daily loss as a signed number.

60. A certain stock loses 31.59 points in 9 days. Express the average daily loss as a signed number.

61. Determine the population of Los Angeles in 1995, 2000, and 2005. Determine the population of your city in 1995, 2000, and 2005. Find the average yearly loss or gain for each 5 years and also for the 10-year period for each city (written as a signed number). List the possible reasons for these changes.

© iStockphoto.com/Christopher Hudson

62. Central Electronics lost $967,140 during one 20-month period. Determine the average monthly loss (written as a signed number). If there are 30 stockholders in this company, determine the total loss per stockholder (written as a signed number).

Exercises 63–64 refer to the chapter application. See page 569.

63. Which continent has a high point that is approximately one-fourth of the height of Mt. Everest?

64. Which continent has a low point that is approximately one-fourth the low point of Africa?

65. Over the 16-month period from October 2007 to February 2009, the Standard & Poor 500 Index lost $814.29 (−$814.29) in value. What was the average monthly loss over this time?

66. In October 2007, the Dow Jones Industrial Average hit an all-time high of 13,930.01. During the financial crisis that followed, the Dow lost value until hitting a low of 7062.93 in February 2009. What was the average monthly loss the Dow over the 16-month period?

STATE YOUR UNDERSTANDING

67. The sign rules for multiplication and division of signed numbers may be summarized as follows:

If the numbers have the same sign, the answer is positive.

If the numbers have different signs, the answer is negative.

Explain why the rules for division are the same as the rules for multiplication.

68. When dividing signed numbers, care must be taken not to divide by zero. Why?

CHALLENGE

Simplify.

69. $[-|-10|(6-11)] \div [(8-13)(11-10)]$

70. $[(14-20)(-5-9)] \div [-(-12)(-8+7)]$

71. $\left(-\dfrac{5}{6} - \dfrac{1}{2}\right)\left(-\dfrac{2}{3} + \dfrac{1}{6}\right) \div \left(\dfrac{1}{3} - \dfrac{3}{4}\right)$

72. $\left(-\dfrac{1}{3} - \dfrac{1}{4}\right)\left(-\dfrac{1}{3} + \dfrac{1}{6}\right) \div \left(\dfrac{1}{3} - \dfrac{3}{4}\right)$

73. $(-0.82 - 1.28)(1.84 - 2.12) \div [3.14 + (-3.56)]$

MAINTAIN YOUR SKILLS

Simplify.

74. $16 \div 4 + 8 - 5$

75. $75 \div 3 \cdot 5 + 3 - (5 - 2)$

76. $14 - 3^2 + 8 - 3 \cdot 2 + 11$

77. $(17 - 3 \cdot 5)^3 + [16 - (19 - 2 \cdot 8)]$

78. Find the volume of a cylinder that has a radius of 8 in. and a height of 24 in. (Let $\pi \approx 3.14$.)

79. Find the volume of a cone that has a radius of 12 in. and a height of 9 in. (Let $\pi \approx 3.14$.)

80. An underground gasoline storage tank is a cylinder that is 72 in. in diameter and 18 ft long. If there are 231 in^3 in a gallon, how many gallons of gasoline will the tank hold? (Let $\pi \approx 3.14$.) Round the answer to the nearest gallon.

81. A swimming pool is to be dug and the dirt hauled away. The pool is to be 27 ft long, 16 ft wide, and 6 ft deep. How many cubic yards of dirt must be removed?

82. To remove the dirt from the swimming pool in Exercise 81, trucks that can haul 8 yd^3 per load are used. How many truckloads will there be?

83. A real-estate broker sells a lot that measures 88.75 ft by 180 ft. The sale price is $2 per square foot. If the broker's commission is 8%, how much does she make?

Order of Operations: A Review

SECTION 8.6

HOW & WHY

◼ **OBJECTIVE** Do any combination of operations with signed numbers.

The order of operations for signed numbers is the same as that for whole numbers, fractions, and decimals.

> **OBJECTIVE**
> Do any combination of operations with signed numbers.

> **To evaluate an expression with more than one operation**
>
> **Step 1. Parentheses**—Do the operations within grouping symbols first (parentheses, fraction bar, etc.), in the order given in steps 2, 3, and 4.
> **Step 2. Exponents**—Do the operations indicated by exponents.
> **Step 3. Multiply** and **Divide**—Do multiplication and division as they appear from left to right.
> **Step 4. Add** and **Subtract**—Do addition and subtraction as they appear from left to right.

EXAMPLES A–G

DIRECTIONS: Perform the indicated operations.

STRATEGY: Follow the order of operations.

WARM-UP

A. Perform the indicated operations:
$-56 + (-32) \div 4$

WARM-UP

B. Perform the indicated operations:
$(-8)(5) - 72 \div (-12)$

WARM-UP

C. Perform the indicated operations:
$9 \div (-0.15) + 6(-2.15)$

WARM-UP

D. Perform the indicated operations:
$15 - \left(\dfrac{7}{12}\right)(-36)$

WARM-UP

E. Perform the indicated operations:
$(-6)(-3)^2 + 43 - (-5)^2$

WARM-UP

F. Perform the indicated operations:
$(-28)(14) - 225 \div (-3)$

WARM-UP

G. How many degrees Celsius is $-4°F$?

A. Perform the indicated operations: $-63 + (-21) \div 7$

$$-63 + (-21) \div 7 = -63 + (-3) \quad \text{Divide first.}$$
$$= -66$$

B. Perform the indicated operations: $(-15)(-3) - 44 \div (-11)$

$$(-15)(-3) - 44 \div (-11) = 45 - (-4) \quad \text{Multiply and divide first.}$$
$$= 45 + 4 \quad \text{Add the opposite of } -4.$$
$$= 49$$

C. Perform the indicated operations: $12 \div (-0.16) + 3(-1.45)$

$$12 \div (-0.16) + 3(-1.45) = -75 + (-4.35) \quad \text{Multiply and divide first.}$$
$$= -79.35$$

D. Perform the indicated operations: $10 - \left(\dfrac{3}{5}\right)(-20)$

$$10 - \left(\dfrac{3}{5}\right)(-20) = 10 - (-12)$$
$$= 10 + 12 \quad \text{Add the opposite of } -12.$$
$$= 22$$

E. Perform the indicated operations: $3(-5)^2 - 3^2 + 7(-4)^2$

$$3(-5)^2 - 3^2 + 7(-4)^2 = 3(25) - 9 + 7(16) \quad \text{Do exponents first.}$$
$$= 75 - 9 + 112$$
$$= 178$$

CALCULATOR EXAMPLE:

F. Perform the indicated operations: $(-18)(23) - (-84) \div (-7)$

The result is -426.

G. Hilda keeps the thermostat on her furnace set at 68°F. Her pen pal in Germany says that her thermostat is set at 20°C. They wonder whether the two temperatures are equal.

STRATEGY: To find out whether 68°F = 20°C, substitute 68 for F in the formula.

$$C = \dfrac{5}{9}(F - 32)$$

$$C = \dfrac{5}{9}(68 - 32) \quad \text{Formula}$$

$$C = \dfrac{5}{9}(36)$$

$$C = 20$$

Therefore, 68°F equals 20°C.

■ **OBJECTIVE**   Do any combination of operations with signed numbers. (See page 603.)

A   *Perform the indicated operations.*

1. $2(-7) - 10$

2. $13 + 4(-5)$

3. $(-2)(-4) + 11$

4. $15 + (-3)(-5)$

5. $3(-6) + 12$

6. $(-5)(6) + 19$

7. $-7 + 3(-3)$

8. $-14 + (-6)2$

9. $(-3)8 \div 4$

10. $(-8)6 \div 3$

11. $(-4)8 \div (-4)$

12. $(-7)12 \div (-6)$

13. $(-8) \div 4(-2)$

14. $(-18) \div 3(2)$

15. $(-3)^2 + (-2)^2$

16. $6^2 - 4^2$

17. $(11 - 3) + (9 - 6)$

18. $(4 - 9) + (5 - 7)$

19. $(3 - 5)(6 - 10)$

20. $(8 - 5)(11 - 15)$

21. $(-3)^2 + 4(-2)$

22. $(-2)^2 - 4(-2)$

23. $-5 + (6 - 8) - 5(3)$

24. $-7 + (5 - 11) - 4(2)$

B

25. $(-13)(-2) + (-16)2$

26. $(-16)(-5) + (-14)5$

27. $(9 - 7)(-2 - 5) + (15 - 9)(2 + 7)$

28. $(10 - 15)(-4 - 3) + (12 - 7)(3 + 2)$

29. $7(-11 + 5) - 44 \div (-11)$

30. $8(-7 + 4) - 54 \div (-9)$

31. $18(-2) \div (-6) - 14$

32. $(-4)(-9) \div (-12) + 11$

33. $-120 \div (-20) - (9 - 11)$

34. $-135 \div (-15) - (12 - 17)$

35. $-2^3 - (-2)^3$

36. $-4^3 - 4^3$

37. $-35 \div 7(-5) - 7^2$

38. $-28 \div 7(-4) - 7^2$

39. $2^2(5 - 4)(7 - 3)^2$

40. $3^2(8 - 6)(6 - 8)^2$

41. $(9 - 13) + (-5)(-2) - (-2)5 - 3^3$

42. $(8 - 11) - (7)(-3) + (-4)(3) - 2^2$

43. $(-3)(-2)(-3) - (-4)(-3) - (3)(-5)$

44. $(-5)(-6)(-1) - (-3)(-6) - (-5)(2)$

45. $(-1)(-6)^2(-2) - (-3)^2(-2)^3$

46. $(-1)(-3)^3(-4) - (-4)^2(-2)^2$

C

47. Find the sum of the product of 12 and −4 and the product of −3 and −12.

48. Find the difference of the product of 3 and 9 and the product of −8 and 3.

Exercises 49–52 refer to the following table, which shows temperatures a satellite recorded during a 5-day period at one location on the surface of Mars.

Temperatures on the Surface of Mars

	Day 1	Day 2	Day 3	Day 4	Day 5
5:00 A.M.	−92°C	−88°C	−115°C	−103°C	−74°C
9:00 A.M.	−57°C	−49°C	−86°C	−93°C	−64°C
1:00 P.M.	−52°C	−33°C	−46°C	−48°C	−10°C
6:00 P.M.	−45°C	−90°C	−102°C	−36°C	−42°C
11:00 P.M.	−107°C	−105°C	−105°C	−98°C	−90°C

49. What is the average temperature recorded during day 5?

50. What is the average temperature recorded at 6:00 P.M.?

51. What is the average high temperature recorded for the 5 days?

52. What is the average low temperature recorded for the 5 days?

Simplify.

53. $[-3 + (-6)]^2 - [-8 - 2(-3)]^2$

54. $[-5(-9) - (-6)^2]^2 + [(-8)(-1)^3 + 2]^2$

55. $[46 - 3(-4)^2]^3 - [-7(1)^3 + (-5)(-8)]$

56. $[30 - (-5)^2]^2 - [-8(-2) - (-2)(-4)]^2$

57. $-15 - \dfrac{8^2 - (-4)}{3^2 + 3}$

58. $-22 + \dfrac{9^2 - 6}{6^2 - 11}$

59. $\dfrac{12(8 - 24)}{5^2 - 3^2} \div (-12)$

60. $\dfrac{15(12 - 45)}{6^2 - 5^2} \div (-9)$

61. $-8|125 - 321| - 21^2 + 8(-7)$

62. $-9|482 - 632| - 17^2 + 9(-9)$

63. $-6(8^2 - 9^2)^2 - (-7)20$

64. $-5(6^2 - 7^2)^2 - (-8)19$

65. Find the difference of the quotient of 28 and −7 and the product of −4 and −3.

66. Find the sum of the product of −3 and 7 and the quotient of −15 and −5.

67. Jason is starting his own company making wooden chairs. For the first six months of operation, his costs were $27,438, and he produced 102 chairs. He sold all of them at a craft show for $175 each. Calculate Jason's total profit for the six-month period, and his profit per chair.

© iStockphoto.com/Adam korzckwa

68. At the end of one full year of operation, Jason's chair company had costs of $34,459. He produced 215 chairs and sold all of them for $175 each. Calculate Jason's profit for the year and his profit per chair.

69. During an "early bird" special, K-Mart sold 25 fishing poles at a loss of $3.50 per pole. During the remainder of the day, they sold 9 poles at a profit of $7.25 per pole. What was the profit on the sale of the fishing poles for the day?

70. Fly America sells 40 seats on Flight 402 at a loss of $52 per seat ($-$52). Fly America also sells 67 seats at a profit of $78 per seat. Express the profit or loss on the sale of the seats as a signed number.

Exercises 71–74 refer to the chapter application. See page 569.

71. For each continent, calculate the average of the highest and lowest points. Which continent has the largest average, and which has the smallest average?

The following table gives the altitudes of selected cities around the world.

Altitudes of Cities

City	Altitude (ft)	City	Altitude (ft)
Athens, Greece	300	Mexico City, Mexico	7347
Bangkok, Thailand	0	New Delhi, India	770
Berlin, Germany	110	Quito, Ecuador	9222
Bogota, Columbia	8660	Rome, Italy	95
Jakarta, Indonesia	26	Tehran, Iran	5937
Jerusalem, Israel	2500	Tokyo, Japan	30

72. Find a group of five cities with an average altitude of less than 100 ft.

73. Find a group of three cities with an average altitude of approximately 350 ft.

74. Find a group of four cities with an average altitude of approximately 7000 ft.

75. The treasurer of a local club records the following transactions during one month:

Opening Balance	$4756
Deposit	$345
Check #34	$212
Check #35	$1218
New check cost	$15
Deposit	$98
National dues paid	$450
Electric bill	$78

What is the balance at the end of the month?

76. The Chicago Bears made the following plays during a quarter of a game:

3 plays lost 8 yd each
8 plays lost 5 yd each
1 quarterback sack lost 23 yd
1 pass for 85 yd
5 plays gained 3 yd each
2 plays gained 12 yd each
1 fumble lost 7 yd
2 passes for 10 yd each

Determine the average movement per play during this quarter. Round to the nearest tenth.

77. During 2009 Tiger Woods won seven tournaments. The tournaments are listed along with his final scores.

JBWere Masters	−14
BMW Championship	−19
WGC Bridgestone Invitational	−12
Buick Open	−20
AT&T National	−13
The Memorial Tournament	−12
Arnold Palmer Invitational	−5

What was his average winning score for the seven tournaments? Round to the nearest tenth.

78. Try this game on your friends. Have them pick a number. Tell them to double it, then add 20 to that number, divide the sum by 4, subtract 5 from that quotient, square the difference, and multiply the square by 4. They should now have the square of the original number. Write a mathematical representation of this riddle.

Locate the error in Exercises 79 and 80. Indicate why each is not correct. Determine the correct answer.

79. $2[3 + 5(-4)] = 2[8(-4)]$
$$= 2[-32]$$
$$= -64$$

80. $3 - [5 - 2(6 - 4^2)^3] = 3 - [5 - 2(6 - 16)^3]$
$$= 3 - [5 - 2(-10)^3]$$
$$= 3 - [5 - (-20)^3]$$
$$= 3 - [5 - (-8000)]$$
$$= 3 - [5 + 8000]$$
$$= 3 - 8005$$
$$= -8002$$

81. Is there ever a case when exponents are not computed first? If so give an example.

CHALLENGE

Simplify.

82. $\dfrac{3^2 - 5(-2)^2 + 8 + [4 - 3(-3)]}{4 - 3(-2)^3 - 18}$

83. $\dfrac{(5 - 9)^2 + (-6 + 8)^2 - (14 - 6)^2}{[3 - 4(7) + 3^3]^2}$

84. $\dfrac{3(4 - 7)^2 + 2(5 - 8)^3 - 18}{(6 - 9)^2 + 6}$

MAINTAIN YOUR SKILLS

Add.

85. $(-17.2) + (-18.6) + (-2.7) + 9.1$

86. $(28.31) + (-8.14) + (-21.26) + (-16)$

Subtract.

87. $48 - (-136)$

88. $-62.7 - (-78.8)$

Multiply.

89. $(-36)(84)(-21)$

90. $(-62)(-22)(-30)$

Divide.

91. $(-800) \div (-32)$

92. $(-25{,}781) \div 3.5$

93. The four Zapple brothers form a company. The first year, the company loses $5832 ($-$5832). The brothers share equally in the loss. Represent each brother's loss as a signed number.

94. AVI Biopharma stock recorded the following gains and losses for the week:

Monday   loss 0.34
Tuesday   loss 0.54
Wednesday   gain 1.32
Thursday   gain 0.67
Friday   loss 0.672

Use signed numbers to find out whether the stock gains or loses for the week.

SECTION

8.7 Solving Equations

OBJECTIVE

Solve equations of the form $ax + b = c$ or $ax - b = c$, where a, b, and c are signed numbers.

VOCABULARY

Recall that the **coefficient** of the variable is the number that is multiplied times the variable.

HOW & WHY

■ **OBJECTIVE**   Solve equations of the form $ax + b = c$ or $ax - b = c$, where a, b, and c are signed numbers.

Note: Before starting this section, you may want to review Getting Ready for Algebra sections in earlier chapters.

The process of solving equations that are of the form $ax + b = c$ and $ax - b = c$, using signed numbers, involves two operations to isolate the variable. To isolate the variable is to get an equation in which the variable is the only symbol on a particular side of the equation.

> **To find the solution of an equation of the form**
> $ax + b = c$ **or** $ax - b = c$
>
> 1. Add (subtract) the constant to (from) each side of the equation.
> 2. Divide both sides by the coefficient of the variable.

EXAMPLES A–C

DIRECTIONS: Solve.

STRATEGY: First, add or subtract the constant to or from both sides of the equation. Second, divide both sides of the equation by the coefficient of the variable.

A. Solve: $-6x + 23 = -7$

$$
\begin{aligned}
-6x + 23 &= -7 && \text{Original equation}\\
-6x + 23 - 23 &= -7 - 23 && \text{Subtract.}\\
-6x &= -30\\
\frac{-6x}{-6} &= \frac{-30}{-6} && \text{Divide.}\\
x &= 5
\end{aligned}
$$

CHECK: Substitute 5 for x in the original equation.

$$
\begin{aligned}
-6(5) + 23 &= -7\\
-30 + 23 &= -7\\
-7 &= -7
\end{aligned}
$$

The solution is $x = 5$.

B. Solve: $-43 = -14x - 71$

$$
\begin{aligned}
-43 &= -14x - 71 && \text{Original equation}\\
-43 + 71 &= -14x - 71 + 71 && \text{Add.}\\
28 &= -14x\\
\frac{28}{-14} &= \frac{-14x}{-14} && \text{Divide.}\\
-2 &= x
\end{aligned}
$$

CHECK: Substitute -2 in the original equation.

$$
\begin{aligned}
-43 &= -14(-2) - 71\\
-43 &= 28 - 71\\
-43 &= 28 + (-71)\\
-43 &= -43
\end{aligned}
$$

The solution is $x = -2$.

C. Solve: $7x - 32 = -88$

$$
\begin{aligned}
7x - 32 &= -88 && \text{Original equation}\\
\underline{+32 \quad\; +32} && \text{Add 32 to both sides.}\\
7x\phantom{{}-32} &= -56\\
\frac{7x}{7} &= \frac{-56}{7} && \text{Divide both sides by 7.}\\
x &= -8 && \text{The check is left for the student.}
\end{aligned}
$$

The solution is $x = -8$.

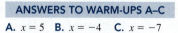

EXERCISES 8.7

■ **OBJECTIVE** Solve equations of the form $ax + b = c$ or $ax - b = c$, where a, b, and c are signed numbers. (See page 610.)

A *Solve.*

1. $-3x + 25 = 4$

2. $-4y + 11 = -9$

3. $-6 + 3x = 9$

4. $-11 + 5y = 14$

5. $4y - 9 = -29$

6. $3x - 13 = -43$

7. $2a - 11 = 3$

8. $5a + 17 = 17$

9. $-5x + 12 = -23$

10. $-11y - 32 = -65$

11. $4x - 12 = 28$

12. $9y - 14 = 4$

B

13. $-14 = 2x - 8$

14. $26 = 3x - 4$

15. $-40 = 5x - 10$

16. $-30 = -5x - 10$

17. $-6 = -8x - 6$

18. $9 = -5x + 9$

19. $-10 = -4x + 2$

20. $20 = -8x + 4$

21. $-14y - 1 = -99$

22. $-16x + 5 = -27$

23. $-3 = -8a - 3$

24. $-12 = 5b + 18$

C

25. $-0.6x - 0.15 = 0.15$

26. $-1.05y + 5.08 = 1.72$

27. $0.03x + 2.3 = 1.55$

28. $0.02x - 2.4 = 1.22$

29. $-135x - 674 = 1486$

30. $94y + 307 = -257$

31. $-102y + 6 = 414$

32. $-63c + 22 = 400$

33. $\dfrac{1}{2}a - \dfrac{3}{8} = \dfrac{1}{40}$

34. $-\dfrac{2}{3}x + \dfrac{1}{2} = \dfrac{3}{4}$

35. If 98 is added to 6 times some number, the sum is 266. What is the number?

36. If 73 is added to 11 times a number, the sum is -158. What is the number?

37. The difference of 15 times a number and 181 is -61. What is the number?

38. The difference of 24 times a number and 32 is -248. What is the number?

39. A formula for distance traveled is $2d = t^2a + 2v$, where d represents distance, v represents initial velocity, t represents time, and a represents acceleration. Find a if $d = 244$, $v = -20$, and $t = 4$.

40. Use the formula in Exercise 39 to find a if $d = 240$, $v = -35$, and $t = 5$.

41. The formula for the balance of a loan (D) is $D = B - NP$, where P represents the monthly payment, N represents the number of payments, and B represents the money borrowed. Find N when $D = \$575$, $B = \$925$, and $P = \$25$.

42. Use the formula in Exercise 41 to determine the monthly payment (P) if $D = \$820$, $B = \$1020$, and $N = 5$.

43. The formula $G = 78 - 6t$ gives G, the number of gallons of water remaining in a bathtub t minutes after the plug has been pulled. Use the formula to find out how many minutes it will take for the bathtub to have 30 gallons of water remaining.

44. The formula $T = P + Prt$ gives the total payback of loan, T, in dollars, for which P in dollars is the amount borrowed and simple interest is charged at the rate of $r\%$ per year for t years. Kyle has borrowed $2600 from his uncle. In two years time he paid off the loan with $2808. What interest rate did Kyle pay his uncle?

Exercises 45–47 refer to the chapter application. See page 569, Use negaive numbers to represent feet below sea level.

45. The high point of Australia is 12,558 ft more than 4 times the lowest point of one of the continents. Write an algebraic equation that describes this relationship. Which continent's lowest point fits the description?

46. The lowest point of Antarctica is 2183 ft less than 12 times the lowest point of one of the continents. Write an algebraic equation that describes this relationship. Which continent's lowest point fits this description?

© iStockphoto.com/David Cannings-Bushell

47. The Mariana Trench in the Pacific Ocean is about 2000 ft deeper than twice one of the other ocean's deepest parts. Write an algebraic equation that describes this relationship. Which ocean's deepest part fits this description? See Exercise 63, Section 8.1.

Solve.

48. $5x + 12 + (-9) = 18$

49. $8z - 12 + (-6) = 38$

50. $-3b - 12 + (-4) = 11 + (-6)$

51. $-5z - 15 + 6 = -21 - 18$

STATE YOUR UNDERSTANDING

52. Explain what it means to solve an equation.

53. Explain how to solve the equation $-3x + 10 = 4$.

CHALLENGE

Solve.

54. $8x - 9 = 3x + 6$

55. $7x + 14 = 3x - 2$

56. $9x + 16 = 7x - 12$

57. $10x + 16 = 5x + 6$

KEY CONCEPTS

SECTION 8.1 Opposites and Absolute Value

Definitions and Concepts	Examples
Positive numbers are greater than zero.	Positive numbers: $4, 7.31, \dfrac{5}{9}$
Negative numbers are written with a dash (−) and are less than zero.	Negative numbers: $-4, -7.31, -\dfrac{5}{9}$
The opposite of a signed number is the number that is the same distance from zero but has the opposite sign.	Opposites: 4 and -4, 7.31 and -7.31, $\dfrac{5}{9}$ and $-\dfrac{5}{9}$
The absolute value of a signed number is its distance from zero on a number line.	$\|-5\| = 5$ $\|5\| = 5$

SECTION 8.2 Adding Signed Numbers

Definitions and Concepts	Examples
To add signed numbers, • If the signs are alike, add their absolute values and use the common sign. • If the signs are unlike, subtract the smaller absolute value from the larger absolute value. The sum has the sign of the number with the larger absolute value.	$4 + 9 = 13$ $(-4) + (-9) = -13$ $-4 + 9 = 9 - 4 = 5$ $4 + (-9) = -(9 - 4) = -5$

SECTION 8.3 Subtracting Signed Numbers

Definitions and Concepts	Examples
To subtract signed numbers, • Rewrite as an addition problem by adding the opposite of the number to be subtracted. • Find the sum.	$5 - 8 = 5 + (-8) = -3$ $-5 - 8 = -5 + (-8) = -13$ $5 - (-8) = 5 + 8 = 13$

SECTION 8.4 Multiplying Signed Numbers

Definitions and Concepts	Examples
To multiply signed numbers, • Find the product of the absolute values. • If there is an even number of negative factors, the product is positive. • If there is an odd number of negative factors, the product is negative.	$3(-9) = -27$ $(-3)(-9) = 27$ $(-2)(-2)(-2)(-2) = 16$ $(-4)(-3)(-2) = -24$

SECTION 8.5 Dividing Signed Numbers

Definitions and Concepts	Examples
To divide signed numbers, • Find the quotient of the absolute values. • If the signs are alike, the quotient is positive. • If the signs are unlike, the quotient is negative.	$-12 \div 6 = -2$ $(-12) \div (-6) = 2$ $12 \div (-6) = -2$

SECTION 8.6 Order of Operations: A Review

Definitions and Concepts	Examples
The order of operations for signed numbers is the same as that for whole numbers: • Parentheses • Exponents • Multiplication/Division • Addition/Subtraction	$(-5)(-2)^3 + (-24) \div (6 - 8)$ $(-5)(-2)^3 + (-24) \div (-2)$ $(-5)(-8) + (-24) \div (-2)$ $40 + 12$ 52

SECTION 8.7 Solving Equations

Definitions and Concepts	Examples
To solve equations of the form $ax + b = c$ or $ax - b = c$, • Add (or subtract) the constant to (from) both sides. • Divide both sides by the coefficient of the variable.	$3x + 8 = -6$ $3x + 8 - 8 = -6 - 8$ $3x = -14$ $\dfrac{3x}{3} = -\dfrac{14}{3}$ $x = -\dfrac{14}{3}$

REVIEW EXERCISES

SECTION 8.1

Find the opposite of the signed number.

1. -39

2. 57

3. -0.91

4. -0.134

Find the absolute value of the signed number.

5. $|-16.5|$

6. $|-386|$

7. $|71|$

8. $|3.03|$

9. Find the opposite of $|-6.4|$.

10. If 93 miles north is represented by $+93$ miles, how would you represent 93 miles south?

SECTION 8.2

Add.

11. $-75 + (-23)$

12. $-75 + 23$

13. $75 + (-23)$

14. $65 + (-45) + (-82)$

15. $-7.8 + (-5.3) + 9.9$

16. $-24 + 65 + (-17) + 31$

17. $-6.8 + (-4.3) + 7.12 + 3.45$

18. $-\dfrac{7}{12} + \dfrac{7}{15} + \left(-\dfrac{7}{10}\right)$

19. The Chicago Bears made the following consecutive plays during a recent football game: 9-yd gain, 5-yd loss, 6-yd loss, and 12-yd gain. Did they get a first down?

20. Intel stock has the following changes in one week: up 0.78, down 1.34, down 2.78, up 3.12, and down 0.15. What is the net change for the week?

SECTION 8.3

Subtract.

21. $19 - (-3)$

22. $-45 - 81$

23. $16 - (-75)$

24. $-134 - (-134)$

25. $-4.56 - 3.25$

26. $-4.56 - (-3.25)$

27. Find the difference between -127 and -156.

28. Subtract -56 from -45.

29. At the beginning of the month, Maria's bank account had a balance of \$562.75. At the end of the month, the account was overdrawn by \$123.15 ($-\$123.15$). If there were no deposits during the month, what was the total amount of the checks Maria wrote?

30. Microsoft stock opened the day at 25.62 and closed the day at 24.82. Did the stock gain or lose for the day? Express the gain or loss as a signed number.

SECTION 8.4

Multiply.

31. $-6(11)$

32. $5(-28)$

33. $-1.2(3.4)$

34. $7.4(-5.1)$

35. $-3(-17)$

36. $-7(-21)$

37. $-4.03(-2.1)$

38. $(-1)(-4)(-6)(5)(-2)$

39. Kroger's promotes a gallon of milk as a loss leader. If Kroger loses 45¢ per gallon, what will be the total loss if they sell 632 gallons? Express the loss as a signed number.

40. Pedro owns 723.5 shares of Pfizer. If the stock loses \$0.32 ($-\$0.32$) a share, what is Pedro's loss expressed as a signed number?

SECTION 8.5

Divide.

41. $-18 \div 6$

42. $153 \div (-3)$

43. $-45 \div (-9)$

44. $-4.14 \div (-1.2)$

45. $-2448 \div 153$

46. $-8342 \div (-97)$

47. Find the quotient of -84.3 and 1.5.

48. Divide -712 by -32.

49. A share of UPS stock loses 15.5 points in 4 days. Express the average daily loss as a signed number.

50. Albertsons grocery store lost \$240.50 on the sale of Wheaties as a loss leader. If the store sold 650 boxes of the cereal during the sale, what is the loss per box, expressed as a signed number?

SECTION 8.6

Perform the indicated operations.

51. $5(-9) - 11$

52. $-3(17) + 45$

53. $-18 + (-4)(5) - 12$

54. $-84 \div 4(7)$

55. $-72 \div (-12)(3) + 6(-7)$

56. $(7 - 4)(4)(-3) + 6(7 - 9)$

57. $(-4)^2(-1)(-1) + (-6)(-3) - 17$

58. $(-1)(3^2)(-4) + 4(-5) - (-18 - 5)$

59. Find the difference of the quotient of 71 and -2.5 and the product of 3.2 and -2.4.

60. A local airline sells 65 seats for $324 each and 81 seats for $211 each. If the break-even point for the airline is $256 a seat, express the profit or loss for the airline for this flight as a signed number.

SECTION 8.7

Solve.

61. $7x + 25 = -10$

62. $-6x + 21 = -21$

63. $71 - 5x = -54$

64. $-55 = 3x + 41$

65. $-43 = 7x - 43$

66. $12y - 9 = -45$

67. $78a + 124 = -890$

68. $-55b + 241 = -144$

69. A formula for relating degrees Fahrenheit (F) and degrees Celsius (C) is $9C = 5F - 160$. Find the degrees Fahrenheit that is equal to $-22°C$.

70. Using the formula in Exercise 69, find the degrees Fahrenheit that is equal to $-8°C$.

TRUE/FALSE CONCEPT REVIEW

Check your understanding of the language of basic mathematics. Tell whether each of the following statements is true (always true) or false (not always true). For each statement you judge to be false, revise it to make a statement that is true.

Answers

1. Negative numbers are found to the left of zero on the number line.

1. _____

2. The opposite of a signed number is always positive.

2. _____

3. The absolute value of a number is always positive.

3. _____

4. The opposite of a signed number is the same distance from zero as the number on the number line but in the opposite direction.

4. _____

5. The sum of two signed numbers is always positive or negative.

5. _____

6. The sum of a positive signed number and a negative signed number is always positive.

6. _____

7. To find the sum of a positive signed number and a negative signed number, subtract their absolute values and use the sign of the number with the larger absolute value.

7. _____

8. To subtract two signed numbers, add their absolute values.

8. _____

9. If a negative number is subtracted from a positive number, the difference is always positive.

9. _____

10. The product of two negative numbers is never negative.

10. _____

11. The sign of the product of a positive number and a negative number depends on which number has the larger absolute value.

11. _____

12. The sign of the quotient, when dividing two signed numbers, is the same as the sign obtained when multiplying the two numbers.

12. _____

13. The order of operations for signed numbers is the same as the order of operations for positive numbers.

13. _____

14. Subtracting a number from both sides of an equation results in an equation that has the same solution as the original equation.

14. _____

TEST

Perform the indicated operations.

1. $-32 + (-19) + 39 + (-21)$

1. _____

2. $(45 - 52)(-16 + 21)$

2. _____

3. $\left(-\dfrac{3}{8}\right) \div \left(\dfrac{3}{10}\right)$

3. _____

4. $\left(-\dfrac{7}{15}\right) - \left(-\dfrac{3}{5}\right)$

4. _____

5. $(-11 - 5) - (5 - 22) + (-6)$

5. _____

6. $-5.78 + 6.93$

6. _____

7. **a.** $-(-17)$
   b. $|-33|$

7. _____

8. $-65 - (-32)$

8. _____

9. $(-18 + 6) \div 3 \cdot 4 - (-7)(-2)(-1)$

9. _____

10. $-110 \div (-55)$

10. _____

11. $(-6)(-8)(2)$

11. _____

12. $(|-7|)(-3)(-1)(-1)$

12. _____

13. $-63.2 - 45.7$

13. _____

14. $(-2)^2(-2)^2 + 4^2 \div (2)(3)$

14. _____

15. $21.84 \div (-0.7)$

15. _____

16. $\left(-\dfrac{1}{3}\right) + \dfrac{5}{6} + \left(-\dfrac{1}{2}\right) + \left(-\dfrac{1}{6}\right)$

16. _____

17. $-112 \div (-8)$

17. _____

18. $-56 - 24$

18. _____

19. $(-7)(4 - 13)(-2) - 5(-2 - 6)$

19. _____

20. $\left(-\dfrac{3}{8}\right)\left(\dfrac{12}{15}\right)$

20. _____

21. $-45 + (-23)$

21. _____

22. $6(-7) + 37$

22. _____

Solve.

23. $-16 = 5x + 14$

23. _____

24. $7x - 32 = 17$

24. _____

25. $24 - 6a = 45$

25. _____

26. Ms. Rosier lost an average of 1.05 lb per week (-1.05 lb) during her 16-week diet. Express Ms. Rosier's total weight loss during the 16 weeks as a signed number.

26. _____

27. The temperature in Chicago ranges from a high of 12°F to a low of -9°F within a 24-hr period. What is the drop in temperature, expressed as a signed number?

27. _____

28. A stock on the New York Stock Exchange opens at 17.65 on Monday. It records the following changes during the week: Monday, $+0.37$; Tuesday, -0.67; Wednesday, $+1.23$; Thursday, -0.87; Friday, $+0.26$. What is its closing price on Friday?

28. _____

29. What Fahrenheit temperature is equal to a reading of -10°C? Use the formula $F = \dfrac{9}{5}C + 32$.

29. _____

30. Find the average of -11, -15, 23, -19, 10, and -12.

30. _____

CLASS ACTIVITY 1

Meteorologists define the average temperature as the mean of the average high temperature and the average low temperature. Consider the following temperature data.

TABLE 1 Average Temperatures for Fairbanks, Alaska, in Degress Fahrenheit

	Jan.	Feb.	Mar.	Apr.	May	June	July	Aug.	Sept.	Oct.	Nov.	Dec.
Average Maximum	−1.6	7.2	23.8	41.0	59.3	70.1	72.3	66.3	54.8	32.0	10.9	1.8
Average Minimum	−18.5	−14.4	−1.7	20.4	38.0	49.5	52.6	47.2	36.2	18.1	−5.6	−14.8
Average												

SOURCE: climatezone.com

1. Calculate the average temperature for each month and enter them in Table 1.

2. What is the average temperature in Fairbanks over the entire year?

TABLE 2 Average Temperatures for Murmansk, Russia, in Degress Celsius

	Jan.	Feb.	Mar.	Apr.	May	June	July	Aug.	Sept.	Oct.	Nov.	Dec.
Average Maximum	−6	−8	−2	1	6	13	16	15	10	3	−3	−6
Average Minimum	−14	−17	−11	−6	0	5	8	6	3	−1	−10	−14
Average												

SOURCE: climatezone.com

3. Calculate the average temperature for each month and enter them in Table 2.

4. What is the average temperature in Murmansk over the entire year?

5. Which city has the colder average temperature for January?

6. Which city has the colder average temperature for the entire year?

CLASS ACTIVITY 2

Table 1 gives the change in price of Ford Motor Company stock during 2007 and 2008. The starting price of the stock was $7.51.

TABLE 1 Monthly Change in Value of Ford Motor Company Stock

1/07	2/07	3/07	4/07	5/07	6/07	7/07	8/07	9/07	10/07	11/07	12/07
+0.62	−0.22	−0.02	+0.15	+0.30	+1.08	−0.91	−0.70	+0.68	+0.38	−1.36	−0.78

1/08	2/08	3/08	4/08	5/08	6/08	7/08	8/08	9/08	10/08	11/08	12/08
−0.09	−0.11	−0.81	+2.54	−1.46	−1.99	−0.01	−0.34	+0.74	−3.01	+0.50	−0.40

1. There are two different methods of calculating the value of the Ford stock at the end of the 2007. Explain each method.

2. Calculate the price of Ford stock at the end of 2007, using both methods. Which method was easier?

3. Calculate the price of Ford stock at the end of 2008.

4. What was the average monthly change for 2008?

5. Multiply the average monthly change for 2008 by 12 and add this value to the beginning value for 2008. How does this compare to the ending value of 2008?

GROUP PROJECT (4 Weeks)

On three consecutive Mondays, locate the final scores for each of the three major professional golf tours in the United States: the Professional Golf Association, PGA; the Ladies Professional Golf Association, LPGA; and the Champions Tour. These scores can usually be found on the summary page in the sports section of the daily newspaper.

1. Record the scores, against par, for the 30 top finishers and ties on each tour. Display the data using bar graphs for week 1, line graphs for week 2, and bar graphs for week 3. Which type of graph best displays the data?

2. Calculate the average score, against par, for each tour for each week. When finding the average, if there is a remainder and it is half of or more than the divisor, round up; otherwise, round down. Now average the average scores for each tour. Which tour scored the best? Why?

3. What is the difference between the best and worst scores on each tour for each week?

4. What is the average amount of money earned by the player whose scores were recorded on each tour for each week? Which tour pays the best?

5. How much did the winner on each tour earn per stroke under par in the second week of your data? Compare the results. Is this a good way to compare the earnings on the tour? If not, why not?

MIDTERM EXAMINATION

CHAPTERS 1–4

Answers

1. Write the place value of the digit 8 in 389,440.

1. _____

2. Add:
   ```
        289
       4675
         52
     78,612
   +    555
   ```

2. _____

3. Write the word name for 67,509.

3. _____

4. Subtract: 78,329
             −69,543

4. _____

5. Add: $703 + 25{,}772 + 1098 + 32$

5. _____

6. Multiply: $(367)(95)$

6. _____

7. Estimate the product and multiply: 803
                                      $\times 906$

7. _____

8. Divide: $347\overline{)73{,}911}$

8. _____

9. Divide: $63\overline{)45{,}981}$

9. _____

10. Perform the indicated operations: $19 - 3 \cdot 4 + 18 \div 3$

10. _____

11. Find the sum of the quotient of 72 and 9 and the product of 33 and 2.

11. _____

12. Find the average of 245, 175, 893, 660, and 452.

12. _____

13. Find the average, median, and mode of 52, 64, 64, 97, 128, 97, 82, and 64.

13. _____

14. The graph shows the number of Honda vehicles sold at a local dealership by model.
    a. Which model has the highest sales?
    b. How many more Accords are sold than Passports?

14. _____

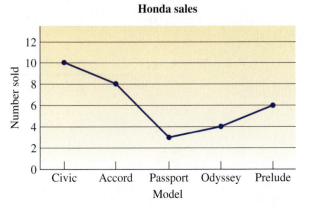

Honda sales

_____

15. Write the least common multiple (LCM) of 30, 24, and 40.

15. _____

16. Is 1785 divisible by 2, 3, or 5?

16. _____

17. Is 107 a prime number or a composite number?

17. _____

18. Is 7263 a multiple of 9?

18. _____

19. List the first five multiples of 32.

19. _____

20. List all the factors of 304.

20. _____

21. Write the prime factorization of 540.

21. _____

22. Change to a mixed number: $\dfrac{83}{7}$

22. _____

23. Change to an improper fraction: $9\dfrac{3}{7}$

23. _____

24. Which of these fractions are proper? $\dfrac{3}{4}, \dfrac{7}{6}, \dfrac{8}{8}, \dfrac{7}{8}, \dfrac{9}{8}, \dfrac{5}{4}, \dfrac{4}{4}$

24. _____

25. List these fractions from the smallest to the largest: $\dfrac{5}{9}, \dfrac{5}{8}, \dfrac{7}{12}, \dfrac{2}{3}$

25. _____

26. Simplify: $\dfrac{114}{150}$

26. _____

27. Multiply and simplify: $\dfrac{3}{16} \cdot \dfrac{4}{9} \cdot \dfrac{5}{6}$

27. _____

28. Multiply. Write the answer as a mixed number. $\left(3\dfrac{3}{5}\right)\left(12\dfrac{6}{7}\right)$

28. _____

29. Divide and simplify: $\dfrac{45}{70} \div \dfrac{9}{14}$

29. _____

30. What is the reciprocal of $3\dfrac{4}{9}$?

30. _____

31. Add: $\dfrac{7}{15} + \dfrac{11}{18}$

31. _____

32. Add: $\begin{aligned} & 6\dfrac{2}{3} \\ + & 11\dfrac{7}{8} \\ \hline \end{aligned}$

32. _____

33. Subtract: $23 - 11\dfrac{4}{7}$

33. _____

34. Subtract: $\quad 54\dfrac{1}{9}$

$\qquad -36\dfrac{4}{7}$

34. _____

35. Find the average of $4\dfrac{5}{6}, 6\dfrac{2}{3}$, and $11\dfrac{3}{4}$.

35. _____

36. Perform the indicated operations: $\dfrac{3}{5} - \dfrac{1}{3} \cdot \dfrac{5}{6} \div \dfrac{5}{6}$

36. _____

37. Write the place value name for eighty thousand two hundred forty and one hundred twenty-two thousandths.

37. _____

38. Round 7.95863 to the nearest tenth, hundredth, and thousandth.

38. _____

39. Write 0.26 as a simplified fraction.

39. _____

40. List the following numbers from smallest to largest: 1.034, 1.109, 1.044, 1.094, 1.02, 1.07

40. _____

41. Add: $134.76 + 7.113 + 0.094 + 5.923 + 25.87$

41. _____

42. Subtract: $\quad 8.3$

$\qquad -5.763$

42. _____

43. Multiply: $\quad 42.765$

$\qquad \times \quad 8.34$

43. _____

44. Multiply: $0.046 \times 100{,}000$

44. _____

45. Divide: $0.902 \div 1{,}000$

45. _____

46. Write 0.00058 in scientific notation.

46. _____

47. Divide: $7.85\overline{)445.88}$

47. _____

48. Change $\dfrac{15}{23}$ to a decimal rounded to the nearest thousandth.

48. _____

49. Find the average and median of 4.5, 6.9, 8.3, 9.5, 3.1, 10.3, and 2.2.

49. _____

50. Perform the indicated operations: $(5.5)^2 - 2.3(4.1) + 11.8 - 9.3 \div 0.3$

50. _____

51. Find the perimeter of the trapezoid.

51. _____

18 cm

7 cm   10 cm

25 cm

52. Find the area of a rectangle that is $7\frac{3}{8}$ ft wide and $10\frac{2}{5}$ ft long.

52. _____

53. On Thursday, October 4, the following counts of chinook salmon going over the dams were recorded: Bonneville, 1577; The Dalles, 1589; John Day, 1854; McNary, 1361; Ice Harbor, 295; and Little Goose, 274. How many chinook salmon were counted?

53. _____

54. Recently General Systems declared a stock dividend of 14¢ a share. Maria owns 765.72 shares of General Systems. To the nearest cent, what was Maria's dividend?

54. _____

55. Houng paid $5.79 for 6.5 pound of navel oranges. What is the price per pound of the navel oranges?

55. _____

FINAL EXAMINATION

CHAPTERS 1–8

1. Add: $\dfrac{7}{12} + \dfrac{7}{15}$

1. _____

2. Write the LCM (least common multiple) of 15, 18, and 35.

2. _____

3. Subtract: $\begin{array}{r} 93.42 \\ -57.69 \end{array}$

3. _____

4. Add: $78.32 + 4.089 + 0.139$

4. _____

5. Divide: $\dfrac{15}{16} \div \dfrac{5}{8}$

5. _____

6. Multiply: $3\dfrac{5}{7} \cdot 35$

6. _____

7. Which of these numbers is a prime number? 99, 199, 299, 699

7. _____

8. Divide. Round the answer to the nearest hundredth. $0.62\overline{)23.764}$

8. _____

9. Subtract: $21 - 13\dfrac{11}{15}$

9. _____

10. Multiply: $(0.0945)(10{,}000)$

10. _____

11. Round to the nearest hundredth: 314.9278

11. _____

12. Multiply: $(11.6)(4.07)$

12. _____

13. Write as a fraction and simplify: 36%

13. _____

14. Add: $\begin{array}{r} 9\dfrac{11}{15} \\[2mm] + \dfrac{5}{6} \\ \hline \end{array}$

14. _____

15. Solve the proportion: $\dfrac{5}{16} = \dfrac{x}{24}$

15. _____

16. What is the place value of the 6 in 23.11567?

16. _____

17. List these fractions from the smallest to the largest: $\dfrac{3}{4}, \dfrac{4}{5}, \dfrac{7}{10}$

17. _____

18. Change to percent: $\dfrac{23}{25}$

18. _____

19. Divide. Round to the nearest thousandth. $46\overline{)908}$

19. _____

20. Write as an approximate decimal to the nearest thousandth: $\dfrac{17}{33}$

20. _____

21. Write the place value name for seventy-two thousand and five hundredths.

21. _____

22. Divide: $0.043\overline{)0.34787}$

22. _____

23. Write as a decimal: $57\dfrac{5}{8}\%$

23. _____

24. Thirty-nine percent of what number is 19.5?

24. _____

25. Write as a percent: 0.0935

25. _____

26. Write as a decimal: $\dfrac{71}{250}$

26. _____

27. Seventy-three percent of 82 is what number?

27. _____

28. Multiply: $(0.26)(4.5)(0.55)$

28. _____

29. If 7 lb of blueberries cost \$24.78, how much would 18 lb cost?

29. _____

30. An iPod is priced at \$345. It is on sale for \$280. What is the percent of discount based on the original price? Round to the nearest tenth of a percent.

30. _____

31. List the first five multiples of 51.

31. _____

32. Write the word name for 8037.037.

32. _____

33. Is the following proportion true or false?

$\dfrac{1.9}{22} = \dfrac{5.8}{59}$

33. _____

34. Write the prime factorization of 750.

34. _____

35. Simplify: $\dfrac{315}{450}$

35. _____

36. Change to a fraction and simplify: 0.945

36. _____

37. Change to a mixed number: $\dfrac{436}{9}$

37. _____

38. Multiply and simplify: $\dfrac{6}{35} \cdot \dfrac{42}{54}$

38. _____

39. Change to a fraction: 0.145

40. A survey at a McDonald's showed that 27 of 50 customers asked for a Big Mac. What percent of the customers wanted a Big Mac?

40. _____

41. Is 2546 a multiple of 3?

41. _____

42. Subtract: $14\dfrac{3}{5} - \dfrac{11}{15}$

42. _____

43. Change to an improper fraction: $13\dfrac{7}{18}$

43. _____

44. List the following decimals from smallest to largest: 2.32, 2.332, 2.299, 2.322

44. _____

45. List all the divisors of 408.

45. _____

46. Divide: $45.893 \div 10^5$

46. _____

47. Divide: $4\dfrac{7}{12} \div 1\dfrac{9}{16}$

47. _____

48. Write a ratio in fraction form to compare 85¢ to $5 (using common units) and simplify.

48. _____

49. Mildred calculates that she pays $0.67 for gas and oil to drive 5 miles. In addition, she pays 30¢ for maintenance for each 5 miles she travels. How much will it cost her to drive 7500 miles?

49. _____

50. Billy works for a large furniture manufacturer. He earns a fixed salary of $1500 per month plus a 3.5% commission on all sales. What does Billy earn in a month in which his sales are $467,800?

50. _____

51. The sales tax on a $55 purchase is $4.75. What is the sales tax rate, to the nearest tenth of a percent?

51. _____

52. Melissa buys a new vacuum cleaner that is on sale at 30% off the original price. If the original price is $245.50 and there is a 6.5% sales tax, what is the final cost of the vacuum cleaner?

52. _____

53. Perform the indicated operations: $82 - 7 \cdot 6 + 45 \div 5$

53. _____

54. Perform the indicated operations: $79.15 - 5.1(8.3) \div 3.4$

54. _____

55. Add: 5 hr 47 min 32 sec
          +2 hr 36 min 48 sec

55. _____

56. Subtract: 9 m 45 cm
              −5 m 72 cm

56. _____

57. Convert 5¢ per gram to dollars per kilogram.

58. Find the perimeter of a trapezoid with bases of 63.7 ft and 74.2 ft and sides of 21.5 ft and 23.6 ft.

59. Find the area of a triangle with a base of 9.3 m and a height of 7.2 m.

60. Find the area of the following geometric figure (let $\pi \approx 3.14$):

6 yd

4 yd

12 yd

61. Find the volume of a box with length of 4.6 ft, width of 7 in., and height of 5 in. (in cubic inches).

62. Find the square root of 578.4 to the nearest hundredth.

63. Find the hypotenuse of a right triangle with legs of 23 cm and 27 cm. Find to the nearest tenth of a centimeter.

64. Add: $(-34) + (-23) + 41$

65. Subtract: $(-73) - (-94)$

66. Multiply: $(5)(-7)(-1)(-3)$

67. Divide: $(-46.5) \div (-15)$

68. Perform the indicated operations: $(-6 - 4)(-4) \div (-5) - (-10)$

69. Solve: $15a + 108 = 33$

70. Find the Fahrenheit temperature that is equivalent to $-15°C$. Use the formula, $9C = 5F - 160$.

CALCULATORS

The wide availability and economical price of current hand-held calculators make them ideal for doing time-consuming arithmetic operations. Even people who are very good at math use calculators under certain circumstances (for instance, when balancing their checkbooks). You are encouraged to use a calculator as you work through this text. Learning the proper and appropriate use of a calculator is a vital skill for today's math students.

As with all new skills, your instructor will give you guidance as to where and when to use it. Calculators are especially useful in the following instances:

- For doing the fundamental operations of arithmetic (addition, subtraction, multiplication, and division)
- For finding powers or square roots of numbers
- For evaluating complicated arithmetic expressions
- For checking solutions to equations

Several different kinds of calculators are available.

- A basic 4-function calculator will add, subtract, multiply and divide. Sometimes these calculators also have a square root key. These calculators are not powerful enough to do all of the math in this text, and they are not recommended for math students at this level.
- A scientific calculator generally has about eight rows of keys on it and is usually labeled "scientific." Look for keys labeled "sin," "tan," and "log." Scientific calculators also have power keys and parenthesis keys, and the order of operations is built into them. These calculators are recommended for math students at this level.
- A graphing calculator also has about eight rows of keys, but it has a large, nearly square display screen. These calculators are very powerful, and you may be required to purchase them in later math courses. However, you will not need all that power to be successful in this course, and they are significantly more expensive than scientific calculators.

We will assume that you are operating a scientific calculator. (Some of the keystrokes are different on graphing calculators, so if you are using one of these calculators, please consult your owner's manual.) Study the following table to discover how the basic keys are used.

Expression	Key Strokes	Display
$144 \div 3 - 7$	[144] [÷] [3] [−] [7] [=]	41.
$3(2) + 4(5)$	[3] [×] [2] [+] [4] [×] [5] [=]	26.
$13^2 - 2(12 + 10)$	[13] [x^2] [−] [2] [×] [(] [12] [+] [10] [)] [=]	125.
$\dfrac{28 + 42}{10}$	[(] [28] [+] [42] [)] [÷] [10] [=]	7.
	or	
	[28] [+] [42] [=] [÷] [10] [=]	7.
$\dfrac{288}{6 + 12}$	[288] [÷] [(] [6] [+] [12] [)] [=]	16.
$19^2 - 3^5$	[19] [x^2] [−] [3] [x^y] [5] [=]	118.
$2\dfrac{1}{3} + \dfrac{5}{6}$	[2] [$a^b/_c$] [1] [$a^b/_c$] [3] [+] [5] [$a^b/_c$] [6] [=]	3 ⌐ 1 ⌐ 6

Notice that the calculator does calculations when you hit the $\boxed{=}$ or ENTER key. The calculator automatically uses the order of operations when you enter more than one operation before hitting $\boxed{=}$ or ENTER. Notice that if you begin a sequence with an operation sign, the calculator automatically uses the number currently displayed as part of the calculation. There are three operations that require only one number: squaring a number, square rooting a number, and taking the opposite of a number. In each case, enter the number first and then hit the appropriate operation key. Be especially careful with fractions. Remember that when there is addition or subtraction inside a fraction, the fraction bar acts as a grouping symbol. But the only way to convey this to your calculator is by using the grouping symbols $\boxed{(}$ and $\boxed{)}$. Notice that the fraction key is used between the numerator and denominator of a fraction and also between the whole number and fractional part of a mixed number. It automatically calculates the common denominator when necessary.

Model Problem Solving

Practice the following problems until you can get the results shown.

Answers

a. $47 + \dfrac{525}{105}$ 52

b. $\dfrac{45 + 525}{38}$ 15

c. $\dfrac{648}{17 + 15}$ 20.25

d. $\dfrac{140 - 5(6)}{11}$ 10

e. $\dfrac{3870}{9(7) + 23}$ 45

f. $\dfrac{5(73) + 130}{33}$ 15

g. $100 - 2^5$ 68

h. $100 - (-2)^5$ 132

i. $4\dfrac{2}{7} - 3\dfrac{3}{5}$ $\dfrac{24}{35}$

PRIME FACTORS OF NUMBERS 1 THROUGH 100

	Prime Factors		Prime Factors		Prime Factors		Prime Factors
1	none	26	$2 \cdot 13$	51	$3 \cdot 17$	76	$2^2 \cdot 19$
2	2	27	3^3	52	$2^2 \cdot 13$	77	$7 \cdot 11$
3	3	28	$2^2 \cdot 7$	53	53	78	$2 \cdot 3 \cdot 13$
4	2^2	29	29	54	$2 \cdot 3^3$	79	79
5	5	30	$2 \cdot 3 \cdot 5$	55	$5 \cdot 11$	80	$2^4 \cdot 5$
6	$2 \cdot 3$	31	31	56	$2^3 \cdot 7$	81	3^4
7	7	32	2^5	57	$3 \cdot 19$	82	$2 \cdot 41$
8	2^3	33	$3 \cdot 11$	58	$2 \cdot 29$	83	83
9	3^2	34	$2 \cdot 17$	59	59	84	$2^2 \cdot 3 \cdot 7$
10	$2 \cdot 5$	35	$5 \cdot 7$	60	$2^2 \cdot 3 \cdot 5$	85	$5 \cdot 17$
11	11	36	$2^2 \cdot 3^2$	61	61	86	$2 \cdot 43$
12	$2^2 \cdot 3$	37	37	62	$2 \cdot 31$	87	$3 \cdot 29$
13	13	38	$2 \cdot 19$	63	$3^2 \cdot 7$	88	$2^3 \cdot 11$
14	$2 \cdot 7$	39	$3 \cdot 13$	64	2^6	89	89
15	$3 \cdot 5$	40	$2^3 \cdot 5$	65	$5 \cdot 13$	90	$2 \cdot 3^2 \cdot 5$
16	2^4	41	41	66	$2 \cdot 3 \cdot 11$	91	$7 \cdot 13$
17	17	42	$2 \cdot 3 \cdot 7$	67	67	92	$2^2 \cdot 23$
18	$2 \cdot 3^2$	43	43	68	$2^2 \cdot 17$	93	$3 \cdot 31$
19	19	44	$2^2 \cdot 11$	69	$3 \cdot 23$	94	$2 \cdot 47$
20	$2^2 \cdot 5$	45	$3^2 \cdot 5$	70	$2 \cdot 5 \cdot 7$	95	$5 \cdot 19$
21	$3 \cdot 7$	46	$2 \cdot 23$	71	71	96	$2^5 \cdot 3$
22	$2 \cdot 11$	47	47	72	$2^3 \cdot 3^2$	97	97
23	23	48	$2^4 \cdot 3$	73	73	98	$2 \cdot 7^2$
24	$2^3 \cdot 3$	49	7^2	74	$2 \cdot 37$	99	$3^2 \cdot 11$
25	5^2	50	$2 \cdot 5^2$	75	$3 \cdot 5^2$	100	$2^2 \cdot 5^2$

SQUARES AND SQUARE ROOTS (0 TO 199)

n	n^2	$\sqrt{n}$	n	n^2	$\sqrt{n}$	n	n^2	$\sqrt{n}$	n	n^2	$\sqrt{n}$
0	0	0.000	50	2,500	7.071	100	10,000	10.000	150	22,500	12.247
1	1	1.000	51	2,601	7.141	101	10,201	10.050	151	22,801	12.288
2	4	1.414	52	2,704	7.211	102	10,404	10.100	152	23,104	12.329
3	9	1.732	53	2,809	7.280	103	10,609	10.149	153	23,409	12.369
4	16	2.000	54	2,916	7.348	104	10,816	10.198	154	23,716	12.410
5	25	2.236	55	3,025	7.416	105	11,025	10.247	155	24,025	12.450
6	36	2.449	56	3,136	7.483	106	11,236	10.296	156	24,336	12.490
7	49	2.646	57	3,249	7.550	107	11,449	10.344	157	24,649	12.530
8	64	2.828	58	3,346	7.616	108	11,664	10.392	158	24,964	12.570
9	81	3.000	59	3,481	7.681	109	11,881	10.440	159	25,281	12.610
10	100	3.162	60	3,600	7.746	110	12,100	10.488	160	25,600	12.649
11	121	3.317	61	3,721	7.810	111	12,321	10.536	161	25,921	12.689
12	144	3.464	62	3,844	7.874	112	12,544	10.583	162	26,244	12.728
13	169	3.606	63	3,969	7.937	113	12,769	10.630	163	26,569	12.767
14	196	3.742	64	4,096	8.000	114	12,996	10.677	164	26,896	12.806
15	225	3.873	65	4,225	8.062	115	13,225	10.724	165	27,225	12.845
16	256	4.000	66	4,356	8.124	116	13,456	10.770	166	27,556	12.884
17	289	4.123	67	4,489	8.185	117	13,689	10.817	167	27,889	12.923
18	324	4.243	68	4,624	8.246	118	13,924	10.863	168	28,224	12.961
19	361	4.359	69	4,761	8.307	119	14,161	10.909	169	28,561	13.000
20	400	4.472	70	4,900	8.367	120	14,400	10.954	170	28,900	13.038
21	441	4.583	71	5,041	8.426	121	14,641	11.000	171	29,241	13.077
22	484	4.690	72	5,184	8.485	122	14,884	11.045	172	29,584	13.115
23	529	4.796	73	5,329	8.544	123	15,129	11.091	173	29,929	13.153
24	576	4.899	74	5,476	8.602	124	15,376	11.136	174	30,276	13.191
25	625	5.000	75	5,625	8.660	125	15,625	11.180	175	30,625	13.229
26	676	5.099	76	5,776	8.718	126	15,876	11.225	176	30,976	13.266
27	729	5.196	77	5,929	8.775	127	16,129	11.269	177	31,329	13.304
28	784	5.292	78	6,084	8.832	128	16,384	11.314	178	31,684	13.342
29	841	5.385	79	6,241	8.888	129	16,641	11.358	179	32,041	13.379
30	900	5.477	80	6,400	8.944	130	16,900	11.402	180	32,400	13.416
31	961	5.568	81	6,561	9.000	131	17,161	11.446	181	32,761	13.454
32	1,024	5.657	82	6,724	9.055	132	17,424	11.489	182	33,124	13.491
33	1,089	5.745	83	6,889	9.110	133	17,689	11.533	183	33,489	13.528
34	1,156	5.831	84	7,056	9.165	134	17,956	11.576	184	33,856	13.565
35	1,225	5.916	85	7,225	9.220	135	18,225	11.619	185	34,225	13.601
36	1,296	6.000	86	7,396	9.274	136	18,496	11.662	186	34,596	13.638
37	1,369	6.083	87	7,569	9.327	137	18,769	11.705	187	34,969	13.675
38	1,444	6.164	88	7,744	9.381	138	19,044	11.747	188	35,344	13.711
39	1,521	6.245	89	7,921	9.434	139	19,321	11.790	189	35,721	13.748
40	1,600	6.325	90	8,100	9.487	140	19,600	11.832	190	36,100	13.784
41	1,681	6.403	91	8,281	9.539	141	19,881	11.874	191	36,481	13.820
42	1,764	6.481	92	8,464	9.592	142	20,164	11.916	192	36,864	13.856
43	1,849	6.557	93	8,649	9.644	143	20,449	11.958	193	37,249	13.892
44	1,936	6.633	94	8,836	9.659	144	20,736	12.000	194	37,636	13.928
45	2,025	6.708	95	9,025	9.747	145	21,025	12.042	195	38,025	13.964
46	2,116	6.782	96	9,216	9.798	146	21,316	12.083	196	38,416	14.000
47	2,209	6.856	97	9,409	9.849	147	21,609	12.124	197	38,809	14.036
48	2,304	6.928	98	9,604	9.899	148	21,904	12.166	198	39,204	14.071
49	2,401	7.000	99	9,801	9.950	149	22,201	12.207	199	39,601	14.107
n	n^2	$\sqrt{n}$	n	n^2	$\sqrt{n}$	n	n^2	$\sqrt{n}$	n	n^2	$\sqrt{n}$

COMPOUND INTEREST TABLE (FACTORS)

	Years					
	1	**5**	**10**	**15**	**20**	**25**
2%						
Quarterly	1.0202	1.1049	1.2208	1.3489	1.4903	1.6467
Monthly	1.0202	1.1051	1.2212	1.3495	1.4913	1.6480
Daily	1.0202	1.1052	1.2214	1.3498	1.4918	1.6487
3%						
Quarterly	1.0303	1.1612	1.3483	1.5657	1.8180	2.1111
Monthly	1.0304	1.1616	1.3494	1.5674	1.8208	2.1150
Daily	1.0305	1.1618	1.3498	1.5683	1.8221	2.1169
4%						
Quarterly	1.0406	1.2202	1.4889	1.8167	2.2167	2.7048
Monthly	1.0407	1.2210	1.4908	1.8203	2.2226	2.7138
Daily	1.0408	1.2214	1.4918	1.8221	2.2254	2.7181
5%						
Quarterly	1.0509	1.2820	1.6436	2.1072	2.7015	3.4634
Monthly	1.0512	1.2834	1.6470	2.1137	2.7126	3.4813
Daily	1.0513	1.2840	1.6487	2.1169	2.7181	3.4900
6%						
Quarterly	1.0614	1.3469	1.8140	2.4432	3.2907	4.4320
Monthly	1.0617	1.3489	1.8194	2.4541	3.3102	4.4650
Daily	1.0618	1.3498	1.8220	2.4594	3.3198	4.4811
7%						
Quarterly	1.0719	1.4148	2.0016	2.8318	4.0064	5.6682
Monthly	1.0723	1.4176	2.0097	2.8489	4.0387	5.7254
Daily	1.0725	1.4190	2.0136	2.8574	4.0547	5.7536

COMPOUND INTEREST TABLE (FACTORS)

GLOSSARY

Absolute value The number of units between a signed number and zero on the number line

Addends Numbers that are added together

Additive inverse A number's additive inverse is the number on the number line that is the same distance from zero but on the opposite side of it. Also called the opposite.

Amount In the percent formula, $R \times B = A$, A is the amount that is compared to B.

Amount of discount During a sale, the amount subtracted from the original (regular) price

Approximate decimal A decimal that represents a rounded value

Area A measure of the amount of surface or space inside a two-dimensional closed figure

Average The sum of a set of numbers divided by the number of numbers in the set. Also called the mean.

Bar graph A graph that uses solid lines or bars of fixed length to represent data

Base in an exponential expression The number used as a repeated factor

Base of a polygon Any side of the polygon, usually a side parallel to the horizon

Base unit of percent In the percent formula, $R \times B = A$, B is the base.

Braces The grouping symbols { }

Brackets The grouping symbols [ ]

Capacity The amount of liquid (volume) in a container

Celsius scale The metric system scale used for measuring temperature

Circle graph A graph used to show a whole unit divided into parts. Also called a pie chart.

Circumference The distance around a circle

Coefficient A number that is multiplied times a variable

Column A vertical line of a table that reads up or down the page

Commission The money salespeople earn based on the dollar value of goods sold

Common denominator Two or more fractions have a common denominator when they have the same denominator

Composite number A whole number greater than 1 with more than two factors (divisors)

Compound interest Interest that is computed on interest already earned.

Counting numbers The numbers 1, 2, 3, and so on. Also called the natural numbers

Cross multiplication In a proportion, multiplying the numerator of each ratio times the denominator of the other

Cross products The products obtained from cross multiplication

Cube A three-dimensional geometric solid with six sides (faces), each of which is a square

Cube of a number The number raised to the third power

Decimal number A number formed using digits and a decimal point

Decimal point The period used in a decimal number to indicate the place values of the digits in the number

Denominator The lower numeral in a fraction

Diameter A line segment from one point on a circle to another point on the circle and that passes through the center

Difference The result of subtracting two numbers

Digits The whole numbers from 0 through 9

Discount The difference between the marked price and the sale price

Discount rate The percent of the original price that is subtracted to get the discount amount. Also called percent of discount.

Dividend In a division problem, the number being divided

Divisible A whole number is divisible by another whole number if the quotient is a whole number and the remainder is zero.

Divisor In a division problem, the number we are dividing by

English system A measurement system commonly used in the United States

Equation A statement that says that two expressions are equal

Equivalent fractions Fractions that are different names for the same number

Equivalent measurements Measures of the same amount but using different units

Even digits The digits 0, 2, 4, 6, 8

Exact decimal A decimal that shows an exact value

Exponent A number, written as a superscript, that indicates the number of times the base is used as a factor

Exponential Property of One If 1 is used as an exponent, the value is equal to the base.

Factors In a multiplication problem, the numbers or expressions being multiplied

Fahrenheit scale The English system scale used for measuring temperature

Fraction A number written using two numbers separated by a fraction bar

Fraction bar The line that separates the numerator and the denominator in a fraction

Front rounding A method of rounding to the highest place value so that all digits become zero except the first digit

Graph An illustration used to display numerical information

Greater than A symbol (>) used to denote that one number is larger than another

Grouping symbols Symbols that indicate operations inside them are to be performed first. Grouping symbols include parentheses, brackets, braces, and fraction bars.

Height For a geometric figure, the perpendicular (shortest) distance from the base to the highest point on the figure

Horizontal scale On a bar or line graph, the values along the horizontal axis

Hypotenuse In a right triangle, the side opposite the right angle

Improper fraction A fraction in which the numerator is equal to or greater than the denominator

Interest A fee charged for borrowing money or an amount paid by the borrower to use another's money or the money paid to use your money.

Interest rate A percent charged or paid on money borrowed

Least common multiple (LCM) For two or more whole numbers, the smallest whole number that is a multiple of each number

Legs In a right triangle, the sides adjacent to the right angle

Less than A symbol (<) used to denote that one number is smaller than another

Like fractions Fractions with common denominators

Like measurements Measurements that have the same unit of measure

Line graph A graph that uses points connected by lines to represent a set of data

Mean The sum of a set of numbers divided by the number of numbers in the set. Also called the average.

Measurement A number together with a unit of measure

Median When a set of numbers is arranged from smallest to largest, the median is the middle number or the average of the two middle numbers.

Metric system The measurement system used by most of the world. Conversions are based on powers of 10.

Mixed number The sum of a whole number and a fraction, written without the plus sign

Mode The number or numbers that occur most often in a set of numbers

Multiple A multiple of a whole number is a product of that number and a natural number.

Multiplication Property of One Any number times 1 is that number.

Multiplication Property of Zero Any number times zero is zero.

Natural numbers The numbers 1, 2, 3, and so on. Also called the counting numbers.

Negative numbers All the numbers less than zero

Number of decimal places The number of digits to the right of a decimal point

Numerator The upper numeral in a fraction

Odd digits The digits 1, 3, 5, 7, 9, and so on.

Opposite of a number The opposite of a number is the number on the number line that is the same distance from zero but on the opposite side. Also called the additive inverse.

Order of operations An established order in which to perform operations when simplifying a mathematical expression

Original price The price at which a business sets out to sell an article

Parentheses The grouping symbols ( )

Partial quotient When two numbers do not divide evenly, the partial quotient is the quotient not including the remainder.

Percent comparison (percent) A ratio with a base unit of 100

Percent of decrease When a value B is decreased by an amount A, the rate of percent R, or A/B, is called the percent of decrease.

Percent of discount The percent of the original price that is subtracted to get the discount amount. Also called the discount rate.

Percent of increase When a value B is increased by an amount A, the rate of percent R, or A/B, is called the percent of increase.

Perfect square A whole number that is the square of another whole number

Perimeter The total distance around the outside of a polygon

Pi The number, represented by the symbol π, that is the quotient of the circumference of a circle and its diameter. π is approximately equal to 3.14 or 22/7

Pictograph A graph that uses symbols or simple drawings to represent a set of data

Pie chart A chart used to show how a whole unit is divided into parts. Also called a circle graph.

Place value The value determined by the position of a digit in a number.

Place value name A method of naming a number by the positions (place values) of its digits

Polygon A closed figure whose sides are line segments

Positive numbers All the numbers greater than zero

Power of 10 The value that is obtained when 10 is written with an exponent

Prime factorization The indicated product of prime numbers that make up a counting number

Prime number A whole number greater than 1 with exactly two factors (divisors)

Principal The amount of money borrowed

Product The result of multiplying two or more numbers

Proper fraction A fraction in which the numerator is less than the denominator

Proportion A statement that two ratios or rates are equal

Pythagorean Theorem The theorem that states that, in a right triangle, the sum of the squares of the legs is equal to the square of the hypotenuse ($a^2 + b^2 = c^2$).

Quadrilateral A polygon with four sides

Quotient The result of dividing two numbers

Radical The symbol ($\sqrt{}$) used to denote a square root of a number

Radius The distance from the center of a circle to any point on the circle

Rate A comparison of two unlike measurements by division

Rate of percent In the percent formula, $R \times B = A$, R is the rate of percent. It includes the percent symbol, %.

Ratio A comparison of two quantities by division

Reciprocal The fraction that is formed by interchanging the numerator and denominator of the original fraction

Remainder The number remaining in a division where the divisor does not divide the dividend evenly.

Right triangle A triangle with one right (90-degree) angle

Round To round a number means to give an approximate value of the number.

Row A horizontal line of a table that reads left to right across the page

Sales tax The amount charged on a purchase to finance state and/or city programs

Scales The values used on either axis of a bar or line graph

Scientific notation A method of writing a number as a product using a number between 1 and 10 and a power of 10

Short division A shortcut division method used when a number is divided by a single digit

Signed numbers The positive numbers, negative numbers, and zero

Simple interest Interest based on borrowing money for 1 year

Simplifying a fraction The process of renaming a fraction by writing it with a smaller numerator and denominator

Solution A number that makes an equation true

Solve a proportion To find a missing number, usually represented as a letter or variable, that makes the proportion true

Square A four-sided polygon with all sides equal and one right angle

Square root The square root of a positive number is the number that is squared to give the original number.

Square of a number The number raised to the second power

Sum The result of adding two or more numbers

Table A method of displaying data using a horizontal and vertical arrangement

Unit fraction A fraction whose numerator and denominator are equivalent measurements

Unit of measure The name of a fixed quantity that is used as a standard

Unit rate A rate with a denominator of one unit

Unlike fractions Fractions with different denominators

Unlike measurements Measurements having different units of measure

Unpaid balance The amount of money still owed on a credit card or loan

Value An assigned or calculated numerical quantity

Variables Letters used to represent numbers. Also called unknowns.

Vertical scale On a bar or line graph, the values along the vertical axis

Volume The amount of space contained within a three-dimensional object

Weight The heaviness of an object

Whole numbers The numbers 0, 1, 2, 3, and so on

Word names Written words that represent a number

ANSWERS

Chapter 1

SECTION 1.1

1. eight hundred forty-three   **3.** four hundred sixty
5. seven thousand, twenty   **7.** 87   **9.** 9500
11. 101,000,000   **13.** twenty-seven thousand, six hundred eighty   **15.** two hundred seven thousand, six hundred ninety   **17.** fifty-four million   **19.** 243,700

21. 22,570   **23.** 19,000,000,000   **25.** $<$

27. $>$ **29.** $<$ **31.** $<$ **33.** 740   **35.** 2700

	Number	Ten	Hundred	Thousand	Ten Thousand
37.	607,546	607,550	607,500	608,000	610,000
39.	6,545,742	6,545,740	6,545,700	6,546,000	6,550,000

41. The percent who exercise regularly is 40%.

43. The over $50,000 category has the lowest percent of nonexercisers.

45. As income level goes up, so does the percent of regular exercisers.

47. Seattle is the only city with an increase. Boston, Chicago, Minneapolis, New Orleans, and San Diego all had a decrease. Although Atlanta had a slight increase in number of homeless, values rounded to the nearest ten are essentially the same.

49. 6840 > 5979   **51.** 656,732,410

53. eighteen thousand, five hundred three dollars

55. $>$ **57.** 9999   **59.** 74,600,000

61. 63,700; 63,800: Rounding the second time changes the tens digit to 5, so rounding to the hundreds place from 63,750 results in a different answer. If told to round to the nearest hundred, the first method is correct.

63. Kimo wrote "eleven thousand, four hundred seventy-five" on the check.

65. 25,400 < 230,000

67. The place value name for the bid is $36,407.

69. To the nearest thousand dollars, the value of the Income Fund of America shares is $185,000.

71. In 2000 there were about 101 million metric tons of carbon emissions.

73. The per capita income in Maine is thirty-two thousand, ninety-five dollars.

75. Maine has the smallest per capita income of the New England states.

77. To the nearest million miles, the distance from Earth to the sun is 93,000,000 miles.

79.

Month	Number of Marriages	Month	Number of Marriages
June	242,000	December	184,000
May	241,000	April	172,000
August	239,000	November	171,000
July	235,000	February	155,000
October	231,000	March	118,000
September	225,000	January	110,000

81. The rivers in increasing length are Yukon, Colorado, Arkansas, Rio Grande, Missouri, Mississippi.

83. Yes, the motor vehicle department's estimate was correct.

85. Jif has fewer of the following nutrients: fat, saturated fat, sodium, carbohydrates, and sugars.

87. *The Dark Knight* took in five hundred thirty-three million, one hundred eighty-four thousand, two hundred nineteen dollars.

89. *Shrek 2* took in about $437,000,000.

91. "Base ten" is a good name for our number system because each place value in the system is 10 times the previous place and one-tenth the succeeding place.

93. Rounding a number is a method of calculating an approximation of that number. The purpose is to get an idea of the value of the number without listing digits that do not add to our understanding. The number 87,452 rounds to 87,000, to the nearest thousand, because 452 is less than halfway between 0 and 1000. Rounded to the nearest hundred it is 87,500 because 52 is more than halfway between 0 and 100.

95. five trillion, three hundred twenty-six billion, nine hundred one million, five hundred seventy thousand

97. 7   **99.** 0

SECTION 1.2

1. 113   **3.** 942   **5.** 707   **7.** 1   **9.** 3810
11. 5789   **13.** 17,500   **15.** 302   **17.** 334
19. 531   **21.** 10   **23.** 486   **25.** 136
27. 4700   **29.** 1100   **31.** 10,000   **33.** 400
35. 5000   **37.** 19,000   **39.** 161,000   **41.** 5000
43. 34,000   **45.** 44 cm   **47.** 221 in.   **49.** 56 ft
51. 134 in.   **53.** The total number of Fords, Toyotas, and Lexuses sold is 3561.

55. 163 more Hondas than Fords are sold.

57. The total of the three best-selling cars sold is 5320.

59. For the week, 5487 salmon went through the ladder. Tuesday's count was 332 more than Saturday's count.

61. The estimated answer is 2000, so Ralph's answer is reasonable.

63. The estimated cost of the items is $1300.

65. The total number of property crimes is 17,014.

67. Sasha consumes 520 calories.

69. Sasha could have eaten 1180 calories for breakfast and lunch.

71. A total of 22,935 trees can be harvested.

73. Fong's Grocery still owes $14,800.

75. The median family income was probably rounded to the nearest thousand. The San Francisco median income was $23,000 higher than Seattle's median income.

77. On average, 530 more husbands killed their wives than wives killed their husbands.

79. On average, there were 209 people who killed a sibling.

81. There were 23,760 more cars sold than sport utility vehicles.

83. There were 550 more utility vehicles sold than cars.

85. There are 20,200,000 adults with HIV/AIDS.

87. The perimeter of the house is 200 ft.

89. Blanche needs 224 in. of lace.

91. Explanations for subtracting that are aimed at 6-year-olds are usually based on physical objects. So $15 - 9 = 6$ because when 9 circles (or pencils or apples or whatever) are removed from 15 circles, 6 circles remain.

 Count the circles that are empty.

93. A sum is the result of adding numbers. The sum of 8, 8, and 2 is 18. Mathematically, we write $8 + 8 + 2 = 18$.

95. seven hundred thousand, nine hundred

97. The dollar sales for the nine cars is $184,000. The Accords' sales were $29,476 more than the Civics' sales.

99. A = 7, B = 2, C = 3, D = 1

GETTING READY FOR ALGEBRA

1. $x = 12$ **3.** $x = 23$ **5.** $z = 14$ **7.** $c = 39$
9. $a = 151$ **11.** $x = 14$ **13.** $y = 58$
15. $k = 168$ **17.** $x = 11$ **19.** $w = 116$

21. The markup is $348.

23. The length of the garage is 9 meters.

25. Let S represent the EPA highway rating of the Saturn and I represent the EPA highway rating of the Impreza. $I + 5 = S$; the Impreza has a highway rating of 30 mpg.

27. Let B represent the total dollars budgeted in a category, S represent the dollars spent in a category, and R represent the dollars not yet spent in a category. $S + R = B$

SECTION 1.3

1. 581   **3.** 291   **5.** 304   **7.** 651   **9.** 296
11. 0   **13.** 3040   **15.** thousands   **17.** 4522
19. 5628   **21.** 3551   **23.** 3478   **25.** 124,800
27. 66,896   **29.** 35,856   **31.** 21,900   **33.** 15,600
35. 38,471   **37.** 34,686   **39.** 328,396   **41.** 3200
43. 25,000   **45.** 250,000   **47.** 240,000

49. 800,000   **51.** 18,000,000   **53.** 418 square yd

55. 961 square ft   **57.** 1944 square cm

59. 102 square ft   **61.** 52,224 square cm

63. 468 square m   **65.** 294 square ft

67. 198 square ft   **69.** 390,365   **71.** 2,376,000

73. The estimated product is 30,000, which is close to 28,438, so Maria's answer is reasonable.

75. The Rotary Club estimates it will sell 5115 dozen roses.

77. The gross receipts from the sale of the Prius are $555,799.

79. The gross receipts are $1,194,000, rounded to the nearest thousand.

81. Washington County grew by 22,488 people for the year.

83. The comptroller realized $44,275 from the sale of the shares.

85. The estimated cost of the 12 blouses is $400, so Carmella should have enough money in the budget.

87. There are 6 feet in a fathom.

89. There are about 18,228 ft in a league.

91. It is not possible to be 20,000 leagues under the sea. The author was taking literary license.

93. Sirius is about 47,040,000,000,000 miles from Earth.

95. There were 136 pages printed in 17 minutes.

97. A RAM of 256 KB has 262,144 bytes.

99. A tablespoon of olive oil has 126 calories from fat.

101. To the nearest thousand gallons, in a 31-day month the water usage is 16,574,000 gallons.

103. The tires cost Ms. Perta $22,272. The gross income from the sale of the tires is $49,184. The profit from the sale of the tires is $26,912.

105. Yes, *Stars Wars: Episode III* would have grossed $760,541,154 if it had doubled its earnings.

107. Explanations for multiplication that are aimed at 8-year-olds are usually based on physical objects. For example, $3(8) = 24$ because 3 groups, each of which has 8 circles (or pencils or apples or whatever), contain a total of 24 circles.

◯◯◯◯ ◯◯◯◯ ◯◯◯◯
◯◯◯◯ ◯◯◯◯ ◯◯◯◯

Count the total number of circles.

109. A product is the result of multiplying numbers. The product of 11 and 8 is 88. Mathematically, we write $11 \times 8 = 88$.

111. To the nearest thousand, the crop is worth $4,449,000.

113. A = 3, B = 2, C = 4, D = 5, E = 9

SECTION 1.4

1. 9   **3.** 13   **5.** 87   **7.** 91   **9.** 17   **11.** 40
13. 82   **15.** 5 R 1   **17.** 4 R 13   **19.** zero
21. 3051   **23.** 32   **25.** 54   **27.** 87   **29.** 76
31. 391   **33.** 181 R 39   **35.** 15 R 52   **37.** 24
39. 12,802 R 18   **41.** 870   **43.** 500

45. The taxes paid per return in week 1 are $5476.

47. The taxes paid per return in week 3 are $4100, rounded to the nearest hundred.

49. The survey finds that 135 trees per acre are ready to harvest.

51. Ms. Munos will pay $2016 for the radios.

53. You would need to spend $1,700,000 per day.

55. The population density of China was about 139 people per square kilometer.

57. The population density of the United States was about 32 people per square kilometer.

59. The average veterinary cost per cat in veterinary fees is $81.

61. Each California representative represents about 625,380 people.

63. The gross state product per person in California was about $50,300.

65. There are about 45 calories per serving.

67. Juan can take 8 capsules per day.

69. About 50,000,000 tickets at $8 each were sold for *Spider-Man* to earn $407,681,000.

71. The average salary for the Steelers was about $1,436,000.

73. The remainder, after division, is the amount left over after all possible groups of the appropriate size have been formed. When 46 objects are separated into groups of 9 (you can make 5 groups of 9) there is 1 object remaining. So $46 \div 9 = 5$ R 1.

75. The company will get the greater gross return in France. The difference in gross receipts is $7125.

77. A = 2, B = 7, C = 9

GETTING READY FOR ALGEBRA

1. $x = 5$ **3.** $c = 18$ **5.** $x = 4$ **7.** $b = 46$
9. $x = 12$ **11.** $y = 312$ **13.** $x = 24$
15. $b = 24{,}408$ **17.** $x = 61$ **19.** $w = 782$

21. The width of the garden plot is 17 feet.

23. He sells 2340 lb of crab.

25. The wholesale cost of one set is $310.

27. Let L represent the low temperature in July and H represent the high temperature in January. $2H = L$; the high temperature is 30°F.

SECTION 1.5

1. 16^6 **3.** 49   **5.** 8   **7.** 15

9. base; exponent; power or value   **11.** 512

13. 1024   **15.** 1000   **17.** 729   **19.** 6561

21. 8800   **23.** 2,100,000   **25.** 23   **27.** 780

29. exponent   **31.** 20,200,000   **33.** 7270

35. 67,340,000   **37.** 528   **39.** 96,100,000,000

41. 4500   **43.** 10^{11} **45.** 161,051

47. 537,824   **49.** 47,160,000,000,000

51. 68   **53.** The size of Salvador's lot is 10,800 ft^2.

55. The skate park's operating budget is approximately $300,000.

57. It is approximately 25,500,000,000,000 miles from Earth to Alpha Centauri.

59. The distance, 6 trillion miles, can also be written as 6×10^{12} and as 6,000,000,000,000.

61. There are 4^4 or 256 bacteria after 3 hours.

63. The number of bacteria will exceed 1000 during the fourth hour.

65. The surface area of the Pacific Ocean is about 64,200,000 square miles.

67. A gigabyte is 10^9 bytes.

69. *Shrek 2* earned about \$437,000,000, or 437×10^6.

71. 60^5 is larger than the gross earnings of *Titanic*.

73. The expression 4^{10} represents the product of 10 fours, that is, $4 \times 4 \times 4 \times 4 \times 4 \times 4 \times 4 \times 4 \times 4 \times 4$.

75. Mitchell's grandparents will deposit \$1,048,576 on his 10th birthday. They will have deposited a total of \$1,398,100.

77. 8962

SECTION 1.6

1. 39   **3.** 1   **5.** 50   **7.** 28   **9.** 48   **11.** 52
13. 30   **15.** 53   **17.** 48   **19.** 16   **21.** 8
23. 106   **25.** 35   **27.** 71   **29.** 324   **31.** 2757
33. 45

35. There were 595 more mallards and canvasbacks than teals and wood ducks.

37. Four times the number of wood ducks added to the number of mallards would be 165 more than the number of teals and canvasbacks.

39. 44   **41.** 54

43. Eltana's Custodial Service used supplies costing \$1546 for the month.

45. The trucker's average weekly income is \$1291. His yearly income is \$64,550.

47. Marla consumes 1150 calories for breakfast.

49. Sally's total charge is \$190.

51. Clay and Connie have \$38 left on their certificate.

53. Ryan will need to save \$150 for 13 months.

55. Derrick will save \$576 per year making his espresso at home.

57. The result is \$641,052,500.

59. Using the order of operations, division takes precedence over subtraction, so $20 - 10 \div 2 = 20 - 5$, or 15.

61. 196

63. Answers will vary.

GETTING READY FOR ALGEBRA

1. $x = 7$ **3.** $y = 33$ **5.** $x = 6$ **7.** $c = 32$

9. $x = 2$ **11.** $c = 12$ **13.** $a = 196$

15. $b = 21$ **17.** Remy bought seven tickets.

19. Rana made six arrangements.

21. $8v = C$; Jessica can purchase nine visits at B-Fit.

23. $4v + 32 = C$; Jessica can purchase 10 visits from Gym Rats.

SECTION 1.7

1. 8   **3.** 16   **5.** 13   **7.** 13   **9.** 7
11. 9   **13.** 16   **15.** 34   **17.** 38   **19.** 37
21. 172   **23.** 156   **25.** 137   **27.** 59   **29.** 58
31. 34   **33.** 73   **35.** 55   **37.** 126   **39.** 86
41. 17   **43.** 3   **45.** no mode   **47.** 43
49. 44 and 55

51. no mode

53. The average number of points is 102 and the median number of points is 103.

55. The average score is 133; the median score is 134.

57. The average gas mileage for the cars is 32 mpg, the median mileage is 32 mpg, and the mode mileage is 34 mpg.

59. The average weight of the wrestlers is 136 lb. The median weight of the wrestlers is 124 lb.

61. The average cost per mile to build the new lines is \$63 million.

63. The average price of the coffee makers is \$70. One of the models cost less than the average price.

65. The average population was 2,299,212 over the 8-year period.

67. Population by Year in Nevada

Year	Population	Year	Population
2000	1,998,257	2005	2,374,452
2001	2,073,496	2006	2,449,691
2002	2,148,735	2007	2,524,930
2003	2,223,974	2008	2,600,169
2004	2,299,213		

69. The average yearly attendance for the 4 years was about 14,600,000.

71. The average assets for the top three banks were \$1,483 billion.

73. The average assets for the top five banks were \$1,125 billion.

75. The mean percentage of Internet users is 36%.

77. The median percentage of Internet users is 34%.

79. Half of the houses on Jupiter Island cost more than $4 million and half cost less than $4 million.

81. In 1950, about half of all men getting married for the first time were 23 years old or younger. In 2002, about half of all men getting married were 27 years old or younger. This means that men are waiting longer to get married. The 23-, 24-, 25-, and 26-year-old men are now below the average of those getting married for the first time.

83. The average earnings were about $545,636,483.

85. The average earnings of all 10 movies were about $481 million.

87. The average, or mean, of 2, 4, 5, 5, and 9 is their sum, 25, divided by 5, the number of numbers. So, the mean is 5. The average gives one possible measure of the center of the group.

89. Answers will vary.

SECTION 1.8

1. Frontier had the least number.

3. Hawaiian had about 14,000 passengers.

5. These airlines totaled about 375,000 passengers.

7. 40   **9.** Full-size

11. The number of vehicles in for repair is 400.

13. Australia had the least number of TV sets.

15. India had about 63 million (63,000,000) TV sets.

17. The difference in the number of sets was about 33 million.

19. The total paid for paint and lumber is $45,000.

21. Steel casting costs $15,000 less than plastic.

23. The company will pay $40,000 for steel casting to double production.

25.

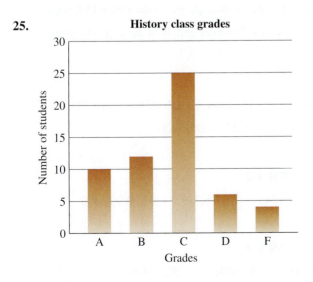

History class grades

27.

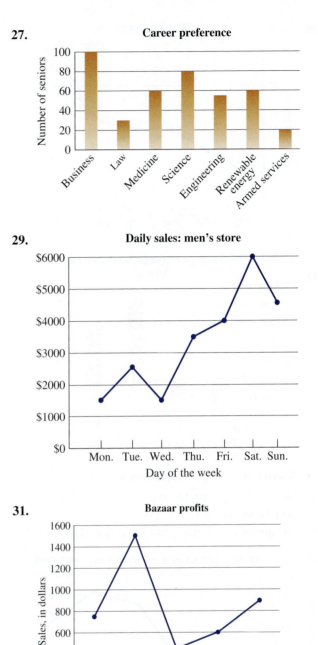

Career preference

29.

Daily sales: men's store

31.

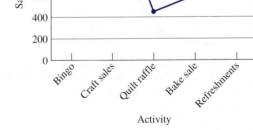

Bazaar profits

33.

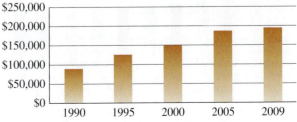

Median cost of a three-bedroom house in Austin, Texas

35.

Full retirement age

37.

Median price of a single-family home in Miami

39. Tokyo had the largest population in 2000.

41. Mexico City is expected to grow the most, 3 million people, during the 15-year period.

43. The greatest amount of growth was from 1960 to 1970.

45.

Overnight camping at Lizard Lake State Park

47.

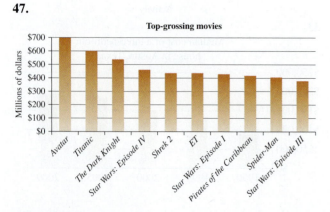

Top-grossing movies

49.

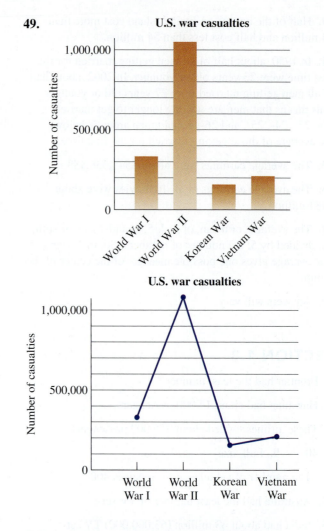

U.S. war casualties

U.S. war casualties

CHAPTER 1 REVIEW EXERCISES

SECTION 1.1

1. six hundred seven thousand, three hundred twenty-one

3. 62,337   **5.** <   **7.** <

9. 183,660, 183,700, 184,000, and 180,000

11. There are 74 more people in the under-15 group.

13. There are 383 more people in the under-15 group.

SECTION 1.2

15. 917   **17.** 1764   **19.** 221   **21.** 1332

23. 102 in.   **25.** 4000, 3200

SECTION 1.3

27. 11,774   **29.** 76,779   **31.** 3,000,000

SECTION 1.4

33. 15   **35.** 86 R 5   **37.** 5300

39. 1331   **41.** 23,000   **43.** 712,000,000

45. $3,400,000,000

47. 65   **49.** 23

51. 57, 64, no mode   **53.** 62, 63, 63

55. The average salary of the siblings, to the nearest hundred, is $87,600.

57. The difference in the temperatures is 15°.

CHAPTER 1 TRUE/FALSE CONCEPT REVIEW

1. false; to write one billion takes 10 digits.   **2.** true
3. true   **4.** false; seven is less than twenty-three.
5. false; $2567 > 2566$ **6.** true   **7.** true   **8.** true
9. false; the sum is 87.   **10.** true   **11.** true
12. false; the product is 45.   **13.** true   **14.** true
15. true   **16.** false; a number multiplied by 0 is 0.
17. true   **18.** false; the quotient is 26.   **19.** true
20. true   **21.** false; division by zero is undefined.

22. false; the value is 49.   **23.** true   **24.** true

25. false; the product is 450,000.   **26.** true

27. false; in $(3 + 1)^2$, the addition is done first.

28. false; in $(11 - 7) \cdot 9$, subtraction is done first.

29. true   **30.** true   **31.** true   **32.** true

33. false; the median of 34, 54, 14, 44, 67, 81, and 90 is 54.

34. true

CHAPTER 1 TEST

1. 212   **2.** 3266   **3.** 28   **4.** 15,836   **5.** $<$
6. 55,000,000   **7.** 238,336   **8.** 730,061   **9.** 1991
10. 372,600   **11.** 39,000   **12.** 230,000
13. 729   **14.** 8687   **15.** 120 ft   **16.** 32,000
17. 6158   **18.** four thousand, five   **19.** 86
20. 259,656   **21.** 7730   **22.** 676,000,000
23. 1160 R 32   **24.** 43   **25.** 345 cm²

26. 514   **27.** average, 697; median, 743; mode, 795
28. It will take the secretary 168 minutes, or 2 hours and 48 minutes, to type 18 pages.

29. Each person will win $13,862,000. Each person will receive $693,100 per year.

30. $120,001–$250,000   **31.** 40   **32.** 5

33. Division B has the most employees.

34. There are 120 more employees in division A.

35. There is a total of 1020 employees.

Chapter 2
SECTION 2.1

1. no   **3.** yes   **5.** yes   **7.** no   **9.** yes
11. no   **13.** yes   **15.** yes   **17.** yes   **19.** 2, 3, 5
21. 2   **23.** 5   **25.** 5   **27.** 3, 5   **29.** yes
31. no   **33.** yes   **35.** yes   **37.** no
39. yes   **41.** no   **43.** yes   **45.** no   **47.** 6, 9, 10
49. 6, 9   **51.** none   **53.** 6, 10   **55.** 6, 9, 10

57. No, it is not possible to divide the collection up evenly, because 439 is not divisible by 3.

59. Yes, 987 is divisible by 3.

61. The band can march in rows of 5, but not rows of 3 or 10, with the same number in each row.

63. No, because 30 is not divisible by 9.

65. They must raise an additional $2.

67. There are 15 animals in the pen.

69. There is a total of 24 elephants and riders in the act.

71. A number is divisible by 5 when the remainder after dividing by 5 is 0. For example, 115 is divisible by 5 because $115 \div 5 = 23$ R 0. However, 116 is not divisible by 5 because $116 \div 5 = 23$ R 1.

73. In the divisibility test for 2, we look at the last digit. If the last digit is even, then the number is divisible by 2. In the divisibility test for 3, we look at the *sum* of the digits. If the sum is divisible by 3, then the number is divisible by 3.

75. yes   **77.** Yes. A number divisible by both 3 and 10 is divisible by 30.

79. 67,000; 70,000

81. 13

83. 72 cm   **85.** 12, 24, 36, 48, 60, and 72

87. 123, 246, 369, 492, 615, and 738

SECTION 2.2

1. 5, 10, 15, 20, 25　　**3.** 23, 46, 69, 92, 115

5. 21, 42, 63, 84, 105　　**7.** 30, 60, 90, 120, 150

9. 50, 100, 150, 200, 250　　**11.** 53, 106, 159, 212, 265

13. 67, 134, 201, 268, 335　　**15.** 85, 170, 255, 340, 425

17. 155, 310, 465, 620, 775

19. 375, 750, 1125, 1500, 1875　　**21.** yes　　**23.** yes

25. no　　**27.** no　　**29.** no　　**31.** yes　　**33.** yes

35. yes　　**37.** yes　　**39.** multiple of 6 and 9

41. multiple of 6 and 15　　**43.** multiple of 6 and 15

45. multiple of 31　　**47.** multiple of 17 and 31

49. No, Jean's car will not be selected, because 14 is not a multiple of 4.

51. The students should work problems 5, 10, 15, 20, 25, 30, 35, 40, 45, 50, and 55.

53. Because goats have 4 feet, the number of goat feet in the pen must be a multiple of 4. But 30 is not a multiple of 4, so Katy counted incorrectly.

55. There could be 6 rows of 130 pots, 10 rows of 78 pots, 12 rows of 65 pots, 13 rows of 60 pots, 15 rows of 52 pots, 20 rows of 39 pots, or 26 rows of 30 pots.

57. Yes, the gear is in its original position because 240 is a multiple of 20.

59. Yes, the number of shares of Yahoo! is a multiple of the number of shares of Hewlett-Packard. There are three times as many shares of Yahoo! as shares of Hewlett-Packard.

61. The triple 3, 4, 5 is a Pythagorean triple because $3^2 + 4^2 = 5^2$ or $9 + 16 = 25$.

63. The third number is 13, so 5, 12, 13 is a Pythagorean triple.

65. If a number is a multiple of 12, then the number is $12 \times$ some whole number. Therefore, the number is also $6 \times 2 \times$ some whole number, so the number is a multiple of 6.

67. First divide 126 by 7 and write the product 7×18. The next multiple is $7 \times 19 = 133$, followed by $7 \times 20 = 140$, and so on. Stop at $7 \times 25 = 175$.

69. no　　**71.** 1333　　**73.** 1, 2, 7, 14

75. 1, 2, 3, 6, 9, 18　　**77.** 1, 2, 3, 5, 6, 10, 15, 30

79. 10　　**81.** 23

83. No, 24 is divisible by both 2 and 8, but not by 16.

SECTION 2.3

1. $1 \cdot 16, 2 \cdot 8, 4 \cdot 4$　　**3.** $1 \cdot 31$　　**5.** $1 \cdot 33, 3 \cdot 11$

7. $1 \cdot 46, 2 \cdot 23$　　**9.** $1 \cdot 49, 7 \cdot 7$

11. $1 \cdot 72, 2 \cdot 36, 3 \cdot 24, 4 \cdot 18, 6 \cdot 12, 8 \cdot 9$

13. $1 \cdot 88, 2 \cdot 44, 4 \cdot 22, 8 \cdot 11$

15. $1 \cdot 95, 5 \cdot 19$

17. $1 \cdot 100, 2 \cdot 50, 4 \cdot 25, 5 \cdot 20, 10 \cdot 10$

19. $1 \cdot 105, 3 \cdot 35, 5 \cdot 21, 7 \cdot 15$

21. $1 \cdot 112, 2 \cdot 56, 4 \cdot 28, 7 \cdot 16, 8 \cdot 14$

23. $1 \cdot 124, 2 \cdot 62, 4 \cdot 31$

25. $1 \cdot 441, 3 \cdot 147, 7 \cdot 63, 9 \cdot 49, 21 \cdot 21$

27. $1 \cdot 459, 3 \cdot 153, 9 \cdot 51, 17 \cdot 27$　　**29.** 1, 3, 5, 15

31. 1, 19　　**33.** 1, 3, 13, 39　　**35.** 1, 2, 23, 46

37. 1, 2, 4, 13, 26, 52　　**39.** 1, 2, 4, 8, 16, 32, 64

41. 1, 2, 3, 4, 6, 8, 9, 12, 18, 24, 36, 72

43. 1, 2, 3, 6, 13, 26, 39, 78

45. 1, 2, 4, 7, 8, 14, 16, 28, 56, 112

47. 1, 2, 67, 134

49. 1, 2, 3, 6, 9, 18, 27, 54, 81, 162

51. $1 \cdot 444, 2 \cdot 222, 3 \cdot 148, 4 \cdot 111, 6 \cdot 74, 12 \cdot 37$

53. $1 \cdot 652, 2 \cdot 326, 4 \cdot 163$　　**55.** $1 \cdot 680, 2 \cdot 340,$ $4 \cdot 170, 5 \cdot 136, 8 \cdot 85, 10 \cdot 68, 17 \cdot 40, 20 \cdot 34$

57.

Number of Programs	Length of minutes
1	120
2	60
3	40
4	30
5	24
6	20
8	15

59. There can be 2 players, 4 players, 13 players, 26 players, or 52 players.

61. The 5 caregivers do not meet the recommendations, because $65 = 5 \cdot 13$, so each caregiver is in charge of 13 preschoolers.

63. We know that the number of band members is not a multiple of 4 or 5. In fact, the number of members is 1 more than a multiple of 4 and also 1 more than a multiple of 5. The multiples of $4 + 1$ are in the list: 5, 9, 13, 17, 21, 25, 29, 33, ...The multiples of $5 + 1$ are in the list: 6, 11, 16, 21, 26, 31, ...The smallest number in both lists is 21.

65. The divisors of 6 are 1, 2, 3, and 6. The sum $1 + 2 + 3 = 6$. Therefore, 6 is a perfect number.

67. The divisors of 28 are 1, 2, 4, 7, 14, and 28. Because $1 + 2 + 4 + 7 + 14 = 28$, 28 is a perfect number.

69. The difference between a factor and a divisor is the operation we use. If we are multiplying, we generally use the word *factor*. If we are dividing, we generally use the word *divisor*.

71. The largest factor is 991.   **73.** 3024   **75.** 35,052   **77.** 33   **79.** 27 R 3   **81.** There is enough wire for 41 speakers, with 16 feet left over.

SECTION 2.4

1. composite   **3.** composite   **5.** composite

7. prime   **9.** composite   **11.** prime

13. composite   **15.** prime   **17.** composite

19. prime   **21.** composite   **23.** composite

25. composite   **27.** composite   **29.** prime

31. prime   **33.** composite   **35.** prime

37. prime   **39.** composite   **41.** composite

43. composite   **45.** prime   **47.** prime

49. composite   **51.** composite   **53.** composite

55. The next year that is a prime number is 2017.

57. The number ends in 8, so it is divisible by 2 and therefore has more than two divisors.

59. This is not true in general, because 2 and 5 are both prime, but 25 and 52 are not.

61. No, 1989 is not a prime number.

63. $M_3 = 2^3 - 1 = 7$; $M_5 = 2^5 - 1 = 31$; and $M_7 = 2^7 - 1 = 127$

65. $M_3 \times (2^{3-1}) = 7 \times (2^2) = 7 \times 4 = 28$

67. A prime number has only two factors, itself and 1. The number 7 is prime. A composite number has three or more factors, itself, 1, and at least one other factor. The number 9 is composite because it has three factors, 1, 3, and 9. A composite number is composed of numbers other than itself and 1.

69. The number 6 has exactly four factors. They are 1, 2, 3, and 6. So does the number 15, with factors 1, 3, 5, and 15. Any number that is the product of exactly two prime numbers has four factors.

71. composite

73. Yes. The final quotient is 15.

75. Yes. The final quotient is 65.

77. No. The quotient of 1029 and 3 is 343, which is not divisible by 3. The final quotient is 343.

79. No. The quotient of 2880 and 5 is 576, which is not divisible by 5. The final quotient is 576.

81. Yes. The final quotient is 11.

SECTION 2.5

1. $2^2 \cdot 3$ **3.** $3 \cdot 5$ **5.** $3 \cdot 7$ **7.** $2^3 \cdot 3$ **9.** $2^2 \cdot 7$

11. $2 \cdot 17$ **13.** $2 \cdot 23$ **15.** $2 \cdot 3 \cdot 7$ **17.** $2 \cdot 3 \cdot 13$

19. $2 \cdot 41$ **21.** $2 \cdot 3^2 \cdot 5$ **23.** $5 \cdot 19$ **25.** $2^2 \cdot 3^3$

27. prime   **29.** $2^2 \cdot 3 \cdot 13$ **31.** $2 \cdot 3 \cdot 29$

33. $2^2 \cdot 3^2 \cdot 5$ **35.** $2^3 \cdot 5^2$ **37.** $2 \cdot 157$

39. $3 \cdot 5 \cdot 23$ **41.** $2^2 \cdot 5 \cdot 23$ **43.** $2^3 \cdot 5 \cdot 13$

45. $3^3 \cdot 17$ **47.** $2 \cdot 3 \cdot 7 \cdot 17$ **49.** 5^4

51. $2 \cdot 3 \cdot 5 \cdot 41$

53. Answers will vary.   **55.** Answers will vary.

57. The numbers 119 and 143 are relatively prime because they have no common factors, as we can see from the prime factorization of each: $119 = 7 \cdot 17$ and $143 = 11 \cdot 13$.

59. The number 97 is relatively prime to 180.

61. A number is written in prime-factored form when (1) it is a prime number or (2) it is written as a product and every factor in the product is a prime number. For example, $18 = 2 \cdot 3^2$ is in prime-factored form, whereas $18 = 3 \cdot 6$ and $18 = 2 \cdot 9$ are not.   **63.** $2^4 \cdot 3 \cdot 47$

65. $17^2 \cdot 23^2$ **67.** The area is 65 in^2.

69. yes   **71.** no   **73.** yes

75. Answers will vary but may include 30, 60, 90, and 120.

SECTION 2.6

1. 36   **3.** 21   **5.** 30   **7.** 40   **9.** 45   **11.** 36
13. 20   **15.** 60   **17.** 30   **19.** 60   **21.** 72
23. 48   **25.** 180   **27.** 48   **29.** 120   **31.** 48
33. 180   **35.** 24   **37.** 48   **39.** 168   **41.** 210
43. 204   **45.** 1400   **47.** 24   **49.** 120

51. The next time they will both have a day off is on May 31.

53. The two gears will return to their original positions in two turns of the 36-tooth gear.

55. They will both be visible in 12 years.

57. The least common multiple of 6 and 10 is 30, so John must run 5 days and his friend 3 days.

59. To find the LCM of 20, 24, and 45, first find the prime factorization of each number. Then build the LCM by using each different prime factor to the highest power that occurs in any single number.

61. 720   **63.** 6300

65. 1, 2, 4, 5, 10, 20, 23, 46, 92, 115, 230, 460

67. 1, 2, 3, 5, 6, 10, 15, 17, 30, 34, 51, 85, 102, 170, 255, 510   **69.** yes   **71.** no

73. They must sell at least 7 more appliances.

CHAPTER 2 REVIEW EXERCISES

SECTION 2.1

1. 44, 50, 478   **3.** 15, 255, 525   **5.** 36, 63

7. 444, 666, 888   **9.** 645   **11.** 7, 9   **13.** 7, 9

15. 4, 5, 10   **17.** 15

19. Yes, the total score is divisible by 4.

SECTION 2.2

21. 7, 14, 21, 28, 35   **23.** 73, 146, 219, 292, 365

25. 85, 170, 255, 340, 425   **27.** 131, 262, 393, 524, 655
29. 211, 422, 633, 844, 1055   **31.** multiple of 6
33. multiple of 6 and 15   **35.** multiple of 9 and 15
37. multiple of 12 and 16   **39.** Yes, as 2691 is a multiple of 23.

SECTION 2.3

41. $1 \cdot 15, 3 \cdot 5$ **43.** $1 \cdot 42, 2 \cdot 21, 3 \cdot 14, 6 \cdot 7$

45. $1 \cdot 236, 2 \cdot 118, 4 \cdot 59$

47. $1 \cdot 354, 2 \cdot 177, 3 \cdot 118, 6 \cdot 59$

49. $1 \cdot 343, 7 \cdot 49$ **51.** 1, 2, 3, 4, 6, 8, 12, 16, 24, 48

53. 1, 2, 3, 4, 5, 6, 10, 12, 15, 20, 30, 60

55. 1, 97   **57.** 1, 2, 3, 4, 6, 9, 12, 18, 27, 36, 54, 108

59. The possible energy costs are $1590 for 1 person, $795 for 2 people, $530 for 3 people, $318 for 5 people, $265 for 6 people, and $159 for 10 people.

SECTION 2.4

61. prime   **63.** prime   **65.** composite

67. composite   **69.** composite   **71.** composite

73. prime   **75.** composite   **77.**   composite

79. The prime number year after 2011 is 2017.

SECTION 2.5

81. 2^5 **83.** $2 \cdot 19$ **85.** $3 \cdot 5^2$ **87.** $2 \cdot 3 \cdot 37$

89. $2^2 \cdot 3^2 \cdot 7$ **91.** $2 \cdot 5 \cdot 29$ **93.** $2^2 \cdot 5 \cdot 17$

95. prime   **97.** $5 \cdot 53$ **99.** $2012 = 2^2 \cdot 503$;
$2013 = 3 \cdot 11 \cdot 61$; $2014 = 2 \cdot 19 \cdot 53$;
$2015 = 5 \cdot 13 \cdot 31$; and so forth.

SECTION 2.6

101. 24   **103.** 360   **105.** 396   **107.** 175

109. 360   **111.** 24   **113.** 72   **115.** 300   **117.** 840

119. Possible answers include 2 and 49; 14 and 49; 2 and 98; and 14 and 98. Other answers are possible.

CHAPTER 2 TRUE/FALSE CONCEPT REVIEW

1. false; not all multiples of 3 end with the digit 3. For example, 15 is a multiple of 3.

2. true   **3.** true   **4.** true   **5.** true   **6.** true

7. false; only one multiple of 300 is also a factor of 300—itself.

8. false; the square of 25 is 625. Twice 25 is 50.

9. false; not all natural numbers ending in 6 are divisible by 6. For example, 16 is not divisible by 6.

10. true

11. false; not all natural numbers ending in 9 are divisible by 3. For example, 29 is not divisible by 3.

12. true

13. false; the sum of the digits $7 + 7 + 7 + 7 + 3 = 31$, which is not divisible by 3.

14. true   **15.** true

16. false; all prime numbers except 2 are odd.

17. true

18. false; every prime number has exactly two factors.

19. true   **20.** true

21. false; the least common multiple (LCM) of three different prime numbers is the product of the three numbers.

22. true

23. false; the largest divisor of the least common multiple (LCM) of three numbers is not necessarily the largest of the three numbers. For example, the LCM of 2, 4, and 6 is 12. The largest divisor of 12 is 12, not 6.

24. true

CHAPTER 2 TEST

1. yes   **2.** 1, 2, 4, 7, 8, 14, 16, 28, 56, 112   **3.** yes
4. no   **5.** 96   **6.** $1 \cdot 75, 3 \cdot 25, 5 \cdot 15$ **7.** $2^3 \cdot 5 \cdot 7$
8. 252   **9.** yes   **10.** 208, 221, 234, 247   **11.** no
12. prime   **13.** composite   **14.** $5 \cdot 11^2$ **15.** 504
16. 2   **17.** 299   **18.** 360

19. No, a set of prime factors has only one product.

20. Any two of the following sets of numbers: 2, 3, 7; or 2, 3, 14; or 2, 3, 21; or 2, 3, 42; or 2, 7, 42; or 3, 7, 14; or 6, 7, 42; or 14, 21, 42.

Chapter 3

SECTION 3.1

1. $\dfrac{5}{8}$ **3.** $\dfrac{2}{5}$ **5.** $\dfrac{4}{5}$ **7.** $\dfrac{5}{4}$ **9.** $\dfrac{11}{8}$ **11.** $\dfrac{14}{11}$

13. Proper fractions: $\dfrac{2}{13}, \dfrac{4}{13}, \dfrac{5}{13}, \dfrac{7}{13}, \dfrac{10}{13}$;

Improper fractions: $\dfrac{13}{13}, \dfrac{15}{13}, \dfrac{17}{13}$

15. Proper fractions: none;

Improper fractions: $\dfrac{17}{16}, \dfrac{18}{17}, \dfrac{29}{19}, \dfrac{30}{21}, \dfrac{23}{23}$

17. Proper fractions: $\dfrac{8}{9}, \dfrac{13}{15}, \dfrac{6}{10}$;

Improper fractions: $\dfrac{12}{10}, \dfrac{10}{9}, \dfrac{19}{19}, \dfrac{16}{15}$

19. Proper fractions: $\dfrac{11}{12}, \dfrac{29}{30}$;

Improper fractions: $\dfrac{3}{3}, \dfrac{6}{5}, \dfrac{8}{8}, \dfrac{15}{14}, \dfrac{19}{19}, \dfrac{21}{20}, \dfrac{24}{23}, \dfrac{32}{31}$

21. $3\dfrac{5}{6}$ **23.** $12\dfrac{5}{7}$ **25.** $11\dfrac{2}{11}$ **27.** $6\dfrac{11}{13}$

29. $15\dfrac{8}{27}$ **31.** $22\dfrac{1}{15}$ **33.** $\dfrac{40}{7}$ **35.** $\dfrac{15}{1}$

37. $\dfrac{76}{9}$ **39.** $\dfrac{217}{5}$ **41.** $\dfrac{299}{7}$ **43.** $\dfrac{493}{8}$

45. The fraction $\dfrac{0}{1}$ is a proper fraction. Any whole number except 0 can be put in the denominator.

47. The error occurred because both numerator and denominator were multiplied by 4: $4\dfrac{2}{3} = \dfrac{14}{3}$.

49. $\dfrac{7}{10}$ **51.** $\dfrac{11}{8}$

53.

55. Of the homes, $\dfrac{23}{35}$ have shake roofs.

57. The mixed numbers $11\dfrac{1}{2}$ and $6\dfrac{1}{8}$ can be written as improper fractions because they represent numbers larger than 1. The dimensions of nonstandard mail are longer than $\dfrac{23}{2}$ in., taller than $\dfrac{49}{8}$ in., and/or thicker than $\dfrac{1}{4}$ in.

59. The tank is $\dfrac{3}{8}$ full.

61. The number 7 is closest to the mark.

63. Jenna can get 104 strips from the wire.

65. There are $\dfrac{357}{510}$ of the bales still in the field.

67. The cup contains $\dfrac{5}{8}$ cup of oil.

69. This is an improper fraction because the numerator is larger than the denominator.

71. To change $\dfrac{34}{5}$ to a mixed number, divide 34 by 5. Write the remainder, 4, over the divisor, 5, to form the fraction part. We have $\dfrac{34}{5} = 6\dfrac{4}{5}$. To change $7\dfrac{3}{8}$ to an improper fraction multiply the denominator, 8, times the whole number, 7. Add the product to 3 to get 59. Write this sum over the denominator. So $7\dfrac{3}{8} = \dfrac{59}{8}$.

73. A mixed number has a value of 1 or more. A proper fraction has a value less than 1.

75. $\dfrac{144}{9}$; $\dfrac{2304}{144}$

77. There will be 87 pieces of rope, 5 pieces for each child. There will be 12 pieces left over.

79. The mixed number is $1234\dfrac{20}{37}$.

81. 18 **83.** 15 **85.** prime **87.** $2 \cdot 3^2 \cdot 23$

89. Bonnie averaged 203 miles per tank and about 51 miles per gallon.

SECTION 3.2

1. $\dfrac{2}{3}$ **3.** $\dfrac{2}{3}$ **5.** $\dfrac{9}{10}$ **7.** $\dfrac{3}{4}$ **9.** $\dfrac{2}{5}$ **11.** $\dfrac{3}{10}$

13. $\dfrac{7}{8}$ **15.** $\dfrac{9}{5}$ **17.** $\dfrac{6}{11}$ **19.** $\dfrac{5}{7}$ **21.** 9

23. $\dfrac{3}{2}$ **25.** $\dfrac{1}{3}$ **27.** $\dfrac{7}{10}$ **29.** $\dfrac{3}{4}$ **31.** $\dfrac{35}{48}$

33. $\dfrac{2}{3}$ **35.** $\dfrac{16}{15}$ **37.** $\dfrac{1}{3}$ **39.** $\dfrac{15}{16}$ **41.** $\dfrac{3}{4}$

43. 5 **45.** $\dfrac{17}{21}$ **47.** $\dfrac{18}{35}$ **49.** $\dfrac{3}{4}$ **51.** $\dfrac{16}{81}$

53. $\dfrac{7}{12}$ **55.** $\dfrac{14}{15}$ **57.** $\dfrac{3}{5}$

59. Four inches is $\dfrac{1}{3}$ of a foot, so 5′4″ is $5\dfrac{1}{3}$ feet.

61. Jimmy Howard saved $\frac{6}{7}$ of the shots on goal. The opposing team scored 4 points on Howard.

63. The fraction of fouled plugs is $\frac{1}{3}$.

65. Gyrid owes $\frac{20}{27}$ of her tuition bill.

67. The food bank was able to meet the needs of $\frac{4}{5}$ of their clients.

69. Of the elk, $\frac{3}{4}$ are cows.

71. 270 in. $= \frac{270}{12}$ ft $= \frac{45}{2}$ ft $= 22\frac{1}{2}$ ft

73. Find the prime factorization of both numerator and denominator. Then eliminate the common factors. The product of the remaining factors is the simplified fraction.

$$\frac{3 \cdot 5 \cdot 5 \cdot 7}{3 \cdot 3 \cdot 5 \cdot 5 \cdot 5} = \frac{7}{15}$$

75. Yes, all simplify to $\frac{5}{12}$. **77.** 252 **79.** 1089

81. 3552 **83.** 16 **85.** 16

SECTION 3.3

1. $\frac{2}{25}$ **3.** $\frac{15}{28}$ **5.** $\frac{10}{33}$ **7.** $\frac{1}{4}$ **9.** $\frac{16}{5}$ or $3\frac{1}{5}$

11. $\frac{5}{96}$ **13.** $\frac{3}{2}$ or $1\frac{1}{2}$ **15.** 1 **17.** $\frac{16}{15}$ or $1\frac{1}{15}$

19. 0 **21.** $\frac{13}{8}$ **23.** $\frac{1}{21}$ **25.** none **27.** $\frac{12}{61}$

29. 17 **31.** $\frac{3}{8}$ **33.** $\frac{32}{21}$ or $1\frac{11}{21}$ **35.** $\frac{1}{2}$

37. $\frac{3}{8}$ **39.** $\frac{4}{3}$ or $1\frac{1}{3}$ **41.** $\frac{49}{30}$ or $1\frac{19}{30}$ **43.** 1

45. $\frac{10}{3}$ or $3\frac{1}{3}$ **47.** $\frac{10}{11}$ **49.** $\frac{7}{8}$ **51.** $\frac{7}{48}$

53. 2 **55.** $\frac{4}{25}$ **57.** $\frac{8}{3}$ or $2\frac{2}{3}$

59. $\frac{3}{4} \cdot \frac{7}{6} = \frac{7}{8}$ because $\frac{3}{4} \cdot \frac{7}{6}$ can be simplified to $\frac{1}{4} \cdot \frac{7}{2}$; then we multiply 1 times 7 to get 7 and 4 times 2 to get 8.

61. The error comes from inverting $\frac{4}{5}$ before multiplying. We invert only in division: $\frac{3}{8} \cdot \frac{4}{5} = \frac{12}{40}$ or $\frac{3}{10}$.

63. There are 8 combinations of four children with a boy as firstborn, and 8 combinations with a girl as firstborn.

65. One combination has a firstborn boy followed by three girls.

67. The container has 60 oz remaining.

69. Becky might save 41 gallons per year.

71. About 12,000 of these patients are expected to survive 5 years.

73. About 29,687,850 did not have access to safe drinking water that year.

75. Multiply $\frac{35}{24}$ by $\frac{14}{40}$ (the reciprocal of $\frac{40}{14}$). Then simplify:
$\frac{35}{24} \div \frac{40}{14} = \frac{49}{96}$. **77.** $\frac{9}{10}$ **79.** $4\frac{2}{5}$ **81.** $3\frac{6}{13}$

83. $\frac{27}{8}$ **85.** $\frac{67}{9}$ **87.** Kevin can haul 15 scoops.

SECTION 3.4

1. $\frac{14}{25}$ **3.** $7\frac{23}{32}$ **5.** $7\frac{1}{3}$ **7.** $22\frac{1}{2}$ **9.** 24

11. $2\frac{4}{5}$ **13.** $18\frac{3}{4}$ **15.** 0 **17.** 28 **19.** $16\frac{1}{2}$

21. $32\frac{1}{15}$ **23.** 105 **25.** $2\frac{4}{7}$ **27.** $1\frac{27}{37}$ **29.** $5\frac{1}{10}$

31. $1\frac{5}{13}$ **33.** $1\frac{1}{3}$ **35.** $\frac{16}{25}$ **37.** $\frac{5}{26}$ **39.** $1\frac{4}{9}$

41. $14\frac{1}{2}$ **43.** $13\frac{1}{2}$ **45.** $17\frac{7}{16}$ **47.** $2\frac{9}{20}$

49. 315 **51.** $\frac{14}{45}$

53. Mixed numbers cannot be multiplied by multiplying the whole number parts and the fraction parts separately. Change them to fractions first: $\frac{5}{3} \cdot \frac{3}{2} = \frac{5}{2} = 2\frac{1}{2}$.

55. The interior of the shed has $42\frac{1}{6}$ ft² of space.

57. The first ring is $1400 per carat and the second ring is $2000 per carat.

59. The iron content is 22 parts per million.

61. The sale price of ground beef is $2 per pound.

63. The more-efficient car emits 800 fewer pounds of CO_2.

65. Shane will need three 2 × 4s.

67. There are exactly $77\frac{5}{7}$ boards in 272 in. Shane will need 78 boards.

69. When we divide by $1\frac{1}{2}$ we are dividing by $\frac{3}{2}$. Dividing by $\frac{3}{2}$ is the same as multiplying by $\frac{2}{3}$. Because $\frac{2}{3}$ is less than 1, the multiplication gives us a smaller number.

71. 1230 **73.** $8\frac{5}{9}$ **75.** $\frac{27}{7}$ **77.** $\frac{133}{9}$ **79.** 1080

81. Yes, because 456,726 is a multiple of 6.

GETTING READY FOR ALGEBRA

1. $x = \dfrac{3}{4}$ **3.** $y = \dfrac{10}{9}$ or $y = 1\dfrac{1}{9}$ **5.** $z = \dfrac{5}{16}$

7. $x = \dfrac{17}{8}$ or $x = 2\dfrac{1}{8}$ **9.** $a = \dfrac{10}{7}$ or $a = 1\dfrac{3}{7}$

11. $b = \dfrac{3}{2}$ or $b = 1\dfrac{1}{2}$ **13.** $z = \dfrac{1}{2}$ **15.** $a = \dfrac{11}{30}$

17. The distance is $\dfrac{3}{4}$ mi.

19. 180 lb of tin were recycled.

SECTION 3.5

1. $\dfrac{4}{6}, \dfrac{6}{9}, \dfrac{8}{12}, \dfrac{10}{15}$ **3.** $\dfrac{10}{12}, \dfrac{15}{18}, \dfrac{20}{24}, \dfrac{25}{30}$

5. $\dfrac{8}{18}, \dfrac{12}{27}, \dfrac{16}{36}, \dfrac{20}{45}$ **7.** $\dfrac{22}{26}, \dfrac{33}{39}, \dfrac{44}{52}, \dfrac{55}{65}$

9. $\dfrac{28}{10}, \dfrac{42}{15}, \dfrac{56}{20}, \dfrac{70}{25}$ **11.** 8   **13.** 18   **15.** 28

17. 4   **19.** 12   **21.** 4   **23.** 66   **25.** 150

27. 20   **29.** 104   **31.** $\dfrac{2}{23}, \dfrac{3}{23}, \dfrac{5}{23}$ **33.** $\dfrac{1}{2}, \dfrac{5}{8}, \dfrac{3}{4}$

35. $\dfrac{1}{4}, \dfrac{3}{10}, \dfrac{2}{5}$ **37.** false   **39.** false   **41.** true

43. $\dfrac{5}{9}, \dfrac{7}{12}, \dfrac{11}{18}$ **45.** $\dfrac{4}{15}, \dfrac{5}{6}, \dfrac{13}{15}, \dfrac{9}{10}$ **47.** $\dfrac{11}{24}, \dfrac{17}{36}, \dfrac{35}{72}$

49. $\dfrac{3}{7}, \dfrac{13}{28}, \dfrac{17}{35}$ **51.** $1\dfrac{9}{16}, 1\dfrac{5}{8}, 1\dfrac{13}{20}$ **53.** false

55. true   **57.** true   **59.** LCM = 24; $\dfrac{12}{24}, \dfrac{16}{24}, \dfrac{4}{24}, \dfrac{15}{24}$

61. From smallest to largest, the strengths are

$\dfrac{3}{32}, \dfrac{1}{8}, \dfrac{1}{4}, \dfrac{5}{16}, \dfrac{3}{8}, \dfrac{1}{2},$ and $\dfrac{9}{16}$.

63. More Americans believe in heaven.

65. More Americans believe in ghosts than believe in black magic.

67. Chang's measurement is heaviest.

69. Moe owns the largest part and Curly owns the smallest.

71. To simplify a fraction is to find a fraction with a smaller numerator and a smaller denominator that is equivalent to the original fraction. To build a fraction is to find a fraction with a larger numerator and a larger denominator that is equivalent to the original.

73. $\dfrac{50}{70}, \dfrac{65}{91}, \dfrac{115}{161}, \dfrac{560}{784}, \dfrac{2905}{4067}$ **75.** 21   **77.** 40   **79.** 288

81. $2^5 \cdot 3$
83. The cost was $3230.

SECTION 3.6

1. $\dfrac{9}{17}$ **3.** $\dfrac{4}{5}$ **5.** 1   **7.** $1\dfrac{7}{15}$ **9.** $\dfrac{16}{19}$ **11.** $1\dfrac{1}{4}$

13. $\dfrac{3}{4}$ **15.** $\dfrac{2}{3}$ **17.** $\dfrac{7}{10}$ **19.** $\dfrac{2}{5}$ **21.** $\dfrac{13}{24}$

23. $\dfrac{1}{2}$ **25.** $\dfrac{13}{16}$ **27.** $\dfrac{5}{9}$ **29.** 1   **31.** $\dfrac{53}{72}$

33. $1\dfrac{17}{20}$ **35.** $2\dfrac{1}{3}$ **37.** $1\dfrac{11}{12}$ **39.** $1\dfrac{32}{63}$ **41.** $1\dfrac{2}{5}$

43. $2\dfrac{17}{40}$ **45.** $\dfrac{5}{9}$ **47.** $1\dfrac{7}{40}$ **49.** $1\dfrac{13}{144}$

51. Only $\dfrac{21}{100}$ of all old tires are not dumped or sent to landfills. This is less than $\dfrac{1}{4}$ of all the old tires.

53. The home and garden users are more than $\dfrac{3}{4}$ of all pesticide users in this country.

55. $\dfrac{1}{5} + \dfrac{2}{5} = \dfrac{3}{5}$. When adding fractions, the parts of the unit must be the same size. So we must keep the common denominator in the result and not add the denominators.

57. The recipe makes $3\dfrac{3}{4}$ gallons of punch.

59. Jonnie needs a bolt that is $1\dfrac{13}{16}$ in. long.

61. The length of the pin is $\dfrac{3}{4}$ in.

63. The length of the rod is $3\dfrac{3}{8}$ in.

65. Shane has $\dfrac{5}{6}$ of the deck boards installed.

67. Each denominator indicates into how many parts a unit has been divided. Unless the units are the same size it would be like adding apples and oranges. You cannot add units of different sizes.

69. $\dfrac{4}{7}$ **71.** Jim consumes 4 g of fat, which is $\dfrac{6}{55}$ of the number of calories.

73. $14\dfrac{25}{48}$ **75.** $27\dfrac{17}{36}$ **77.** $8\dfrac{23}{24}$ **79.** 48

81. The instructor needs $418\dfrac{1}{2}$ in. of wire solder.

SECTION 3.7

1. $12\dfrac{8}{9}$ **3.** $11\dfrac{3}{5}$ **5.** $12\dfrac{2}{15}$ **7.** $12\dfrac{13}{14}$ **9.** $7\dfrac{5}{14}$

11. $10\dfrac{2}{15}$ **13.** $22\dfrac{3}{20}$ **15.** $11\dfrac{1}{2}$ **17.** $25\dfrac{3}{5}$

19. $18\dfrac{5}{18}$ **21.** $21\dfrac{1}{4}$ **23.** $20\dfrac{7}{12}$ **25.** $34\dfrac{14}{15}$

27. $337\frac{11}{36}$ **29.** $107\frac{14}{33}$ **31.** $55\frac{11}{12}$ **33.** $110\frac{43}{72}$

35. $41\frac{11}{30}$ **37.** $50\frac{3}{5}$ **39.** $18\frac{5}{8}$ **41.** $64\frac{3}{8}$

43. Nancy needs $98\frac{1}{2}$ in. of molding for the frame.

45. Elizabeth needs $10\frac{5}{8}$ yd of fabric for the dress and jacket.

47. Jeff lost $25\frac{3}{4}$ lb.

49. The perimeter is $7\frac{17}{40}$ in.

51. The total rainfall for the three months is $12\frac{67}{100}$ inches.

53. There are $61\frac{27}{80}$ mi of road that need to be resurfaced. It will cost about $920,000.

55. It is $17\frac{1}{2}$ in. from point A to point D.

57. I prefer the methods of this section rather than changing to improper fractions first because (a) it avoids fractions with large numerators and (b) requires fewer steps.

59. Yes, the statement is true.

61. 26 **63.** 222 **65.** 35 **67.** $\frac{10}{21}$

69. Millie's score is 35, which is under par.

SECTION 3.8

1. $\frac{2}{3}$ **3.** $\frac{1}{3}$ **5.** $\frac{3}{8}$ **7.** $\frac{1}{2}$ **9.** $\frac{1}{16}$ **11.** $\frac{7}{45}$

13. $\frac{2}{21}$ **15.** $\frac{1}{18}$ **17.** $\frac{7}{12}$ **19.** $\frac{7}{27}$ **21.** $\frac{1}{24}$

23. $\frac{23}{60}$ **25.** $\frac{1}{2}$ **27.** $\frac{19}{48}$ **29.** $\frac{5}{12}$ **31.** $\frac{11}{36}$

33. $\frac{9}{20}$ **35.** $\frac{13}{30}$ **37.** $\frac{7}{24}$ **39.** $\frac{17}{48}$ **41.** $\frac{5}{36}$

43. $\frac{3}{200}$ **45.** $\frac{31}{56}$

47. One-fifth of all Caucasian Americans have naturally blonde or red hair.

49. Willie has $\frac{5}{12}$ of the stimulus, or $200, left.

51. In that year, $\frac{9}{25}$ of the population did not hold stock.

53. The employed accounted for $\frac{22}{25}$ of the work force.

55. For Americans 80 or older, $\frac{8}{25}$ of the population are male.

57. The difference in diameters is $\frac{11}{16}$ in.

59. Discretionary spending accounted for $\frac{39}{100}$ of the total budget.

61. They are not equal, because the denominators (parts of the unit), 4 and 2, are not the same. We cannot get the result by subtracting the numerators and the denominators. We must first find the common denominator. So, $\frac{3}{4} - \frac{1}{2} = \frac{3}{4} - \frac{2}{4} = \frac{1}{4}$. **63.** $\frac{218}{625}$

65. The project will require 101 bricks, but 1 of them will be cut to fit. One brick, $3\frac{3}{4}$ in. wide, plus the mortar joint on one end, $\frac{3}{8}$ in., has a width of $4\frac{1}{8}$ in. The length of the garden is 10 ft, or 120 in. We must add 16 in. to the length to account for the border on the ends. The width of the garden is 72 in. We do not need to add on to the width. Each length requires $\left(136 \div 4\frac{1}{8}\right)$ bricks, or about 33 bricks. Each width requires $\left(72 \div 4\frac{1}{8}\right)$ bricks, or about 18 bricks.

67. 12,200 **69.** $6\frac{2}{5}$ **71.** $7\frac{9}{20}$ **73.** $27\frac{7}{10}$

75. It will take them 8 days to build the wall.

SECTION 3.9

1. $7\frac{2}{13}$ **3.** $12\frac{1}{3}$ **5.** $5\frac{1}{8}$ **7.** $2\frac{3}{7}$ **9.** $16\frac{17}{60}$

11. $2\frac{1}{2}$ **13.** $8\frac{5}{6}$ **15.** $172\frac{7}{9}$ **17.** $6\frac{5}{12}$ **19.** $8\frac{2}{5}$

21. $56\frac{1}{3}$ **23.** $15\frac{23}{60}$ **25.** $22\frac{5}{6}$ **27.** $15\frac{47}{48}$

29. $63\frac{1}{5}$ **31.** $2\frac{7}{32}$ **33.** $9\frac{31}{40}$ **35.** $74\frac{7}{8}$

37. $27\frac{31}{120}$ **39.** $74\frac{35}{48}$ **41.** $16\frac{19}{78}$ **43.** $\frac{27}{35}$

45. If 1 is borrowed from 16 in order to subtract the fraction part, $\frac{1}{4}$, then the whole-number part is 15 − 13, or 2. So $16 - 13\frac{1}{4} = \left(15 + \frac{4}{4}\right) - \left(13 + \frac{1}{4}\right) = 2\frac{3}{4}$.

47. She trims $2\frac{3}{8}$ lb. **49.** He has $12\frac{1}{20}$ tons left.

51. He won by $5\frac{3}{4}$ in.

53. It is $32\frac{5}{6}$ ft longer.

55. They have $19\frac{3}{10}$ mi to go.

57. There will be $5\frac{7}{8}$ inches of the board left over.

59. Haja can buy a bolt that is $\frac{3}{8}$ in. in diameter.

61. First change the fractions to equivalent fractions with a common denominator. Thus, $\frac{1}{3} = \frac{8}{24}$ and $\frac{5}{8} = \frac{15}{24}$. Because $\frac{15}{24}$ is the larger fraction, we must borrow 1 from the 4 and add it to $\frac{8}{24}$. We have $4\frac{8}{24} = 3 + 1\frac{8}{24} = 3\frac{32}{24}$. Now we can subtract: $3\frac{32}{24} - 2\frac{15}{24} = 1\frac{17}{24}$.

63. No, they are not equal.

65. The snail's net distance is $3\frac{2}{3}$ ft. It will take 6 days to gain over 20 ft.

67. 62 **69.** 18 **71.** 7 **73.** $4\frac{1}{2}$

75. The shipping weight is 1160 oz, or $72\frac{1}{2}$ lb.

GETTING READY FOR ALGEBRA

1. $a = \frac{1}{2}$ **3.** $c = \frac{5}{8}$ **5.** $x = \frac{11}{72}$ **7.** $y = 1\frac{38}{63}$

9. $a = 1\frac{11}{40}$ **11.** $c = 3\frac{11}{24}$ **13.** $x = 1\frac{1}{36}$

15. $3\frac{1}{6} = w$ **17.** $a = 36\frac{4}{9}$ **19.** $c = 21\frac{2}{21}$

21. The height 10 years ago was $43\frac{9}{16}$ ft.

23. She bought 46 lb of nails.

SECTION 3.10

1. $\frac{7}{11}$ **3.** $\frac{2}{17}$ **5.** $\frac{1}{2}$ **7.** $\frac{4}{5}$ **9.** $\frac{1}{40}$

11. $\frac{7}{8}$ **13.** $\frac{43}{50}$ **15.** $\frac{5}{8}$ **17.** $1\frac{5}{48}$ **19.** $\frac{41}{90}$

21. $\frac{5}{6}$ **23.** 0 **25.** $\frac{11}{48}$ **27.** 0 **29.** $\frac{6}{11}$

31. $\frac{11}{21}$ **33.** $\frac{16}{45}$ **35.** $\frac{16}{45}$ **37.** $2\frac{5}{9}$

39. $\frac{5}{12}$ **41.** $\frac{49}{90}$ **43.** $3\frac{2}{9}$ **45.** $7\frac{1}{5}$

47. $\frac{67}{75}$ **49.** $2\frac{2}{5}$ **51.** $3\frac{37}{54}$

53. $3\frac{1}{15}$ **55.** The average length is $33\frac{2}{3}$ in.

57. The mothers need 55 cups of cereal, $1\frac{1}{4}$ cups of butter, $33\frac{3}{4}$ cups of chocolate chips, $27\frac{1}{2}$ cups of marshmallows, 40 cups of pretzels, and 10 cups of raisins.

59. The average winning throw was 68 ft $6\frac{3}{40}$ in.

61. Normal body temperature is 37°C.

63. Shane will need three 2×12s, although there will be 6 ft left over.

65. 1. Do operations in parentheses first, following steps 2, 3, and 4. 2. Do exponents next. 3. Do multiplication and division as they occur. 4. Do addition and subtraction as they occur. The order is the same as for whole numbers.

67. $2\frac{57}{1000}$ **69.** The total amount paid is $19,444.

71. $2\frac{13}{14}$ **73.** $5\frac{5}{9}$ **75.** $2\frac{5}{8}$ **77.** $3 \cdot 5^2 \cdot 13$

79. Jason walked 25 miles and raised $250.

CHAPTER 3 REVIEW EXERCISES

SECTION 3.1

1. $\frac{4}{5}$ **3.** $\frac{9}{10}$ **5.** Improper fractions: $\frac{8}{3}, \frac{9}{9}, \frac{22}{19}$

7. $7\frac{5}{11}$ **9.** $82\frac{3}{5}$ **11.** $\frac{77}{12}$ **13.** $\frac{77}{6}$ **15.** $\frac{17}{1}$

17. The wholesaler can pack $2684\frac{19}{24}$ cases of beans.

SECTION 3.2

19. $\frac{3}{4}$ **21.** $\frac{2}{3}$ **23.** $\frac{3}{5}$ **25.** 17 **27.** $\frac{1}{3}$

29. $\frac{13}{16}$ **31.** $\frac{1}{5}$ **33.** $\frac{2}{3}$ **35.** She is $\frac{3}{5}$ years old.

SECTION 3.3

37. $\frac{6}{49}$ **39.** $\frac{18}{55}$ **41.** 1 **43.** $\frac{8}{3}$ **45.** $\frac{9}{14}$

47. $\frac{3}{4}$ **49.** $\frac{77}{90}$

51. Rent accounts for $\frac{1}{7}$ of their income.

SECTION 3.4

53. $2\frac{6}{25}$ **55.** $16\frac{5}{8}$ **57.** $2\frac{13}{16}$ **59.** $18\frac{3}{4}$ **61.** 8

63. $\frac{6}{7}$ **65.** $3\frac{3}{5}$ **67.** $28\frac{1}{5}$

69. She harvests 31,078 bushels of corn.

71. $\dfrac{4}{6}, \dfrac{6}{9}, \dfrac{10}{15}, \dfrac{16}{24}$ **73.** $\dfrac{6}{28}, \dfrac{9}{42}, \dfrac{15}{70}, \dfrac{24}{112}$ **75.** 36

77. 120 **79.** $\dfrac{1}{2}, \dfrac{3}{5}, \dfrac{7}{10}$ **81.** $\dfrac{1}{5}, \dfrac{2}{9}, \dfrac{3}{11}$

83. $\dfrac{5}{14}, \dfrac{3}{8}, \dfrac{11}{28}, \dfrac{3}{7}$ **85.** true **87.** false

89. The smallest is $\dfrac{7}{16}$ ton and the largest is $\dfrac{3}{4}$ ton.

91. $\dfrac{11}{15}$ **93.** $\dfrac{2}{3}$ **95.** $\dfrac{5}{8}$ **97.** $\dfrac{3}{5}$ **99.** $\dfrac{7}{15}$

101. $1\dfrac{5}{8}$ **103.** The bamboo grew $1\dfrac{1}{8}$ in.

105. $9\dfrac{1}{10}$ **107.** $15\dfrac{5}{14}$ **109.** $13\dfrac{5}{24}$

111. $25\dfrac{43}{80}$ **113.** $54\dfrac{3}{4}$ **115.** $60\dfrac{7}{12}$

117. Russ's total weight loss was $9\dfrac{11}{30}$ lb.

119. $\dfrac{9}{16}$ **121.** $\dfrac{5}{8}$ **123.** $\dfrac{13}{24}$ **125.** $\dfrac{1}{30}$

127. $\dfrac{17}{75}$ **129.** $\dfrac{7}{20}$ **131.** She has $\dfrac{5}{12}$ oz left.

133. $85\dfrac{3}{10}$ **135.** $15\dfrac{1}{5}$ **137.** $2\dfrac{5}{6}$ **139.** $6\dfrac{7}{48}$

141. $4\dfrac{41}{60}$ **143.** $10\dfrac{25}{48}$

145. a. During the year, $3\dfrac{5}{8}$ in. more rain falls in Westport.

b. In a 10-yr period, $277\dfrac{1}{12}$ in. more rain falls in Salem.

147. 1 **149.** $1\dfrac{7}{12}$ **151.** $\dfrac{1}{2}$ **153.** $1\dfrac{1}{64}$

155. $\dfrac{15}{32}$ **157.** $1\dfrac{13}{30}$

159. The class average is $\dfrac{7}{10}$ of the problems correct.

CHAPTER 3 TRUE/FALSE CONCEPT REVIEW

1. true **2.** false; written as a mixed number, $\dfrac{7}{8} = 0\dfrac{7}{8}$.

3. false; the numerator always equals the denominator, so the fraction is improper. **4.** true

5. false; when a fraction is completely simplified, its value remains the same. **6.** true **7.** true **8.** true

9. false; the reciprocal of an improper fraction is less than 1. **10.** true **11.** true **12.** true

13. false; like fractions have the same denominators.

14. false; to add mixed numbers, add the whole numbers and add the fractions.

15. true **16.** true **17.** true **18.** true

CHAPTER 3 TEST

1. $20\dfrac{1}{3}$ **2.** $1\dfrac{7}{24}$ **3.** $\dfrac{79}{9}$ **4.** $\dfrac{3}{8}, \dfrac{2}{5}, \dfrac{3}{7}$ **5.** $\dfrac{11}{1}$

6. 27 **7.** $9\dfrac{2}{15}$ **8.** $18\dfrac{20}{27}$ **9.** 0 **10.** $\dfrac{2}{3}$

11. $6\dfrac{4}{5}$ **12.** $\dfrac{1}{2}$ **13.** $\dfrac{2}{9}$ **14.** $\dfrac{1}{3}$ **15.** $\dfrac{12}{35}$

16. $6\dfrac{13}{20}$ **17.** $\dfrac{5}{8}$ **18.** $\dfrac{11}{14}$ **19.** $\dfrac{5}{16}$ **20.** $2\dfrac{5}{8}$

21. $\dfrac{7}{8}, \dfrac{7}{9}, \dfrac{8}{9}$ **22.** $2\dfrac{5}{8}$ **23.** $8\dfrac{13}{40}$ **24.** $\dfrac{7}{12}$

25. $7\dfrac{6}{11}$ **26.** $\dfrac{4}{5}$ **27.** $1\dfrac{7}{8}$ **28.** $\dfrac{9}{50}$

29. true **30.** $\dfrac{5}{5}, \dfrac{6}{6}, \dfrac{7}{7}$

31. There are 22 truckloads of hay in the rail car.

32. She needs to make 25 lb of candy.

Chapter 4
SECTION 4.1

1. twenty-six hundredths **3.** two hundred sixty-seven thousandths **5.** seven and two thousandths
7. eleven and ninety-two hundredths **9.** 0.42
11. 0.409 **13.** 9.059 **15.** 0.0308
17. eight hundred five thousandths **19.** eighty and five hundredths **21.** Sixty-one and two hundred three ten-thousandths **23.** ninety and three thousandths
25. 0.0035 **27.** 1800.028 **29.** 505.005

31. 65.065

	Unit	Tenth	Hundredth
33. 35.777	36	35.8	35.78
35. 729.638	730	729.6	729.64
37. 0.6157	1	0.6	0.62

39. $67.49   **41.** $548.72

	Ten	Hundredth	Thousandth
43. 35.7834	40	35.78	35.783
45. 86.3278	90	86.33	86.328
47. 0.91486	0	0.91	0.915

49. $72   **51.** $7822

53. The tax rate in Gainsville is twenty two and one thousand nine hundred six ten-thousandths.

55. The rounded tax rate for Ft. Lauderdale is 19.81.

57. Carlos writes "fifty-seven and seventy-nine hundredths dollars" or "fifty-seven and 79/100 dollars" on the check.

59. The position of the arrow to the nearest hundredth is 3.74.

61. The position of the arrow to the nearest tenth is 3.7.

63. seven hundred fifty-six and seven thousand one hundred four ten-thousandths

65. The account value to the nearest cent is $3478.59.
67. 213.1101   **69.** The population density of Kuwait is about 131 people per square kilometer.

71. The number appears to have been rounded to the nearest hundredth.

	Thousand	Hundredth	Ten-thousandth
73. 7564.35926	8000	7564.36	7564.3593
75. 78,042.38875	78,000	78,042.39	78,042.3888

77. The speed was rounded to the thousandths place.

79. The speed is 763 mph to the nearest mile per hour and 763.0 to the nearest tenth.

81. In 43.29, the numeral 4 is in the tens place, which is two places to the left of the decimal point. This means there are four tens in the number. In 18.64, the 4 is in the hundredths place, which is two places to the right of the decimal point. This means that there are four hundredths in the number.

83. The rounded value, 8.283, is greater than the original value. $8.283 > 8.28282828$ or $8.28282828 < 8.283$

85. The place value name is 0.00000522.   **87.** Answers will vary.   **89.** 77   **91.** 475   **93.** true

95. 6209, 6215, 6218, 6223, 6227

97. Kobe must hit 34 field goals.

SECTION 4.2

1. $\dfrac{83}{100}$ **3.** $\dfrac{13}{20}$ **5.** $\dfrac{329}{500}$ **7.** $\dfrac{41}{50}$ **9.** $\dfrac{12}{25}$

11. $10\dfrac{41}{100}$ **13.** $\dfrac{1}{8}$ **15.** $12\dfrac{6}{25}$ **17.** $11\dfrac{43}{125}$

19. $\dfrac{3}{4}$ **21.** 0.1, 0.4, 0.7

23. 0.06, 0.17, 0.24   **25.** 3.179, 3.185, 3.26

27. false

29. false   **31.** 0.046, 0.047, 0.047007, 0.047015, 0.0477

33. 0.555, 0.5552, 0.55689, 0.55699   **35.** 25.005, 25.0059, 25.051, 25.055   **37.** false   **39.** true

41. As a simplified fraction, the probability is $\dfrac{1}{16}$.

43. The best (lowest) bid for the school district is $2.6351 made by Tillamook Dairy.

45. As a simplified fraction, the highest percentage of free throws made in a season is $\dfrac{104}{125}$.

47. As a simplified fraction, the lowest percentage of free throws made by both teams in a game is $\dfrac{81}{200}$.

49. $\dfrac{67}{80}$ **51.** $25\dfrac{1}{40}$ **53.** Norado needs less acid.

55. 3.0007, 3.00077, 3.00092, 3.002, 3.00202

57. 82.58, 82.78, 82.80, 82.85, 82.86, 83.01, 83.15, 83.55

59. Maria should choose the 0.725 yd.

61. Belgium had fewer people per square kilometer.

63. The fat grams are $7\dfrac{19}{20}$, $8\dfrac{9}{10}$, and $10\dfrac{17}{20}$, respectively. The least amount of fat is in the frozen plain hash browns.

65. O'Neal made $\dfrac{609}{1000}$ of his attempts and missed $\dfrac{391}{1000}$ of his shots.

67. The table sorted by averages from lowest to highest is

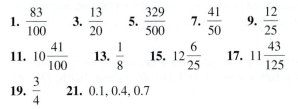

National League

Year	Name	Team	Average
2005	Derrek Lee	Chicago	0.335
2007	Matt Holliday	Colorado	0.340
2006	Freddy Sanchez	Pittsburgh	0.344
2004	Barry Bonds	San Francisco	0.362
2008	Chipper Jones	Atlanta	0.364

American League

Year	Name	Team	Average
2008	Joe Mauer	Minnesota	0.328
2005	Michael Young	Texas	0.331
2006	Joe Mauer	Minnesota	0.347
2007	Maglio Ordonez	Detroit	0.363
2004	Ichiro Suzuki	Seattle	0.372

69. $\dfrac{11}{25}, \dfrac{101}{250}, \dfrac{1011}{25{,}000}$ **71.** 11,509   **73.** 6785

75. $\dfrac{11}{8}$ or $1\dfrac{3}{8}$ **77.** $\dfrac{1}{64}$

79. The attendance at the theater was 4593.

SECTION 4.3

1. 1.4   **3.** 5.9   **5.** 15.8   **7.** 40.09   **9.** 4

11. 55.867   **13.** 31.68   **15.** 7.065   **17.** 9.0797

19. 157.485   **21.** 0.1714   **23.** 1.2635

25. 120.3876   **27.** 795.342   **29.** 38.268

31. 0.3   **33.** 6.1   **35.** 4.1   **37.** 18.04

39. 20.87   **41.** 0.266   **43.** 2.244   **45.** 0.369

47. 3.999   **49.** 81.906   **51.** 0.03086   **53.** 7.2189

55. 0.0727   **57.** 4.24   **59.** 0.189   **61.** 0.2719

63. 0.7966

65. Manuel bought 85.2 gallons of gas on his trip.

67. 1689.4

69. The total gross state products for the six states is $6.138 trillion.

71. The total state gross product for the east coast states is $2.441 trillion dollars.

73. Jack spent $30.88 on gas.

75. The total cost of the groceries is $28.73.

77. The total length of the Seikan and the Dai-shimizu is 76.44 km.

79. The period from 2005 to 2010 is projected as the largest increase, 3.7 million.

81. The top of the tree is 83 ft from ground level.

83. The distance is 6.59375 in.

85. Sera is 0.778 second faster than Muthoni.

87. The relay race was completed in 38.95 sec.

89. The fourth sprinter must have a time of 13.16 sec or less.

91. When subtracting fractions, we must first get common denominators and then subtract the numerators. When subtracting decimals, we align decimal points. This results in columns of like place value, which are then subtracted. The numbers in each column have the same place value, which means if they were written as fractions they would have common denominators. Thus, aligning the decimal points results in subtracting numbers with common denominators.

93. A total of 26 5.83s must be added.

95. The missing number is 0.1877.

97. The difference is 0.4475.   **99.** $20\dfrac{37}{40}$ or 20.925

101. 391,571   **103.** 42,579   **105.** $\dfrac{2}{3}$ **107.** $25\dfrac{1}{5}$

109. Harry and David will need 8280 pears to fill the order.

GETTING READY FOR ALGEBRA

1. $x = 11.1$ **3.** $y = 13.83$ **5.** $t = 0.484$
7. $x = 12.51$ **9.** $w = 7.2$ **11.** $t = 8.45$
13. $a = 3.84$ **15.** $x = 16.17$ **17.** $a = 17.7$
19. $s = 7.273$ **21.** $c = 596.36$

23. Two years ago the price of the water heater was $427.73.

25. The markup is $158.66.

27. Let C represent the cost of the groceries, then $24 + C = $61. The shopper can spend $37 on groceries.

SECTION 4.4

1. 4.5   **3.** 9.5   **5.** 0.72   **7.** 0.64   **9.** 0.036
11. 0.126   **13.** four   **15.** 0.06176   **17.** 11.468
19. 2.8728   **21.** 12.0768   **23.** 22.631
25. 0.023852   **27.** 0.167232   **29.** 0.004455
31. 32.2812   **33.** 1.2672   **35.** 27.1

37. 2221.8   **39.** 532.79   **41.** 108.40

43. Grant purchased a total of 99 gallons of gas.

45. To the nearest cent, Grant paid $48.37 for the fifth fill-up.

47. Grant paid the least for his fill-up at $2.399 per gallon.

49. 288.1675   **51.** 9164.632   **53.** 22,569.207

55. 3958.00   **57.** Sonya paid $30.41 for the steak.

59. It costs $179.25 to rent the midsize car.

61. It costs $31.52 less to rent the full-size car.

63. Store 1 is selling the freezer–refrigerator for the least total cost.

65. Steve burns 2384.25 calories per week.

67. The total linear feet of steel is 744.975 ft.

69. The total weight of beef consumed is 317.6 lb in 1995, 254.8 lb in 2005, and 236 lb in 2015. Greater awareness of potential heart problems associated with the consumption of large quantities of red meat probably accounts for the decline.

71. The number of gallons of water used by each type of toilet is as follows: pre-1970, 1,149,005 gallons; 1970s, 731,185 gallons; low-flow, 323,810.5 gallons. A total of 825,194.5 gallons of water is saved by using the low-flow model.

73. The Sanchez estate owes $4055 in taxes.

75. Bolt's second 100 meters took 9.61 sec.

77. The number of decimal places in the product is the sum of the number of decimal places in the two factors. So the product of 5.73 and 4.2 will have three decimal places.

79. The smallest whole number is 111.

81. The missing number is 0.0252.

83. The missing number is 3.   **85.** 820,000
87. 2200   **89.** 692,000   **91.** 476   **93.** 4210

SECTION 4.5

1. 1.93   **3.** 9260   **5.** 1355.7   **7.** 0.5818
9. 83.25   **11.** 1070   **13.** right   **15.** 78,324
17. 9.97   **19.** 57,900   **21.** 0.90775
23. 2,873,000   **25.** 60.5632

27. 0.0003276   **29.** 7.5×10^5 **31.** 9.1×10^{-5}

33. 4.1953×10^3 **35.** 1,200,000   **37.** 0.004

39. 943,000   **41.** 4.37×10^4 **43.** 5.87×10^{-7}
45. 6.84×10^{-11} **47.** 6.4004×10
49. 0.00007341   **51.** 1,770,000,000
53. 0.0000000311   **55.** 0.0000000148

57. The total cost of the tires is $3968.

59. The total cost of Ms. James's land is $310,000.

61. The land area of Earth is 5.2×10^7 square miles.

63. One nanometer is 1×10^{-9} m.

65. Light travels 1 mile in approximately 5.4×10^{-5} second.

67. The shortest wavelength of visible light is 0.00004 cm.

69. The total energy use of the four families is 1,232,300,000 BTUs.

71. The per capita consumption of milk in Cargill Cove in 2010 was 30.6 gallons.

73. They are multiplying the actual number by 1000.

75. The mathematically calculated average is 0.424.

77. The approximate length of one parsec is 1.92×10^{13} miles.

79. 6.5×10^6 **81.** 1.3365×10^3 or 1336.5

83. 312   **85.** 62 R 1   **87.** 2596   **89.** 9200

91. The width of the rectangle is 21 in. or 1.75 ft.

SECTION 4.6

1. 0.8   **3.** 4.9   **5.** 326.7   **7.** 3030   **9.** 5.53
11. 0.7875   **13.** whole number   **15.** 1.2   **17.** 6.6
19. 0.10   **21.** 8.12   **23.** 1.92   **25.** 452.90
27. 9.153   **29.** 0.016   **31.** 14.444   **33.** 7.1
35. 7.95   **37.** 7.3   **39.** 12.4   **41.** 9.775
43. 6.62, 6.75, no mode   **45.** 20.848, 22.46, 23.64
47. 15.26, 15.4, no mode   **49.** 0.668, 0.6685, no mode
51. 0.56396, 0.5506, 0.5066

53. The average closing price of Microsoft was $24.64.

55. 91.74

57. To the nearest tenth of a cent, the unit price of apples is $0.948, or 94.8¢, per pound.

59. To the nearest tenth of a cent, the unit price of pork chops is $1.991, or 199.1¢ per pound.

61. To the nearest cent, the cost of 1.5 lb of cod fillet is $5.99.

63. To the nearest cent, the average donation was $959.41.

65. Orlando has an average daily temperature of 82.9°F. The median of the average daily temperatures is 84°F, and the modes of the average daily temperatures are 73°F, 78°F, 84°F, and 92°F.

67. The average price per gallon is 2.502. The median price per gallon is 2.489. The mode prices per gallon are 2.489 and 2.599.

69. The average low temperature for the cities was 25.3°F.

71. The mode for the high temperatures was 33°F.

73. To the nearest mile, June got 49 miles per gallon.

75. To the nearest foot, there are 315 ft of cable on the spool.

77. To the nearest tenth, the length of the beam is 19.3 ft.

79. Denmark has the smallest area and Norway has the smallest population.

81. Denmark, the country with the largest population density, is over six times more crowded than Sweden. Finland and Norway are about equal in density.

83. To the nearest hundredth, the ERA is 1.64.

85. The fraction is $\frac{7}{10}$; it means that the runner is successful 7 out of 10 times.

87. The decimal point in a quotient is positioned directly above the decimal point in the dividend when dividing by a whole number. When dividing by a decimal, move the decimal point in the divisor to make it a whole number.

Move the decimal point in the dividend the same number of places. Now the decimal point in the quotient is positioned above the relocated decimal in the dividend. This works because a division problem may be thought of as a fraction $\left(\dfrac{\text{dividend}}{\text{divisor}}\right)$ and moving the decimals in both the same number of places to the right is the same as multiplying a fraction by 1 in the form of some power of 10 over itself.

89. Division Procedure

Operation on Decimals	Procedure	Example
Division	If the divisor is not a whole number, move the decimal point to the right until it is a whole number. Move the decimal point in the dividend the same number of places to the right. Now place the decimal point in the quotient directly above the decimal point in the dividend. Now divide.	$0.25\overline{)3.5675}$ $\dfrac{14.27}{25\overline{)356.75}}$ $\dfrac{25}{106}$ $\dfrac{100}{6\,7}$ $\dfrac{5\,0}{1\,75}$ $\dfrac{1\,75}{0}$

91. 5.95

93. $\dfrac{5}{6}$ **95.** $\dfrac{52}{11}$ **97.** $17\dfrac{11}{12}$ **99.** 68 **101.** $\dfrac{3}{5}$

GETTING READY FOR ALGEBRA

1. $x = 7$ **3.** $y = 308.5$ **5.** $t = 0.23$
7. $m = 0.06$ **9.** $q = 562.5$ **11.** $h = 28.125$
13. $y = 2.66$ **15.** $c = 0.2028$ **17.** $x = 1.04$
19. $s = 0.02415$ **21.** $z = 85.7616$

23. There are 23 servings.

25. The current is 9.5 amps.

27. The length of the rectangle is 18.4 ft.

29. Let S represent the number of students; then $20S = 3500$. The instructor has 175 students in her classes.

SECTION 4.7

1. 0.75 **3.** 0.15 **5.** 0.8125
7. 6.625 **9.** 56.584

	Tenth	Hundredth
11. $\dfrac{3}{7}$	0.4	0.43
13. $\dfrac{5}{12}$	0.4	0.42
15. $\dfrac{11}{13}$	0.8	0.85
17. $\dfrac{2}{15}$	0.1	0.13
19. $7\dfrac{7}{18}$	7.4	7.39

21. $0.8\overline{1}$ **23.** $0.08\overline{3}$

	Hundredth	Thousandth
25. $\dfrac{9}{79}$	0.11	0.114
27. $\dfrac{45}{46}$	0.98	0.978

29. $0.\overline{384615}$ **31.** $0.\overline{857142}$

33. The micrometer reading will be 2.375 in.

35. The measurements are 0.375 in., 1.25 in., and 0.5 in.

	Hundredth	Thousandth	Ten-thousandth
37. $\dfrac{21}{52}$	0.40	0.404	0.4038
39. $16\dfrac{15}{101}$	16.15	16.149	16.1485

41. The cost of the fabric is $4\dfrac{2}{25}$, or $4.08, per yard.

Decimals are easier to use because we write money in decimal form.

43. Michael can run a mile in 6 minutes 27 sec.

45. Anchorage had 19.23 hours of daylight.

47. The winning time was $2\dfrac{101}{240}$ hr, or about 2.42 hours.

49. The larger is $\dfrac{7}{625}$.

51. The fraction is less than 0.1. Yes, the decimal equivalent is 0.015375.

53. 4.69, 0.02   **55.** 9   **57.** 13   **59.** 80
61. 210,000

63. The average sale price of the houses was $257,682.

SECTION 4.8

1. 0.5   **3.** 0.02   **5.** 0.3   **7.** 29.4   **9.** 0.68
11. 10.07   **13.** 50.8   **15.** 2.22   **17.** 10.5
19. 26.264   **21.** 0.20   **23.** 0.4   **25.** 10
27. 1   **29.** 0.0018   **31.** 40   **33.** ten
35. hundred-thousandth

37. The estimate is 0.19. The answer is reasonable.

39. The estimate is 0.0003. The answer is reasonable.

41. Elmer spends $15.55.   **43.** d   **45.** 19.3

47. 25.8145   **49.** 26.539

51. The estimate is 0.0016, so Alex's answer is reasonable.

53. The estimate is $120, but since we rounded the price down, she will not have enough money for the shirts.

55. The perimeter is about 111 yd.

57. Rosalie will save $636.48 a year by bringing her lunch from home.

59. Matthew paid $116.60 for the items.

61. The average earning of the nine players was $473,555,56.

63. The Red Wings averaged 29 shots per game and the Penguins averaged 26.7 shots per game.

65. In the problem $0.3(5.1)^2 + 8.3 \div 5$, we divide 8.3 by 5 and add the result to $0.3(5.1)^2$. We do this because the order of operations requires that we divide before adding. In the problem $[0.3(5.1)^2 + 8.3] \div 5$, the entire quantity $[0.3(5.1)^2 + 8.3]$ is divided by 5. The insertion of the brackets has changed the order of operations so that we add before dividing because the addition is inside grouping symbols.

67. $3.62 \div (0.02 + 72.3 \cdot 0.2) = 0.25$

69. $(1.4^2 - 0.8)^2 = 1.3456$ **71.** Answers will vary.

73. 0.84375   **75.** 2.32

77. $\dfrac{51}{125}$ **79.** $6\dfrac{21}{25}$

81. The clerk should price the TV at $490.38.

GETTING READY FOR ALGEBRA

1. $x = 8.16$ **3.** $x = 3$ **5.** $0.1 = t$
7. $x = 741$ **9.** $x = 0.32$ **11.** $m = 63.1$
13. $y = 23.56$ **15.** $p = 4.375$
17. $24.3 = x$ **19.** $h = 26.8$ **21.** $c = 6.275$

23. The Celsius temperature is 120°C.

25. Gina has made 15 payments.

27. Let H represent the hours of labor; then $36H + 137.50 = 749.50$. She put in 17 hours of labor on the repair.

CHAPTER 4 REVIEW EXERCISES

SECTION 4.1

1. six and twelve hundredths   **3.** fifteen and fifty-eight thousandths   **5.** 21.05   **7.** 400.04

	Tenth	Hundredth	Thousandth
9. 34.7648	34.8	34.76	34.765
11. 0.467215	0.5	0.47	0.467

SECTION 4.2

13. $\dfrac{19}{25}$ **15.** $\dfrac{8}{3125}$ **17.** 0.89, 0.95, 1.01
19. 0.717, 7.017, 7.022, 7.108   **21.** true

SECTION 4.3

23. 34.708   **25.** 13.7187   **27.** 48.1149
29. 32.6 in.
31. Hilda's take-home pay is $3464.69.

SECTION 4.4

33. 28.245   **35.** 0.0243402   **37.** 0.148
39. 0.0019   **41.** Millie pays $1243.31 for 23.75 yd.

SECTION 4.5

43. 0.013765   **45.** 73,210   **47.** 7.8×10^{-3}
49. 1.43×10^{-5} **51.** 70,000,000
53. 0.0641   **55.** The average price of a bat is $37.35.

SECTION 4.6

57. 0.037   **59.** 177.1875   **61.** 0.84   **63.** 70.77

65. To the nearest cent, the average donation was $65.45.

67. 7.912, 8.09   **69.** 0.519, 0.56135

71. The unit price is $1.563 per ounce.

SECTION 4.7

73. 0.5625   **75.** 17.376   **77.** 0.61   **79.** $0.\overline{692307}$

81. The value of the Microsoft share is 24.28.

SECTION 4.8

83. 5.27   **85.** 31.998   **87.** 33.982

89. The estimate is 9.8, so Jose's answer is reasonable.

91. The estimate is 0.024, so Louise's answer is reasonable.

93. The estimate is $270, so Ron should have enough money in the budget.

CHAPTER 4 TRUE/FALSE CONCEPT REVIEW

1. false; the word name is "seven hundred nine thousandths."   **2.** true

3. false; the expanded form is $\frac{8}{10} + \frac{5}{100}$.   **4.** true

5. false; the number on the left is smaller because 732,687 is less than 740,000.

6. true   **7.** true   **8.** true

9. false; the numeral in the tenths place is 7 followed by a 4, so it rounds to 356.7.

10. true

11. false; $9.7 - 0.2 = 9.5$, as we can see when the decimal points are aligned.

12. false; if we leave out the extra zeros in a product such as $0.5(3.02) = 1.51$, the answer has fewer decimal places.

13. true   **14.** false; the decimal moves to the left, resulting in a smaller number.

15. false; move the decimal left when the exponent is negative.   **16.** true

17. false; most fractions have no exact terminating decimal form. For example, the fractions $\frac{1}{3}, \frac{1}{6}, \frac{1}{7}$, and $\frac{1}{9}$ have no exact terminating decimal equivalents.

18. false; $\frac{4}{11} = 0.\overline{36}$ **19.** true   **20.** true

CHAPTER 4 TEST

1. 5.453   **2.** 0.6699, 0.6707, 0.678, 0.6789, 0.682
3. seventy-five and thirty-two thousandths   **4.** 33.7212

5. 0.184   **6.** 57.90   **7.** 72.163   **8.** $18\frac{29}{40}$

9. 7.23×10^{-7} **10.** 0.739   **11.** 73,000

12. 21.82   **13.** 97.115   **14.** 0.0000594

15. 9045.065   **16.** 91.7   **17.** 3.0972×10^5

18. 46.737   **19.** 0.05676   **20.** 0.00037

21. The average monthly offering was $126,774.33.

22. 7895.308   **23.** Grant pays $67.86 for the plants.

24. Allen Iverson had played in 72 games.

25. The player's slugging percentage is 686.

26. Jerry lost 0.03 lb more than Harold.

Chapter 5
SECTION 5.1

1. $\frac{4}{15}$ **3.** $\frac{1}{5}$ **5.** $\frac{4}{5}$ **7.** $\frac{1}{2}$ **9.** $\frac{3}{7}$ **11.** $\frac{7}{15}$

13. $\frac{\$227}{2 \text{ donors}}$ **15.** $\frac{55 \text{ mi}}{1 \text{ hr}}$ **17.** $\frac{35 \text{ mi}}{1 \text{ gal}}$

19. $\frac{15 \text{ rose bushes}}{2 \text{ rows}}$ **21.** $\frac{2 \text{ trees}}{7 \text{ ft}}$

23. $\frac{17 \text{ scholarships}}{48 \text{ applicants}}$

25. $\frac{60 \text{ apples}}{1 \text{ box}}$ **27.** $\frac{15 \text{ pies}}{2 \text{ sales}}$

29. 40 mpg (miles per gallon)   **31.** 4 ft per sec

33. 32 cents per lb   **35.** 0.008 qt per mi

37. 2.4 children per family   **39.** 83.3 ft per sec

41. 187.8 lb per in^2 **43.** 741.7 gal per hr

45. a. The price per rose of the special is $14.24.
b. This is a savings of $14.85 over buying four separate roses.

47. The 8-oz can is 9.88 cents per ounce and the 20-oz can is 10.95 cents per ounce, so the 8-oz can is the better buy.

49. The 10-lb sale bag is the better buy, at a cost of 54.9 cents per pound.

51. a. The salsa costs 22¢ per ounce.   **b.** The sale price of the salsa is 16¢ per ounce.   **c.** Jerry can save $3.84.

53. Answers will vary. For example, meat, fish, fruit, vegetables, and bulk food are usually priced by the pound, so the unit price is the same for any amount purchased. Boxed items such as cereal, dried pasta, cookies, laundry detergent, and pet food are usually cheaper in larger quantities. Occasionally, meat and other perishables may be discounted because they are near their expiration date, so the unit price is lower.

55. a. The ratio of people with type O blood to all people is $\frac{23}{50}$.

b. The ratio of people with type AB blood to people with type A or B blood is $\frac{2}{25}$.

57. The ratio of sale price to regular price is $\frac{4}{7}$.

59. The population density of Dryton City is about 97.6 people per square mile.

61. Answers will vary.

63. a. The ratio of laundry use to toilet use is $\frac{7}{20}$.

b. The ratio of bath/shower use to dishwashing use is $\frac{16}{3}$.

65. A total of 1.25 mg of lead is enough to pollute 25 L of water.

67. The retrieval rate is $\frac{105 \text{ in.}}{5 \text{ turns}} = 21$ in. per turn.

69. The average number of credits per student was $\frac{4020 \text{ credits}}{645 \text{ students}} \approx 6$ credits per student. The FTE for the term was $\frac{4020 \text{ credits}}{15 \text{ credits}} = 268$ FTE.

71. This is a ratio, because no units are given. It means that 1 inch on the map represents 150,000 inches (about 2.37 miles) in the real world.

73. A ratio compares two measurements. The ratio of \$5 to \$1000 is $\frac{5}{1000}$, or $\frac{1}{200}$. A rate compares unlike measurements. The rate of 350 cents per 20 oz is $\frac{350 \text{ cents}}{20 \text{ oz}}$, or $\frac{35 \text{ cents}}{2 \text{ oz}}$. A unit rate compares unlike measurements, where the second measurement is 1 unit.

The unit rate of 350 cents to 20 oz is 17.5¢ per ounce.

75. Examples of 2-to-1 ratios are eyes to people, hands to people, and legs to people. Examples of 3-to-1 ratios are triplets to mothers, petals to trillium flowers, and leaves to poison ivy stems.

77. $\frac{1}{15}$ **79.** 0.464   **81.** 0.525

83. $4\frac{8}{15}$ **85.** $\frac{94}{125}$

SECTION 5.2

1. true   **3.** false   **5.** false   **7.** true   **9.** false

11. false   **13.** true   **15.** $d = 20$ **17.** $c = 6$

19. $y = 10$ **21.** $c = 21$ **23.** $x = \frac{9}{4}$ **25.** $p = \frac{5}{3}$

27. $x = 11.5$ **29.** $z = 4.4$ **31.** $w = 4\frac{1}{3}$

33. $a = 12.8$ **35.** $d = 100$ **37.** $b = \frac{3}{8}$

39. $x = 0.45$ **41.** $w = 0.75$ **43.** $y = 250$

45. $t = 8$ **47.** $w \approx 1.4$ **49.** $y \approx 7.8$

51. $a \approx 0.33$ **53.** $c \approx 1.24$

55. The value in the box must be 10 so that the cross products are the same.

57. The error is multiplying straight across instead of finding cross products. Multiplying straight across goes with fraction multiplication, but cross multiplication goes with proportions.

59. Fran's minimum allowable level of HDL is about 39.

61. About 20 of the rivers in New York would be expected to show an increase in pollution.

63. About 93 girls in the school could be expected to be on a diet.

65. The proportion $\frac{1}{196,416} = \frac{1}{200,000}$ is not mathematically true, but it is close to being true. The atlas probably used 3.1 miles as a rounded number.

67. $3.5(y) = \left(\frac{1}{4}\right)7$ To solve the proportion, first cross multiply.

$3.5y = \frac{7}{4}$

$y = \frac{7}{4} \div 3.5$ Rewrite as division. We can simplify using decimals or fractions.

If we choose fractions, we get $y = \frac{7}{4} \div \frac{7}{2}$, or $y = \frac{1}{2}$.

If we choose decimals, we get $y = 1.75 \div 3.5$, or $y = 0.5$.

69. $a = 9$ **71.** $w \approx 8.809$ **73.** 175.1804

75. 6.23   **77.** 0.0004835   **79.** 103.1475   **81.** 5

SECTION 5.3

1. 6 **2.** 4 **3.** 15 **4.** h **5.** $\dfrac{6}{4} = \dfrac{15}{h}$
6. 10 in. **7.** 30

8. 18 **9.** h **10.** 48 **11.** $\dfrac{30}{18} = \dfrac{h}{48}$ **12.** 80 ft

13. 9 **14.** 12

15. N **16.** 108 **17.** $\dfrac{9}{12} = \dfrac{N}{108}$

18. The manufacture will make 81 footballs.

19. T **21.** $\dfrac{3}{65} = \dfrac{T}{910}$

23. The school will need 4 additional teachers.

25. $\dfrac{18}{2} = \dfrac{75}{x}$

27. The bank requires a payment of $239.38.

29. a. There were about 400,282 males in Bahrain in 2009.

b. The ratio of males to females in Bahrain was $\dfrac{11}{9}$, or about 1.22 males per female.

31. The family would spend $9000 on food.

33. Fifty pounds of fertilizer will cover 2500 ft^2.

35. $\dfrac{11}{15} = \dfrac{x}{30}$ **37.** About 121,813,780 people in China cannot read.

39. A 42-pound child should receive 280 mg of Tylenol.

41. It will take 17 hr to go 935 mi.

43. The second room will cost $350.90.

45. Ida needs 0.4 cm^3 (0.4 cc) of the drug.

47. Abbey's waist should be no more than 28″.

49. Betty needs 12 lb of cashews and 28 lb of peanuts.

51. Debra needs 42 quarts of blue paint.

53. Lucia needs 65 lb of ground round for the meatballs.

55. The cost of the computer is about £555.13.

57. The comparable price for a 60-month battery is $141.65.

59. 12 m^3 of air must have less than 2820 mg of ozone.

61. The store must reduce the price of each box by 17¢ in order to beat the warehouse outlet's price.

63. The actual bridges are about 3.09 mi apart.

65. A proportion is composed of two equal ratios or rates. The speed of a vehicle can be measured in different units: $\dfrac{60 \text{ mi}}{1 \text{ hr}} = \dfrac{88 \text{ ft}}{1 \text{ sec}}$. Food recipes can be cut or expanded by keeping the ingredient measures in proportion. Prices for many items, especially food and clothing, are proportional to the number purchased.

67. From a consumer's viewpoint, it is better if there is a price break when buying larger quantities. Many products have a lower unit price when bought in bulk or larger packages.

69. The drive shaft speed is 1120 revolutions per minute.

71. 167.85; 200 **73.** $0.01399 > 0.011$

75. Mrs. Diado will pay $17.88 for 12 lb of the ground beef.

77. $\dfrac{127}{200}$ **79.** 0.615

CHAPTER 5 REVIEW EXERCISES

SECTION 5.1

1. $\dfrac{1}{5}$ **3.** $\dfrac{6}{5}$ **5.** $\dfrac{3}{4}$ **7.** $\dfrac{2}{3}$ **9.** $\dfrac{9 \text{ people}}{10 \text{ chairs}}$

11. $\dfrac{8 \text{ applicants}}{3 \text{ jobs}}$ **13.** $\dfrac{14 \text{ books}}{3 \text{ students}}$ **15.** $\dfrac{85 \text{ people}}{3 \text{ committees}}$

17. 25 mph **19.** 9¢ per pound of potatoes

21. 37.5 mpg **23.** $0.35 per croissant

25. No, the first part of the country has a rate of 3.5 TVs per household and the second part of the country has a rate of about 3.3 TVs per household.

SECTION 5.2

27. true **29.** false **31.** false **33.** $r = 11$

35. $t = 9$ **37.** $f = \dfrac{3}{5}$ **39.** $r = 10\dfrac{2}{3}$

41. $t = 5\dfrac{22}{25}$ **43.** $a \approx 10.6$ **45.** $c \approx 27.4$

47. The store brand must be cheaper than $6.24 for a 50-load box.

SECTION 5.3

49. Merle will study 37.5 hr per week.

51. Juan must work 180 hr to pay for his tuition.

53. Larry makes $43.75.

CHAPTER 5 TRUE/FALSE CONCEPT REVIEW

1. true **2.** true **3.** false;
$\dfrac{18 \text{ miles}}{1 \text{ gallon}} = \dfrac{54 \text{ miles}}{3 \text{ gallons}}$, or $\dfrac{18 \text{ miles}}{1 \text{ hour}} = \dfrac{54 \text{ miles}}{3 \text{ hours}}$

4. false; to solve a proportion, we must know the values of three of the four numbers.

5. false; if $\dfrac{8}{5} = \dfrac{t}{2}$, then $t = 3\dfrac{1}{5}$.

6. false; in a proportion, two ratios are *stated* to be equal but they may not be equal.   **7.** true   **8.** true

9. true   **10.** false; the table should look like this:

	First Tree	Second Tree
Height	18	x
Shadow	17	25

CHAPTER 5 TEST

1. $\dfrac{4}{5}$ **2.** Ken would answer 45 questions correctly.

3. $w = 0.9$ **4.** false   **5.** $y = 6.5$ **6.** true

7. Mary should be paid $84.24.   **8.** $\dfrac{1}{9}$

9. The price of 10 cans of peas is $8.20.

10. One would expect 17.5 lb of bones.

11. $x = 0.625$ **12.** They will catch 24 fish.

13. Jennie will need 10 gal of gas.   **14.** $a = 5.3$

15. yes   **16.** The tree will cast a 10.5-ft shadow.

17. The population density is 37.2 people per square mile.

18. $y \approx 1.65$ **19.** The crew could do 26 jobs.

20. There are 30 females in the class.

Chapter 6
SECTION 6.1

1. 29%   **3.** 72%   **5.** 125%   **7.** 62%   **9.** 32%

11. 56%   **13.** 48%   **15.** 55%   **17.** 130%

19. 100%   **21.** 250%   **23.** 42.5%   **25.** 37.5%

27. 87.5%   **29.** 37.5%   **31.** $66\dfrac{2}{3}\%$ **33.** 2

35. The percent of eligible voters who exercised their right to vote was 82%.

37. $32\dfrac{1}{4}\%$ **39.** 350%   **41.** $766\dfrac{2}{3}\%$

43. The luxury tax is 11%.

45. The annual interest rate is 1.95%.

47. The tax rate is 1.48%.

49. In 1980, 0.8% of the juvenile population was arrested for burglary. In 2007, the percent was about 0.225%.

51. In 2007, $33\dfrac{1}{3}\%$ of women 25 to 29 had a bachelor's degree.

53. Carol spends 82% of her money on the outfit.

55. The percent paid in interest was 20%.

57. Percent is the amount per hundred. It is the numerator of a fraction with a denominator of 100. It is the number of hundredths in a decimal.

59. $\dfrac{109}{500}$; 21.8%   **61.** $\dfrac{776}{500}$ or $\dfrac{194}{125}$; $1\dfrac{69}{125}$, 155.2%

63. 4733.5   **65.** 207,800

67. 0.0000672   **69.** 0.009003

71. Ms. Henderson earned $3172.05 during the month.

SECTION 6.2

1. 47%   **3.** 232%   **5.** 8%   **7.** 496%

9. 1900%   **11.** 0.83%   **13.** 95.2%   **15.** 59.2%

17. 7.31%   **19.** 2000%   **21.** 1781%

23. 0.044%   **25.** 710%   **27.** 88.67%

29. $81.1\overline{6}\%$ or $81\dfrac{1}{6}\%$

31. 24.09%   **33.** 0.96   **35.** 0.73

37. 0.0135   **39.** 9.08   **41.** 6.525   **43.** 0.000062

45. 0.00071   **47.** 39.4   **49.** 0.00092   **51.** 1

53. 6.62   **55.** 0.005   **57.** 0.001875   **59.** 0.7375

61. 0.00125   **63.** 4.13773

65. The tax on income in Colorado was 4.63%.

67. The Girl Scout sold 36% of her quota of cookies on the first day.

69. The employees will get a raise of 0.0315.

71. The Credit Union will use 0.0634 to compute the interest.

73. 0.005   **75.** 0.886

77. The river pollution due to agricultural runoff is 0.6 of the total pollution.

79. The golf team won 87.5% of their matches.

81. Kurt Warner's passing completion rate was 67%.

83. In 2016, the number of physical therapy jobs will be 127.2% of the number of jobs in 2006.

85. One mile is about 1.609 kilometers.

87. The New York City subway system has 0.406 of the riders of the Moscow system.

89. The diameter of Pluto is about 0.274 times the diameter of Earth.

91. The percent of income paid for both Social Security and Medicare is 9.1%.

93. The sale price is 89% of the original price. The store will advertise "11% off."

95. The new box is 1.25 times the size of the old box.

97. The decimal will be greater than 1 when the % is greater than 100%. For example, 235% written as a decimal is 2.35 and 3.4% written as a decimal is 0.034.

99. 1,800,000% **101.** 42.5% or $42\frac{1}{2}$%; $42\frac{5}{9}$%

103. 0.6, 0.569 **105.** Answers will vary.

107. 0.140625 **109.** 1.8

111. $\frac{41}{400}$ **113.** 350,000

115. A total of $22.44 is saved on the trip.

SECTION 6.3

1. 67% **3.** 74% **5.** 85% **7.** 50% **9.** 85%

11. 105% **13.** 187.5% **15.** 6.3% **17.** 460%

19. $33\frac{1}{3}$% **21.** $483\frac{1}{3}$% **23.** $491\frac{2}{3}$%

25. 61.5% **27.** 55.6% **29.** 78.6%

31. 3237.9% **33.** $\frac{3}{25}$ **35.** $\frac{17}{20}$

37. $\frac{13}{10}$ or $1\frac{3}{10}$ **39.** 2 **41.** $\frac{21}{25}$

43. $\frac{1}{4}$ **45.** $\frac{9}{20}$ **47.** $\frac{3}{2}$ or $1\frac{1}{2}$ **49.** $\frac{91}{200}$

51. $\frac{17}{250}$ **53.** $\frac{121}{200}$ **55.** $\frac{1}{400}$ **57.** $\frac{1}{40}$

59. $\frac{1}{550}$ **61.** $\frac{177}{400}$ **63.** $\frac{199}{60}$ or $3\frac{19}{60}$

65. Kobe Bryant made 85% of his free throws.

67. President Barack Obama received 52.9% of the votes.

69. Barack Obama received 67.8% of the Electoral College votes.

71. Gilmore made 59.9% of his field goals.

73. 18.61% **75.** 137.88% **77.** $\frac{2}{45}$ **79.** $\frac{11}{4000}$

81. Miguel is taking 162.5% of the recommended allowance.

83. It represents $\frac{63}{50}$, or $1\frac{13}{50}$, of last year's enrollment.

85. The fraction of residents between the ages of 25 and 40 is $\frac{7}{40}$.

87. St. Croix is $\frac{3}{5}$, or 60%, of the area of the Virgin Islands.

89. In 2004, 57.8% of the days were smoggy. In 2009, 45.8% of the days were smoggy. Possible reasons include better emission controls on cars and trucks and more pollution controls on factories.

91. During the 10 years, $\frac{29}{50}$ of the salmon run was lost.

93. In Kenya, $\frac{17}{10,000}$ of the population is white, meaning that 17 out of every 10,000 people in Kenya are white.

95. The percent taken off the original price is 48%.

97. The grill was 33% off.

99. Sales can be described either by fractions or percents. For example, you might hear about a sale of $\frac{1}{3}$ off or of 33% off.

Common statistics are also given in either form. You might hear that $\frac{1}{5}$ of the residents of a town are over 65 years old or that 20% of the residents are over 65 years old. It is important to note that in such circumstances, fractions are generally approximations of the actual statistics.
Percent forms of statistics may be approximations, but they are generally considered more accurate than fraction forms.

101. 2.3% **103.** $\frac{1}{4000}$ **105.** $1\frac{2001}{2500}$

107. Answers will vary. **109.** $\frac{21}{25}$

111. $4\frac{13}{200}$ **113.** 0.825 **115.** 56.7%

117. 0.0813

SECTION 6.4

	Fraction	Decimal	Percent
1.	$\frac{1}{10}$	0.1	10%
3.	$\frac{3}{4}$	0.75	75%
5.	$1\frac{3}{20}$	1.15	115%

#	Fraction	Decimal	Percent
7.	$\dfrac{1}{1000}$	0.001	0.1% or $\dfrac{1}{10}$%
9.	$4\dfrac{1}{4}$	4.25	425%
11.	$\dfrac{11}{200}$	0.055	$5\dfrac{1}{2}$%
13.	$\dfrac{27}{200}$	0.135	13.5% or $13\dfrac{1}{2}$%
15.	$\dfrac{5}{8}$	0.625	$62\dfrac{1}{2}$%
17.	$\dfrac{17}{50}$	0.34	34%
19.	$\dfrac{2}{25}$	0.08	8%
21.	$\dfrac{24}{25}$	0.96	96%
23.	$\dfrac{7}{20}$	0.35	35%
25.	$\dfrac{1}{8}$	0.125	12.5% or $12\dfrac{1}{2}$%

27. The swim suit is 25% off the regular price.

29. Hank's Super Market offers the better deal.

31. Seller B offers the better deal and both offers fall within the supervisor's authorized amount.

33. Stephanie had the best batting average.

35. Peru has the higher literacy rate.

37. The TV sells for $\dfrac{2}{5}$ off the regular price.

39. Melinda gets the best deal from Machines Etc.

41. Fractions are commonly used to describe units with subdivisions that are not powers of 10. For example, inches are subdivided into fourths and eighths, so it is common to have measures of $14\dfrac{3}{8}$ inches. Rates are usually stated using decimal forms. For example, unit pricing is generally given in tenths of a cent, as in Cheerios costing 19.4¢ per ounce. Statistics are most often given using percent forms. For example, 52% of total orchestra expenses are salaries of the musicians.

	Fraction	Decimal	Percent
43.	$\dfrac{23}{7}$	3.3	328.6%
	$\dfrac{4}{5}$	0.8	80.0%
	$\dfrac{421}{200}$	2.1	210.5%

45. $y = 41.4$ **47.** $x = 3.735$ **49.** $B = 200$

51. $A \approx 9.5$ **53.** The Bacons should collect \$62,800 in insurance money.

SECTION 6.5

1. 54   **3.** 75   **5.** 300%   **7.** 75%   **9.** 40
11. 27.3   **13.** 80%   **15.** 100   **17.** 2   **19.** 117
21. 12.25   **23.** 5.5%   **25.** 655   **27.** 36.4
29. 270   **31.** 265   **33.** 76.8%   **35.** 21.49
37. 62.5%   **39.** 205   **41.** 48.2%   **43.** 330.65
45. 227.265   **47.** 23.7%

49. $\dfrac{41}{96}$ **51.** $114\dfrac{3}{10}$ **53.** 132.0%

55. The problem is in equating a 40¢ profit with a 40% profit. A 40¢ profit means that the company makes 40¢ on each part sold. A 40% profit means that the company makes 40% of the cost on each part. But 40% of 30¢ is only 12¢. So a 40% profit would mean selling the parts for 42¢. The company is making a much higher profit than that. The 40¢ profit is actually 133% of the cost.

57. $\dfrac{17}{75}$ **59.** 0.1   **61.** $x = 45$

63. $a = 0.25$ or $a = \dfrac{1}{4}$

65. $w = 20.2$ **67.** The number of miles is 145.

69. The number of inches needed is $2\dfrac{5}{8}$.

SECTION 6.6

1. There are about 838 undergraduate students from low-income families.

3. The Hispanic or Latino population of Los Angeles County was about 4,657,000.

5. Alexa's score was 79%.

7. The tip was 16% of the check.

9. Sand is $33\dfrac{1}{3}$% of the mixture.

11. Delplanche Farms has about 31% of its acreage in soybeans.

13. The General Fund Budget was about 41.7% of the total budget.

15. Mickey Mantle got a hit 29.8% of the time.

17. The population of Japan is about 125 million.

19. The fruit stand had 24 boxes of bananas in stock.

21. She must sell four refrigerators, or \$4580 worth of refrigerators.

23. Mary Ann lost about 14.9% of her original weight.

25. The recommended daily values are as follows: fat, 63 g; sodium, 2444 mg; potassium, 3600 mg; and dietary fiber, 21 g.

27. In 1900, the U.S. population was about 75.7 million people.

29. Viewers will watch about 8.35 hours per day, or about 8 hours 21 minutes.

	New Amount	Increase or Decrease	Percent of Increase or Decrease
	Amount		
31.	345	415	Increase of 70	20.3%
33.	764	888	Increase of 124	16.2%
35.	2900	3335	Increase of 435	Increase of 15%

37. The Denver Nuggets' average home attendance decreased by 0.8%

39. The percent of increase in the number of persons was about 5.2%.

41. The percent increase for the year was 47.1%

43. The percent of increase in the number of families was about 87.9%.

45. The percent of decrease in production was about 6.0%.

47. The percent of increased mileage is 7.6%.

49. The median home size increased 34.8% over the 20-year period.

51. Asians are the smallest identifiable population in Texas.

53. The second largest ethnic group in Texas is Hispanic.

55. Chrysler sold more cars than Nissan.

57.

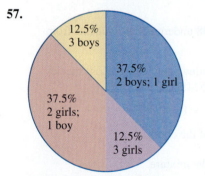

12.5% 3 boys
37.5% 2 boys; 1 girl
37.5% 2 girls; 1 boy
12.5% 3 girls

Possibilities of boy–girl combinations

59.

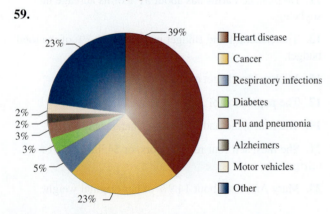

39% Heart disease
23% Cancer
2% Respiratory infections
2% Diabetes
3% Flu and pneumonia
3% Alzheimers
5% Motor vehicles
23% Other

Major causes of death

61. The percent of increase is 100%.

63. Carol's baby's weight increased by 215% during the first year.

65. Nonfood items accounted for 30.4% of the cost. Meat products accounted for 23.5%.

67. Sally receives $7560 and Rita receives $5040.

69. Joe pays a total of $17,508 for the car.

71. The engine develops 113.75 horsepower.

73. Peter attended 90% of the classes.

75. The score on the algebra test is 85%.

SECTION 6.7

	Marked Price	Sales Tax Rate	Amount of Tax	Total Cost
1.	$238	5.8%	$13.80	$251.80
3.	$90.10	3.8%	$3.42	$93.52
5.	$628	4.5%	$28.26	$656.26
7.	$1499	3.6%	$53.96	$1552.96

9. The sales tax is $60.92.

11. The suit costs George $506.47.

13. The sales tax rate is 6.3%.

15. Jim pays $654.99 for the TV set.

17. The sales tax rate is about 3.6%.

19. The sales tax rate is about 7.5%.

	Original Price	Rate of Discount	Amount of Discount	Sale Price
21.	$75.82	18%	$13.65	$62.17
23.	$320	30%	$96	$224
25.	$15.95	3.1%	$0.49	$15.46
27.	$1798	5.1%	$91.70	$1706.30

29. Joan pays $161.46 for the painting.

31. Larry pays $48,129.90 for the SUV.

33. The discount is $274.40. Melvin pays $705.59 for the generator.

35. The discount price is $334.88 and the sales tax rate is 4.8%.

37. Heather will save $27 off the original price.

39. The TV is on sale for $402.71.

41. The rebate on the canopy is $26.74.

43. The sale price of the overalls is $32.06.

45. The price of the coat is $104.08. She receives about 58.2% off.

47. The mother pays $77.48 for the two pairs of shoes. This is a 21.7% savings off the original price.

49. The price of the clubs will be $928.13.

	Sales	Rate of Commission	Commission
51.	$4890	9%	$440.10
53.	$67,320	8.5%	$5722.20
55.	$1100	4%	$44
57.	$346,800	9%	$31,212

59. Grant's commission for the week was $625.57.

61. The commission on the sale of the tickets is $25.48.

63. He earned $741.07 last week.

65. Carlita earned $421.75 last week.

67. Perry's sales for the week totaled $5933.33.

69. Ms. James earned commissions of $1433.50 for the week.

71. A sales tax is a percent increase in the price paid by the consumer, because it is an additional amount of money paid.

73. In July Marlene earned $3183 and in August she earned $5175.

75. $\dfrac{59}{72}$ **77.** $\dfrac{23}{48}$ **79.** $\dfrac{1}{6}$ **81.** $\dfrac{13}{9}$ or $1\dfrac{4}{9}$

83. The perimeter is $18\dfrac{19}{20}$ in.

SECTION 6.8

	Principal	Rate	Time	Interest	Total Amount Due
1.	$10,000	5%	1 year	$500	$10,500
3.	$5962	8%	1 year	$476.96	$6438.96
5.	$32,000	8%	4 years	$10,240.00	$42,240.00
7.	$4500	5.5%	9 months	$185.63	$4685.63

9. Maria earns $180 in interest.

11. Nancy has earned $1402.50 in interest over the 3-year period.

13. Tyra earns $172.95 in interest.

15. Janna pays $2424.58 to pay off the loan.

17. a. The interest paid was $5038. **b.** The interest rate was 11%.

19. The ending balance is $29,710.80.

21. The ending balance is $198,612.

23. The amount of interest earned is $10,850.56.

25. The amount of interest earned is $138,643.

27. Juanita's account will earn $81 more than Jose's account.

29. Carl's account earns $177 more interest.

31. a. Mark pays $57.85 in interest. **b.** $142.15 goes to pay off the balance. **c.** The unpaid balance at the end of the June is $3824.95.

33. a. The minimum monthly payment is $171. **b.** The interest is $52.71. **c.** $118.29 goes to pay off the balance. **d.** The unpaid balance at the end of one month is $4158.71.

35. a.

	Beginning balance	Minimum payment	Interest paid	Balance reduction	Ending balance
Month 1	$10,000	$400	$108.33	$291.67	$9708.33
Month 2	$9708.33	$388	$105.17	$282.83	$9425.50
Month 3	$9425.50	$377	$102.11	$274.89	$9150.61

b. The amounts are decreasing because they are percentages of a decreasing amount (the balance). **c.** Since the balance reduction amounts are decreasing by less than $10 per month, Joe will probably make his $8000 balance by the end of the year.

37. The minimum monthly payments are primarily interest owed on the balance, and very little goes to reducing the amount of the balance. A late fee or additional charges could easily be more than the reduction of the outstanding balance, so the balance actually increases for the month.

39. It will take 14 months to pay off the balance and the 14th payment will be for $82.32. Linn will have paid $146.72 in interest. **41.** 315.028 **43.** 36.325

45. 1.30456 **47.** 57.015 **49.** Each of Mary's nieces and nephews received $197,600.

CHAPTER 6 REVIEW EXERCISES
SECTION 6.1

1. 75% **3.** 78% **5.** $57\dfrac{1}{7}\%$

SECTION 6.2

7. 65.2%   **9.** 0.017%   **11.** The Phoenix Suns won 75.6% of their games.   **13.** 0.48   **15.** 0.000625

17. The decimal number used to compute the interest is 0.1845.

SECTION 6.3

19. 68.75%   **21.** 48.1%   **23.** The Classic League basketball team won about 80.43% of their games.

25. $1\frac{13}{20}$ or $\frac{33}{20}$ **27.** $\frac{13}{40}$

29. The Bears defense was on the field $\frac{29}{50}$ of the game.

SECTION 6.4

31–38.

Fraction	Decimal	Percent
$\frac{17}{25}$	0.68	68%
$\frac{37}{50}$	0.74	74%
$\frac{3}{200}$	0.015	1.5%
$3\frac{11}{40}$	3.275	327.5%

SECTION 6.5

39. 100.1   **41.** 21.25%   **43.** 12.58   **45.** 18.2%

SECTION 6.6

47. Melinda's yearly salary was $24,035.

49. Omni Plastics currently employs 429 people.

51. The Hummer H2 decreased in value by 12%.

53. Most students received a B grade.

SECTION 6.7

55. The clubs cost Mary $495.33.

57. The cost of the suit to the consumer is $405.54.

59. The New Balance shoes are sold at about 28% off.

SECTION 6.8

61. Wanda owes her uncle $6215.

63. Minh's credit card balance will be $1311.46.

CHAPTER 6 TRUE/FALSE CONCEPT REVIEW

1. true   **2.** true

3. false; rewrite the fraction with a denominator of 100 and then write the numerator and write the percent symbol.

4. false; move the decimal point two places to the right and write the percent symbol.

5. true   **6.** false; the base unit is always 100.

7. true   **8.** true   **9.** false; the proportion is $\frac{A}{B} = \frac{X}{100}$.

10. true   **11.** true   **12.** false; $4\frac{3}{4} = 475\%$

13. false; 0.009% = 0.00009   **14.** false; $B = 43{,}000$

15. true   **16.** true

17. false; it is the same as a decrease of 27.75%.

18. false; her salary is 99% of what it was Monday.

19. true   **20.** true   **21.** true

22. false; $\frac{1}{2}\% = 0.005$

23. true   **24.** false; the sales tax rate is 6.5%

CHAPTER 6 TEST

1. The percent of discount on the computer is 27.8%.

2. 3.542%   **3.** The population is 56% female.

4. 74.6%   **5.** 84.375%   **6.** 45   **7.** $\frac{54}{325}$

8. 0.78%   **9.** 11.7   **10.** 681.8%

11. $\frac{68}{25}$ or $2\frac{18}{25}$ **12.** 0.0789

13. The Adams family spends $1120 per month on rent.

14–19.

Fraction	Decimal	Percent
$\frac{13}{16}$	0.8125	81.25%
$\frac{78}{125}$	0.624	62.4%
$\frac{37}{200}$	0.185	18.5%

20. 75.6%   **21.** 0.04765   **22.** The total price of the tires is $380.39.

23. The sale price of the Hilfiger jacket is $149.07.

24. The percent of increase in population is 20.6%.

25. The percent of B grades is 23.4%.

26. Jerry earned $5151 last month.

27. Ukiah will owe his aunt $6955 at the end of 1 year.

28. The percent of calories from fat is 47.4%.

29. Greg's investment is worth $14,823.

30. Loraine's credit card balance at the end of May is $6801.79

Chapter 7

SECTION 7.1

1. feet or meters   **3.** inches or millimeters

5. inches or centimeters   **7.** feet or meters

9. feet or meters   **11.** inches or centimeters

13. feet or meters   **15.** yards, feet, or meters

17. 108 in.   **19.** 12,500 cm   **21.** 36 ft

23. 0.9685 km   **25.** 600,000 cm   **27.** 2.60 mi

29. 2.25 ft   **31.** 6.22 m   **33.** 74.57 mi

35. 57.41 ft   **37.** 143 yd   **39.** 4.94 ft   **41.** 200 mm

43. 116 in   **45.** 2600 mi   **47.** 5 yd 1 ft

49. 19.69 mi   **51.** 24,075 mm   **53.** 5 ft 9 in.

55. 2461 km   **57.** 19 ft 7 in.   **59.** 8 yd 3 in.

61. Maria needs to buy 34 ft 8 in. of garland.

63. Their average height is 1.34 m.

65. The Millau Viaduct is about 1125 ft tall.

67. Nine lanes will require 8 lengths of dividers, which is a total of 400 m.

69. Figures for the Amazon and the Congo Rivers appear to be rounded, because of the two and three terminal zeros. This implies they are probably estimates.

71. The Amazon River is just about twice as long as the São Francisco River.

73. The yard is 20 ft wide and 28 ft long.

75. The main walkway is 4 ft wide, and the others are 2 ft wide.

77. Only like measures may be added or subtracted. Add (subtract) the numbers and keep the common unit of measure.
79. 3,476,000

81. $\frac{7}{12}$ **83.** $2\frac{1}{2}$ **85.** 4.5369   **87.** 84.4

SECTION 7.2

1. 107 oz   **3.** 25 mL   **5.** 4 gal   **7.** 8.5 qt

9. 84 qt   **11.** 126 lb   **13.** 82 oz   **15.** 30 gal

17. 88 c   **19.** 12.79 gal   **21.** 55 kg   **23.** 80 lb

25. 558 lb   **27.** 8000 g   **29.** 8 g   **31.** 85 oz

33. 620 lb   **35.** 223 g   **37.** 72,000 mg

39. 368.54 g   **41.** 0°C   **43.** 40°C   **45.** 20°C

47. 480 sec   **49.** 133 days   **51.** 66.7°C

53. 8.9°C   **55.** 159.8°F   **57.** 57.2°F

59. 22,776 hr   **61.** 0.2 day

63. 26 lb 6 oz   **65.** 7 gal 2 qt

67. The grocery sold 55 lb 7 oz of hamburger.

69. Each student receives 12.5 mL of acid.

71. The average temperature is 28.9°C.

73. Each serving is 0.05 gal or 6.4 oz of juice.

75. Four bags contain 1588 g of potato chips.

77.

Kilograms	54	58	63	69	76	85	97	130
Pounds	119	128	139	152	168	187	214	287

79. The *Jahre Viking* weighs about 673 million kg fully loaded.

81. He needs 11 servings of cereal and milk to get sufficient protein.

83. Because grams measure weight and inches measure length, it is not possible to add these measures.

85. There are 480 grains in 1 oz t and 5760 grains in 1 lb t.

87. Each will receive 109.5 dwt, or 5 oz t 9 dwt 12 grains of platinum.

89. 848   **91.** 41.94

93. $43\frac{11}{18}$ **95.** 153.45   **97.** 0.05684 km

SECTION 7.3

1. 48 ft   **3.** 130 in.   **5.** 74 mm   **7.** 60 m

9. 83 mm   **11.** 800 ft   **13.** 90 ft   **15.** 258 cm

17. 40 yd   **19.** 168 m   **21.** 170 ft   **23.** 21.98 in.

25. 37.68 mm   **27.** 50.24 km   **29.** 49.13 cm

31. 74.55 yd   **33.** 110.27 mm   **35.** 214 in.

37. 75.70 in.

39. June will need 168 ft of railroad ties.

41. The track is about 440 yd long.

43. The inside dimensions of the frame are 20 in. by 26 in.

45. They need 117 ft of fence.

47. The football player ran 19,376 yd, which is about 11 mi.

49. Marylane needs 72 ft of lumber for the project.

51. The total cost of the lumber is $139.41.

53. Yolande needs about 170 in. of lace for the second row.
55. The perimeter of the patio is 60 ft.   **57.** The center line divides the yard into identical but mirror-image halves.

59.

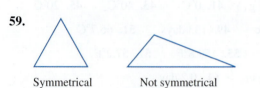

Symmetrical   Not symmetrical

61. Perimeter is the distance around a flat surface. To find the perimeter of the figure, add the lengths of the four sides. If the figure were a park, the sides could be measured in miles or kilometers. If the figure were a room, the sides could be measured in feet or meters. If it were a piece of paper, the sides could be measured in inches or centimeters.

63. A square has all four sides equal. This doesn't have to be true for a rectangle.

65. 850,085   **67.** 32,570,000

69. 277,276   **71.** 6 lb 1 oz

73. Fred's average score was 87.

SECTION 7.4

1. 121 ft^2 **3.** 17.5 m^2 **5.** 54 ft^2 **7.** 60 yd^2

9. $346\frac{1}{2}$ cm^2 **11.** 255 m^2 **13.** 170 ft^2

15. 1620 in^2 **17.** 396 cm^2 **19.** 135 in^2

21. 680 in^2 **23.** 289.3824 cm^2 **25.** 47.08 ft^2

27. 15,552 in^2 **29.** 27,878,400 ft^2 **31.** 3.13 yd^2

33. 38.33 yd^2 **35.** 396 ft^2 **37.** 0.54 mi^2

39. 248.00 in^2 **41.** 129.17 ft^2 **43.** 13,935.46 cm^2

45. 237 ft^2 **47.** 320 m^2 **49.** 683 yd^2

51. 346.87 cm^2 **53.** Zakaria needs about 3.4 gal of stain.

55. The 12-in. dinner plate has about 35 in^2 more area than the 10-in. plate.

57. Debbie needs 562.5 oz of weed killer.

59. The doors require 36 ft^2 of glass.

61. The gable requires 162 ft^2 of sheathing.

63. Maureen needs 32 ribbons of fabric and 3840 in^2 of fabric.

65. April and Larry need about 151 ft^2 of tile.

67. The region can be covered with six bags of seed.

69. The den can be paneled with six sheets of paneling.

71. The patio and walkways have an area of 326 ft^2.

73. Area is usually measured in square units. For example, square feet and square meters can be used. Acres are also used to measure land area.

75. Cover the area with as many whole squares as possible. For the leftover area, estimate half or quarter squares. Then find the total of the measures of all of the squares.

77. 3100 ft^2 will be walkway and the remaining 20,900 ft^2 will be lawn.

79. 670,800   **81.** 46   **83.** 17

85. average: 50; median: 17.5; mode: 16

87.

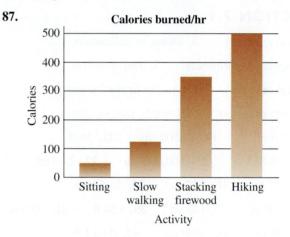

SECTION 7.5

1. 120 ft^3 **3.** 4719 cm^3 **5.** 168 cm^3

7. 50.24 ft^3 **9.** 27,000 mL

11. 3.5325 ft^3 or 6104.16 in^3

13. 4.19 ft^3 **15.** 4725 in^3 **17.** 38,250 cm^3

19. 1.5 ft^3 **21.** 1,000,000 cm^3 **23.** 46,656 in^3

25. 8 ft^3 **27.** 10 cm^3 **29.** 7.85 ft^3

31. 0.35 m^3 **33.** 2.14 mi^3

35. The volume of the granary is 4396 yd^3.

37. 5520 cm^3 **39.** 315 ft^3 **41.** 177,120 cm^3

43. The volume of the pasta box is $94\frac{45}{64}$ in^3.

45. The bricklayer needs 21 ft^3 of mortar.

47. It is necessary to round up in this case because we cannot allow ourselves to run short of needed materials. Numbers rounded according to the rule in Section 1.1 could be less than the actual amount necessary.

49. The garden is approximately 33,696 in^2, so they will need 269,568 in^3 = 156 ft^3 ≈ 6 yd^3 of topsoil.

51. The can is about 18.4 in. tall.

53. The aquarium has an actual capacity of about 74.8 gal.

55. The aquarium will support 3 angelfish.

57. Consider the pointed end as a solid on its side with a triangular base ($V = Bh$). The remainder of the figure is a rectangular solid ($V = \ell wh$). Calculate the volume of each piece and add the two volumes.

59. The pool holds 2400 ft^3 of water.

61. 729   **63.** 197   **65.** 7225   **67.** 3600

69. 72 in^2 or 0.5 ft^2

SECTION 7.6

1. 7   **3.** 6   **5.** 10   **7.** 30   **9.** $\dfrac{7}{10}$ **11.** 0.9

13. 13.23   **15.** 28.65   **17.** 3.67   **19.** 0.78

21. 107.24   **23.** $c = 10$ **25.** $c = 35$ **27.** $a = 25$

29. $b = 12$ **31.** $b \approx 77.46$ **33.** $c = 28.86$

35. $b \approx 58.09$ **37.** $a = 42$

39. The hypotenuse measures 75 cm.

41. The side of the square measures 12 ft.

43. The TV is advertised as a 48-in. screen.

45. The top of the ladder is about 11.62 ft up the house.

47. Lynzie is about 38.48 mi from her starting point.

49. It is about 127.28 ft from first base to third base.

51. Scott needs 12 ft of wire for the supports.

53. A square root of a number is a number that, when squared, gives back the original number. To find $\sqrt{289}$, we look for the number whose square is 289. Because $20^2 = 400$, we know that the number is less than 20. Working backward, we have $18^2 = 324$, which is too large but close. The number $17^2 = 289$. Therefore, $\sqrt{289} = 17$.

55. The rod will fit in the second box but not in the first one.   **57.** 611.231   **59.** $80 = 2^4 \cdot 5$

61. 1806   **63.** 72.5 m$^2 \approx 780.36$ ft^2.

65. The volume of the can is about 502.65 in^3.

CHAPTER 7 REVIEW EXERCISES

SECTION 7.1

1. 46 mm   **3.** 1 yd or 3 ft   **5.** 1 ft 4 in.   **7.** 144 in.

9. 381 mm

SECTION 7.2

11. 36 qt   **13.** 86 gal   **15.** 10 hr 20 min
17. 224 oz   **19.** 35°C

SECTION 7.3

21. $P = 98$ cm   **23.** $P = 69$ ft   **25.** $P = 40.82$ yd

27. The perimeter is 54 m.

29. The city will need 1010 ft of fencing.

SECTION 7.4

31. $A = 2075$ mm^2 **33.** $A = 231$ cm^2

35. $A = 62.13$ yd^2

37. The area of the square is 360,000 mm^2.

39. One coat must cover 206 ft^2, so two coats must cover 412 ft^2; therefore, she does not have enough paint.

SECTION 7.5

41. $V = 4608$ cm^3 **43.** $V = 100$ in^3 **45.** $V = 336$ m^3

47. The volume is about 47,689 ft^3.

49. There are 46,656 in^3 in a cubic yard.

SECTION 7.6

51. 23   **53.** $\dfrac{9}{7}$ or 1.29   **55.** $b = 18$

57. $b \approx 14.42$

59. The top of the ladder is about 17.75 ft high.

CHAPTER 7 TRUE/FALSE CONCEPT REVIEW

1. true   **2.** false; equivalent measures have different units of measurement, as in 6 ft and 2 yd.   **3.** true

4. false; the perimeter of a square can be found in inches or meters or other measures of length.   **5.** true

6. false; the volume of a cube is $V = s^3$, or the area of a square is $A = s^2$, or a square has no volume.

7. false; the area of a circle is $A = \pi r^2$, or the area of a circle is $A \approx 3.14\,r^2$.   **8.** true   **9.** true   **10.** true

11. false; $3^2 = 9$ or $\sqrt{9} = 3$ **12.** true   **13.** true

14. false; one milliliter is equivalent to 1 cubic centimeter.

15. false; measurements can be added or subtracted only if they are expressed with the same unit of measure.

16. true

17. false; volume can also be measured in cubic units.

18. false; the Pythagorean Theorem states that the sum of the squares of the lengths of the legs of a right triangle is equal to the square of the length of the hypotenuse.

19. false; 1 yd$^2 = 9$ ft^2 **20.** true

CHAPTER 7 TEST

1. 8329 mm   **2.** 118 in.   **3.** 1440 in^3

4. The room can be covered by 184 ft^2 of tile.

5. 17.607   **6.** 29 cm   **7.** 40 in^2 **8.** 2 yd 2 ft 7 in.

9. In the English system, there are pounds, ounces, and tons. In the metric system, there are grams, milligrams, and kilograms.   **10.** 48 pt

11. The windows require 88 ft of molding.

12. 260 in^3 **13.** 48 cm^2 **14.** The volume is 24 ft^3.

15. Each student will receive 8 oz of grape juice.

16. Lining the garden plots will cost $184.80.   **17.** 6 cm

18. 96.2 km^2 **19.** Each player runs 5120 yd.

20. Area is a measure of two-dimensional space, whereas volume is a measure of three-dimensional space.

21. John's Meat Market averaged sales of 489.4 lb of steak each day.

22. Li needs about 282.6 in. of lace. She must buy 8 yd.

23. The box has a larger volume than the bag, so all the dog food will fit into the box.

24. Macy is 3.7 mi from her starting point.

25. Neela needs more than 544 in^2 of wrapping paper.

Chapter 8
SECTION 8.1

1. 4   **3.** −5   **5.** 2.3   **7.** $-\dfrac{3}{5}$ **9.** $\dfrac{6}{7}$

11. −61   **13.** 42   **15.** 3.78   **17.** $-\dfrac{17}{5}$

19. −0.55   **21.** −113.8   **23.** 0.0123

25. 4   **27.** 24   **29.** 3.17   **31.** $\dfrac{5}{11}$ **33.** $\dfrac{3}{11}$

35. 32 or −32   **37.** 0.0065   **39.** 355   **41.** $\dfrac{11}{3}$

43. $\dfrac{33}{7}$ **45.** 0   **47.** 0.341   **49.** $-\dfrac{5}{17}$

51. −78

53. The opposite of a gain of 1.54 is −1.54.

55. The opposite of a reading of 14°C is −14°C.

57. The opposite of 1915 B.C. is A.D. 1915.

59. As signed numbers, the distance up the mountain is +1415 ft and the distance down the mountain is −1143 ft.

61. Eighty miles south would be represented by −80 miles.

63.

Ocean	Deepest Part (ft)	Average Depth (ft)
Pacific	Mariana Trench, −35,840	−12,925
Atlantic	Puerto Rico Trench, −28,232	−11,730
Indian	Java Trench, −23,376	−12,598
Arctic	Eurasia Basin, −17,881	−3,407
Mediterranean	Ionian Basin, −16,896	−4,926

65. 24   **67.** −14

69. On the first play, the Seahawks made −8 yd. The opposite of the loss of eight yards is +8 yd.

71. The distances driven by the Golden family expressed as signed numbers are +137 mi, +212 mi, −98 mi, and −38 mi.

73. The positive real numbers and zero each have an absolute value that is the number itself. This is because the absolute value of a number is its distance (distance is positive) from zero and the positive numbers are all a positive distance from zero.

75. 7   **77.** positive   **79.** 59   **81.** 5   **83.** 20.1 m

85. 69 in.   **87.** Each part of the coil is 1.5 m.

SECTION 8.2

1. 2   **3.** 2   **5.** −15   **7.** −14   **9.** 0

11. −12   **13.** −46   **15.** −14   **17.** −26

19. 7   **21.** −6   **23.** −108   **25.** 0   **27.** 27

29. −10   **31.** −7   **33.** −8.3   **35.** 2.8

37. $-\dfrac{7}{12}$ **39.** $-\dfrac{7}{6}$ **41.** −121   **43.** −189

45. −30   **47.** 81   **49.** −757   **51.** −114

53. The sum of the temperatures recorded at 9:00 A.M. is −349°C.

55. The sum of the temperatures recorded at 11:00 P.M. is −505°C.

57. The net change in weight the cargo master should report is −108 lb.

59. The altitude of the airplane at 4 P.M. is 29,870 ft.

61. The inventory at the end of the month is 30,370 volumes.

63. No, the Patriots were 2 yd short of a first down.

65. a. The net change for the Nordstrom stock is down 1.01. **b.** The closing price is $33.01.

67. The net gain for the month is $5077.

69. Adding signed numbers can be thought of as moving along a number line according to the absolute value and sign of the numbers. Starting at zero, move as many spaces as the absolute value of the first number. Move right if the number is positive and left if it is negative. Then move according to the second number. The sign of the sum is determined by which side of zero you end up on. So if you have moved right for more spaces than you have moved left, the sum will be positive. And if you have moved left more spaces than you have moved right, the sum will be negative. Symbolically, compare the absolute values of the numbers being added. The sign of the one with the largest absolute value (corresponding to the most spaces moved in a particular direction) is the sign of the answer.

71. -124 **73.** 32   **75.** 78   **77.** 83

79. The perimeter is 142 in.

81. The perimeter is 124 in.

83. One pound of bananas costs $0.72.

SECTION 8.3

1. 4   **3.** -12 **5.** -5 **7.** 3   **9.** -20

11. -9 **13.** 32   **15.** -34 **17.** -28 **19.** 1

21. 0   **23.** -18 **25.** -80 **27.** 118

29. -119 **31.** 11   **33.** -182 **35.** 204

37. 0   **39.** 144   **41.** 173   **43.** -9.98

45. -1.44 **47.** -56.14 **49.** 116   **51.** -687

53. The change in temperature is 85°C.

55. Joe wrote checks totaling $796.32.

57. The range of altitude for each continent is Africa, 19,852 ft; Antarctica, 25,191 ft; Asia, 30,340 ft; Australia, 7362 ft; Europe, 18,602 ft; North America, 20,602 ft; and South America, 22,965 ft. The smallest range is in Australia. It means it is a rather flat continent.

59. The depth of the ocean floor at Mauna Kea is $-19,680$ ft, or 19,680 ft below sea level.

61. The difference in temperatures recorded at 11:00 P.M. on day 3 and day 5 is -15°C.

63. The difference between the highest and lowest temperatures recorded during the 5 days is 105°C.

65. Thomas's bank balance is $-$6.41, or $6.41 overdrawn.

67. Janna wrote checks totaling $477.14.

69. The national debt increased by $6.24 trillion over the 9-year period.

71. The difference in the scores was -6.

73. On a number line, start at zero and move 12 units to the right, positive. Now if we were adding -8, we would move 8 units to the left of $+12$, but because we are subtracting we reverse the direction and move an additional 8 units to the right, ending at $+20$.

75. The first problem is addition, and the second is subtraction. The answer to the two problems is the same, -3.

77. -12.68 **79.** -41 **81.** 112   **83.** 5772

85. The area of the circle is about 615.44 in^2

87. The area of the rectangle is 127.84 cm^2.

89. To carpet the floor, you need 76.25 yd^2.

SECTION 8.4

1. -8 **3.** -10 **5.** -80 **7.** -77 **9.** 11

11. -306 **13.** -255 **15.** -9 **17.** -350

19. $-\dfrac{1}{4}$ **21.** 3   **23.** 28   **25.** 99   **27.** 60

29. -11 **31.** 210   **33.** 5.4   **35.** 24.2   **37.** $\dfrac{1}{6}$

39. 1.645   **41.** -2520 **43.** 465   **45.** 7.14

47. $-\dfrac{3}{10}$ **49.** -14.035 **51.** 0.5423   **53.** 0

55. 21   **57.** -240 **59.** $\dfrac{1}{3}$

61. The Celsius measure is -15°C.

63. Ms. Riles's loss expressed as a signed number is -20.8 lb.

65. The Dow Jones loss over 12 days is -20.88.

67. -49 **69.** -42 **71.** -252

73. Safeway's loss expressed as a signed number is $-$65.62.

75. Winn Dixie's loss expressed as a signed number is $-$31.38.

77. The low temperature was about -89°C.

79. The trader paid $6257.80 for the stock. She received $4321.50 when she sold the stock. She lost $1936.30, or $-$1936.30, on the sale of the stock.   **81.** Asia

83. An even number of negatives may be grouped in pairs. Because the product of each pair of negatives is positive, the next step is multiplying all positive numbers. So the final product is positive.

85. -70 **87.** 56   **89.** 6   **91.** 67   **93.** 143.41

95. The cost of the wall-to-wall carpeting is $2502.

97. It takes 39.75 ft^2 of sheet metal to make the box.

SECTION 8.5

1. -2 **3.** -4 **5.** -3 **7.** -8 **9.** 8

11. -6 **13.** -2.02 **15.** -35 **17.** -3

19. -1.2 **21.** 2   **23.** 3   **25.** 7   **27.** 6

29. −5   **31.** 7   **33.** 8   **35.** 4.04

37. $\dfrac{1}{2}$ **39.** 5   **41.** −45   **43.** 116   **45.** −4.5

47. 0   **49.** −0.0026   **51.** −452   **53.** 16

55. Each member's share of the loss is −$125.65.

57. Mr. Harkness loses an average of −4.6 lb per week.

59. The average daily loss for the stock is −3.82 points.

61. Answers will vary.

63. The continent is Australia.

65. The average monthly loss of the Standard & Poor 500 Index was $50.89, or −$50.89.

67. The division rules are the same as a consequence of the fact that division is the inverse of multiplication. Every division fact can be rewritten as a multiplication fact.

So we have: $a \div b = a \times \dfrac{1}{b}$.

69. −10   **71.** $-\dfrac{8}{5}$ **73.** −1.4   **75.** 125

77. 21   **79.** The volume of the cone is about 1356.48 in³.

81. The amount of dirt to be removed is 96 yd³.

83. The broker makes $2556.

SECTION 8.6

1. −24   **3.** 19   **5.** −6   **7.** −16   **9.** −6

11. 8   **13.** 4   **15.** 13   **17.** 11   **19.** 8

21. 1   **23.** −22   **25.** −6   **27.** 40   **29.** −38

31. −8   **33.** 8   **35.** 0   **37.** −24   **39.** 64

41. −11   **43.** −15   **45.** 144   **47.** −12

49. The average temperature recorded during day 5 is −56°C.

51. The average high temperature recorded for the 5 days is −34°C.

53. 77   **55.** −41   **57.** $-\dfrac{62}{3}$ **59.** 1

61. −2065   **63.** −1594   **65.** −16

67. Jason's total profit for the six months was −$9588, or a loss of $94 per chair.

69. K-Mart had a profit of −$22.25 on the fishing poles.

71. The average of the highest and lowest points are Africa, 9414 ft; Antarctica, 4268.5 ft; Asia, 13,858 ft; Australia, 3629 ft; Europe, 9209 ft; North America, 10,019 ft; and South America, 11,351.5 ft. The largest average is in Asia. The smallest average is in Australia.

73. The three cities are Athens, Bangkok, and New Delhi.

75. The balance at the end of the month is $3226.

77. Tiger Woods's average winning score for the seven tournaments was about −13.6.

79. The problem is worked according to the order of operations. So we work inside the brackets first, $3 + 5(-4)$. This is where the error has been made. We must multiply first, so $2[3 + 5(-4)] = 2[3 + (-20)]$
$$= 2[-17]$$
$$= -34$$

81. Yes, if a different operation is inside parentheses. For example, in $(3 - 6)^2$ the subtraction is done first. So, $(3 - 6)^2 = (-3)^2 = 9$.

83. −11   **85.** −29.4   **87.** 184   **89.** 63,504

91. 25   **93.** Each brother has a loss of $1458, or −$1458.

SECTION 8.7

1. $x = 7$ **3.** $x = 5$ **5.** $y = -5$ **7.** $a = 7$

9. $x = 7$ **11.** $x = 10$ **13.** $x = -3$ **15.** $x = -6$

17. $x = 0$ **19.** $x = 3$ **21.** $y = 7$ **23.** $a = 0$

25. $x = -0.5$ **27.** $x = -25$ **29.** $x = -16$

31. $y = -4$ **33.** $a = \dfrac{4}{5}$

35. The number is 28.   **37.** The number is 8.

39. The acceleration is 33.

41. There will be 14 payments.

43. It will take 8 minutes for the tub to have 30 gal remaining.

45. The equation is $7310 = 4L + 12{,}558$, where L represents the lowest point. Asia's lowest point fits the description.

47. The equation is $-35{,}840 = 2D - 2000$, where D represents the deepest part. The Ionian Basin's deepest part fits this description.

49. $z = 7$ **51.** $z = 6$

53.
$$
\begin{array}{rl}
-3x + 10 = & 4 \\
\underline{-10 \qquad} & \underline{-10} \\
-3x = & -6 \\
\dfrac{-3x}{-3} = & \dfrac{-6}{-3} \\
x = & 2
\end{array}
$$
Add −10 to both sides.
Simplify.
Divide both sides by −3.
Simplify.

55. $x = -4$ **57.** $x = -2$

CHAPTER 8 REVIEW EXERCISES

SECTION 8.1

1. 39 **3.** 0.91 **5.** 16.5 **7.** 71 **9.** −6.4

SECTION 8.2

11. −98 **13.** 52 **15.** −3.2 **17.** −0.53

19. Yes, the Bears gained 10 yd.

SECTION 8.3

21. 22 **23.** 91 **25.** −7.81 **27.** 29

29. Maria wrote checks totaling $685.90.

SECTION 8.4

31. −66 **33.** −4.08 **35.** 51 **37.** 8.463

39. Kroger's loss expressed as a signed number is −$284.40.

SECTION 8.5

41. −3 **43.** 5 **45.** −16 **47.** −56.2

49. The average daily loss of the stock is −3.875 points.

SECTION 8.6

51. −56 **53.** −50 **55.** −24 **57.** 17

59. −20.72

SECTION 8.7

61. $x = -5$ **63.** $x = 25$ **65.** $x = 0$

67. $a = -13$

69. The Fahrenheit temperature is −7.6°F.

CHAPTER 8 TRUE/FALSE CONCEPT REVIEW

1. true **2.** false; the opposite of a positive number is negative. **3.** false; the absolute value of a nonzero number is always positive. **4.** true **5.** false; the sum of two signed numbers can be positive, negative, or zero.

6. false; the sum of a positive signed number and a negative signed number may be positive, negative, or zero.

7. true **8.** false; to subtract two signed numbers, add the opposite of the number to be subtracted. **9.** true

10. true **11.** false; the sign of the product of a positive number and a negative number is negative.

12. true **13.** true **14.** true

CHAPTER 8 TEST

1. −33 **2.** −35 **3.** $-\dfrac{5}{4}$ **4.** $\dfrac{2}{15}$ **5.** −5

6. 1.15 **7. a.** 17 **b.** 33 **8.** −33 **9.** −2

10. 2 **11.** 96 **12.** −21 **13.** −108.9

14. 40 **15.** −31.2 **16.** $-\dfrac{1}{6}$ **17.** 14

18. −80 **19.** −86 **20.** $-\dfrac{3}{10}$ **21.** −68

22. −5 **23.** $x = -6$ **24.** $x = 7$ **25.** $a = -3.5$

26. Ms. Rosier lost −16.8 lb.

27. The drop in temperature is −21°F.

28. The closing price on Friday is 17.97.

29. The Fahrenheit temperature is 14°F.

30. The average is −4.

CHAPTERS 1–4 MIDTERM EXAMINATION

1. Ten-thousand **2.** 84,183

3. sixty-seven thousand, five hundred nine

4. 8786 **5.** 27,605 **6.** 34,865
7. 720,000; 727,518 **8.** 213 **9.** 729 R 54 **10.** 13

11. 74 **12.** 485

13. average, 81; median, 73; mode, 64

14. a. Civic has the highest sales. **b.** There are 5 more Accords sold than Passports.

15. 120 **16.** It is divisible by 3 and 5.

17. prime number **18.** yes

19. 32, 64, 96, 128, 160

20. 1, 2, 4, 8, 16, 19, 38, 76, 152, 304

21. $2 \cdot 2 \cdot 3 \cdot 3 \cdot 3 \cdot 5 = 2^2 \cdot 3^3 \cdot 5$

22. $11\dfrac{6}{7}$ **23.** $\dfrac{66}{7}$ **24.** $\dfrac{3}{4}, \dfrac{7}{8}$ **25.** $\dfrac{5}{9}, \dfrac{7}{12}, \dfrac{5}{8}, \dfrac{2}{3}$

26. $\dfrac{19}{25}$ **27.** $\dfrac{5}{72}$ **28.** $46\dfrac{2}{7}$ **29.** 1 **30.** $\dfrac{9}{31}$

31. $\dfrac{97}{90}$ or $1\dfrac{7}{90}$ **32.** $18\dfrac{13}{24}$ **33.** $11\dfrac{3}{7}$

34. $17\dfrac{34}{63}$ **35.** $7\dfrac{3}{4}$ **36.** $\dfrac{4}{15}$ **37.** 80,240.122

38. 8.0, 7.96, 7.959 **39.** $\dfrac{13}{50}$

40. 1.02, 1.034, 1.044, 1.07, 1.094, 1.109

41. 173.76   **42.** 2.537   **43.** 356.6601

44. 4600   **45.** 0.000902   **46.** 5.8×10^{-4}

47. 56.8   **48.** 0.652

49. average, 6.4; median, 6.9   **50.** 1.62

51. The perimeter is 60 cm.   **52.** $76\frac{7}{10}$ ft^2

53. A total of 6950 chinook were counted.

54. Maria received $107.20 as a dividend.

55. The navel oranges cost $0.89 per pound.

CHAPTERS 1–8 FINAL EXAMINATION

1. $\frac{21}{20}$ or $1\frac{1}{20}$ **2.** 630   **3.** 35.73   **4.** 82.548

5. $\frac{3}{2}$ or $1\frac{1}{2}$ **6.** 130   **7.** 199   **8.** 38.33

9. $7\frac{4}{15}$ **10.** 945   **11.** 314.93   **12.** 47.212

13. $\frac{9}{25}$ **14.** $10\frac{17}{30}$ **15.** $x = \frac{15}{2}$ or $7\frac{1}{2}$

16. ten-thousandths   **17.** $\frac{7}{10}, \frac{3}{4}, \frac{4}{5}$

18. 92%   **19.** 19.739   **20.** 0.515   **21.** 72,000.05

22. 8.09   **23.** 0.57625   **24.** 50   **25.** 9.35%

26. 0.284   **27.** 59.86   **28.** 0.6435

29. The blueberries would cost $63.72.

30. The percent of discount is about 18.8%.

31. 51,102, 153, 204, 255

32. eight thousand, thirty-seven and thirty-seven thousandths

33. false   **34.** $2 \cdot 3 \cdot 5 \cdot 5 \cdot 5 = 2 \cdot 3 \cdot 5^3$

35. $\frac{7}{10}$ **36.** $\frac{189}{200}$ **37.** $48\frac{4}{9}$ **38.** $\frac{2}{15}$

39. $\frac{29}{200}$

40. Of the customers surveyed, 54% ordered a Big Mac.

41. no   **42.** $13\frac{13}{15}$ **43.** $\frac{241}{18}$

44. 2.299, 2.32, 2.322, 2.332

45. 1, 2, 3, 4, 6, 8, 12, 17, 24, 34, 51, 68, 102, 136, 204, 408

46. 0.00045893

47. $2\frac{14}{15}$ **48.** $\frac{17}{100}$

49. It will cost Mildred $1455 to drive 7500 miles.

50. Billy earns $17,873 for the month.

51. The sales tax rate is 8.6%.

52. The final cost of the vacuum cleaner is $183.02.

53. 49   **54.** 66.7   **55.** 8 hr 24 min 20 sec

56. 3 m 73 cm   **57.** $50 per kilogram

58. The perimeter is 183 ft.   **59.** 33.48 m^2

60. 62.13 yd^2 **61.** 1932 in^3 **62.** 24.05

63. The hypotenuse of the triangle is about 35.5 cm.

64. -16 **65.** 21   **66.** -105 **67.** 3.1

68. 2   **69.** $a = -5$

70. The Fahrenheit temperature is 5°F.

INDEX OF APPLICATIONS

INDEX

D